Clayey Barrier Systems for Waste Disposal Facilities

OTHER TITLES AVAILABLE FROM
E & FN SPON

Landfilling of Waste: Leachate
T.H. Christensen, R. Cossu and R. Stegmann

Landfilling of Waste: Barriers
T.H. Christensen, R. Cossu and R. Stegmann

Water Quality Assessments: A guide to the use of biota, sediments and water in environmental monitoring
D. Chapman

Microbiology for Environmental and Public Health Engineers
R.M. Sterritt and J.N. Lester

Ecological Effects of Waste Water
E.B. Welch

Comprehensive Water Analysis
T.R. Crompton

For more information on these and other titles please contact The Promotion Department, E & FN Spon, 2–6 Boundary Row, London SE1 8HN, Telephone 071 865 0066

Clayey Barrier Systems for Waste Disposal Facilities

R. Kerry Rowe

Department of Civil Engineering, University of Western Ontario, London, Canada

Robert M. Quigley

Geotechnical Research Centre, University of Western Ontario, London, Canada

and

John R. Booker

School of Civil Engineering, University of Sydney
Australia

E & FN SPON

An Imprint of Chapman & Hall

London · Glasgow · Weinheim · New York · Tokyo · Melbourne · Madras

Published by E & FN Spon, an imprint of Chapman & Hall, 2–6 Boundary Row, London SE1 8HN, UK

Chapman & Hall, 2–6 Boundary Row, London SE1 8HN, UK

Blackie Academic & Professional, Wester Cleddens Road, Bishopbriggs, Glasgow G64 2NZ, UK

Chapman & Hall GmbH, Pappelallee 3, 69469 Weinheim, Germany

Chapman & Hall USA, One Penn Plaza, 41st Floor, New York NY 10119, USA

Chapman & Hall Japan, ITP-Japan, Kyowa Building, 3F,
2-2-1 Hirakawacho, Chiyoda-ku, Tokyo 102, Japan

Chapman & Hall Australia, Thomas Nelson Australia, 102 Dodds Street, South Melbourne, Victoria 3205, Australia

Chapman & Hall India, R. Seshadri, 32 Second Main Road, CIT East, Madras 600 035, India

---

First edition 1995

Typeset in 10½/12½pt Sabon by Cambrian Typesetters, Frimley, Surrey

Printed in England by Clays Ltd, St Ives plc, Bungay, Suffolk

ISBN 0 419 19320 0

A catalogue record for this book is available from the British Library

♾ Printed on permanent acid-free text paper, manufactured in accordance with ANSI/NISO Z39.48-1992 and ANSI/NISO Z39.48-1984 (Permanence of Paper).

Contents

Preface

The design of waste disposal facilities typically involves some form of 'barrier' that separates the 'waste' from the general groundwater system. This barrier is intended to minimize the migration of contaminants from the facility, thus the environmental impact of the facility is intimately related to its design and long-term performance. Natural clayey deposits or recompacted clayey liners frequently represent a key component of these barriers. However, there are many practical situations where neither a low permeability clay nor plastic liner (geomembrane) alone is sufficient to prevent unacceptable long-term environmental impact. In these circumstances, the barrier is a designed system which may involve numerous components (e.g. multiple or composite liners, multiple leachate collection systems etc.) which work together to provide protection to the groundwater environment.

This book deals with the design and performance of barrier systems, which may include natural clayey deposits, recompacted clayey liners or even porous rock as part of the barrier system. Some consideration is also given to other components of the barrier system, such as leachate collection systems and geomembrane liners. The book was originally written as a set of notes for short courses which the authors have given for practicing engineers, educators and hydrogeologists. It was then extended as a text for a senior or graduate level course. We have attempted to explain all key concepts, assuming little background knowledge other than introductory courses in soil mechanics, chemistry and hydrogeology. In particular, Chapter 1 is essential to the reader not familiar with the basic concepts. Appendix A also presents a glossary of terms which should assist the reader not familiar with some of the terminology.

The objectives of this book are two-fold. Firstly, to focus on the clayey component of the barrier. Contaminants can potentially pass through a barrier by advection (e.g. the movement of contaminants due to groundwater movement) and diffusion (i.e. the movement of contaminants from locations of high concentration to locations of lower concentration). The potential impact of a given contaminant will generally depend on both the hydraulic conductivity and diffusion coefficient for the barrier. Thus the book deals with the construction/compaction of clayey liners, the significance of clay mineralogy in the design of barriers and the determination of appropriate hydraulic conductivity values and diffusion coefficients. In particular, the field studies to be discussed in Chapter 9 demonstrate conclusively that diffusive contaminant transport can and does occur, even in the absence of advection.

The second objective is to discuss the role which theoretical modeling can play in the design of barriers. This role may include the determination of some relevant design parameters from laboratory tests and the evaluation of the potential impact of a proposed design on groundwater quality both under working conditions and in the event of a failure of part of the engineered system (e.g. the leachate collection system).

The state-of-the-art has now advanced to a point where modeling can be used as an aid to engineering judgement in most landfill designs. However, the results of an analysis are only as good as the assumptions and parameters on which it is based. There will always be some uncertainty associated with parameters and design assumptions, and for this reason, analysis will be most useful for performing sensitivity studies to examine the implications of different design scenarios and for determining the potential significance of uncertainty regarding key parameters.

The first chapter deals with basic concepts, including the types of barrier being considered, the nature and relative importance of different transport mechanisms and the choice of suitable boundary conditions. It also includes a discussion of methods of analysis, ranging from simple hand methods to simple finite layer techniques which can be readily implemented on a microcomputer.

The second chapter discusses a number of design considerations for barrier systems which may include clayey liners, geomembrane (plastic) liners, composite liners, leachate collection systems, leak detection systems and hydraulic control layers. Factors discussed include leachate composition, leachate collection systems, leakage through liners, hydraulic containment and consideration of the service life of components of barrier systems.

Chapters 3 and 4 examine issues of permeability and clay–leachate compatibility for clayey liners, while Chapter 5 discusses the modeling of flow though liner systems.

Chapter 6 provides a discussion of the process of molecular diffusion and highlights the many factors involved in the migration of contaminant through soil due to a difference in chemical potential (concentration).

Chapter 7 describes a finite layer technique for the analysis of contaminant migration in intact and fractured porous media.

Chapter 8 discusses the evaluation of a number of key parameters such as the diffusion coefficient and sorption parameters (i.e. distribution or partitioning coefficient), and shows that both parameters can often be estimated from a single test using a simple finite layer analysis to match calculated and observed migration of a chemical species of interest through a site specific clayey soil. Chapter 9 then illustrates the important role of molecular diffusion with reference to a number of field case histories.

Chapter 10 discusses the importance of the finite mass of contaminant available for transport into the groundwater system together with factors such as the effect of landfill size, sorption, liner thickness and advective velocity. Examples show how analysis may improve the designer's 'feel' for the effectiveness of potential contaminant attenuation mechanisms and proposed design both for natural and engineered systems involving intact clayey barriers.

Chapter 11 examines the migration of contaminants in fractured porous media and highlights the important role of matrix diffusion (i.e. diffusion from the fractures into the intact matrix adjacent to the fractures). Consideration is given to both fractured porous rock and fractured clayey deposits and to the benefits to be derived from installing an intact (e.g. compacted clay) liner over fractured media.

Finally, Chapter 12 discusses the integration of hydrogeology and engineering in the design of barrier systems and the assessment of potential containment impact. Topics discussed include the effect of the natural hydrogeology on engineering design, the service life of engineered systems, the contaminating lifespan of a landfill, the implications of failure of leachate collection systems and the development of action level concentrations for use in monitoring.

Much of the research reported in this book was supported by operating and strategic funds from the Natural Sciences and Engineering Research Council of Canada. In addition, major funding was received from the Ontario Ministry of the Environment for clay–organic liquid compatibility assessment and Leda clay–MSW leachate compatibility assessment.

Many graduate students, postdoctoral fellows and research associates have also been involved. These include: Drs F. Barone, F. Fernandez, D. Goodall, C. Ohikere, P. Nadarajah and E.K. Yanful, Messrs K. Badv, C.J. Caers, A. Hammoud, T. Helgason, J. Mucklow, E. San, J.D. Smith, Ms L. Eggleston and V.E. Crooks. Our technical staff of G. Lusk and W. Logan have expedited the laboratory work. Ms J. Lemon, E. Milliken and J. Lapp were responsible for the careful typing of the manuscript.

April 1994 R. Kerry Rowe, Robert M. Quigley and John R. Booker

Chapter one

Basic concepts

1.1 Introduction

This book was written as a text for short courses for professional engineers and hydrogeologists, and for senior level and graduate level courses, dealing with the design, construction and performance of barrier systems. The objective of this first chapter is to describe the basic concepts that will be expanded on in later chapters. Different types of barrier systems will be discussed. The different factors associated with contaminant transport through barrier systems are explained and the relative importance of different transport mechanisms is examined. Finally, some of the basic concepts associated with modeling of contaminant movement through barrier systems and assessing impact on groundwater quality are introduced and illustrated for some simple cases.

1.2 Overview of barrier systems

The impact of a waste disposal facility on groundwater quality will depend on the nature of the site, the climate, the type of waste, the local hydrogeology and the presence of a dominant flow path, and, perhaps most importantly, on the nature of the barrier which is intended to limit and control contaminant migration. These days, barriers will usually include one or more of the following components:

1. natural clayey soils such as lacustrine clay or clayey till;
2. recompacted clayey liners;
3. cut-off walls;
4. natural bedrock;
5. geomembranes in composite liner systems.

The following subsections provide an introduction to some of the design considerations associated with these components. In this discussion terms such as 'diffusion', 'advection' and 'sorption' are used. The reader unfamiliar with these terms may wish to review section 1.3 prior to reading the following. A glossary defining key terminology is given in Appendix A.

1.2.1 Natural clayey deposits

(a) Diffusion-controlled systems

The simplest barrier systems involve thick natural deposits of clayey soil where the water table is near the surface. For the unusual case of a landfill which does not mound above the original ground surface, it may be possible to design the landfill such that, after closure, the liquid (leachate) levels in the landfill rise to the natural water table, as shown schematically in Figure 1.1, and there is no significant gradient or advective transport. The primary consideration in the evaluation of this type of barrier design is to ensure that the hydrogeologic environment is such that one can reasonably expect low hydraulic gradients over the long term. Under these circumstances the movement of contaminants will be slow and controlled primarily by diffusion. The clay itself can also act as an important medium for the attenuation of some contaminants due to processes such as sorption, precipitation and biodegradation. Nevertheless, migration will occur and it is important to check that contaminant impact at the site boundary, due to molecular diffusion from the landfill, will meet regulatory requirements.

The diffusive movement of contaminants through saturated clayey deposits is relatively well understood and, as discussed in Chapter 9, research has shown that natural, diffusion-controlled, chemical profiles established over thousands of years are consistent with simple theoretical predictions.

(b) Advective–diffusive transport

While situations such as that shown in Figure 1.1 do exist, they are not common and a more typical situation involving mounded waste is shown in Figure 1.2. Prior to landfilling, there are downward gradients from the water table to the underlying aquifer. The clayey soil provides a natural barrier; however, in order to design a suitable landfill, careful consideration must be given to the selection of the landfill base elevations and to the design of the leachate collection system.

Figure 1.2 shows a situation where the base elevations are below the groundwater level but above the potentiometric surface (i.e. water levels) in the aquifer; hence there will be downward advective flow from the landfill to

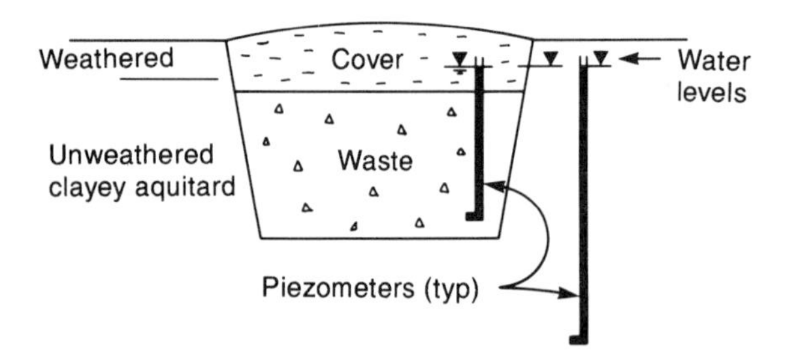

Figure 1.1 Natural barrier design involving diffusive transport (but no advection) in deep clayey deposit. (Reproduced from Rowe, 1988, with permission of the *Canadian Geotechnical Journal*.)

the aquifer. Figure 1.3 shows a situation where the base elevations are such that the design leachate mound is below the potentiometric surface in the aquifer; hence there is upward advective flow from the aquifer to the landfill. This design relies on hydraulic containment of contaminants and may be referred to as a 'hydraulic trap' (as discussed in section 1.2.1(d)). A discussion of modeling of these types of situations for intact soil is given in Chapters 10 and 12.

(c) Fractured clays

A major consideration for a landfill design such as that shown in Figure 1.2 is whether the clayey soil or till is fractured. Recent research by D'Astous *et al.* (1989), Herzog *et al.* (1989) and in unpublished reports has provided evidence to suggest that clayey till beneath the obviously weathered and fractured zone may be fractured to depths of 6–15 m.

It is important to recognize that fractures with very thin (e.g. 10–15 μm) openings at relatively wide spacings of 1–3 m can still have a significant effect in terms of increasing the bulk hydraulic conductivity (compared to that of the intact soil). Furthermore, even if the fractures are closed prior to construction of the landfill, stress relief due to excavation of the landfill cell may, under some circumstances, cause opening of fractures in the underlying clays.

If present, fractures are likely to control the hydraulic conductivity of the clay and hence the downward advective velocity, and it may be necessary to construct a clay liner by reworking the existing soil (if it is suitable) to reduce outward advective transport from the landfill to acceptable levels. Relatively simple semi-analytical models have now been developed which will allow the designer (Chapter 11) to estimate potential impact associated with fractured clay, or a liner overlying fractured clay, for situations similar to that shown in Figure 1.2.

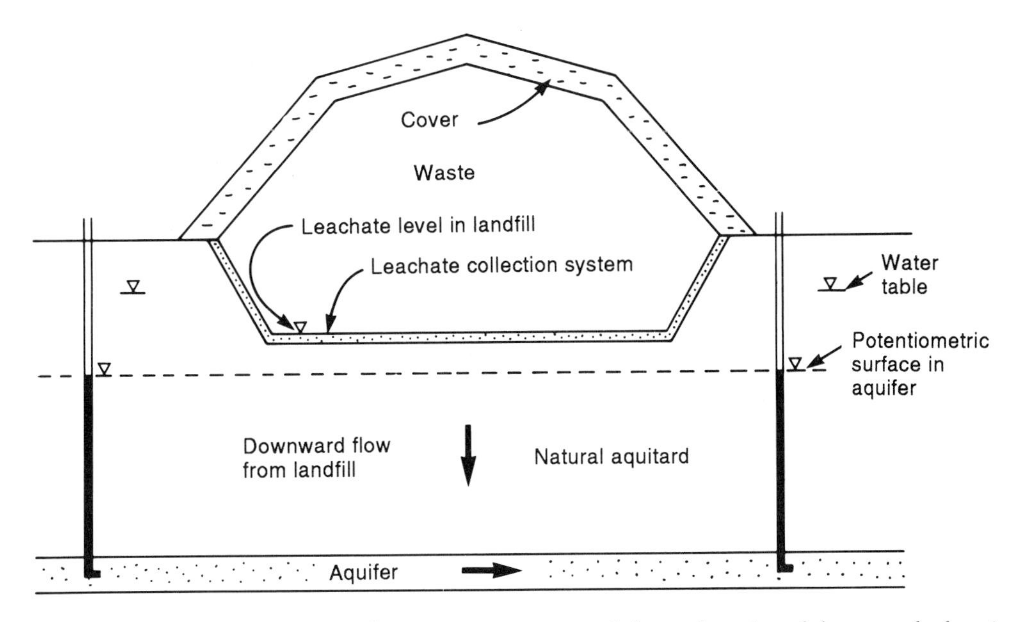

Figure 1.2 Barrier design involving a leachate collection system, a natural clayey deposit and downward advective–diffusive transport.

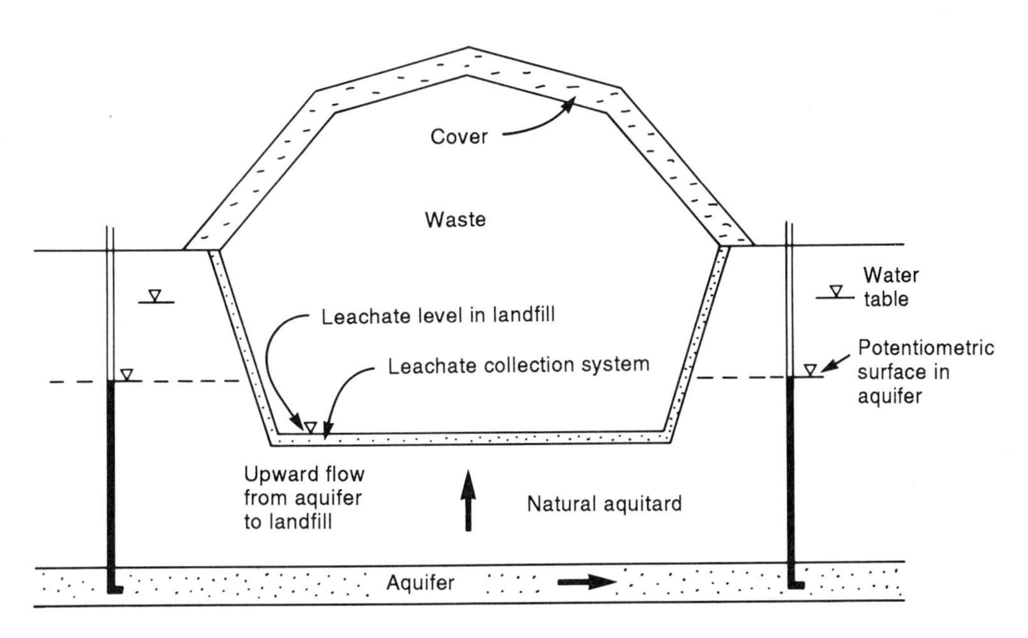

Figure 1.3 Barrier design involving a leachate collection system, a natural clayey deposit, upward advection and downward diffusion – a hydraulic trap.

(d) Hydraulic traps

The hydraulic trap shown in Figure 1.3 is very attractive from a contaminant impact perspective since the inward advective flow of groundwater from the aquifer tends to inhibit the outward diffusion of contaminants. However, as illustrated in Chapter 10, there may still be a potential impact on the underlying aquifer, even with an operating hydraulic trap. Appropriate contaminant transport calculations are required to assess whether the impact will be significant or negligible.

The design of landfills intended to operate as a hydraulic trap is far from straightforward. Very careful consideration must be given to a number of conflicting criteria. For example, lowering the landfill base elevations increases the inward hydraulic gradient and hence inward advective flow. This reduces the amount of outward diffusion (which is good); however, by lowering the base contours, one also reduces the thickness of the barrier which is separating the waste from the underlying aquifer (which is not so good). Thus, if there were to be a failure of the leachate collection system and mounding of leachate which reversed the hydraulic gradient, the attenuation afforded by the barrier would be reduced. Furthermore, the lower the base of the landfill the greater the potential for opening of fractures in the clay due to uplift pressure and the greater the potential for blowout of the base of the landfill. One can reduce the likelihood of blowout by depressurizing the aquifer during construction; however, this option requires careful evaluation of the number of wells required to gain adequate drawdown (i.e. lowering) of water levels in the aquifer over large areas and the potential effect on off-site water users.

The operation of the hydraulic trap shown schematically in Figure 1.3 implicitly assumes that the construction of the landfill will not significantly affect the water levels in the underlying aquifer. It cannot be assumed that this will necessarily be the case. Construction of a landfill as shown schematically in Figure 1.3 would eliminate any recharge of the aquifer that had previously occurred over the area of the landfill footprint. In addition, the leachate collection system is removing groundwater from the aquifer for as long as the hydraulic trap is maintained. The combination of these two effects can be expected to result in a reduction in head (water pressure) in the aquifer unless the aquifer is very permeable or the site is over a natural groundwater discharge area. The extent to which the head in the aquifer is reduced will depend on the hydrogeological characteristics of the aquifer and overlying clayey soils, as well as on the size of the landfill (i.e. the larger the landfill, the greater the shadow it casts on the aquifer; hence the greater the potential for a decrease in head within the aquifer). The implications of this shadow effect should be considered carefully in the design of a hydraulic trap and are discussed in more detail in Chapter 12.

Consideration should also be given to the impact of potential pumping of the aquifer for use as water supply and the impact of potential long-term changes in climate over the contaminating lifespan of the landfill (i.e. the period of time during which the landfill contains contaminants which could have an unacceptable impact if released to the environment). In some cases these factors alone may necessitate a more elaborate design such as that discussed subsequently with respect to Figure 1.8 (section 1.2.2(b)).

1.2.2 Compacted clay liners

Compacted clay liners have been the subject of debate with respect to both the hydraulic conductivity which can be achieved in the field (e.g. Day and Daniel, 1985 and related discussion) and the potential impact of soil–leachate interaction on hydraulic conductivity (e.g. Green, Lee and Jones, 1981; Brown and

Anderson, 1983; Brown, Green and Thomas, 1983, 1984; Fernandez and Quigley, 1985; Bowders and Daniel, 1987). However, experience to date would suggest that with good engineering practice and quality control, good quality, low hydraulic conductivity liners can be constructed of compacted inactive clays, as discussed in Chapter 9. Furthermore, it would appear that while some very concentrated organic wastes may increase the hydraulic conductivity of clay, provided that one considers clay–leachate compatibility in the selection of the liner material, large increases in hydraulic conductivity can be avoided, as discussed in Chapter 4.

Inactive clayey liners which are compacted at a water content higher than the Standard Proctor optimum moisture content and not allowed subsequently to dry out, will be nearly saturated, compressible on loading and should behave in a satisfactory manner similar to natural unfractured clayey barriers. Provided that the design minimizes outward gradients, the primary transport mechanism through a well-designed compacted clayey barrier will be diffusion.

The present state-of-the-art has not clarified the susceptibility of sodium bentonite–sand mixtures to damage by saline and organic leachates, and particular care should be taken in the design of this type of compacted liner.

(a) Single liner

Figure 1.4 shows a recompacted clayey barrier constructed on natural soils. The liner may be required for one of two reasons. Firstly, if the natural soil is a fractured clayey soil then the liner may be required to retard movement (or potential movement) of contaminant along the fractures. For situations where there is outward flow from the landfill (e.g. water level (a) in Figure 1.4), the assessment of potential impact is likely to consider advective–diffusive migration through the liner and the potential for attenuation due to matrix diffusion as contaminant migrates along the fractures, as discussed in Chapter 11. For situations where there is a hydraulic trap (e.g. water level (b) in Figure 1.4), the assessment of potential impact will require consideration of outward diffusion. The effect of the inward advective flow needs to be considered carefully. For example, if the inward flow is primarily through the fractures, then the outward diffusion of contaminant through the matrix between fractures may not be significantly reduced by the inward flow if the fracture spacing is relatively wide. On the other hand, if the fractures are closely spaced and/or nonconductive, then the situation may be similar to that discussed in the previous section where the outward diffusion is resisted by inward advective flow. Clearly, each specific case should be carefully examined on its own merits.

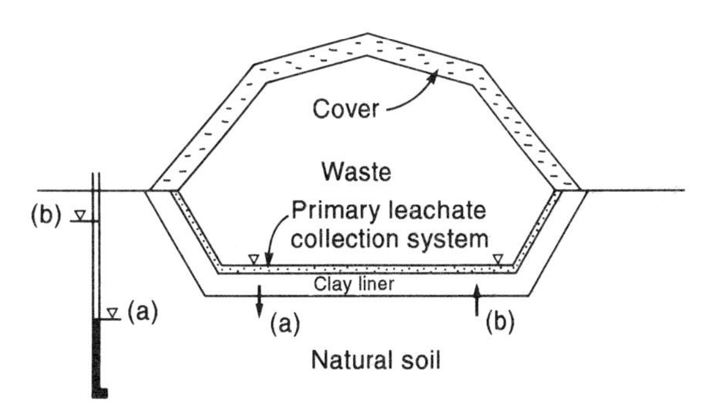

Figure 1.4 A compacted clayey liner used as the primary barrier.

The second reason for constructing a clay liner is that, while intact, the surrounding natural soil does not have a low enough hydraulic conductivity (permeability) to provide an adequate barrier. A typical example is Metropolitan Toronto's Keele Valley landfill in Ontario, Canada. The available data would suggest that, to date, the liner is performing well. Based on the results of the shallow lysimeters and the majority of conductivity sensor sets, it would appear that it is performing better than would be expected for the specified 1.2 m thick liner with a 1×10^{-10} m/s hydraulic conductivity (Reades *et al.*, 1989; King *et al.*, 1993). Even though the outward hydraulic gradient through the liner exceeds unity, the data suggest that the rate of contaminant migration through the liner is governed by diffusion. No evidence has been found for bulk field hydraulic conductivities higher than those measured on undisturbed samples in the laboratory at an effective consolidation pressure of about 160 kPa (Reades *et al.*, 1989).

It is noted that Reades *et al.* (1989) report that the diffusion coefficient inferred from the observed diffusion profile through the Keele Valley compacted clay liner of 6.5×10^{-10} m^2/s (0.02 m^2 per annum, or m^2/a) is very close to that of 7×10^{-10} m^2/s (0.022 m^2/a) obtained earlier in laboratory testing. Furthermore, as discussed in Chapter 9, the data would appear to suggest that the 0.3 m thick sand blanket constructed over the liner is acting more like part of the barrier than as a drain, based on the observed change in the upper portion (5–10 cm) of the sand (i.e. the presence of a biologically produced black slime) (King *et al.*, 1993) and the fact that a very good diffusion profile can be traced through the sand and clay liner. In the Keele Valley landfill the sand blanket is primarily intended as protection for the liner, and so clogging of the sand is desirable, since it increases the effective thickness of the liner with respect to diffusion. However, the clogging of the sand observed at Keele Valley is also a warning for other projects where one might be considering using a sand blanket as a drainage layer, since it raises concerns regarding the effectiveness of sand blankets as an essential part of a leachate collection system.

(b) Liner with secondary leachate collection/hydraulic control system

There are some situations where the conceptual designs indicated in Figures 1.2–1.4 may not provide sufficient confidence that there will be a negligible effect on groundwater quality. Under these circumstances, an additional level of engineering in the form of a secondary leachate collection system or hydraulic control layer may be appropriate as shown in Figures 1.5–1.8.

For situations where the potentiometric surface in the underlying aquifer is at or below the base of the landfill (Figure 1.5, water level (a)), the construction of a permeable drainage system, which is located beneath the compacted clay liner, serves two purposes. Firstly, the drainage layer functions as a secondary leachate collection system which can remove a portion of the leachate which passes through the liner (and some passage is to be expected through any liner system where there are outward gradients). Secondly, this layer serves to reduce the hydraulic gradient through the underlying soil for aquifer water levels below the landfill base

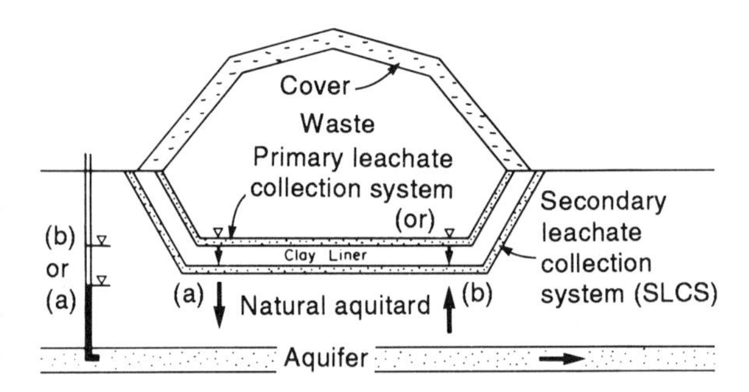

Figure 1.5 A compacted clayey liner used in conjunction with a secondary leachate collection/detection system.

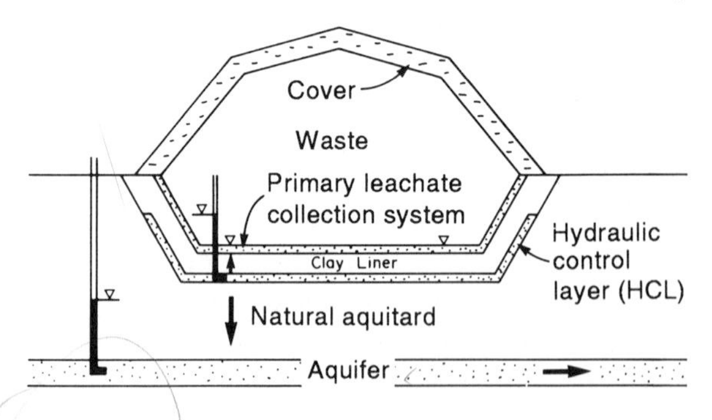

Figure 1.6 A compacted clayey liner used in conjunction with a hydraulic control layer (HCL) and engineered hydraulic trap above the HCL.

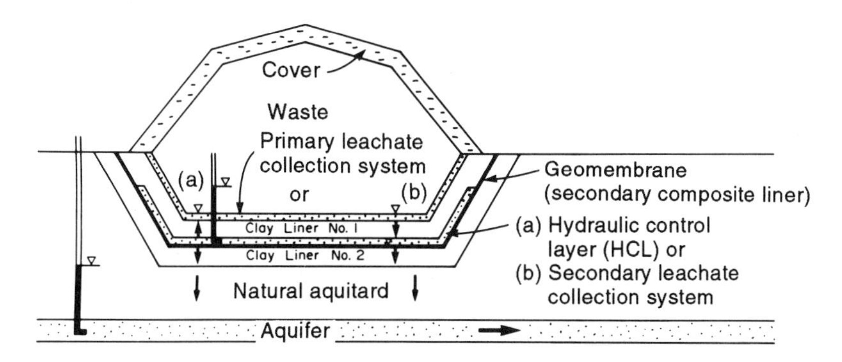

Figure 1.7 A compacted clayey primary liner used in conjunction with an engineered hydraulic control layer and hydraulic trap to minimize contaminant impact together with a composite secondary liner (geomembrane and clayey liner) used to minimize volume of fluid needed to maintain the hydraulic trap. By pumping the hydraulic control layer, this can also be used as a secondary leachate collection system.

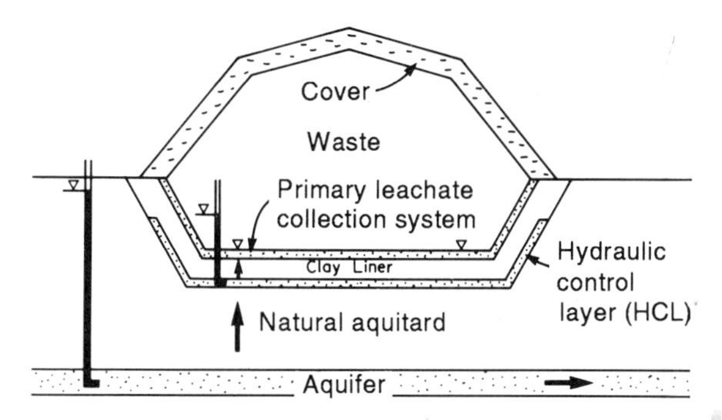

Figure 1.8 A compacted clayey liner used in conjunction with a primary leachate collection system and a hydraulic control layer to create a 'natural' hydraulic trap.

(case (a) in Figure 1.5) or to induce a hydraulic trap to the secondary system when the potentiometric surface in the aquifer is above the level of the secondary leachate collection system (water level (b) in Figure 1.5).

If the natural soil has a relatively low hydraulic conductivity (e.g. 1×10^{-9} m/s or less), then the permeable layer beneath the primary liner may be used as a hydraulic control layer, as shown schematically in Figure 1.6. In this concept the permeable layer is saturated and maintained at a pressure above that in the landfill. This creates an inward gradient and an engineered hydraulic trap. The inward advective flow will resist outward diffusion of contaminants. Some fluid from the hydraulic control layer will also move downward (i.e. there are downward gradients from the hydraulic control layer for the case shown in Figure 1.6). The volume of fluid required to maintain the hydraulic trap can be reduced by introducing a geomembrane liner (and/or a second clay liner) below the hydraulic control layer as shown in Figure 1.7.

In environments where the water table is at or near the ground surface, or where the potentiometric surface in an underlying aquifer is near the surface, a design such as that shown in Figure 1.8 may be warranted. In this case, there is both a natural hydraulic trap (i.e. water flows from the natural soil into the hydraulic control layer) and an engineered hydraulic trap (i.e. water flows from the hydraulic control layer into the landfill). Where practical, this design has the following advantages. Firstly, since there is inward flow to the hydraulic control layer and a relatively impermeable clay liner, it may be possible to design the system such that the engineered hydraulic trap is entirely passive. That is, all water required to maintain an inward gradient from the hydraulic control layer to the landfill is provided by the natural hydrogeologic system and no injection of water to the hydraulic control layer is required.

Secondly, because of the two-level hydraulic trap, there will be substantially greater attenuation of any contaminants that do migrate through the primary liner. Thus, the impact at the site boundary can be expected to be substantially less than that which would be calculated for the case shown in Figure 1.6. Thirdly, since fluid can be injected and withdrawn from the hydraulic control layer it is possible to control the concentration of contaminant in the layer, and hence the impact at the boundary, in the event of a major failure of either the liner or primary leachate collection system. The designs shown in Figure 1.7 (water level (a)) and Figure 1.8 have the same advantage.

The design of all landfills should involve careful consideration of the design life of the engineered components of the system. The conceptual designs shown in Figures 1.6, 1.7 and 1.8 provide a high level of redundancy in the event that there is a major failure of the primary leachate collection system, followed by an increase in leachate levels to above the potentiometric surface in the underlying aquifer.

When designing hydraulic traps involving an engineered hydraulic control layer, consideration should be given to the implications of failure of the primary leachate collection system, the effect of the landfill shadow on the long-term performance of the hydraulic trap and the potential for blowout. A more detailed discussion of the design of hydraulic control layers is given in Chapter 12.

1.2.3 Cut-off walls and permeable surrounds

Cut-off walls are most commonly used to limit contaminant migration from existing sites which have not been adequately designed; however, they can also be used in controlling migration from new sites where it may be desirable to isolate the (potentially contaminated) groundwater in a relatively thin and shallow aquifer beneath the landfill. For example, in the case shown schematically in Figure 1.9(a), the thickness of the natural clay barrier may not be enough to prevent potential contamination of water flowing along the underlying minor aquifer. By constructing cut-off walls around the site and hence reducing the flow in the aquifer locally, it is possible to change an advection-controlled system beneath the landfill into a diffusion-controlled system, thereby substantially reducing the impact on off-site groundwater quality. It is, of course, still necessary to consider diffusive migration through the cut-off wall and into the aquifer. This can be achieved using similar techniques to those which will be discussed for natural or compacted clayey barriers in Chapter 10.

An interesting variation of this concept developed by Matich and Tao (1984), and referred to as the 'pervious surround concept', involves minimizing advective transport through a waste pit by surrounding it with a multilayered pervious envelope, with less permeable material adjacent to the waste and more permeable material outside of this, as shown schematically in Figure 1.9(b). In this

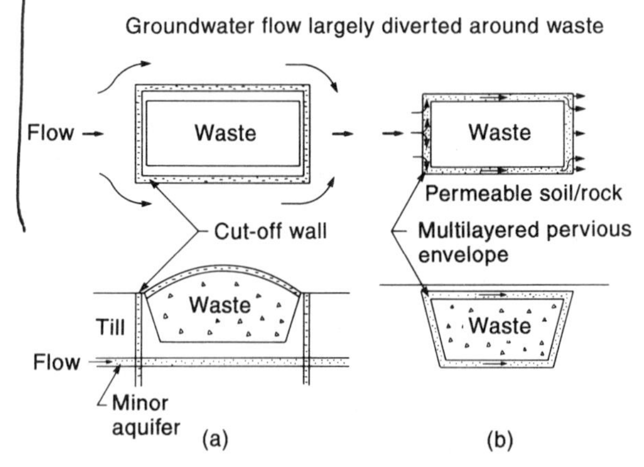

Figure 1.9 (a) A cut-off wall is used to divert groundwater flow from beneath a natural clayey barrier. (b) Pervious material is placed outside the waste to divert groundwater flow around (rather than through) the waste. (After Rowe, 1988; reproduced with permission of the *Canadian Geotechnical Journal*.)

way, water flow is directed around the outside of the pit rather than through the pit, and contaminant migration would be predominantly by diffusion from the waste through the less permeable material, together with advective–dispersive transport within the more permeable outer zone. Thus, from the standpoint of modeling, determination of contaminant loading of the groundwater for this case is very similar to that for waste sites separated from an underlying aquifer or drainage system by a clayey barrier, as shown in Figure 1.2.

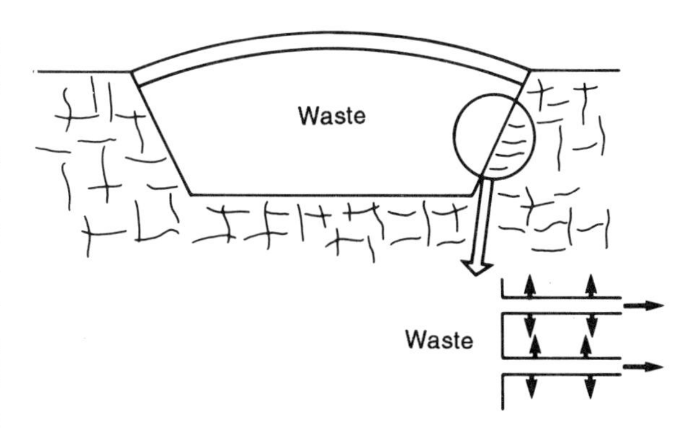

Figure 1.10 Landfill located in fractured shale. Contaminant transport along the fractures is attenuated by diffusion into the matrix of the shale adjacent to the fractures. (After Rowe, 1988, reproduced with permission of the *Canadian Geotechnical Journal.*)

1.2.4 Natural bedrock

A topic of particular interest in some regions is the migration of contaminants from existing or proposed landfills excavated into, or sitting on top of, fractured rock. Typically, the intact rock has a very low hydraulic conductivity and contaminant migration will primarily involve advective–dispersive transport along the fractures in the rock (Figure 1.10). In these cases, the primary mechanism limiting the movement of contaminant is the process of matrix diffusion, whereby contaminant is removed from the fracture as it diffuses into the matrix of the rock. For example, monitoring of an existing landfill at Burlington, Ontario (Gartner Lee Ltd., 1986) suggests that after 15 years' migration, contaminant movement in fractured shale downgradient of the Burlington landfill is probably not more than 25 m, and that substantial attenuation has occurred. The migration of contaminants through fractured porous rock is discussed in Chapter 11.

1.2.5 Geomembrane liners as primary barriers

There are situations where abundant, local, natural or recompacted clays can provide a cost effective and safe primary barrier for landfills. However, there are also situations where the type of waste, landfill size, local hydrogeology and geotechnical conditions are such that the natural soil liner alone is not sufficient to prevent unacceptable contaminant impact at some time in the future. In these cases, an appropriate design involving a geomembrane as part of a primary (and, if necessary, secondary) composite liner may provide a cost effective means of gaining the environmental protection required.

The US EPA has funded a large number of studies that have led to the development of technical guidelines for landfills designed with geomembrane liners (e.g. US EPA, 1988, 1989). An overview of the design of geosynthetic containment systems has been published by Koerner (1990a). These documents provide guidance for many practical applications. However, it must also be recognized that they have been developed for very specific regulatory conditions, and while much of the information is directly transferable to other conditions, particular care is required to ensure that the implications of different regulatory, climatic, geotechnical and hydrogeologic conditions in other regions and countries are recognized when designing landfills.

Figure 1.11 is a schematic detail showing the use of a 1–2 mm thick geomembrane as the

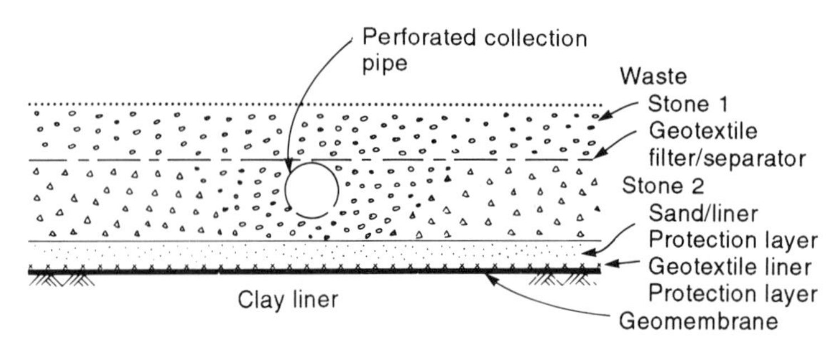

Figure 1.11 Schematic of a primary leachate collection system and geomembrane primary liner.

primary liner in a liner system. This detail could be coupled with the scenario shown in Figure 1.4 (water level (a)) to provide a composite liner. This type of system might be necessary if the clay liner itself could not meet regulatory requirements. For water table conditions (a) shown in Figure 1.4 or the situations shown in Figures 1.2 and 1.5(a), there is an outward hydraulic gradient from the landfill. The installation of a geomembrane above the clay liner would be expected to reduce substantially the outward movement of contaminant by advective processes; however, some migration would still be expected due to leakage through the geomembrane (primarily due to holes, as discussed in Chapter 2). Furthermore, diffusion can also occur through the geomembrane (as discussed in Chapter 12). Consideration should be given to both mechanisms.

Groundwater conditions shown in Figure 1.3, or as implied by water level (b) in Figure 1.4, pose a somewhat different problem if a geomembrane were proposed as part of the primary liner. Without the geomembrane, the landfill would operate as a hydraulic trap, with small quantities of groundwater flowing into the landfill and being collected by a leachate collection system. If a geomembrane is used as part of the primary liner, careful consideration must be given to the effect of uplift pressure on the geomembrane and it may be necessary either to install a granular groundwater control layer beneath a liner (as discussed in the next paragraph), or to depressurize the underlying aquifer during construction in order to prevent problems due to uplift. Also, with the geomembrane in place, one cannot assume that there is any inward advective velocity, and contaminant migration by pure diffusion through the geomembrane and soil must be considered.

A design such as that shown in Figure 1.8 could be used to limit pressures on the liner by managing the pumping of the hydraulic (groundwater) control layer. In the limit where there is full drawdown of fluid levels at the pump, the hydraulic control layer would act as a secondary leachate collection system (like that shown in Figure 1.5(b)) but would also collect groundwater. This would be beneficial from the perspective of contaminant transport since there would be a hydraulic trap to resist outward movement of any contaminant which did escape through the composite (geomembrane–clay) liner. The disadvantage would be the difficulty in distinguishing the level of leakage through the composite liner from the groundwater; hence monitoring of volumes of fluid would probably not provide a good indication of failure of the primary leachate collection system. Monitoring of the chemistry of fluid collected from the secondary collection system would provide an indication of a failure of the primary system, but there is the potential for considerable dilution.

Geomembranes have also found application in double liner systems such as that shown in Figure 1.12. Here the leachate collection system overlies the primary geomembrane liner which, in turn, overlies a leak detection system (secondary leachate collection system) and secondary composite geomembrane and clay liner.

The leachate collection system above the geomembrane is primarily intended to minimize the head drop (driving force) across the geomembrane (Geoservices, 1987a,b). Recognizing that it is almost impossible to construct a liner which does not allow some contaminant escape, the leak detection system (which may be a granular layer (Figure 1.12) and/or a geosynthetic drainage layer for some situations (Figure 1.13)) is intended to detect and collect most of the leachate which does escape through the primary geomembrane. A clay liner is often used as a backup in these systems to provide additional containment of any leachate that does escape through the geomembrane. The clayey liner also provides long-term attenuation of contaminant which may still remain after the landfill has been decommissioned and has gone beyond the post-closure maintenance period (which is often a 30-year period but may vary considerably depending on local regulations). This may, in fact, be the most important role of the clayey barrier for systems such as that shown in Figure 1.12, and calculations can be performed to determine the likely extent of impact following the post-closure maintenance period.

For composite liner systems (e.g. Figure 1.12), particular care is required to prevent desiccation of the underlying clay liner (especially on side slopes) due to potentially high temperatures at the underside of the geomembrane while it is exposed. The risk of this desiccation can be reduced by covering the geomembrane with some insulating material (e.g. the protection layer and the leachate collection system) as quickly as possible. Consideration should also be given to the potential for clay liner desiccation due to high temperatures in landfill leachate (e.g. Collins, 1993). However, it would appear that in a number of practical situations in Canada, neither the temperature nor the temperature gradient have been as high as examined by Collins. Furthermore, the soil tested by Collins was particularly susceptible to desiccation cracking. Nevertheless, consideration should be given to the potential for desiccation cracking due to landfill-induced temperature effects when designing composite liners.

1.2.6 Significance of the cover

While the primary focus of this book is on the migration of contaminants from waste disposal facilities, it should be noted that the generation of mobile contaminants is often related to the movement of fluid or gas into the waste. For example, the volume of leachate generated in a

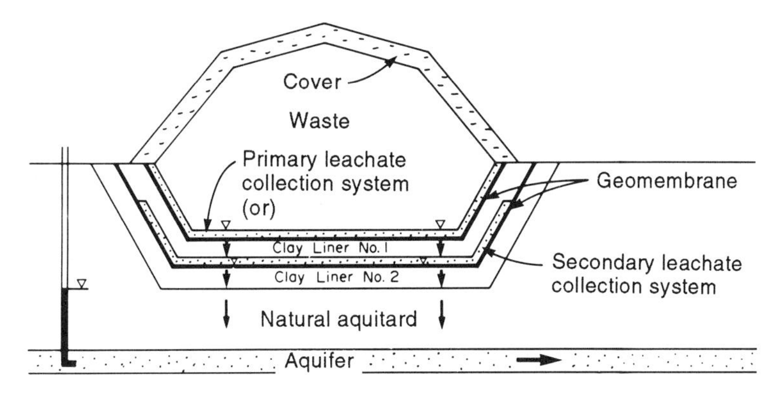

Figure 1.12 Geosynthetic liner system with primary and secondary composite liners each involving a geomembrane and compacted clayey liner.

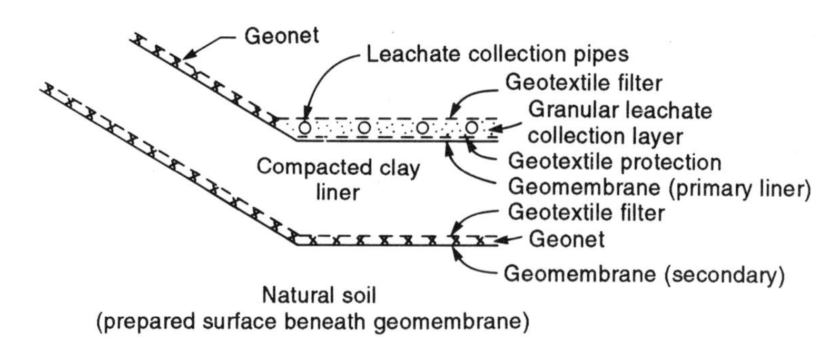

Figure 1.13 Schematic of a primary leachate collection system, composite liner and secondary geosynthetic leachate collection/detection system.

landfill is directly related to the movement of water through the cover. As will be discussed in subsequent chapters, the leachate concentrations and the contaminating lifespan of a landfill (defined fully in Chapter 2) may be related to the infiltration through the cover and this, in turn, may influence the performance of the underlying natural clayey deposit as a barrier.

The generation of contaminants may also be related to the movement of gas through a cover. For example, sulphide-rich mine tailings may pose little problem if kept in a reduced state but may generate considerable acidified leachate if oxygen is allowed to migrate through the cover to the waste. In these instances, the primary design criterion for a cover may be to minimize the diffusion of oxygen, from the atmosphere, through the cover and into the mine tailings (Yanful, 1993).

1.3 Transport mechanisms and governing equations

Evaluation of the design for a waste disposal facility involves making a quantitative prediction of potential impact of the waste on groundwater quality, keeping in mind that under most circumstances involving contaminant movement through a barrier and into an aquifer, the best one can expect to do is to predict trends and a likely range of concentrations at any given point in space and time.

There are four aspects of any attempt to make quantitative predictions, namely the need to:

1. identify the controlling mechanisms;
2. formulate or select a theoretical model;
3. determine the relevant parameters;
4. solve the governing equations.

The controlling mechanisms for contaminant transport are discussed in the following subsections, together with the development of the governing differential equation. The relevant boundary conditions are discussed in sections 1.5 and 1.6. The determination of relevant parameters is discussed in Chapters 3, 4 and 8. Section 1.7 illustrates some simple 'hand solutions' to the governing equations. More elaborate solution techniques are examined in Chapter 7 and their application is illustrated in Chapters 10, 11 and 12.

When dealing with contaminant transport through saturated clayey barriers or the matrix of an intact rock (e.g. shale), the primary transport mechanisms are advection and diffusion. When dealing with transport through aquifers the key transport mechanisms are usually advection and dispersion. In fractured materials, transport is generally controlled by advection and dispersion along the fractures and diffusion from the fractures into the matrix for the adjacent porous media.

1.3.1 Advective transport

Advection involves the movement of contaminant with flowing water. This may be thought of as being analogous to a moving walkway at an airport (Figure 1.14). If it is assumed that once people step onto the walkway they remain standing, then they will be transported along at the speed of the walkway. In this analogy, the speed of the walkway corresponds to the groundwater (seepage) velocity, v. The movement of contaminants at a speed corresponding to the groundwater velocity is often referred to as plug flow. The time required for a plug of contaminant to move a given distance, d, is equivalent to d divided by v (Figure 1.14). The concentration, c, of people on the walkway is the number of people per unit area, and the total number of people being transported past any given checkpoint in a unit of time is defined as the flux.

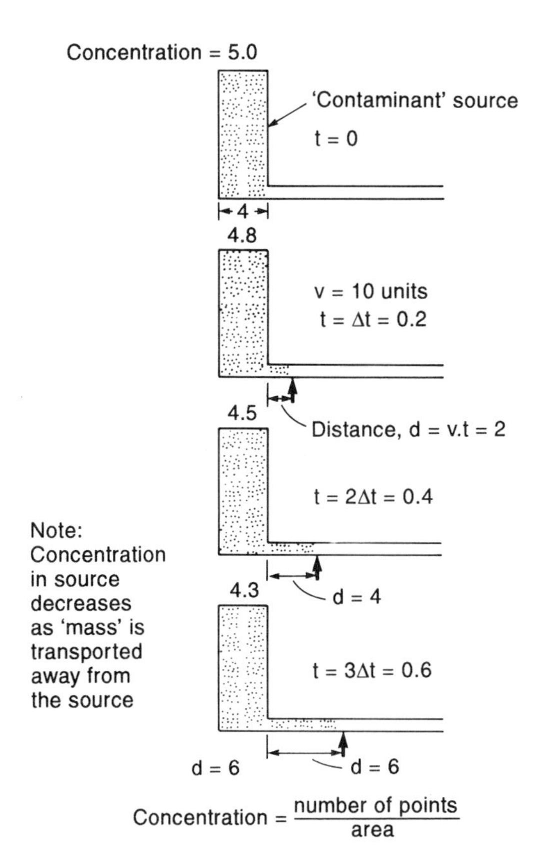

Figure 1.14 Schematic showing advective transport (e.g. people standing on a moving walkway) and a decrease in source concentration with time (e.g. in an airport holding bay) as individuals move out of the holding bay.

Thus when dealing with contaminants in groundwater (rather than people in airports) the mass of contaminant transported by advection per unit area per unit time (i.e. the mass flux, f) is given by

$$f = nvc \qquad (1.1a)$$

$$= v_a c \qquad (1.1b)$$

where n is the effective porosity of the soil [–]; v is the groundwater velocity (seepage velocity) $[LT^{-1}]$; v_a is the Darcy velocity $= nv$ $[LT^{-1}]$; c is the concentration of contaminant at the point and time of interest; concentration is the mass of contaminant per unit volume of fluid $[ML^{-3}]$. L, T and M in square brackets indicate the basic dimensions of the quantity, L, being length, T, time and M, mass.

Thus, just as the total number of people passing out of the holding bay of a terminal or moving walkway (Figure 1.14) is obtained by summing (counting) the number of people passing the checkpoint at the door of the holding bay, so too the total mass transported from a contaminant source into a barrier up to some specific time, t, is obtained by summing (integrating) the mass flux (equation 1.1) with respect to time, τ:

$$m_a = A_0 \int_0^t nvc \, d\tau \qquad (1.2)$$

where m_a = mass of contaminant transported from the landfill by advection [M]; A_0 is the cross-sectional area of the landfill through which contaminant is passing $[L^2]$; and all other terms are as defined above.

If there were no flow, then there would be no movement of contaminant into the barrier **by advection.** It should be noted that the velocity at which contaminant moves **by advection** is the groundwater velocity. In the analogy, this is the velocity of the moving walkway since people are assumed to remain standing; if people walk on the walkway then they can move faster still, and this is analogous to diffusion, as discussed in the next section.

1.3.2 Diffusive transport

Diffusion involves the movement of contaminant from points of high chemical potential (concentration) to points of low chemical potential (concentration). In terms of analogy, this corresponds to the desire of people who have been locked up in a hot, crowded, airport holding bay to spread out once the doors of the holding bay are opened. Even if the moving walkway is stopped, people will walk at different rates away from the holding bay, with the fastest walkers out in front having the most space around them (i.e. the lowest concentration).

The mass flux transported by diffusion alone can be written as

$$f = -nD_e \frac{\partial c}{\partial z} \tag{1.3}$$

where n is the effective porosity of the soil [–] (as previously defined); D_e is the effective diffusion coefficient $[L^2T^{-1}]$; $\partial c/\partial z$ is the concentration gradient (i.e. the change in concentration with distance). The negative sign arises from the fact that contaminants move from high to low concentrations; hence the gradient $\partial c/\partial z$ will usually be negative.

The total mass of contaminant, m_d, transported out of a landfill **by diffusion** up to some specific time, t, is obtained by integrating equation 1.3 with respect of time τ, to give

$$m_d = A_0 \int_0^t \left(-nD_e \frac{\partial c}{\partial z}\right) d\tau \tag{1.4}$$

where all terms are as previously defined.

1.3.3 Advective–diffusive transport

As discussed in section 1.3.1, in the absence of diffusion, contaminant would be transported out of a landfill at the groundwater (seepage) velocity. Again, in the absence of diffusion, there would be no outward contaminant transport through a landfill barrier if the advective flow were into the landfill. However, diffusion cannot be neglected. If the direction of diffusive

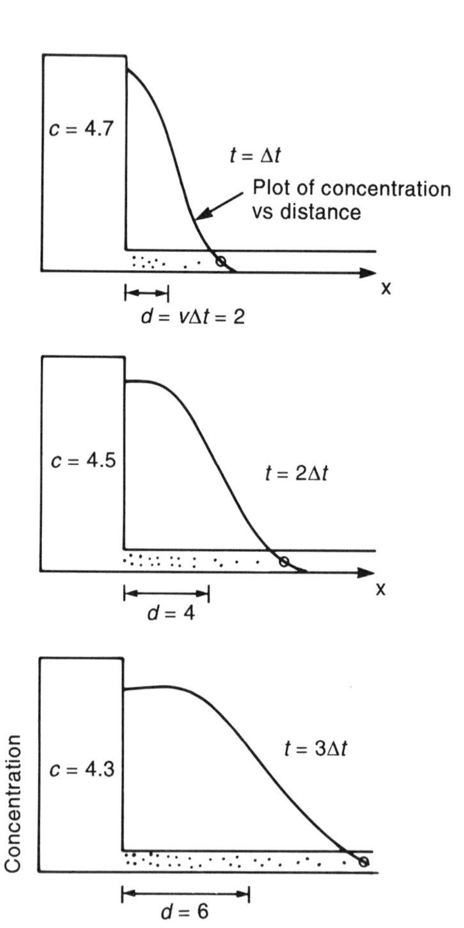

Figure 1.15 Schematic showing advective–diffusive transport (e.g. people walking at different rates on a moving walkway) and a decrease in source concentration with time (e.g. an airport holding bay) as individuals move out of the holding bay.

transport is the same as the direction of advective flow, then it will increase the amount of contaminant transport and decrease the time it takes for contaminant to move to a given point away from the source. In terms of analogy, this corresponds to people spreading out from our airport holding bay by walking on the moving walkway. Suppose that the first and most eager person to leave the lounge drops yellow paint on the moving walkway as soon as he/she steps onto it. If the walkway is moving at a velocity v, then after some time Δt the yellow paint will have moved by advection a distance $d = v\Delta t$ (Figure 1.15). However, if that individual was also walking at a speed v relative to the walkway, then in the same time Δt they would have moved a distance $2d$. Thus the location of the first person out front marks the location of the contaminant plume. Other slower walkers will be spread out behind our lead walker. Thus we get a change in concentration (i.e. number of people per unit area) starting at zero ahead of the location of our lead walker, with the concentration increasing as we move back to behind the point where our lead walker first stepped onto the walkway (i.e. where there is yellow paint). Meanwhile, back in the holding bay, there had originally been five people per unit area (Figure 1.15). The fact that many of them have moved out of the holding bay by stepping onto the walkway has reduced the number of people in the holding bay; so the concentration (number of people per unit area of holding bay) drops from the original value of five people per unit area as time passes and more people leave the lounge.

Diffusion can also occur in the direction opposite to advective transport; so, as will be demonstrated in Chapter 10, it is possible for contaminant to escape from a landfill, even though the groundwater flow is directed into the landfill. The level of impact can be calculated, and often the landfill can be designed to maintain an acceptable level of impact (which may be no significant impact). This situation is again analogous to our airport holding bay, but where the moving walkway is moving into the holding bay; for people to escape they must walk along the walkway in the opposite direction to the walkway's movement. This is the hydraulic trap discussed in section 1.2. If the walkway is moving slowly (i.e. low advective velocity into the holding bay/landfill), then it is not difficult for people to walk out of the holding bay. As the inward speed of the walkway is increased it becomes more and more difficult for people to escape, and eventually one could set the inward velocity of the walkway so high that even the fastest runner cannot move outward along the inward moving walkway.

For the case of advective–diffusive transport the mass flux, f, is given by

$$f = nvc - nD_{\mathrm{e}}\frac{\partial c}{\partial z} \qquad (1.5)$$

and the total mass, m, transported from the landfill up to a specific time, t, is given by

$$m = A_0 \int_0^t \left(nvc - nD_{\mathrm{e}}\frac{\partial c}{\partial z}\right)\mathrm{d}\tau \qquad (1.6)$$

where the velocity, v, is positive if it is out of the landfill and negative if it is into the landfill, and all other terms are as previously defined.

1.3.4 Dispersion

When contaminant migration is associated with relatively high flows (as in many aquifers), there is a third transport mechanism to be considered; namely, mechanical dispersion. This process involves mixing that occurs due to local variations in the flow velocity of the groundwater. It too can be illustrated in terms of analogy. Picture a busload of school children wearing bright red uniforms arriving at a crowded fair. In the bus they have a high concentration (i.e. there are a large number of children per unit area); once let out of the bus the children will disperse (spread out) in the crowd, and before long an aerial view of the fairground would show that the red uniforms were widely distributed in the crowd, not necessarily because the students wanted to be separated, but rather because of mixing that takes place when trying to move through a large crowd.

The dispersion of contaminant also involves a mixing and spreading of the contaminant due to nonhomogeneity in the aquifer (Freeze and Cherry, 1979). Although this mechanism is totally different from the diffusion process, for most practical purposes, it can be mathematically modeled in the same way; hence the two processes are often lumped together as a composite parameter, D, called the coefficient of hydrodynamic dispersion:

$$D = D_e + D_{md} \tag{1.7}$$

where D_e is the effective (molecular) diffusion coefficient for the contaminant species of interest $[L^2T^{-1}]$; D_{md} is the coefficient of mechanical dispersion $[L^2T^{-1}]$.

The mass flux is then given by

$$f = nvc - nD\frac{\partial c}{\partial z} \tag{1.8}$$

where D is defined by equations 1.7 and 1.9a (discussed below) and all other terms are as previously defined.

When dealing with transport through intact clayey soil, diffusion will usually control the parameter D and dispersion is negligible (Gillham and Cherry, 1982; Rowe, 1987 and section 1.3.8). In aquifers, the opposite tends to be true and dispersion tends to dominate. It is often convenient to model the dispersive process as a linear function of velocity (Bear, 1979; Freeze and Cherry, 1979):

$$D_{md} = \alpha v \tag{1.9a}$$

where α is dispersivity [L].

The dispersivity, α, tends to be scale dependent (Gillham and Cherry, 1982). A number of values of dispersivity back-calculated for a number of different cases and summarized by Anderson (1979) are given in Table 1.1. This table shows the estimated longitudinal dispersivity α_L in the direction of groundwater flow and the ratio of transverse dispersivity α_T (in a direction perpendicular to the direction of flow) to longitudinal dispersivity. Typically, the longittudinal and transverse directions relate to two horizontal directions (Figure 1.16). For sand and gravel deposits the transverse dispersivity is commonly less than 30% of the longitudinal value (i.e. $\alpha_T/\alpha_L \leqslant 0.3$ in many cases) and is often of the order of 10% of the longitudinal value ($\alpha_T/\alpha_L \approx 0.1$) (Domenico and Schwartz, 1990). When there is pronounced horizontal stratification the vertical mechanical dispersivity may be small, and the vertical coefficient of hydrodynamic dispersion may be similar to the diffusion coefficient (Sudicky, 1986).

Gelhar *et al.* (1985) examined field experiments at 55 sites. Based on the data, a number of approximate relationships can be established. A reasonable fit to the available data gives

$$\begin{aligned} \alpha_L &\approx \frac{x^2}{100}\ \text{m} \qquad x \leqslant 100\ \text{m} \\ \alpha_L &= 100\ \text{m} \qquad x > 100\ \text{m} \end{aligned} \tag{1.9b}$$

where α_L is the longitudinal dispersivity in meters; x is the scale of the observation in meters.

In experiments with high quality data (all of which involved scales of 100 m or less), the dispersivity was less than 3 m and typically lay between the dispersivity given by equation 1.9b and that given by equation 1.9c below. Looking at all the available data it would also appear that some of the high quality data defines a typical lower bound which is given by

$$\begin{aligned} \alpha_L &\approx \frac{x}{100}\ \text{m} \qquad x \leqslant 1000\ \text{m} \\ \alpha_L &= 10\ \text{m} \qquad x > 1000\ \text{m} \end{aligned} \tag{1.9c}$$

Table 1.1 Regional dispersivities (after Anderson, 1979)

Aquifer	*Location*	*Porosity*	*Longitudinal dispersivity (m)*	α_T/α_L
Alluvial	Rocky Mountain	0.30	30.5	1.0
	Colorado	0.20	30.5	0.3
	California	NR	30.5	0.3
	Lyons, France	0.2	12	0.33
	Barstow, CA	0.40	61	0.3
	Sutter Basin, CA	0.5–0.2	80–200	0.1
Glacial	Long Island, NY	0.35	21.3	0.2
Limestone	Brunswick, GA	0.35	61	0.3
Fractured basalt	Idaho	0.10	91	1.5
	Hanford Site, WA	NR	30.5	0.6
Alluvial	Barstow, CA	0.40	61	0.003
	Alsace, France	NR	15	0.067
Glacial till over shale	Alberta, Canada	0.001 and 0.053	3.0 and 6.1	0.2
Limestone	Cutler Area, FL	0.25	6.7	0.1

Note: α_L, longitudinal dispersivity; α_T, transverse dispersivity; NR, not reported.

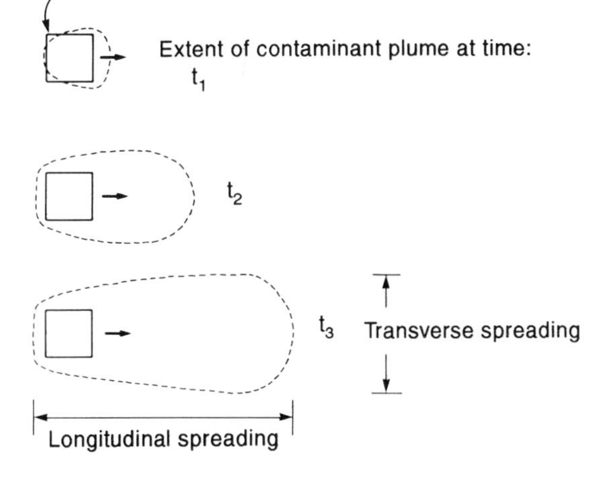

Figure 1.16 Schematic showing plan view of longitudinal spreading of a contaminant plume due to advection and longitudinal dispersion parallel to the average direction of groundwater flow and transverse spreading perpendicular to groundwater flow due to transverse dispersion.

Examining all the available data studied by Gelhar *et al.* (1985), it would appear that a reasonable upper estimate of dispersivity is given by

$$\alpha_L \approx \frac{x^4}{100}\ \text{m} \qquad x \leqslant 10\ \text{m}$$
$$\alpha_L \approx 200\ \text{m} \qquad x > 10\ \text{m} \qquad (1.9d)$$

It should be noted that for scales exceeding 1000 m a limited number of dispersivities have been reported which are less than that given by equation 1.9c, and that a limited number exceed that given by equation 1.9d. The range for data for lengths exceeding 1000 m is 5 m to 5600 m; none of this is high quality data.

Since mechanical dispersion is related to local changes in groundwater velocity, and since this is related to local changes in hydraulic conductivity, it seems reasonable that there should be some relationship between dispersivity and a geostatistical description of the hydraulic conductivities in an aquifer, and such relationships have been developed (e.g. Gelhar and Axness, 1983). These approaches may be useful when there are sufficient data to characterize the geostatistics of the hydraulic conductivities meaningfully; frequently, this is not possible.

When dealing with fractured porous media, the apparent dispersion inferred from tracer tests can be affected significantly by the phenomenon of matrix diffusion. Thus when considering migration of contaminants in this medium, the effects of matrix diffusion (to be discussed in section 1.4.2) and mechanical dispersion (discussed above) should be examined separately. The modeling of matrix diffusion from fractures is discussed in Chapter 7 and illustrated in Chapter 11.

It is evident from the foregoing (and especially by the spread of estimates of α implied by equations 1.9b to 1.9d) that there will always be considerable uncertainty regarding dispersivity in granular or fractured media; hence in the design of barriers it will usually be necessary to perform sensitivity studies to evaluate the effects of this uncertainty on predicted impact.

1.3.5 Sorption

In the previous sections, various transport mechanisms have been discussed. These represent means by which contaminant species in solution move in the groundwater. Equally important, however, are the mechanisms which remove contaminant from solution. These processes may include cation exchange whereby cations such as K^+, Na^+, Pb^{2+}, Cd^{2+}, Fe^{2+}, Cu^{2+} etc. replace other cations (e.g. Ca^{2+}, Mg^{2+}) on the surface of the clay, and hence are removed from solution (this is usually accompanied by an increase in the pore fluid concentration of the desorbed cations Ca^{2+}, Mg^{2+} etc.). Other mechanisms include precipitation of heavy metals (e.g. Yanful, Nesbit and Quigley, 1988b) and removal of organic contaminants from solution by interaction with solid organic matter in the soil.

In the simplest case, the sorption processes can be modeled as being linear and reversible, and so the mass of contaminant removed from solution, S, is proportional to the concentration in solution, c:

$$S = K_d c \qquad (1.10)$$

where S is the mass of solute removed from solution per unit mass of solid [–]; K_d is the partitioning or distribution coefficient $[M^{-1}L^3]$; c is the equilibrium concentration of solute (mass of solute per unit volume of pore fluid) $[ML^{-3}]$.

A plot of the variation in solid-phase concentration S versus the solution-phase concentration under equilibrium is called an isopleth or isotherm. The case represented by equation 1.10 is a linear isopleth (Figure 1.17) and is often regarded as a reasonable approximation for low concentrations of contaminant; the assumption that the process giving rise to sorption is reversible is generally safe even if the process is not reversible. At high concentrations, sorption is nonlinear and more complex relationships between the solid-phase concentration S and the solution concentration have been devised. Two that are more commonly used are the Langmuir and Freundlich isopleths. The Langmuir isopleth is given by

$$S = \frac{S_m bc}{1 + bc} \qquad (1.11a)$$

where S_m is the solid-phase concentration corresponding to all available sorption sites being occupied; b is a parameter representing the rate of sorption; S is the mass of solute removed from solution (sorbed) per unit mass of solid; c is the equilibrium concentration of solute (mass per unit volume of pore fluid).

The Langmuir isopleth is plotted in Figure 1.18(a). The parameters S_m and b are best obtained by performing batch tests in which a known volume of contaminant at a known concentration is mixed with a known volume of dry soil. The solute concentration of species which interact with the soil will decrease to an equilibrium value, c. From this test, the equilibrium solid-phase and solute concentrations (S and c) can be deduced for a range of solute concentrations. The parameters S_m and b can then be estimated by plotting the data in the form $1/S$ versus $1/c$ as shown in Figure 1.18(b).

The Langmuir isopleth has a good theoretical basis but, unfortunately, is not always adequate for describing sorption processes. The Freundlich isopleth provides an empirically based alternative model which sometimes provides a better quantitative description of sorption. It can be written as

$$S = K_f c^{\varepsilon} \tag{1.12a}$$

where K_f and ε are empirically determined constants, and the relationship between S and c is plotted as in Figure 1.19(a) for $\varepsilon > 1$ and $\varepsilon < 1$. As was the case for the Langmuir parameters S_m and b, the Freundlich parameters are also best determined by performing batch tests. Since, in this case, equation 1.12a can be rewritten as

$$\ln S = \ln K_f + \varepsilon \ln c \tag{1.12b}$$

it is most convenient to plot $\ln S$ versus $\ln c$ as shown in Figure 1.19(b) and this readily allows the evaluation of K_f and ε.

At the beginning of this section it was indicated that the linear isopleth is the simplest representation of sorption, and indeed it is. It should not be inferred, however, that this is simply a mathematically convenient representation. Examination of equation 1.11a and Figure 1.18(a) shows that at low concentrations the Langmuir isopleth is essentially linear. Clearly, also, the Freundlich isopleth reduces to equation 1.10 for $\varepsilon = 1$. The linear isopleth is, in fact, often a good representation for the sorption of contaminants found in leachate from municipal waste disposal sites where the concentrations are often relatively low. However, unfortunately this is not always the case. Furthermore, it should be noted that the sorption of a given

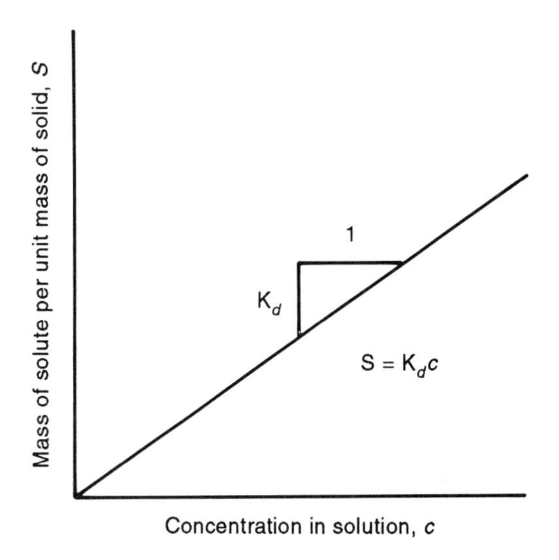

Figure 1.17 Linear isopleth.

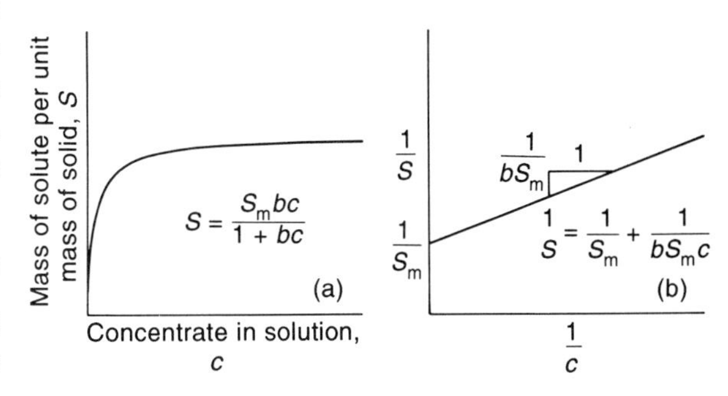

Figure 1.18 Langmuir isopleth: (a) plot showing sorption as a function of concentration; (b) plot used to determine parameters.

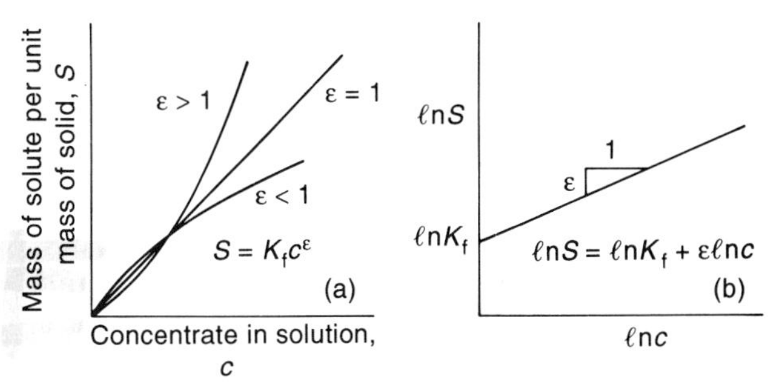

Figure 1.19 Freundlich isopleth: (a) plot showing sorption as a function of concentration; (b) log plot used to determine parameters.

species is often dependent on the presence of other competing species, and hence cannot be determined in isolation.

Analyses of contaminant transport assuming linear sorption (equation 1.10) can readily be performed using analytical, semi-analytical (e.g. finite layer) or numerical (e.g. finite element or finite difference) techniques. Nonlinear sorption poses more of a problem for analysis.

Numerical techniques (finite element) can be used in a time-marching algorithm in which the concentration at time $t_{i+1} = t_i + \Delta t$ is equal to the concentration, c_i, at the previous time, t_i, plus the change in concentration, Δc, due to contaminant migration over the time interval Δt. With this approach, nonlinear sorption can be modeled using tangent approximations to the isopleth over the time period Δt. Thus, in the analysis, the increment in sorption, ΔS, over this period, Δt, is given by

$$\Delta S = \frac{S_{\mathrm{m}} b}{(1 + bc_{\mathrm{r}})^2} \Delta c \qquad (1.11\mathrm{b})$$

for the Langmuir isopleth and

$$\Delta S = (\varepsilon K_{\mathrm{f}}\, c_{\mathrm{r}}^{\varepsilon - 1})\, \Delta c \qquad (1.12\mathrm{c})$$

for the Freundlich isopleth, where ΔS is the increment in mass removed from solution in the time Δt; Δc is the increase in solute concentration in the time Δt; c_{r} is the reference concentration which may be the concentration of solute at time t_i, or if an iterative procedure is adopted, c_{r} may be taken at the average of c_i and c_{i+1}.

Equations 1.11b or 1.12c can be implemented readily in standard finite element contaminant transport codes (essentially the distribution coefficient K_{d} varies with time, and for any time increment is given by equations 1.11b and 1.12c, since $K_{\mathrm{d}} = \Delta S/\Delta c$). However, it is not a trivial matter to obtain accurate results. Since the sorption is dependent on the reference concentration c_{r}, and since the procedure is incremental (viz. $c_{i+1} = c_i + \Delta c$), significant errors can accumulate unless a small time step and fine finite element mesh are used to capture the high concentration gradient at small times.

An alternative approach is to adopt semi-analytical formulations (e.g. the finite layer technique) where an iterative approach is adopted to determine an appropriate secant approximation to the isotherm. Using semi-analytical techniques (to be discussed in detail in Chapter 7), the concentration at time t_{i+1} can be determined directly (i.e. without determining the concentration at time t_i) provided that the nonlinear isopleth can be approximated by an appropriate secant distribution coefficient. For a Langmuir isopleth, the secant K_{d} is given by

$$K_{\mathrm{d}} = \frac{S_{\mathrm{m}} b}{1 + bc_{*}} \qquad (1.11\mathrm{c})$$

For the Freundlich isopleth, the secant K is given by

$$K_{\mathrm{d}} = K_{\mathrm{f}}\, c_{*}^{\varepsilon - 1} \qquad (1.12\mathrm{d})$$

where c_{*} is the estimated concentration at the point and time of interest.

In order to implement this approach in finite layer codes it is necessary to split the deposit into sublayers and use an iterative technique to determine the secant K_{d} for each sublayer, as follows.

1. Calculate the concentration at the top and bottom of each sublayer based on an estimated K_{d} value for each sublayer (an initial value of zero may be used or a value of some user-specific starting estimate c_{r} may be used).
2. Determine a new secant value of K_{d} for each sublayer using equation 1.11c or equation 1.12d, where c_{*} is the average concentration in the sublayer determined from the concentration at the top and bottom of the sublayer in step 1.
3. Repeat step 1 using the new estimate of K_{d} for each sublayer until the process converges (i.e. the value of K_{d} used in step 1 is

essentially the same as the new value of K_d determined in step 2).

With this approach, the accuracy of the solution will depend on the number of sublayers. However, the number of sublayers required is not particularly large (compared with the number of finite elements required in a numerical analysis), provided that they are arranged to provide a reasonable cover in the zone where there is a significant concentration profile. There is no accumulation of error in the approach (i.e. the accuracy at time t_2 is not in any way dependent on the accuracy at time t_1) and one can directly calculate the concentrations at any particular time of interest without determining them at earlier times. This approach is implemented in the program POLLUTE (version 6) (Rowe and Booker, 1994).

Unless otherwise noted, it will be assumed in this book that the sorption processes are linear and can be represented in terms of the partitioning or distribution coefficient K_d. The product ρK_d (where ρ is the dry density of the soil) is a dimensionless measure of the amount of sorption which is likely to occur. A contaminant species is said to be conservative if there is no sorption (i.e. $\rho K_d = 0$). A typical example of a conservative contaminant would be the chloride ion (Cl^-). Typical examples of contaminants whose migration and impact may be greatly retarded by sorption processes are the heavy metals Pb^{2+}, Cd^{2+}, Fe^{2+} and Cu^{2+}, and, in the presence of soil organics, hydrocarbons (e.g. benzene, toluene) and halogenated aliphatic compounds (dichloromethane etc.).

1.3.6 Radioactive and biological decay

Some elements undergo radioactive decay to lighter elements and many organic compounds undergo biological decay into other simpler compounds. The time it takes for the concentration of the particular species to be reduced to half of the original concentration may be referred to as its half-life. For example, phenol is a benchmark chemical for biodegradability studies and there is a large body of information on its time rates of degradation in sewage, soil and fresh water (Howard, 1989).

For substances which undergo first order decay, the rate of reduction of concentration is proportional to the current concentration, so that

$$\frac{\partial c}{\partial t} = -\lambda c \tag{1.13a}$$

where λ is the first order decay constant $[T^{-1}]$, which has three components due to radioactive decay, biological decay and fluid withdrawal respectively:

$$\lambda = \Gamma_R + \Gamma_B + \Gamma_S$$

Γ_R (= ln 2/(radioactive half-life)) is the radioactive decay constant; Γ_B (= ln 2/(biological decay half-life)) is the biological decay constant; Γ_S is the volume of fluid removed per unit volume of soil per unit time from beneath a landfill (to be discussed below).

This equation has the solution

$$c(t) = c_0 e^{-\lambda t} \tag{1.13b}$$

where c_0 is the initial concentration $[ML^{-3}]$.

For radioactive decay which is controlled by an element's atomic structure and is essentially independent of environment, there are substantial available data that can be used to estimate the decay constant Γ_R. Since biological decay depends on many factors such as the presence of appropriate bacteria, substrate, temperature etc., the rate of decay will be specific to a given environment. Much more research is required to allow quantification of the decay, but lack of clear quantification should not mean that the process is overlooked. For example, examination of data from an Ontario landfill, over a period of one decade, suggests that phenol concentrations have degraded with a half-life of less than one year. However, because of potential differences from one landfill to another, care is required in extrapolating this experience.

In addition to the decay of naturally occurring organics (e.g. phenol, benzene, toluene), there is evidence to indicate that synthetic chemicals found in waste (e.g. halogenated aliphatic compounds) can undergo biologically mediated breakdown. Table 1.2 summarizes published half-lives of various halogenated aliphatic compounds in water (at 20°C) in the absence of bacteria. As indicated by Vogel *et al.* (1987), these processes can be accelerated by micro-organisms which mediate (i.e. enhance/promote) substitution reactions (e.g. for monohalogenated or dihalogenated aliphatic compounds such as dichloromethane, chloroethane, 1,1-dichloroethane, 1,2-dichloroethane etc.). Thus the half-lives given in Table 1.2 should represent upper bounds to half-life in a biologically active environment, and there is growing evidence that there can be significant biological breakdown of at least some of these contaminants under anaerobic conditions.

Table 1.2 Environmental half-lives from abiotic hydrolysis or dehydrohalogenation of halogenated aliphatic compounds at 20°C (modified from Vogel, Criddle and McCarty, 1987)

Compound	*Half-life (years)*
Methanes	
Dichloromethane	1.5, 704[a]
Trichloromethane	1.3, 3500[a]
Tetrachloromethane (carbon tetrachloride)	7000
Bromomethane	0.10
Dibromomethane	183
Tribromomethane	686
Bromochloromethane	44
Bromodichloromethane	137
Dibromochloromethane	274
Ethanes	
Chloroethane	0.12
1,2-Dichloroethane	50
1,1,1-Trichloroethane	0.5, 2.5
1,1,2,-Trichloroethane	170
1,1,1,2-Tetrachloroethane	384
1,1,2,2-Tetrachloroethane	0.8
1,1,2,2,2-Pentachloroethane	0.01
Bromoethane	0.08
1,2-Dibromoethane	2.5
Ethenes	
Trichloroethene	0.9, 2.5
Tetrachloroethene	0.7, 6
Propanes	
1-Bromopropane	0.07
1,2-Dibromopropane	0.88
1,3-Dibromopropane	0.13
1,2-Dibromo-3-chloropropane	35

[a]Very large and questionable range of reported values.

A decrease in concentration may also occur if fluid is removed from the soil by some mechanism (as discussed below). In this case $\lambda = \Gamma_S$, where Γ_S equals the volume of fluid removed per unit volume of soil per unit time in the region where fluid is being removed from below the landfill. There are a number of possible ways in which this may occur. For example, it may arise in a landfill barrier system because leachate is being withdrawn at a constant rate from the landfill by a collection system (e.g. a secondary leachate collection system or active hydraulic control layer as discussed in section 1.2.2(b)). Alternatively, it may occur if there is a horizontal flow in a soil deposit where the transport of contaminant is predominantly vertical. In this case, water entering below the upgradient edge of the overlaying landfill will generally bring negligible contaminant into the aquifer beneath the landfill while on the down-gradient edge of the landfill contaminant will be removed from beneath the landfill by the flowing water.

1.3.7 Governing differential equations

As previously noted (e.g. equation 1.8), the mass flux, f, which is transported per unit area per unit time due to advective–diffusive–dispersive transport can be given by

$$f = nvc - nD\frac{\partial c}{\partial z} \qquad (1.8)$$

Thus, by considering conservation of mass within any small region, the change in concentration with time is given by

$$n\frac{\partial c}{\partial t} = -\frac{\partial f}{\partial z} - \varrho\frac{\partial S}{\partial t} - n\lambda c \qquad (1.14a)$$

which simply says that the increase in contaminant concentration within a small volume is equal to the increase in mass due to advective–diffusive transport ($-\partial f/\partial z$, where the negative sign implies a net increase in mass within the element) minus the decrease in mass due to sorption ($-\partial S/\partial t$) minus the decrease in mass due to first order decay processes ($-n\lambda c$).

Substituting Equation 1.8 for the mass flux f and equation 1.10 for the sorbed concentration S into equation 1.14a then gives

$$n\frac{\partial c}{\partial t} = \left(nD\frac{\partial^2 c}{\partial z^2} - nv\frac{\partial c}{\partial z}\right) - \varrho K_d\frac{\partial c}{\partial t} - n\lambda c \qquad (1.14b)$$

where n is the effective porosity of the soil [–]; c is the concentration at depth z and time t [ML^{-3}]; $D = D_e + D_{md}$ is the coefficient of hydrodynamic dispersion [L^2T^{-1}]; v is the seepage or (average linearized) groundwater velocity [LT^{-1}]; ϱ is the dry density of the soil [ML^{-3}]; K_d is the distribution or partitioning coefficient [$M^{-1}L^3$]; $v_a = nv$ is the Darcy or discharge velocity [LT^{-1}]); $\lambda = \Gamma_R + \Gamma_B + \Gamma_S$ is the first order decay constant [T^{-1}] defined in section 1.3.6.

It is worth noting that in some texts, equation 1.14b is rearranged in the form

$$(n + \varrho K_d)\frac{\partial c}{\partial t} = nD\frac{\partial^2 c}{\partial z^2} - nv\frac{\partial c}{\partial z} - n\lambda c \qquad (1.14c)$$

Neglecting first order decay and dividing throughout by $(n + \varrho K_d)$, this becomes

$$\frac{\partial c}{\partial t} = D^*\frac{\partial^2 c}{\partial z^2} - v^*\frac{\partial c}{\partial z} \qquad (1.14d)$$

where $D^* = D/R$; $v^* = v/R$; $R = 1 + \varrho K_d/n$ is the retardation coefficient.

This is a mathematically correct procedure but can lead to severe difficulties if the boundary conditions are flux-controlled (e.g. boundary conditions like those discussed in sections 1.5 and 1.6). For this reason, the use of parameters D^* and v^* and equation 1.14d is to be discouraged.

Equation 1.14 describes advective–diffusive–dispersive transport under one-dimensional conditions. This can be generalized readily to three-dimensional conditions and rewritten as

$$n\frac{\partial c}{\partial t} = nD_x\frac{\partial^2 c}{\partial x^2} + nD_y\frac{\partial^2 c}{\partial y^2} + nD_z\frac{\partial^2 c}{\partial z^2} - nv_x\frac{\partial c}{\partial x} - nv_y\frac{\partial c}{\partial y} - nv_z\frac{\partial c}{\partial z} - \varrho K_d\frac{\partial c}{\partial t} - n\lambda c \qquad (1.15)$$

where D_x, D_y, D_z are the coefficients of hydrodynamic dispersion in the x, y and z cartesian directions respectively [L^2T^{-1}]; and v_x, v_y, v_z are the components at groundwater velocity in the three cartesian directions [LT^{-1}].

1.3.8 Relative importance of different transport mechanisms

A considerable number of laboratory tests have been performed to verify the applicability of the advection–dispersion model (equations 1.7, 1.8 and 1.14). The available data suggest that for the majority of cases the model was quite adequate for practical purposes (e.g. Fried, 1976). The laboratory tests indicate that at relatively low velocities the dispersion coefficient is equal to the effective diffusion coefficient, while at high velocities, the dispersion coefficient increases as a linear function of velocity.

Perkins and Johnston (1963) published an empirical relationship which provides some insight as to what constitutes low and high velocities. Based on the results of a number of tests on homogeneous samples, the coefficient of hydrodynamic dispersion, D, was given by

$$D = D_e + 1.75dv \text{ (m}^2\text{/a)} \qquad (1.16)$$

where d is the mean grain diameter of the soil (m).

The effective (molecular) diffusion coefficient D_e often lies in the range 0.005 to 0.05 m^2/a. Adopting these two values, equation 1.16 can be plotted to give the variation in the coefficient of hydrodynamic dispersion, D, with velocity as shown in Figure 1.20 for the case where the mean grain size is 200 μm.

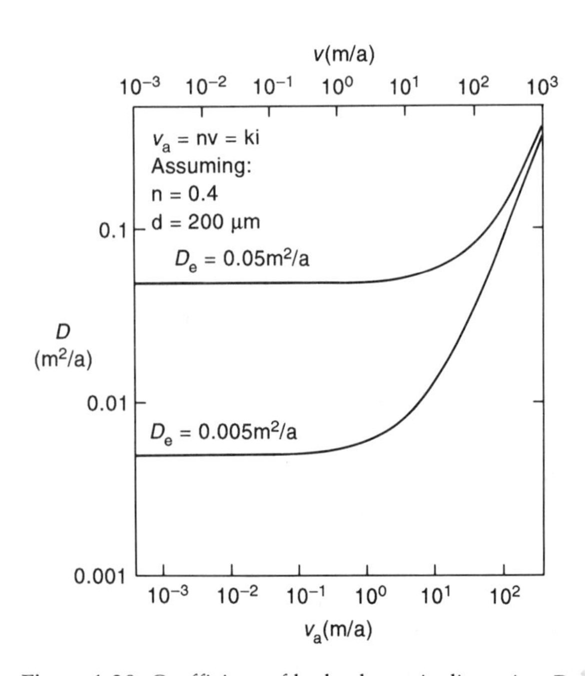

Figure 1.20 Coefficient of hydrodynamic dispersion D, as a function of Darcy velocity v_a and seepage velocity v based on data from Perkins and Johnston (1963). (Modified from Rowe, 1987.)

Assuming that equation 1.16 is applicable to saturated, homogeneous, unfractured, silts, silty clays or clayey soils with hydraulic conductivity (permeability, k) less than 10^{-7} m/s, mechanical dispersion may be neglected for hydraulic gradients less than 1 (i.e. in most of such cases). For a saturated homogeneous sand with hydraulic conductivity of 10^{-5} m/s or less (and $d \leqslant$ 200 μm), these results also suggest that diffusion will generally dominate over mechanical dispersion for hydraulic gradients less than 0.01. For coarser sands or where hydraulic conductivities or gradients are higher, mechanical dispersion may be significant.

It is evident from Figure 1.20 that for $d \leqslant$ 200 μm, the coefficient of hydrodynamic dispersion is quite independent of the groundwater velocity for Darcy velocities v_a less than 10^{-1} m/a. This then raises the question as to how important advective transport is for these velocities (i.e. $v_a < 10^{-1}$ m/a). To provide some answers to this question, analyses were performed to determine the peak chemical flux exiting from beneath a 1.2 m thick clay liner ($n = 0.4$) as a function of the advective velocity v_a. For purposes of illustration the diffusion coefficient was taken to be 0.018 m^2/a (see Chapter 8 for a detailed discussion of diffusion coefficients). Unless otherwise noted, the leachate concentration was assumed to be constant at $c_0 = 1$ g/L (i.e. $c_0 = 1000$ g/m^3) and the liner was assumed to be totally washed (i.e. such that the base concentration $c_b = 0$).

Figure 1.21(a) shows the final (steady state) variation in concentration with depth beneath the landfill for the case of pure diffusion ($v_a = 0$) and for a Darcy velocity of 0.006 m/a. Notice that the concentrations for $v_a > 0$ are greater than those for $v_a = 0$ throughout the layer. Furthermore, the concentration gradient ($\partial c/\partial z$) at the bottom of the liner for $v_a > 0$ is also greater than that for $v_a = 0$. Thus it follows that for a given soil and contaminant (i.e. given value of D), the mass of contaminant (per unit area, per unit time) passing through the barrier and

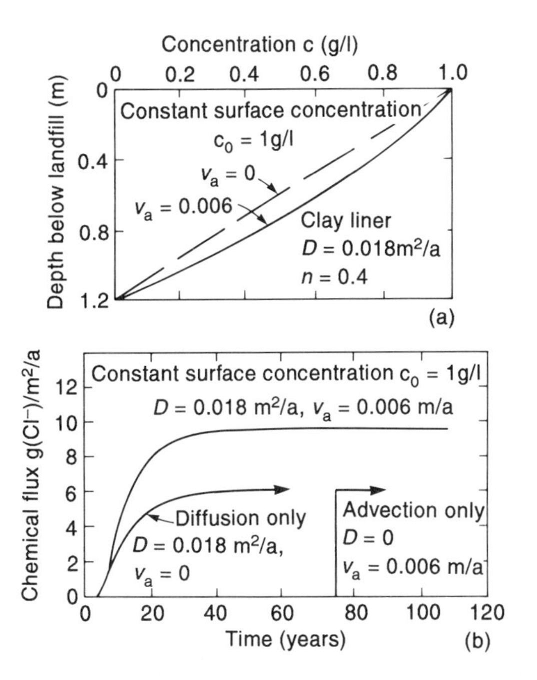

Figure 1.21 Effect of assumptions used to model contaminant migration through a liner: (a) final (steady-state) concentration profile through the liner; (b) chemical flux passing out of the liner (exit flux) assuming constant contaminant concentration in the liner.

into the underlying aquifer (i.e. the exiting chemical flux or exit flux), will increase with increasing Darcy velocity v_a.

It may be tempting to estimate the peak flux loading, f, on the aquifer by performing two simple hand calculations:

$$f = -nD\frac{\partial c}{\partial z} = \frac{nDc_0}{H} \quad (\text{g/m}^2/\text{a})$$
(assuming $v_a \ll nD$) (1.17a)

$$f = nvc_0 = v_a c_0 \quad (\text{g/m}^2/\text{a})$$
(assuming $v_a \gg nD$) (1.17b)

where H is the thickness of the liner (m) and c_0 is the constant leachate concentration (1 g/L = 1000 g/m^3). However, for cases where neither v_a nor D dominate, this approach can give misleading results. For example, Figure 1.21(b) shows the variation in exit flux with time for the case of pure diffusion ($D = 0.018$ m^2/a, $v_a = 0$), pure advection ($D = 0$, $v_a = 0.006$ m/a) and advective–diffusive transport ($D = 0.018$ m^2/a, $v_a = 0.006$ m/a). Coincidentally, in this example the maximum flux of 6 g/m^2/a is identical for both the pure diffusion and pure advection cases. Conventional calculations performed neglecting diffusion and assuming plug flow ($D = 0$, $v_a = 0.006$ m/a) suggest that no contaminant would escape into the aquifer until the seepage front arrived at the base after 75 years. However, diffusion is important and due to diffusion alone an exit flux exceeding 10% of the maximum flux would be expected after only five years. Indeed, the maximum flux of 6 g/m^2/a would be attained after only 50 years (compared to 75 years for plug flow). Consideration of both advection and diffusion gives a substantially higher flux at any time with the peak flux being 55% higher than the peak flux obtained by considering either diffusion or advection independently.

In many practical situations involving clay liners, the hydraulic conductivity will be less than 10^{-9} m/s and the hydraulic gradient less than 0.2. These cases will involve Darcy velocities $v_a = 0.006$ m/a or less.

By performing calculations such as those for which the results are shown in Figure 1.21 for a range of velocities v_a, it is possible to establish when advection dominates over diffusion (and *vice versa*), while from equation 1.16 it is also possible to estimate the range of velocities over which diffusion dominates over mechanical dispersion (and *vice versa*), and the results are summarized in Figure 1.22. At velocities likely to be encountered with a well-constructed composite liner (geomembrane plus compacted clay), diffusion is likely to dominate. For velocities often encountered with outward gradients and a good compacted clay liner ($k \approx 10^{-10}$ m/s), diffusion and advection may both be of significance, but not mechanical dispersion. On the other hand, in sand and gravel aquifers, advection *will* dominate over diffusion and dispersion *may* dominate over diffusion.

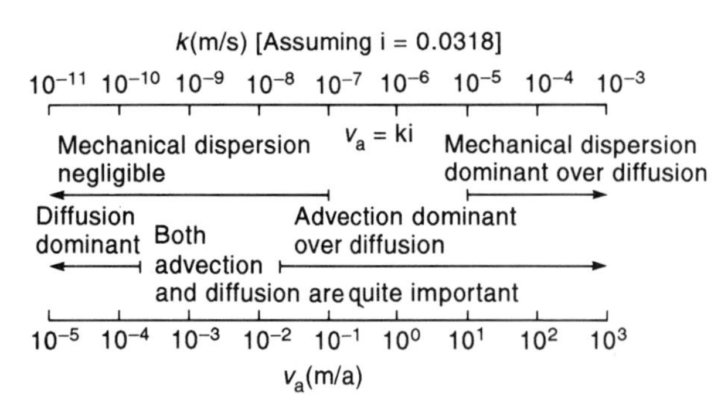

Figure 1.22 Range of Darcy velocities over which diffusion or mechanical dispersion may control the coefficient of hydrodynamic dispersion, and the range of velocities over which diffusion or advection may control the magnitude of the exit flux for a 1.2 m thick liner. (Modified from Rowe, 1987.)

1.4 Complicating factors

1.4.1 Unsaturated soils

The advective–diffusive movement of contaminants through unsaturated soils is somewhat more complicated than for saturated soils. The partial differential equation governing one-dimensional movement is given by

$$\frac{\partial}{\partial t}(\theta c) = \frac{\partial}{\partial z}\left(\theta D \frac{\partial c}{\partial z}\right) - \frac{\partial}{\partial z}(v_a c) - \theta\lambda c \qquad (1.18)$$

where θ is the volumetric water content (equal to porosity for a saturated soil) [–]; λ is a term which takes account of sorption and biological, chemical and radioactive decay [T^{-1}]. All other terms are as previously defined.

This equation bears a marked similarity to equation 1.14b, but this similarity may be deceptive. For an unsaturated soil the volumetric water content θ, the coefficient of hydrodynamic dispersion D and the Darcy velocity may vary in both space and time.

The movement of contaminant through unsaturated soils is a very complex phenomenon, as demonstrated by a number of laboratory and field studies (e.g. de Smedt, 1981; Gerhardt, 1984). The simplest case is that in which there is negligible advective transport through the unsaturated soils. This situation can only arise when the net infiltration is negligible (e.g. below an intact composite liner). Under these circumstances, the migration of contaminant in solution will be very slow since the migration will be purely by diffusion, and it has been shown (e.g. Klute and Letey, 1958; Porter *et al.*, 1960; Rowe and Badv, 1994a,b,c) that the effective diffusion coefficient in unsaturated soils may (at least in some circumstances) be substantially lower than in similar saturated soils.

In humid climates the unsaturated soil will usually be hydraulically active, and advective transport (which may vary with time) must be considered. As noted above, the diffusion coefficient is dependent on the volumetric water content, and hence will vary both spatially and temporally in a hydraulically active region. The advective transport will depend, in part, on the hydraulic conductivity of the soil. This tends to increase with the volumetric water content of the soil up to a maximum value for a saturated soil (e.g. Gardner, 1958; van Genuchten, 1978). Thus, the hydraulic conductivity of an unsaturated soil will be far more sensitive to point-to-point variations in grain size distribution than that of saturated soils, and this alone makes the determination of representative hydraulic conductivities substantially more difficult. Additional uncertainty arises from the effects of seasonal variations in infiltration and assumptions concerning the expected long-term weather pattern which may influence calculated contaminant transport through unsaturated soils.

Various investigators have questioned the direct application of equation 1.18 for unsaturated soils (e.g. Gaudet *et al.*, 1977; de Smedt, 1981). The problem tends to be manifest as an apparent dispersion well in excess of what would be expected for a saturated soil. In an attempt to explain this phenomenon, various researchers (e.g. Rao *et al.*, 1974; Gaudet *et al.*, 1977; de Smedt, 1981) have proposed multiple water phase models which involve advective–dispersive transport through the mobile water (typically in the smaller saturated pores) together with diffusion of contaminant into (or from) the immobile water (generally in the larger unsaturated pores). This approach appears to give reasonable agreement with experiments, although the parameters used are generally selected by matching the experimental and theoretical behavior. De Smedt (1981) has shown that a reasonable fit to his experimental data could also be obtained using equation 1.18 and an effective dispersion coefficient D given by

$$D = \frac{\theta_m}{\theta} D_m + \frac{\theta_{im}^2 v^2}{\theta\, \theta_m \beta} \qquad (1.19)$$

where D_m is the dispersion coefficient in the mobile water $[L^2T^{-1}]$; θ_{im}, θ_m, θ are volumetric water content in the immobile and mobile phases, and the bulk soil respectively [–]; v is the seepage velocity (v_a/θ) $[LT^{-1}]$; β is the coefficient for solute transport between phases $[T^{-1}]$.

An inspection of equation 1.19 indicates the difficulties of using this approach in practice, since the parameters β, v, θ_m, θ_{im}, θ may be expected to vary both temporally and spatially. Furthermore, the parameter β must be determined by curve-fitting laboratory results for a particular situation.

It may be concluded that even though some significant progress has been made concerning the prediction of contaminant transport through unsaturated soils, this is still a formidable undertaking.

1.4.2 Fractured porous media

When contaminants move through fractured porous media (e.g. fractured clay, shale, sandstone) the movement of contaminants along the fractures is typically by advection and dispersion. However, as contaminants move along these fractures, there is usually a difference between the concentration of a given species in the fracture at a given point and the concentration in the pore water of the adjacent intact material. This concentration difference (gradient) gives rise to diffusion of contaminants between the fluid in the fractures and the pore water in the matrix of the adjacent material (the matrix pore water). For example, if the concentration of contaminant in the fractures is higher than in the matrix pore water, then contaminant will diffuse into the matrix, thereby reducing the concentration in the fracture. This phenomenon is referred to as matrix diffusion, and can have a significant effect on the movement of contaminants through fractured media, as discussed in Chapter 11.

The one-dimensional movement of contaminants in a fractured system consisting of a set of parallel fractures (more complex cases are considered in Chapter 7) can be written as

$$n_f \frac{\partial c_f}{\partial t} = n_f D \frac{\partial^2 c_f}{\partial z^2} - v_a \frac{\partial c_f}{\partial z} - \Delta K_f \frac{\partial c_f}{\partial t} - n_f \lambda c_f - \dot{q} \qquad (1.20)$$

where n_f is the fracture porosity [–]; c_f is the concentration in a fracture at depth z and time t $[ML^{-3}]$; D is the coefficient of hydrodynamic dispersion in the fractures $[L^2T^{-1}]$; v_a is the Darcy velocity $[LT^{-1}]$; Δ is the surface area of fracture per unit volume within the fracture medium $[L^{-1}]$; K_f is the fracture distribution coefficient for sorption onto the fracture surface, defined by Freeze and Cherry (1979) as the mass of solute adsorbed per unit area of surface divided by the concentration of solute in solution [L]; λ is the first order decay constant $[T^{-1}]$; $\dot{q}$ is the rate at which contaminant is being transported into the matrix (per unit volume) by diffusion from the fractures $[ML^{-3}T^{-1}]$.

The rate of contaminant transport into the matrix, $\dot{q}$, depends on the porosity of the intact material, the diffusion coefficient of the matrix material and fracture spacing. It is evaluated for one-, two- and three-dimensional conditions in Chapter 7 and the effect of varying these parameters is examined in Chapter 11.

1.5 Modeling the finite mass of contaminant

In many practical situations, the mass of contaminant within a landfill (in the field) or a source reservoir (in the laboratory) is limited and mass will be reduced as contaminant is transported into the soil. In terms of the analogy shown in Figures 1.14 and 1.15, the number of people in the airport holding lounge (after disembarkation from their plane) is limited. As people move out of the lounge, the number of people remaining in the lounge drops, and hence so does the concentration (i.e. number of people per unit area). The only way that the concentration can increase again is if another plane arrives and more people (mass) are added to the holding lounge. While this may happen in an airport holding lounge it does not normally happen in landfills after the landfill is closed.

Experience has shown that the concentration of potential contaminants generally increases during operation of the disposal facility, reaches a peak and then declines. The increase in concentration may be related to:

1. the physical processes of leaching of contaminant from solid waste as water infiltrates through the waste; and/or
2. chemical and biological processes which generate the chemical species of interest from the synthesis, or breakdown, of existing chemical species in the waste (e.g. due to biological action).

Likewise, the decrease in concentration with time may be related to:

1. the physical process of removal of contaminant, in the form of leachate, from the landfill, and/or
2. chemical and biochemical processes which result in precipitation and/or the synthesis or breakdown of the chemical species of interest into other chemical forms.

In the design of barrier systems it is generally not practical to model the details of the leaching processes or of any associated chemical or biological processes. However, reasonable engineering approximations can be made which will allow the designer to obtain some insight into the potential impact of the finite mass of contaminant. Thus, for the purposes of performing design calculations, it is often conservative to assume that:

1. the concentration of a contaminant of interest reaches the peak concentration, c_0, instantaneously;
2. all of the mass of this contaminant species, m_{TC}, is in solution at the time that the peak concentration occurs.

The mass of contaminant available for transport into the soil can be represented in terms of the peak concentration, c_0, and the reference height of leachate, H_r, or the equivalent height of leachate, H_f. In the case of laboratory diffusion tests, H_r and H_f are identical and correspond to the actual height of source fluid (leachate) directly above the soil. In the case of the landfill, H_r may be defined for each contaminant species of interest and corresponds to the volume of fluid (per unit area of landfill) that, at a concentration c_0, would contain the total mass, m_0, of that contaminant species which could be released either for transport or collection. It does not include contaminant that is, and is expected always to be, in a solid immobile form, or contaminant that is released in the gas phase. The equivalent height of leachate, H_f, has a similar definition except that it only corresponds to that portion of the mass that is available for transport into the hydrogeologic system. Thus the essential difference between H_r and H_f is that H_r includes the mass collected by the leachate collection system, while H_f excludes this mass.

Considering conservation of mass within the source solution, one can write:

$$\begin{bmatrix}\text{Mass of contaminant}\\ \text{within source at}\\ \text{time } t\end{bmatrix} = \begin{bmatrix}\text{Initial mass of}\\ \text{contaminant}\\ \text{with source}\end{bmatrix} - \begin{bmatrix}\text{Mass of}\\ \text{contaminant}\\ \text{transported}\\ \text{into the soil}\end{bmatrix} - \begin{bmatrix}\text{Mass of contaminant}\\ \text{lost due to}\\ \text{first order decay}\\ \text{processes}\end{bmatrix}$$

$$m_t = m_{TC} - m - m_{DC} \tag{1.21}$$

Substituting equation 1.6 for the mass of contaminant transported into the soil gives

$$m_t = m_{TC} - A_0 \int_0^t \times \left(nvc(\tau) - nD \frac{\partial c(\tau)}{\partial z} \right) d\tau - m_{DC} \tag{1.22}$$

where $m_t = A_0 H_r c(t)$ is the mass of contaminant in source at time t [M]; $m_{TC} = A_0 H_r c_0$ is the initial mass of contaminant in the landfill [M]; $m_{DC} = A_0 H_r \int_0^t \lambda_T c(\tau)\, d\tau$ is the mass lost due to first order decay processes [M]; $c(t)$ is the concentration in the landfill at time t; A_0 is the area of landfill through which contaminant can pass into the soil; $\lambda_T = \Gamma_R + \Gamma_{BT} + \Gamma_{ST}$ is the first order decay constant [T^{-1}]; Γ_R is the decay constant for radioactive decay within the source [T^{-1}]; Γ_{BT} is the decay constant for biological decay within the source [T^{-1}]; Γ_{ST} is the volume of leachate withdrawn from unit volume of the source during unit time (e.g. from the leachate collection system) [T^{-1}]; H_r is the reference height of leachate [L].

Thus, if f_T is the flux entering the surface of the deposit and c_T is the concentration at the surface, then

$$c_T(t) = c_0 - \frac{1}{H_r} \int_0^t f_T(\tau)\, d\tau - \int_0^t \lambda_T c_T(\tau)\, d\tau \tag{1.23a}$$

If $\lambda_T = 0$ and $H_r \to \infty$, equation 1.23a leads to the boundary condition $c_T(t) = c_0$.

Equation 1.23a explicitly models both the full mass of a given contaminant in the landfill ($m_{TC} = A_0 H_r c_0$) and removal of contaminant from the landfill (e.g. by a leachate collection system) in terms of Γ_{ST}. As discussed in more detail in Chapter 10, an alternative approach is to reduce the total mass to the amount available for transport into the groundwater ($m_0 = A_0 H_f c_0$) (i.e. excluding that portion of the mass which will be collected by a primary leachate collection system). For this case the finite mass boundary condition can be written in terms of the equivalent height of leachate, H_f, as indicated in Figure 1.23. This gives rise to the boundary condition

$$c_T(t) = c_0 \frac{1}{H_f} \int_0^t f_T(\tau)\, d\tau \tag{1.23b}$$

where H_f = equivalent height of leachate [L].

For the case where $\lambda_T = 0$ (no mass removed from landfill except by migration into the underlying barrier system or groundwater) then $H_r = H_f$ and equations 1.23a and 1.23b are identical.

1.6 Modeling a thin permeable layer as a boundary condition

In many situations (e.g. Figures 1.2 to 1.8) it may be desirable to be able to model advective–diffusive transport through a relatively thin clayey barrier and along an aquifer of thickness h and porosity n_b (Figure 1.24). If there is (say) advective transport through the barrier and into the aquifer, then, strictly speaking, continuity requires that the velocity in the aquifer should vary with horizontal position below the barrier. However, in many practical situations where interest is focused on impact at the downgradient edge of the landfill, or a boundary compliance point close to the edge of the landfill, the aquifer velocity, v_b, may be assumed to be uniform at a value equal to that at the downgradient edge of

$$\begin{bmatrix}\text{Mass of Contaminant}\\ \text{in the landfill at time } t\\ m_t = A_0 H_f c_T(t)\end{bmatrix} = \begin{bmatrix}\text{Initial Mass}\\ \text{of contaminant}\\ m_0 = A_0 H_f c_0\end{bmatrix} - \begin{bmatrix}\text{Mass which has}\\ \text{passed into the}\\ \text{soil up to time } t\\ A_0 \int_0^t f_T(c,\tau)\mathrm{d}\tau\end{bmatrix}$$

$$\therefore m_t = m_0 - A_0 \int_0^t f_T(c,\tau)\mathrm{d}\tau$$

and

$$A_0 H_f\, c_T(t) = A_0 H_f c_0 - A_0 \int_0^t f_T(c,\tau)\mathrm{d}\tau$$

where A_0 is the area of landfill through which contaminant can pass into the soil;
H_f is the equivalent height of leachate;
$f_T(c,\tau)$ is the surface flux (mass per unit area per unit time) passing into the soil.

Thus dividing by the equivalent volume of leachate ($A_0 H_f$) gives

$$c_T(t) = c_0 - \frac{1}{H_f}\int_0^t f_T(c,\tau)\mathrm{d}\tau$$

Figure 1.23 Finite mass boundary condition to simulate the landfill on top of the liner.

the landfill. If it is also assumed that the concentration in the aquifer is uniform across its thickness, then consideration of conservation of mass within a small element of the aquifer between $(x, x + \mathrm{d}x)$ gives

$$\begin{aligned} n_b h\,\mathrm{d}x\, c_b(x,t) &= \mathrm{d}x \int_0^t f_b(x,c,\tau)\mathrm{d}\tau \\ &- h \int_0^t [f_x(x + \mathrm{d}x,c,\tau) - f_x(x,c,\tau)]\mathrm{d}\tau \\ &- n_b h\,\mathrm{d}x \int_0^t \lambda^*_b\, c_b(x,\tau)\mathrm{d}\tau \end{aligned} \tag{1.24}$$

which follows from the observation that the mass of contaminant in the element at a specific time, t, is equal to the total mass transported into the element from the clayey barrier less the net mass transported out of the element in the x direction and the mass loss due to first order decay. Assuming that the mass flux, f_x, into the aquifer is governed by equation 1.8, dividing throughout by the pore volume, $n_b h\,\mathrm{d}x$, and taking limits as $\mathrm{d}x$ tends to zero gives

$$\begin{aligned} c_b(x,t) &= \int_0^t \left(\frac{f_b(x,c,\tau)}{n_b h} - \frac{v_b}{n_b}\frac{\partial c_b(x,\tau)}{\partial x} + D_H \frac{\partial^2 c_b(x,\tau)}{\partial x^2} \right)\mathrm{d}\tau \\ &- \int_0^t \lambda^*_b c_b(x,\tau)\mathrm{d}\tau \end{aligned} \tag{1.25}$$

where n_b is the porosity of the aquifer [–]; h is the thickness of the aquifer [L]; v_b is the Darcy velocity in the aquifer (in the x direction) (LT^{-1}]; D_H is the coefficient of hydrodynamic dispersion in the aquifer (x direction) (L^2T^{-1}]; $c_b(x,\tau)$ is the concentration in the aquifer at position x at time τ [ML^{-3}]; $f_b(x,\tau)$ is the flux into the aquifer at position x at time τ [$ML^{-2}T^{-1}$]; λ^*_b is the first order decay constant for the base, and includes the effect of radioactive decay Γ_R and biological decay Γ_{Bb}, but does not include Γ_{Sb}, the effect of 'washing out' due to horizontal flow since this is explicitly considered by another term in the equation [T^{-1}].

For two-dimensional problems an aquifer can be modeled as a boundary condition by invoking equation 1.25. As will be discussed in Chapters 7 and 10, there are many practical situations where additional simplification is possible by considering one-dimensional (1-D) advective–diffusive transport down to the aquifer (e.g. vertically) combined with 1-D advective (e.g. horizontal) transport out from beneath the landfill. Thus, for a landfill of width W and length L (where L is the dimension in the

[Mass in base beneath the landfill at time t] = [Mass transported into the base up to time t] − [Mass transported out from beneath the base up to time t]

$$\therefore m(t) = \int_0^t WLf_b(c,\tau)d\tau - \int_0^t Wh\, v_b c_b d\tau$$

where

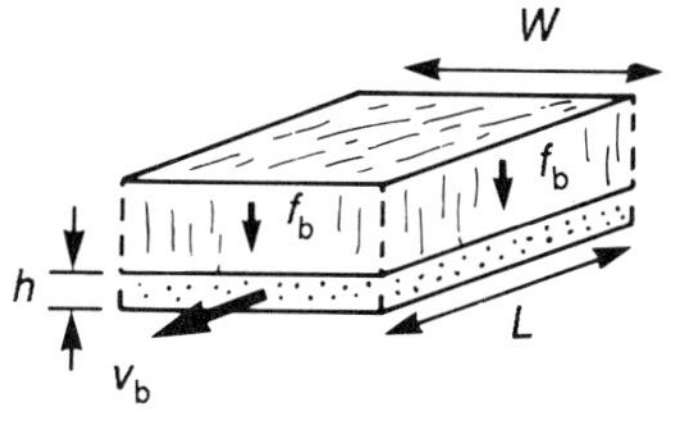

$f_b(c,\tau)$ is the flux into the base ($z = H$) at time τ;
W is the width of the landfill;
L is the length of the landfill;
c_b is the concentration in the base at time τ ($= m(t)/(WLhn_b)$);
h is the thickness of permeable layer;
n_b is the porosity of the permeable layer;
v_b is the Darcy velocity in permeable layer.

Dividing by the volume of fluid in the base ($WLhn_b$),

$$c_b(t) = \int_0^t \frac{f_b(c,\tau)}{n_b h} d\tau - \int_0^t \frac{v_b c_b}{n_b L} d\tau$$

Figure 1.24 Boundary conditions for modeling a thin aquifer beneath a liner.

direction of groundwater flow in the underlying aquifer), as shown in Figure 1.24, the equation for concentration in the aquifer (equation 1.25) reduces to

$$c_b(t) = \int_0^t \left(\frac{f_b(c_b,\tau)}{n_b h} - \lambda_b c_b(\tau) \right) d\tau \qquad (1.26)$$

where $\lambda_b = \lambda_b^* + \Gamma_{Sb} = \Gamma_R + \Gamma_{Bb} + v_b/(n_b L)$.

1.7 Hand solutions to some simple problems

Analytical solutions can be obtained for simplified cases typically involving a single homogeneous layer (barrier) subject to simplified boundary conditions. Numerous solutions have been reported in the literature (e.g. Lapidus and Amundson, 1952; Ogata and Banks, 1961; Ogata, 1970; Lindstrom *et al.*, 1967; Selim and Mansell, 1976; Rowe and Booker, 1985a; Booker and Rowe, 1987). Although these analytical solutions have sometimes been expressed in graphical form suitable for hand computation (e.g. Ogata, 1970; Booker and Rowe, 1987), the evaluation of the analytical solution generally requires the use of a computer. The primary uses of analytical solutions are to:

1. perform quick sensitivity studies and preliminary design calculations;
2. check the results of more sophisticated analyses.

1.7.1 Ogata–Banks equation

Probably the best known analytical solution is that for the concentration c at time t and depth z beneath the surface of a barrier which is assumed to be infinitely deep and subject to a constant surface concentration c_0 (e.g. Ogata, 1970):

$$c(z,0) = 0 \qquad z > 0 \qquad (1.27a)$$
$$c(0,t) = c_0 \qquad t \geqslant 0 \qquad (1.27b)$$
$$c(\infty,t) = 0 \qquad t \geqslant 0 \qquad (1.27c)$$

This solution can be written in the form

$$\frac{c}{c_0} = \frac{1}{2}\left[\operatorname{erfc}\left(\frac{z - vt}{2(Dt)^{1/2}}\right) + \exp\frac{vz}{D}\operatorname{erfc}\left(\frac{z + vt}{2(Dt)^{1/2}}\right)\right] \quad (1.28)$$

i.e. $\operatorname{erfc}(x) = 1 - \operatorname{erf}(x)$. For contaminant species which are retarded due to sorption and have a retardation coefficient $R = 1 + \varrho K_d/n$, equation 1.28 may be used by scaling the time of interest τ such that $t = \tau/R$. Thus when one assumes a constant source concentration, sorption simply slows the rate of advance of the contaminant plume.

Equation 1.28 may be readily programmed or may be solved graphically using Figure 1.25, as illustrated in Figures 1.26 and 1.27.

The arrows in Figure 1.25 are located by the calculation in Figure 1.26 that for a nonretarded species with $v_a = 0.002$ m/a ($v_s = 0.005$ m/a)

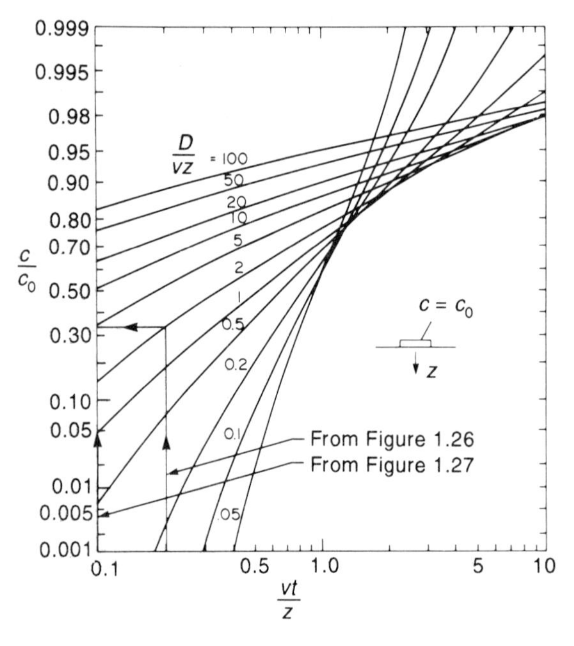

Figure 1.25 Ogata and Banks graphical solution to the advection–dispersion equation for an infinitely deep deposit with a constant surface concentration. (Modified from Ogata and Banks, 1961.)

Assume:	Downward Darcy velocity	$v_a = 0.002$ m/a
	Diffusion coefficient	$D = 0.02$ m²/a
	Porosity	$n = 0.4$
	No sorption	$\rho K_d = 0$
	Source concentration	$c_0 = 1500$ mg/l
	Time of interest	$t = 80$ years
	Depth of interest	$z = 2$ m

Deduce: $v = v_a/n = 0.002/0.4 = 0.005$ m/a

$$\frac{D}{vz} = \frac{0.02}{0.005 \times 2} = 2$$

$$\frac{vt}{z} = \frac{0.005 \times 80}{2} = 0.2$$

From graphical solution: $c/c_0 = 0.33$

$\therefore c = 0.33 \times 1500 = 500$ mg/l

Figure 1.26 Example problem A1 using the Ogata and Banks solution for a conservative contaminant: $H_f = \infty$.

and $D = 0.02$ m²/a, at a depth of 2 m the concentration would have increased to a third of the source value (i.e. 500 mg/L in this case) after 80 years' migration.

Figure 1.27 shows an example calculation assuming $v_a = 0.002$ m/a, $D = 0.01$ m²/a, $\varrho K_d = 1.2$. In particular, it is noted that a modest level of sorption as implied by $\varrho K_d = 1.2$ gives rise to a retardation coefficient $R = 4$ and this reduced the effective time used in the calculation by a factor of four (i.e. from $\tau = 160$ years to $t = 40$ years). Hence, for the example, the impact at $z = 2$ m after 160 years is 75 mg/L (i.e. 5% of the initial concentration).

It should be noted that in the example shown in Figure 1.27 the value of $D = 0.01$ is half that used in Figure 1.25. If the same value of D had been used (i.e. $D = 0.02$ m²/a) this would give $D/vz = 2$ and hence $c/c_0 = 0.15$ or $c \approx 225$ mg/L.

1.7.2 Booker–Rowe equation

In section 1.5 it was indicated that the finite mass of contaminant within a landfill could be represented in terms of an equivalent height of leachate H_f. An analytical solution for one-dimensional migration in an infinitely deep

Assume: Downward Darcy velocity $v_a = 0.002$ m/a
Diffusion coefficient $D = 0.01$ m²/a
Porosity $n = 0.4$
Sorption $\rho K_d = 1.2$
Source concentration $c_0 = 1500$ mg/l
Time of interest $\tau = 160$ years
Depth of interest $z = 2$ m

Deduce: $v = v_a/n = 0.005$ m/a

$$R = 1 + \rho K_d/n = 1 + \frac{1.2}{0.4} = 4$$

$$\frac{D}{vz} = \frac{0.01}{0.005 \times 2} = 1$$

$$t = \frac{\tau}{R} = \frac{160}{4} = 40$$

$$\frac{vt}{z} = \frac{0.005 \times 40}{2} = 0.1$$

From graphical solution: $c/c_0 \approx 0.05$
$c = 0.05 \times 1500 = 75$ mg/l

(A more rigorous analysis gives $c = 61$ mg/l)

Figure 1.27 Example problem B1 using the Ogata and Banks solution for a reactive contaminant: $H_f = \infty$.

deposit where the source concentration varies with time (as mass is transported into the barrier), has been given by Booker and Rowe (1987) and the concentration at any depth z and time t can be written as

$$c(z,t) = c_0 \exp(ab - b^2t)(bf(b,t) - df(d,t))/(b-d) \quad (1.29a)$$

where

$$f(b,t) = \exp(ab + b^2 t)\text{erfc}\left(\frac{a}{2t^{1/2}} + bt^{1/2}\right) \quad (1.29b)$$

$$f(d,t) = \exp(ad + d^2t)\text{erfc}\left(\frac{a}{2t^{1/2}} + dt^{1/2}\right) \quad (1.29c)$$

$$a = z\left(\frac{(n + \varrho K_d)}{nD}\right)^{1/2}$$

$$b = v\left(\frac{n}{4D(n + \varrho K_d)}\right)^{1/2}$$

$$d = \frac{nD}{H_f}\left(\frac{n + \varrho K_d}{nD}\right)^{1/2} - b$$

and all other terms are as previously defined. The function $f(p,t)$ (for $p = b$ or $p = d$) may be evaluated directly by computer or using a hand calculator by observing that

$$f(p,t) = \exp(-a^2/4t)\phi(x)$$

where $x = (pt^{1/2} + a/(2t^{1/2}))$ and the function $\phi(x)$ is given in Figure 1.28. With some algebra, equation 1.29 reduces to equation 1.28 for $H_f \rightarrow \infty$ (i.e. $d = -b$).

Figures 1.29 and 1.30 illustrate the use of the graphical method. The cases analyzed are the same as those examined in Figures 1.26 and 1.27 respectively except that the equivalent height of leachate, H_f, is taken to be 1 m in Figures 1.29 and 1.30 compared to $H_f = \infty$ for Figures 1.26 and 1.27. Comparing the results obtained for Figures 1.26 and 1.29 (or 1.27 and

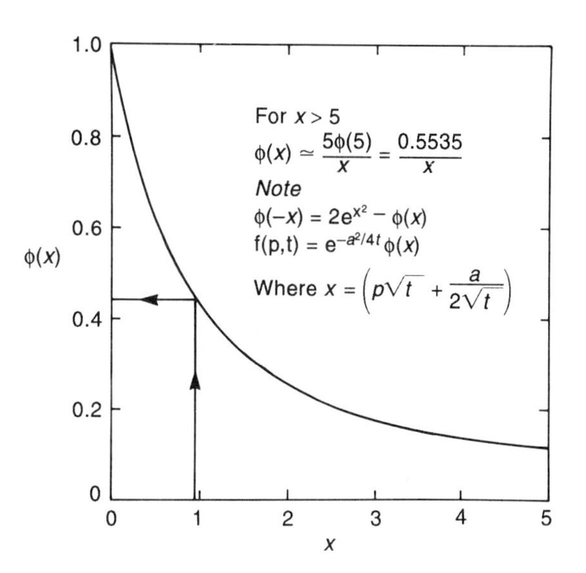

Figure 1.28 Booker and Rowe graphical solution to the advection–dispersion equation for an infinitely deep deposit with a finite mass of contaminant in the source. (Modified from Booker and Rowe, 1987; reproduced with permission of the *International Journal for Numerical and Analytical Methods in Geomechanics*.)

Deduce: $v = 0.002/0.4 = 0.005$ m/a

$$a = z\left(\frac{n + \rho K_d}{nD}\right)^{1/2} = 2\left(\frac{0.4 + 0}{0.4 \times 0.02}\right)^{1/2} = 14.1$$

$$b = v\left(\frac{n}{4D(n + \rho K_d)}\right)^{1/2} = 0.005\left(\frac{0.4}{4 \times 0.02(0.4 + 0)}\right)^{1/2} = 0.0177$$

$$d = \frac{nD}{H_f}\left(\frac{n + \rho K_d}{nD}\right)^{1/2} - b = \frac{0.4 \times 0.002}{1}\left(\frac{0.4 + 0}{0.4 \times 0.02}\right)^{1/2} - 0.0177$$

$$= 0.039$$

$f(b,t) = \exp(-a^2/4t)\ \phi(bt^{1/2} + 0.5a/t^{1/2})$
$= 0.537\ \phi(0.946)$

From graphical solution: $\phi(0.946) = 0.44$
$\therefore f(b,t) = 0.537 \times 0.44 = 0.236$

$f(d,t) = \exp(-a^2/4t)\ \phi(dt^{1/2} + 0.5a/t^{1/2})$
$= 0.537\ \phi(1.137)$

From graphical solution: $\phi(1.137) = 0.39$
$\therefore f(d,t) = 0.537 \times 0.39 = 0.209$

$c = c_0 \exp(ab - b^2t)\ [bf(b,t) - df(d,t)]/(b - d)$
$= 1500 \exp(0.22)\ [0.0177 \times 0.236 - 0.039 \times 0.209]/(0.0177 - 0.039)$
$= 350$ mg/l

(cf. $c = 500$ mg/l if $H_f = \infty$)

Figure 1.29 Example problem A2 using the Booker and Rowe solution for a conservative contaminant: $H_f = 1$ m.

Deduce: $v = 0.002/0.4 = 0.005$ m/a

$$a = z\left(\frac{n + \rho K_d}{nD}\right)^{1/2} = 2\left(\frac{0.4 + 1.2}{0.4 \times 0.01}\right)^{1/2} = 40$$

$$b = v\left(\frac{n}{4D(n + \rho K_d)}\right)^{1/2} = 0.005\left(\frac{0.4}{4 \times 0.01(0.4 + 1.2)}\right)^{1/2} = 0.0125$$

$$d = \frac{nD}{H_f}\left(\frac{n + \rho K_d}{nD}\right)^{1/2} - b = \frac{0.4 \times 0.001}{1}\left(\frac{0.4 + 1.2}{0.4 \times 0.02}\right)^{1/2} - 0.0125$$

$$= 0.675$$

$f(b,t) = \exp(-a^2/4t)\ \phi(bt^{1/2} + 0.5a/t^{1/2})$
$= 0.082\ \phi(1.74)$

From graphical solution: $\phi(1.74) = 0.28$
$\therefore f(b,t) = 0.082 \times 0.28 = 0.023$

$f(d,t) = \exp(-a^2/4t)\ \phi(dt^{1/2} + 0.5a/t^{1/2})$
$= 0.082\ \phi(2.43)$

From graphical solution: $\phi(2.43) \approx 0.2$
$\therefore f(d,t) = 0.082 \times 0.2 = 0.0164$

$c = c_0 \exp(ab - b^2t)\ [bf(b,t) - df(d,t)]/(b - d)$
$= 1500 \times 0.024$
$= 36$ mg/l

(cf. $c = 60$ mg/l if $H = \infty$)

Figure 1.30 Example problem B2 using the Booker and Rowe solution for a reactive contaminant: $H_f = 1$ m.

1.30) it is seen that when one considers the finite mass of contaminant, the concentration at the point and time of interest is substantially reduced. This matter will be discussed in greater detail in Chapter 10.

1.8 Summary

This chapter has provided an introduction to some of the design considerations associated with a number of different barrier systems. This has included a discussion which has dealt with clay and geomembrane liners, leachate collection and leak detection systems, hydraulic control layers and hydraulic traps.

The contaminant transport processes of advection, diffusion and dispersion have been described and it has been demonstrated that for low permeability unfractured clayey barriers, diffusion may be a significant and, in many cases, the dominant transport mechanism.

The roles of sorption, biodegradation, partial saturation and matrix diffusion as retardation mechanisms have also been discussed briefly. Based on this introduction to concepts, the governing differential equations for contaminant transport have been developed for simple cases. The finite mass of contaminant has been discussed and simple boundary conditions for a landfill (or other source) having a finite mass of contaminant and for liners underlain by a thin aquifer have been presented. Finally, the solution of the governing equations for contaminant transport through a clayey barrier has been illustrated by means of a number of simple examples.

Chapter two

Design considerations

2.1 Introduction

Attitudes regarding the design of municipal waste landfills and other disposal facilities for industrial and mining waste have changed significantly over the last couple of decades. It is not so long ago that waste was simply dumped in old rock quarries or gravel pits, and it was assumed that the impact would be reduced to acceptable levels, due to dilution of the leachate or chemical waste by the groundwater. In response to the environmental problems that this approach has caused, many governments are now regulating the design of waste disposal facilities. In the USA, for example, the Congress has mandated regulations relating to the minimum design for waste disposal sites which involve a single composite liner and a landfill cover no more permeable than the liner as the standard design. Alternative designs may be permitted in approved states. In general, interest is only focused on a 30-year period after closure of the landfill. While this approach is much better than uncontrolled disposal of waste, it does not necessarily ensure that 'good' designs are implemented and that a site which meets the Environmental Protection Agency (EPA) requirements will not be an environmental hazard for the future (Rowe, 1991b).

In the Province of Ontario, Canada, the proponent of a disposal site is required to demonstrate that the proposed facility will not impact on the present or future 'reasonable use' of groundwater at the site boundary (MoEE, 1993a). In essence, this recognizes that over the long term it is almost impossible to guarantee that there will be no impact on groundwater due to a waste disposal facility. While it is impossible to prove that any facility will never cause an environmental problem, it is possible to design a facility which, based on what we know today, should not impact on the reasonable use of groundwater, and hence on the environment. Similarly, it is also possible to identify designs which, based on what we know today, have a reasonable likelihood of contaminating groundwater and the environment beyond the site. It may be that it will be tens or even hundreds of years before there is a serious impact; however, a long time to impact does not justify a design in the context of modern attitudes towards protecting the environment, not only for ourselves, but also for future generations.

The previous chapter provided an introduction to barrier systems and a number of design considerations associated with these systems. This chapter provides a background discussion concerning approaches to contaminant impact assessment, the contaminating lifespan of a landfill, and some factors to be considered when assessing the service life of the components of an engineered system. The objective of the design will be to design a system such that even when allowance is made for the uncertainty concerning the service life of some components of the system, the system itself can be designed such that the service life of the system exceeds the contaminating lifespan of the landfill. The material presented in this and subsequent chapters provides the basis for developing designs which meet this objective.

2.2 Impact assessment

As noted in Chapter 1, the design of a barrier system is often intimately related to the environmental impact assessment of the proposed landfill. Environmental impact assessments are generally driven by regulatory requirements and, as a consequence, typical 'acceptable' barrier systems can be expected to vary regionally as a result of both variations in hydrogeologic conditions and variations in regulatory requirements. The fundamental question underlying most impact assessments is whether the proposed landfill will have no more than a negligible effect on groundwater quality at the site boundary. However, this perfectly reasonable question raises two subsidiary questions – what is a 'negligible effect' and over what period of time must the effect be negligible? The answer to the latter question has very important implications since it is intimately related to the design life of the engineering features of the facility. The design of a barrier system for a landfill that is only required to have 'negligible' impact for a 30-year post-closure period is likely to be different from the design for a 100-year period which, in turn, may be quite different from that required to have negligible effect on groundwater quality in perpetuity.

Typically, environmental regulations fall into one of the following categories:

1. essentially no regulation;
2. prescriptive regulations which specify minimum requirements such as 'two liners of which at least one is a synthetic liner';
3. regulations requiring 'no impact' or 'negligible impact' for a prescribed period of time (e.g. 30 years or 100 years post-closure);
4. regulations requiring negligible impact in perpetuity.

2.2.1 Nonexistent regulations

The situation where there is no regulation provides considerable latitude to the landfill proponent and designer in terms of barrier system adopted. It also provides little assurance that the environment will be protected unless the design is subjected to a rigorous preconstruction review.

2.2.2 Prescriptive regulations

Prescriptive regulations are simple. They are typically based on the perception of the regulators as to what constitutes a safe design, and will implicitly have negligible effects on groundwater quality. Unfortunately, prescriptive regulations may not recognize that potential impact is not only related to the details of the barrier, but may also be highly dependent on many other factors including (but not limited to) the local hydrogeologic conditions, the size of the landfill (both in areal extent and thickness of waste), the infiltration into the landfill and the detailed design of the leachate collection system. Prescriptive regulations may be easy to administer, but they create a situation where for one landfill the design may be overly conservative, while for a second landfill the prescriptive design may provide no assurance that the long-term impact will be negligible.

2.2.3 No impact versus negligible impact

Regulations which require no impact are emotionally desirable but involve practical difficulties. Typically, some form of model is required to demonstrate that there will be no impact, and a monitoring program is required to assess the performance of the facility. In terms of modeling, the no-impact requirement is often impossible to achieve, since for most designs the process of molecular diffusion will result in some contaminant migration from the landfill, even for a 30 m thick clay deposit, or a composite liner system composed of a geomembrane and compacted clay liner. The requirement of no impact is also impossible to enforce since, at the very least, the impact that can actually occur is controlled by the detection limits used in the chemical analysis of the groundwater. Thus no impact in practice becomes a negligible impact condition where negligible is related to sampling procedure and the analytical detection limits, which vary depending on the chemicals being considered and the analytical technique used.

In short, **no-impact** regulations may place unrealistic restrictions on the design of facilities since, if properly modeled, the vast majority of facilities would not meet this requirement. Furthermore, this approach creates an arbitrary *de facto* limit on allowable potential impact, which is directly related to field sampling and laboratory techniques, and not to considerations of health and safety.

A more meaningful approach is to require that the facility has negligible effect on groundwater, where negligible is quantified on the basis of considerations of background chemistry, the chemical species and the potential aesthetic and health-related implications of an increase in concentration in the groundwater. An example of this type of approach is contained in the Ontario Ministry of the Environment and Energy (MoEE) guidelines associated with its 'Reasonable Use Policy' (MoEE, 1993a) which provides a means of quantifying a negligible effect in a rational and consistent manner.

2.2.4 Regulatory period

The regulated time period over which a landfill is required to have no, or negligible, effect on groundwater quality, or on a neighbor's reasonable use of groundwater drawn from near the property boundary, has a significant effect on the design. For example, one can readily design a barrier (and buffers) such that there is negligible impact at the site boundary for a 30-year period, simply by designing the system such that the travel time for contaminants to reach the boundary (at detectable levels) is greater than 30 years (or 100 years or whatever arbitrary limit is set). However, this does not mean that there will not be a significant environmental impact. It may simply mean that the impact of that facility is being passed on to future generations. As discussed in Chapter 12, one can readily encounter situations where the proposed design is not expected to have more than a negligible impact for 100 years but can be predicted to have a significant and unacceptable impact after 150–200 years.

In the design of barriers, it is important to recognize that the clay and geosynthetic–clay barrier being commonly used today will, if properly constructed, greatly retard the migration of contaminants from a waste disposal facility but, unless carefully designed, may still ultimately result in significant long-term contamination of groundwater. The likelihood of this occurring increases with the trend to larger and larger landfills.

2.2.5 Service life and contaminating lifespan

A second factor intimately related to the regulated period of no or negligible impact is the service life of the facility. If properly designed, specified and constructed, the service life of many of the key components of a landfill barrier system, such as leachate collection systems and geomembrane liners, is likely to exceed 30 years. Under these circumstances, it is a relatively straightforward matter to design a landfill to meet requirements for no or negligible impact for a 30-year post-closure period. Monitoring and contingency measures should, of course, be established to detect and rectify any unpredictable failure that might occur during this period.

However, when one is required to ensure negligible impact for periods exceeding 30 years, careful consideration must be given to the effect of degradation of the engineering components of the system.

The most stringent regulations with respect to environmental assessment require that the landfill have negligible effects on groundwater quality (typically at the site boundary) in perpetuity. For example, the Ontario Ministry of the Environment and Energy's (MoEE) 'Reasonable Use Policy' places no time constraint on the period during which the landfill is to have a negligible effect on a neighbor's reasonable use of groundwater.

The MoEE's Engineered Facilities Policy (MoEE, 1993b) addresses the service life of the engineered facility, and requires that the service life exceed the contaminating lifespan, where the contaminating lifespan may be defined as the period of time during which the landfill will produce contaminants at levels that could have unacceptable impacts if they were discharged to the environment.

The contaminating lifespan will depend on the contaminant transport pathway, the leachate strength, the mass of waste (and, in particular, the thickness of waste) and the infiltration through the cover. The contaminating lifespan generally increases with increasing thickness of waste and decreasing infiltration into the landfill. The more hydrogeologically suitable the site (i.e. the greater the potential for natural attenuation of contaminants), the shorter the contaminating lifespan. For the limiting case of a site that has sufficient natural attenuation that no engineered barrier system or leachate collection system is required, the contaminating lifespan is zero. Chapter 12 provides a more detailed discussion of contaminating lifespan.

2.3 Waste and leachate composition

This book focuses on two types of leachate:

1. domestic or municipal solid waste leachate (MSWL);
2. liquid hydrocarbons, soluble and insoluble, in simple aqueous systems.

2.3.1 Domestic waste leachate

(a) Domestic solid refuse

The composition of average municipal refuse (picked up at the curbside) may vary from country to country, and will depend on the local cultural background of the generating community. In North America, the composition of refuse should normally be similar to that presented in Table 2.1 (Ham *et al.*, 1979). Another examination of waste composition by Hughes, Landon and Farvolden (1971) is summarized in Table 2.2. A summary of this type is particularly useful in that it quantifies the proportion of the total waste mass of a number of contaminant species. For example, Table 2.2 indicates that chloride represents 0.097% (say 0.1%) of the total dry mass of the waste examined by Hughes *et al.* While the composition may vary from one region or country to another, information such as that presented in Tables 2.1 and 2.2 can be compiled for each such region/country. This type of information is used in Chapters 10, 11 and 12 when modeling the finite mass of contaminant.

(b) Leachate generation

Percolation of rainwater through solid waste leaches out soluble salts and biodegraded organic products to form a foul-smelling, gray leachate. Fine-grained soils used as daily cover may also be incorporated as suspended solids in the leachate.

Bacterial growth starts immediately under oxidizing conditions generating temperatures which have been recorded to be in excess of 60°C in some wastes (Collins, 1993). Since optimum growth takes place at C:N:P ratios of

Table 2.1 Municipal solid waste refuse composition

Component	Percent of all refuse by weight	Moisture percent by weight[b]	Analysis (percent dry weight)[a] Volatile matter	Carbon	Hydrogen	Oxygen	Nitrogen	Sulfur	Noncombustibles
Rubbish, 64%									
Paper	42.0	10.2	84.6	43.4	5.8	44.3	0.3	0.20	6.0
Wood	2.4	20.0	84.9	50.5	6.0	42.4	0.2	0.05	1.0
Grass	4.0	65.0	–	43.3	6.0	41.7	2.2	0.05	6.8
Brush	1.5	40.0	–	42.5	5.9	41.2	2.0	0.05	8.3
Greens	1.5	62.0	70.3	40.3	5.6	39.0	2.0	0.05	13.0
Leaves	5.0	50.0	–	40.5	6.0	45.1	0.2	0.05	8.2
Leather	0.3	10.0	76.2	60.0	8.0	11.5	10.0	0.40	10.1
Rubber	0.6	1.2	85.0	77.7	10.4	–	–	2.0	10.0
Plastic	0.7	2.0	–	60.0	7.2	22.6	–	–	10.2
Oils, paints	0.8	0.0	–	66.9	9.7	5.2	2.0	–	16.3
Linoleum	0.1	2.1	65.8	48.1	5.3	18.7	0.1	0.40	27.4
Rags	0.6	10.0	93.6	55.0	6.6	31.2	4.6	0.13	2.5
Street sweepings	3.0	20.0	67.4	34.7	4.8	35.2	0.1	0.20	25.0
Dirt	1.0	3.2	21.2	20.6	2.6	4.0	0.5	0.01	72.3
Unclassified	0.5	4.0	–	16.6	2.5	18.4	0.05	0.05	62.5
Food wastes, 12%									
Garbage	10.0	72.0	53.3	45.0	6.4	28.2	3.3	0.52	16.0
Fats	2.0	0.0	–	76.7	12.1	11.2	0.0	0.00	0.0
Noncombustibles, 24%									
Metals	8.0	3.0	0.5	0.8	0.04	0.2	–	–	99.0
Glass and ceramics	6.0	2.0	0.4	0.6	0.03	0.1	–	–	99.3
Ashes	10.0	10.0	3.0	28.0	0.5	0.8	–	0.5	70.2
Composite refuse, as received									
All refuse	100	20.7	–	28.0	3.5	22.4	0.33	0.16	24.9

Source: Ham *et al.* (1979). (EPA, 1983, Table 2–3)

[a]Analysis of the respective components.

[b]Moisture content of the respective component in the waste.

about 100:5:1, it is probably food wastes which fuel early biological reactions. Low average nitrogen levels in bulk waste (only 0.5%) and rapid depletion of oxygen soon render the system anaerobic, cooler and far less reactive. Farquhar (1987) reports representative temperatures in the waste in landfills in Ontario as typically between 10 and 20°C, which is well below the 35°C considered to be optimal for methanogenesis (Zehnder, 1978). In other parts of the world (e.g. UK, Germany) internal temperatures in landfills of 30–40°C are common.

MSW leachate is a complex liquid which changes in characteristics as one passes from the early acetic phase of young leachate to the methanogenic phase of older leachate. Because of the time period over which landfills may be constructed, the leachate that is collected may be a mixture of young and older leachate from different parts of the landfill. Typically, the acetic phase is characterized by high organics

Table 2.2 Municipal solid waste refuse composition (after Hughes, Landon and Farvalden, 1971)

Component	*Proportion relative to dry weight of refuse or percent by weight*
Crude fiber	38.3%
Moisture content	18.2%
Ash	20.2%
Free carbon	0.57%
Nitrogen	
free	0.02 mg/g
organic	1.23 mg/g
Water solubles	
sodium	2.33 mg/g
chloride	0.97 mg/g
sulfate	2.19 mg/g
COD	42.29 mg/g
Phosphate	0.15 mg/g
Hardness	10.12 mg/$CaCO_3$/g
Major metals	
aluminum, iron, silicon	>5.00% (by spectrographic analysis)
Minor metals	
calcium, magnesium, potassium	1.0–5.0% (by spectrographic analysis)

with BOD_5/COD ratios greater than 0.4 (Ehrig, 1989). Table 2.3 summarizes some typical characteristics of young and older leachate.

Much of the organic phase is extremely biodegradable, resulting in high initial BOD values. Chian (1977) reports original total organic carbon values of ~17 g/l (~1.7%) in a two-month-old Illinois leachate with the following fatty acid composition (g/l; 49% of TOC = 0.8% of leachate):

1. Acetic acid 1.75
2. Propionic acid 0.51
3. Isobutyric acid 0.31
4. Butyric acid 3.08
5. Isovaleric acid 0.72
6. Valeric acid 0.56
7. Hexanoic acid 1.49

These acids are very important since they render leachate weakly acidic and mobilize heavy metals. Table 2.4 illustrates the nature of these low molecular weight aliphatic acids. The remaining organics appear to be predominantly humic-carbohydrates, carboxyl compounds, carbonyl compounds etc. (Chian, 1977).

With these comments in mind, Table 2.5 illustrates the composition of three US, three Italian and two German leachates, and Table 2.6 summarizes results for a number of Canadian leachates.

Experience would indicate that the concentration of key contaminants in leachate reaches a peak value some time after landfilling begins and then decreases with time. Even for conservative species there is a sound theoretical basis for a decrease in concentration with time due to dilution as new leachate is generated and other leachate is removed by the leachate collection system (Rowe, 1991a; Ehrig and Scheelhaase, 1993). For nonconservative species, this decay can be accelerated by both chemical and biological breakdown. To illustrate the basic trends, Figure 2.1 shows an empirical curve for the change in chloride concentration based on the age of MSW landfills in North America. Figure 2.2 shows the observed decay of phenol in a landfill based on a number of years of monitoring data.

An examination of Table 2.6 shows that the concentration of chloride in landfill leachate often lies between 30 and 12 000 mg/l. Based on available data, a chloride concentration of 1500 to 2500 mg/l would appear to be a reasonable source concentration for impact calculations for the MSW leachate in Ontario. An average concentration for chloride of 2100 mg/l has been reported by Ehrig and Scheelhaase (1993) based on European data. In areas where there may be less leachate dilution, higher values may be more appropriate. Similarly, if the landfill is used to dispose of road salt, sewage sludge or industrial salty wastes then higher values may be appropriate.

There is some question concerning the potential effect of recycling on leachate quality. Preliminary German experience (Collins, 1991) suggests that the removal of paper, glass, metals, plastics and textiles (about 36% of the refuse stream) had no significant effect on chloride concentrations, but significantly reduced the concentration and mass of iron and

Table 2.3 Characteristics of young and older leachate

	Young leachate	*Older leachate*
Composition		
Water	~95%	~99%
Dissolved and suspended inorganics	~3%	~1%
Dissolved and suspended organics	~2%	~½%
Chemical characteristics		
Chemical oxygen demand (COD)	~23 000 mg/l	~3000 mg/l
Biological oxygen demand (BOD_5)	~15 000 mg/l	~180 mg/l
pH	5.2–6.1	7.2–8

Table 2.4 Short chain carboxylic acids

Acetic — CH_3COOH

```
    H
    |    //O
H - C - C
    |    \O - H
    H
```

Propionic (carboxylic acid C_3) — CH_3CH_2COOH

```
    H   H
    |   |    //O
H - C - C - C
    |   |    \O - H
    H   H
```

Butyric (carboxylic acid C_4) — $CH_3CH_2CH_2COOH$

```
    H   H   H
    |   |   |    //O
H - C - C - C - C
    |   |   |    \O - H
    H   H   H
```

Valeric (carboxylic acid C_5) — $CH_3CH_2CH_2CH_2COOH$

Hexanoic (carboxylic acid C_6) — $CH_3(CH_2)_4COOH$

Isobutyric (2-methylpropionic acid) — $(CH_3)_2CH_2COOH$

Table 2.5 Composition of MSW landfill leachates. Concentration of constituents (mg/l) and pH

	USA[a]			*Italy*[a]			*Germany*[a]	
Consituent	*1*	2	3	4[b]	5[b]	6[c]	7[d]	8[e]
BOD_5	–	13 400	–	3 000	10 400	2 125	57–2 700	400–45 900
COD	42 000	18 100	1 340	38 520	28 060	7 750	1 450–6 340	1 630–63 700
Total solids	36 250	12 500	–	–	–	–	–	–
Total volatile fatty acids	–	930	333	1 574	435	–	–	–
Organic nitrogen as N	–	107	–	60	554	125	–	–
Ammonia nitrogen as N	950	117	862	1 293	1 203	1 040	620–2 080	620–3 500
pH	6.2	5.1	6.9	6.0	6.3	8.5	7.2–7.9	5.7–8.1
Total alkalinity as $CaCO_3$	8 965	2 480	–	5 125	4 250	8 250	–	–
Chemicals and metals								
Arsenic	–	–	0.11	–	–	–	–	–
Boron	–	–	29.9	–	–	–	–	–
Cadmium	–	–	1.95	–	–	–	–	–
Calcium	2 300	1 250	354.1	175	–	–	70–290	130–4 000
Chloride	2 260	180	1.95	2 231	1 868	3 650	1 490–3 550	1 490–21 700
Chromium	–	–	<0.1	–	–	–	–	–
Copper	–	–	<0.1	–	–	–	–	–
Iron	1 185	185	4.2	47	330	–	8–79	8–870
Lead	–	–	4.46	–	–	–	–	–
Magnesium	410	260	233	1 469	827	–	100–270	100–840
Manganese	58	18	0.04	42	27	–	0.2–4.0	0.3–28
Mercury	–	–	0.008	–	–	–	–	–
Nickel	–	–	0.3	–	–	–	–	–
Phosphate	82	1.3	–	–	–	2.3	–	–
Potassium	1 890	500	–	1 200	1 200	–	–	–
Sodium	1 375	160	748	1 400	1 300	–	–	–
Sulfate	1 280	–	<0.01	1 600	1 860	219	1–115	1–121
Zinc	67	–	18.8	7	5	–	–	–

[a]Sources: 1, Wigh (1979); 2, Breland (1972); 3, Griffin and Shimp (1978); 4–6, Cancelli and Cazzuffi (1987); 7, 8, Brune *et al.* (1991)
[b]Young landfill.
[c]Old landfill.
[d]Weak concentration.
[e]Strong concentration.

Table 2.6 MSW leachate from Canada

		Ontario								Alberta
	Note	1	2	3	4	5	6	7	8	9
BOD_5	a	3 500	2 330–12 500	450–12 500	–	1 850–3 875	23 500	30	>5 000	–
COD	a	4 700	53–21 200	28–21 800	11 000–82 000	–	10 600	–	–	–
TOC	a	1 650	–	6 190	–	–	3 060	–	5 710	1–10 500
DOC	a	–	18–7 090	10–7 360	270–25 000	–	4 600	20	–	–
TDS	a	6 100	–	–	–	–	–	–	–	2 610–29 500
Ammonia (as N)	a	–	–	355	–	–	715	–	295	–
TKN	a	–	–	370	–	–	750	30	365	11–508
pH		6.6	6.2	6.7	–	–	7.0	7.0	6.6	5.3–7.1
Total alkalinity (as $CaCO_3$)	a	2 600	–	6 330	–	–	5 490	770	6 720	1 280–7 620
Aluminum	a	1	–	–	–	–	–	–	–	0.06–18.2
Boron	a	5	–	–	–	–	–	–	–	1–24.9
Cadmium	a	–	0.006	0.8–52	–	–	0.035	0.005	0.001	0.003
Calcium	a	540	60–2 500	74–1 700	60–4 580	–	535	480	1 740	157–2 910
Chloride	a	–	(750)	(505)	(1 915)	(1 400)	–	–	–	–
		670	30–3 800	28–3 130	30–12 000	730–2 230	1 270	240	3 400	795–18 000
Chromium	a	0.15	0.06	0.28	–	0.09	0.05	0.02	0.001	0.006–0.75
Copper	a	–	0.04	0.036	–	0.1	–	–	–	0.01
Iron	a	80	4–540	0.4–560	6–1 670	45–160	270	2	520	1–2 650
Lead	a	–	0.03	0.07	–	0.22	<0.05	–	0.014	0.04–1.7
Magnesium	a	240	10–465	45–560	21–1 750	–	190	240	640	270–630
Manganese	a	7	8	6	3	2	–	–	–	0.001–1.9
Mercury	a	–	0.06	0.2	–	–	–	–	–	0.0006
Nickel	a	0.15	0.22	0.50	–	0.39	–	–	–	0.03–28.6
Phosphate	a	<0.8	0.8	–		–	–	–	–	–
Potassium	a	240	227	37	18–1 680	–	800	40	545	20–1 660
Sodium	a	–	(1 095)	(580)	(3 140)	–	–	–	–	–
		430	90–10 400	14–3 650	210–16 000	–	1 330	210	3 110	284–8 420
Sulphate	a	–	(103)	(74)	(290)	–	–	–	–	–
		380	1–490	2–4 200	23–2 530	–	510	1 715	25	58–2 490
Zinc	a	2.4	(1.5)	(2.2)	–	(3.8)	1.4	0.3	20	0.005–0.4
Phenol	b	3 900	(2 750)	(970)	(1 130)	(1 280)	5 750	16	64 000	–
1,1-Dichloroethane	b	60	–	14	–	–	–	–	–	ND–120
1,2-Dichloroethane	b	6	–	18	–		–	–	–	–
Benzene	b	7	–	18	–	–	7	–	–	26–750
Dichloromethane	b	–	–	1	3 500	–	130	–	–	9–2 000
Ethylbenzene	b	4	–	80	–	–	30	–	–	40–180
Toluene	b	80	–	(90)	–	–	270	1.6	–	58–1 400
Trichloroethylene	b	–	–	90	–	–	–	–	–	–
Meta- and *para*- xylene	b	–	–	120	–	–	50	–	–	80–900
Ortho-xylene	b	–	–	70	–	–	30	–	–	–

[a]In mg/l or ppm. [b]In µg/l or ppb. Number in brackets represents geometric mean of data. ND: not detected.

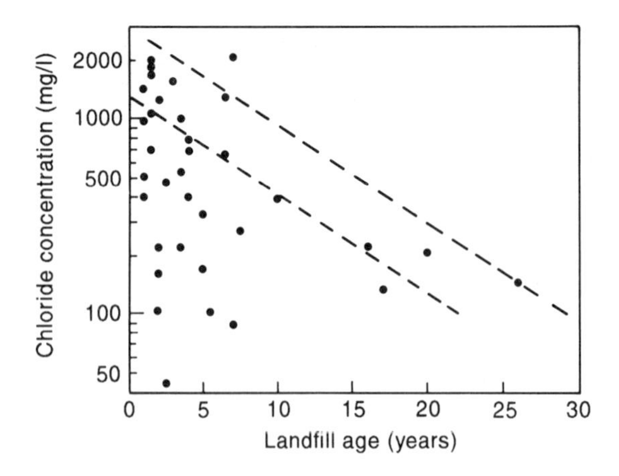

Figure 2.1 Chloride concentrations observed in landfills of different age. (Note that no correlation has been made with size, climate or initial source concentration.) (Based on data from Lu, Eichenberger and Stearns, 1985.)

calcium collected in leachate. The removal of an additional 32% of the waste as organic material to be composted lead to additional reductions in concentrations of iron and calcium and did not significantly affect chloride. This is beneficial with respect to extending the longevity of leachate collection systems (section 2.4). Additional research is required to identify whether these German findings are applicable in other countries/regions.

Tables 2.6 and 2.7 provide data concerning organic compounds detected at a number of landfills. In Table 2.6, Ontario landfills 2, 5 and 7 are considered to be reasonable representatives of municipal solid waste. Some industrial waste is expected at the other landfills listed in Tables 2.6 and 2.7. It should be noted that the concentrations of organic compounds are low. The concentrations of various potential contaminants existing in waste (e.g. Tables 2.5–2.7) may be compared with typical drinking water standards as listed in Table 2.8.

(c) MSW leachate stability

The dissolved organics in raw municipal solid waste leachate are highly biodegradable so that biochemical alteration continues in the sample bottles brought back to the laboratory for testing. Bacteria growth results in a rapid rise in pH, reduction in E_h (oxygen depletion as in reduction of SO_4^{2-} to S^{2-}) and apparently production of CO^{2-}. Since most leachates seem to have a high concentration of Ca^{2+}, flocs of calcite ($CaCO_3$) appear within a couple of days of laboratory storage if stored at room temperature (20°C) and Ca^{2+} levels in the leachate experience a rapid reduction. Similarly, Fe^{2+} which may be abundant in acidic leachate declines rapidly as it forms an amorphous black slime of FeS_2. Typical curves showing these phenomena are presented in Figures 2.3 and 2.4.

Although these reactions are greatly inhibited by storage at 4°C, this temperature is not suitable for long duration hydraulic conductivity testing. This lack of stability may cause chemical

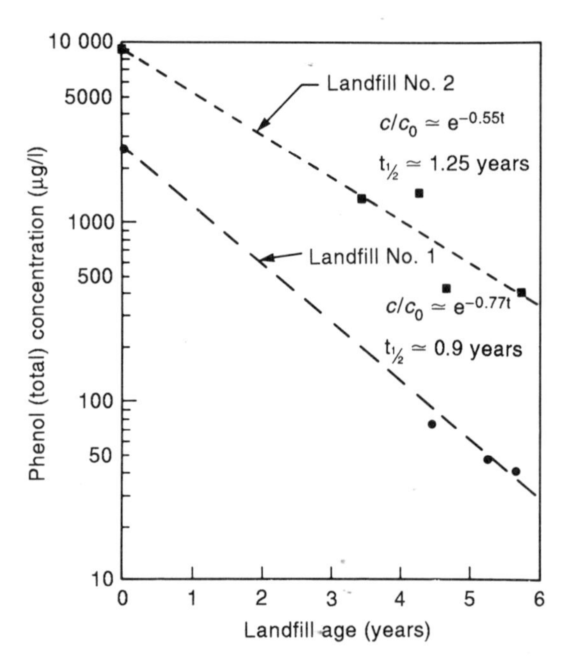

Figure 2.2 Observed decrease in phenol concentration with time at two landfills. (Note that there was no significant decay in chloride concentration during this time period. These landfills do not have a leachate collection system.)

Table 2.7 Liquid hydrocarbon contaminants found in leachate contaminated groundwater at six landfill sites (μg/l) (after Barker *et al.*, 1987)

	Landfill sites					
Contaminant	*Borden*	*Woolwich*	*North Bay*	*New Borden*	*Upper Ottawa Street*	*Tricil*
Aliphatic, aromatic and carboxylic acids	20	>10 000	>300	–	>1 000	–
Carbon tetrachloride	<1	5	<1	9	P	ND
Chloroform	<1	20	<1	25	5	ND
Trichloroethylene	1	37	2	750	P	ND
Trichloroethane	ND	7	<1	90	20	8.44
Tetrachloroethylene	ND	2	<1	<1	<1	ND
Acetone	<1	–	6	–	6	–
Tetrahydrofuran	P	–	9	–	200	–
1,4-Dioxane	ND	–	<1	–	P	–
Benzene	3	70	51	50	60	7 920
Toluene	1	7 500	60	1 400	2 600	9 520
Xylenes	<1	700	140	500	3 500	–
Ethylbenzene	<1	1 100	64	120	700	3 320
Tetramethylbenzene	ND	10	250	70	450	–
Chlorobenzene	ND	ND	105	ND	110	–
Dichlorobenzenes	<1	ND	13	ND	5	–
Naphthalene	ND	50	15	260	P	2 350
Phenols	–	1 100	10	P	P	–
Benzothiozoles	<1	30	10	ND	P	–
PAHs	–	–	ND	–	ND	–
Phthalates	<1	P	110	P	P	–

Note: Many of these landfills are known to have accepted industrial waste.
P: detected but concentration not estimated.
ND: not detected.
–: not determined.

control and interpretation problems in laboratory tests. For example, the Ca^{2+}, Fe and pH values in the influent permeant must be continuously monitored along with the effluent chemistry during hydraulic conductivity testing for clay–leachate compatibility assessment, as discussed in Chapter 4. Also, the potential for similar decreases should also be considered during diffusion testing. A preliminary review of these problems has been published by Quigley, Yanful and Fernandez (1990).

2.3.2 Liquid hydrocarbons

The very complex subject of hydrocarbon liquids defies succinct summarization for purposes of a book such as this. Nevertheless, a useful summary of the classes of organic compounds and

Table 2.8 Some drinking water standards

Constituent	*Recommended concentration limit*[a] *(mg/l)*
Inorganic	
Total dissolved solids	500
Chloride	250
Copper	1
Hydrogen sulfide	0.05
Iron	0.3
Manganese	0.05
Sodium	200
Sulfate	250
Zinc	5
Organic	
Acetone	1
Phenols	0.002
Toluene	0.024
	Maximum permissible concentration (μg/l)
Inorganic	
Antimony	10
Arsenic	50
Barium	1 000
Boron	5 000
Cadmium	5
Chromium (Cr^{VI})	50
Selenium	10
Lead	10
Mercury	1
Nitrate as nitrogen	10 000
Silver	50
Organic	
Cyanide	100
DDT	30
Endrine	0.2
Lindane	4
Methoxychlor	100
Toxaphene	2
2,4-D	100
2,4,5-TP silvex	10
Synthetic detergents	500
1,2-Dichloroethene	70
1,2-Dichloroethane	5
1,4-Dichlorobenzene	5
1,4-Dioxane	412
Dichloromethane (methylene chloride)	50
Tetrachloroethylene	500
Trichloroethylene	50
Trichloromethane (chloroform)	100
Xylenes	300
Vinyl chloride	2
Radionuclides and radioactivity	*Maximum permissible activity (pCi/l)*
Radium 226	1
Strontium 90	8
Gross beta activity	1 000
Gross alpha activity	15
Bacteriological	
Total coliform bacteria	1 per 100 ml

Sources: US Environmental Protection Agency, 1987; World Health Organization, 1984; European Standards, 1981; Ontario Ministry of the Environment, 1990; Canadian Drinking Water Quality Guidelines, 1988.
[a]Recommended concentration limits for these constituents are mainly to provide acceptable esthetic and taste characteristics.

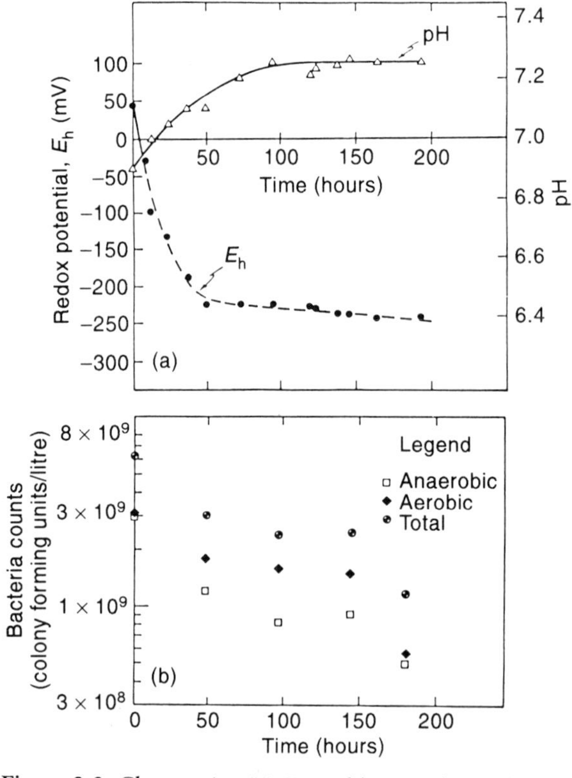

Figure 2.3 Changes in pH, E_h and bacterial population during eight-day experiment on the stability of MSW leachate from London, Canada. (Modified from Quigley, Yanful and Fernandez, 1990.)

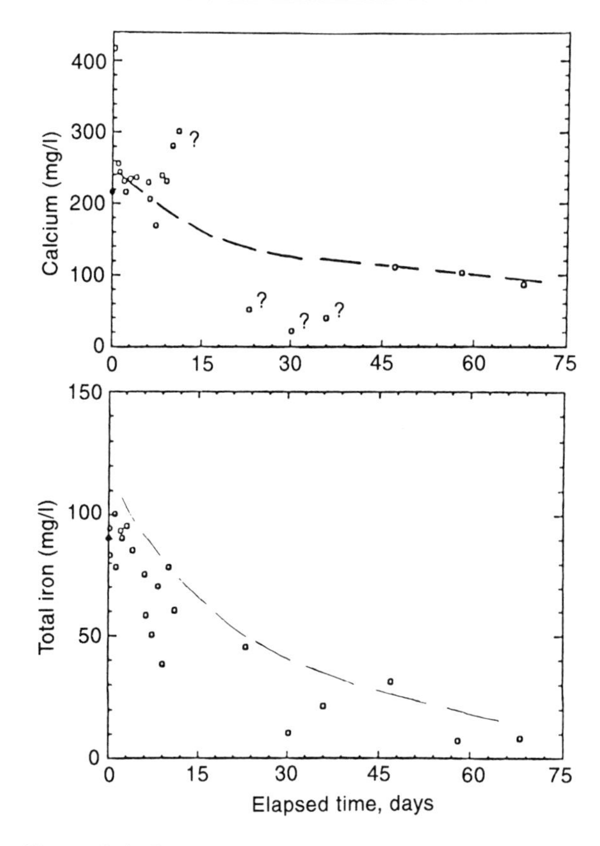

Figure 2.4 Concentration versus time curves for calcium and iron in a typical raw MSW leachate stored in sealed jars at 21°C.

their properties has been prepared by Mitchell and Madsen (1987). The components considered by them are shown in Figure 2.5 and their properties are summarized in Table 2.9.

For purposes of this book, two classes of liquid hydrocarbons will be considered:

1. organics which are insoluble in water; this includes dense nonaqueous phase liquids (DNAPLs) and light nonaqueous phase liquids (LNAPLs);
2. organics which are soluble in water.

Since the solubility decreases rapidly with increasing size of a soluble organic molecule, consideration will be focused on small molecules consisting of up to about eight carbon atoms.

The solubility of organic liquids is generally controlled by their polarity. Nonpolar or weakly polar substances dissolve in nonpolar or weakly polar solvents. Highly polar compounds dissolve in highly polar solvents including water. Fortunately, the polarity or dipole moment (in debyes) is usually closely proportional to the dielectric constant, ε, which figures so significantly in the Gouy–Chapman theory of potential (to be discussed in Chapter 3).

Methanol, CH_3OH, is soluble in water because the hydrogen on a methanol molecule interchanges easily with hydrogen on water molecules producing nearly infinite miscibility. As shown in Table 2.10, the solubility decreases rapidly for alcohols consisting of chains longer than three carbons.

In water–wet clay barriers, solubility plays a critical role in the interrelationships between the barrier and a retained organic liquid. Several terms are useful to describe solubility:

1. hydrophilic is used to describe water-loving or soluble organics;
2. hydrophobic is used to describe water-hating or insoluble organics;
3. lyophilic is a general term describing mutual attraction (solubility) between two liquids;
4. lyophobic is a general term describing mutual repulsion between two liquids.

When large volumes of contaminant are available, the source concentration may be controlled by the solubility limit and the equivalent height of leachate may be quite large.

Finally, a few words about the dielectric constant, ε. Liquids with a high dielectric constant (water at 80, methanol at 34, etc.) dissolve ionic compounds, not only because they are polar and efficiently hydrate the dissociated species, but also because they have insulating properties. In the case of NaCl in water, the force of attraction between the hydrated Na^+ and Cl^- atoms is greatly reduced by the high dielectric solvent. This applies

Type of compound		Functional group	Specific example	
Saturated hydrocarbons	alkanes	$\equiv C - C \equiv$	$H_3C - CH_3$	ethane
Unsaturated hydrocarbons	(alkenes)	$= C = C =$	$H_2C = CH_2$	ethylene
	(alkynes)	$- C \equiv C -$	$HC \equiv CH$	acetylene
	(polyolefins)	$= C = C - C = C =$	$H_2C = CH - CH = CH_2$	butadiene
Aromatic hydrocarbons			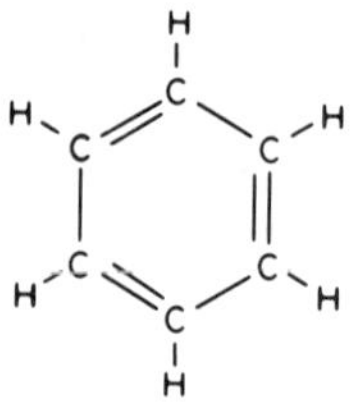	benzene
Alcohols		$- OH$	$CH_3CH_2 - OH$	ethyl alcohol
Phenols		$- OH$	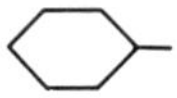 $CHCH_3$ (with OH)	I – phenyl – ethanol
Ethers		$- O -$	$CH_3CH_2 - O - CH_2CH_3$	ethyl ether
Aldehydes		$H-C=O$	$CH_3 - C(=O) - H$	acetaldehyde
Keytones		$>C = O$	$CH_3 - C(=O) - CH_3$	acetone
Organic acids (carbolic acids)		$- C(=O)OH$	$CH_3 - C(=O) - OH$	acetic acid
			$CH_3 - CH_2 - C(=O) - OH$	proprionic acid
Organic bases (amines)		$- NH_2$	$CH_3 - NH_2$	methylamine
Halogenated hydrocarbons		$- C\ell$	$H - CH(C\ell) - C\ell$	Dichloromethane
		$- F$	 $C\ell$	Chlorobenzene
		$- Br$	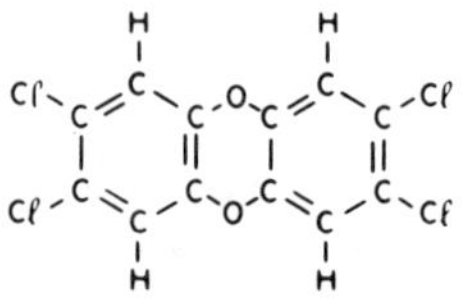 tetrachlorodibenzo – *p* – dioxin	

Figure 2.5 Classes of organic compounds. (Modified from Mitchell and Madsen, 1987.)

Table 2.9 Properties of organic chemicals used in hydraulic conductivity testing of clays (after Mitchell and Madsen, 1987)

Class of compound	*Compound*	*Formula*	*Solubility in in water (g/l)*	*Dielectric constant*	*Dipole moment (debye)*	*Density (g/cm^3)*
Hydrocarbons	Heptane	C_7H_{16}	<0.3	1.9	0	0.684
	Cyclohexane	C_6H_{12}	<0.3	2.0	0	0.779
	Benzene	C_6H_6	0.7	2.3	0	0.879
	Xylene (di-			2.27 (*para*)	0	
	methylbenzene)	C_8H_{10}	<0.3	2.37 (*meta*)		0.880
				2.57 (*ortho*)	0.62	
	Tetrachloro-methane (carbontetra-chloride)	CCl_4	0.8	2.2	0	1.594
	Trichloro-ethylene (TCE)	C_2HCl_3	1	3.4	0	1.464
	Nitrobenzene	$C_6H_5NO_2$	2	35.7	4.22	1.204
Alcohols and phenols	Methanol	CH_3OH	∞	33.6	1.70	0.791
	Ethanol	C_2H_5OH	∞	25.0	1.69	0.789
	Ethyleneglycol	$C_2H_6O_2$	∞	37.7	2.28	1.119
	Phenol	C_6H_5OH	86	13.1	1.45	1.072
Ethers	1,4-Dioxane	$C_4H_8O_2$	∞	2.2	0	1.034
Aldehydes and ketones	Acetone	C_3H_6O	∞	21.5	2.9	0.79
Organic acids	Acetic acid	$C_2H_4O_2$	∞	6.15	1.74	1.049
Organic bases	Aniline	$C_6H_5NH_2$	36	6.89	1.55	1.02
Mixed chemicals	Xylene	C_8H_{10}	∞ in	2.27–2.57	0–0.62	0.880
	Acetone	C_3H_6O	acetone	21.5	2.9	0.79
	Sodium acetate	CH_3COONa				
	Glycerol	$C_3H_8O_2$	∞	42.5	–	1.261
	Acetic acid	$C_2H_4O_2$	∞	6.15	1.74	1.049
	Salicylic acid	$C_7H_6CO_3$	Slight	–	–	1.443
	5% ammonia	NH_3	–	17.3	–	–
	copper tetramine, nickel hexamine					

equally in the double layer (discussed in Chapter 3) where the presence of water reduces the force of attraction between the negative clay particle and positive cations. If organic liquids of lower dielectric constant enter the double layer they effect an increase in the force of attraction and contract the double layer.

For purposes of this book, organic liquids are subdivided into two groups:

1. insoluble (lyophobic) DNAPLs and LNAPLs which normally have values for dielectric constant of 2 to 4;
2. soluble (lyophilic) compounds which normally have dielectric constant values of about 10 or greater.

The use of the dielectric constant is very convenient since it plays a prominent role in double layer theory. The special case of low dielectric liquids which are soluble in water (e.g. dioxane) will be considered in Chapter 4.

Table 2.10 Solubility of alcohols in water (from Morrison and Boyd, 1983)

Alcohol		*Solubility, g/100 g H_2O*
Methanol	CH_3OH	∞
Ethanol	CH_3CH_2OH	∞
Propanol	$CH_3CH_2CH_2OH$	∞
Butanol	$CH_3CH_2CH_2CH_2OH$	7.9
Pentanol	$CH_3CH_2CH_2CH_2CH_2OH$	2.3
Hexanol	$CH_3CH_2CH_2CH_2CH_2CH_2OH$	0.6
Heptanol	$CH_3CH_2CH_2CH_2CH_2CH_2CH_2OH$	0.2
Octanol	$CH_3CH_2CH_2CH_2CH_2CH_2CH_2CH_2OH$	0.05

2.4 Leachate mounding and collection

2.4.1 Function of the collection system

The movement of contaminant through a barrier will, as discussed in Chapter 1, depend on both the advective (Darcy) velocity and diffusion. The Darcy velocity will, in turn, depend on the hydraulic conductivity of the barrier and the hydraulic gradient.

The hydraulic gradient will often depend on the engineering of the landfill, since landfill construction will frequently change the hydraulic conditions at a site. For example, modern landfills are commonly designed to have a leachate underdrain system. These collection systems may serve several functions. Firstly, by lowering the height of leachate mounding, leachate seeps (and consequent contamination of surface waters) can be minimized. Secondly, by reducing the leachate head acting on the base of the landfill, the hydraulic gradient through the underlying barrier and the Darcy velocity out of the landfill, can be reduced to acceptable levels in many cases. Thirdly, by removing contaminant from the landfill, the mass of contaminant available for transport into the hydrogeologic system will be reduced.

Of the functions listed above, the second and third are of greatest significance to the impact of the landfill on groundwater quality. Various methods of estimating the height of the leachate mound (and hence the head within the landfill) have been proposed. Once the height of leachate mounding has been calculated, it is a relatively straightforward calculation to estimate the average hydraulic gradient and Darcy velocity through the barrier. As demonstrated in the following section, it is then possible to estimate the impact of the leachate collection system on the mass of contaminant available for transport into the barrier. The following paragraphs will discuss a number of methods available for estimating the height of leachate mounding.

2.4.2 Leachate mounding

If the barrier beneath the waste is flat and of low permeability compared to the waste, then the height, h, of the mound between two drains separated by a distance (Figure 2.6(a)) may be estimated from an equation given by Harr (1962) which (on simplification) can be written as

$$h = \Omega^{1/2}[(l - x)x]^{1/2} \quad (2.1a)$$

$$\Omega = \frac{q}{k_w} \quad (2.1b)$$

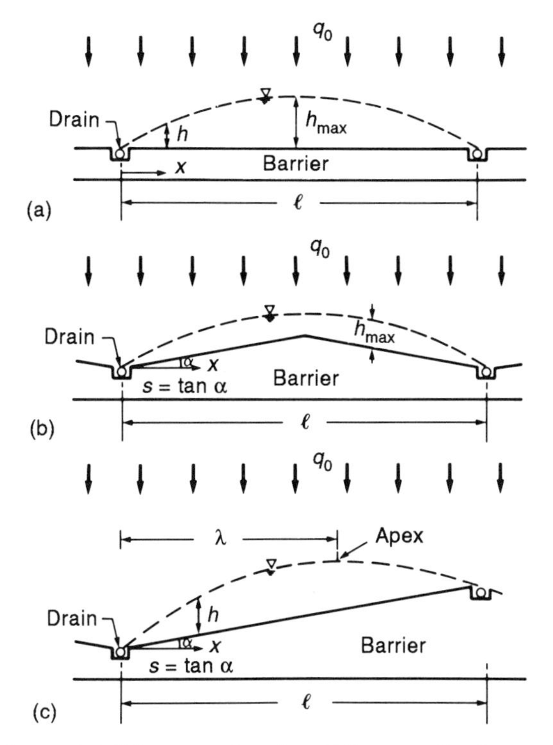

Figure 2.6 Schematic showing different collection systems. (From Rowe, 1988; reproduced with permission of the *Canadian Geotechnical Journal*.)

where h is measured relative to the head at the leachate collection drains [L]; l is the spacing between drains [L]; x is the distance from one of these points [L]; q is the portion of the steady-state infiltration that is being collected by the drains [LT^{-1}]; and k_w the hydraulic conductivity of the waste (or other material) between the drains [LT^{-1}]. At the midpoint between draw-down points, this equation reduces to

$$h_{max} = 0.5l\Omega^{1/2} \qquad (2.2)$$

where h_{max} is the maximum height of mounding above the barrier. Based on the variation in h with position x, it can be shown that the average value of h is given by

$$\bar{h} = 0.785h_{max} = 0.393l\Omega^{1/2} \qquad (2.3)$$

Using the same basic assumptions as adopted in the development of equation 2.2, Moore (1983) developed a solution for the case of a sloping geometry, as indicated in Figure 2.6(b), and derived an equation for the maximum height of mounding above the barrier, h_{max}:

$$h_{max} = 0.5l[(\Omega + s^2)^{1/2} - s] \qquad (2.4)$$

where $s = \tan\alpha$ is the slope of the barrier.

McBean *et al.* (1982) considered a different sloping collection system, as shown in Figure 2.6(c). This case is somewhat more complicated than the previous two cases, and one cannot write a simple explicit equation for the height h of leachate above the barrier. However, assuming zero pressure head within the drains, the height h and the distance at which it occurs can be related by the equation

$$x = \lambda(1 - A\exp B) \qquad (2.5a)$$

where

$$A = \frac{\Omega^{1/2}}{\left(\dfrac{h^2}{(\lambda - x)^2} - \dfrac{sh}{(\lambda - x)} + \Omega\right)^{1/2}} \qquad (2.5b)$$

$$B = \frac{s}{(4\Omega + s^2)^{1/2}}\left[\tan^{-1}\left(\frac{-s}{(4\Omega + s^2)^{1/2}}\right) - \tan^{-1}\frac{\left(\dfrac{2h}{(\lambda - x)} - s\right)}{(4\Omega + s^2)^{1/2}}\right] \qquad (2.5c)$$

The leachate mounding between drains can be readily calculated by determining the location x for an assumed height of mounding h using a successive substitutions algorithm. However, to do so, one must first know the value of λ. This can be done by a process of trial and error:

1. estimate λ;
2. calculate the leachate mound to the right of the lower drain (Figure 2.6(c)) over a distance λ;
3. calculate the leachate mound to the left of the upper drain over a distance (−λ) (noting that both positive and negative slopes s are permitted);

4. if the height h at the location λ (calculated in steps 2 and 3) do not agree, then revise the estimate of λ and repeat the procedure until the calculated mound is continuous between the two adjacent drains.

Equations 2.1–2.5 have been developed assuming that the infiltration q is equal to the flow passing into the collection points. When a portion of the infiltration water moves down through the barrier, the actual flow to the collection points q is equal to the difference between the infiltration q_0, and the flow into the barrier q_i (i.e. $q = q_0 - q_i$). These equations provide an approximate estimate of the head above the barrier.

More elaborate equations for estimating the height of leachate mounding have been proposed (e.g. Wong, 1977; Dematrocopoulis *et al.*, 1984). These equations have been developed to consider explicitly leakage through the barrier, and in this sense are more realistic than equations 2.1–2.5. However, to make the problem tractable, a number of other assumptions have been made. For example, Wong assumes that the leachate collection system is above the water table; the effects of groundwater flow beneath the barrier are negligible; and the leachate instantly saturates a rectilinear volume above the liner, and retains this shape while draining towards the collection drain and through the barrier. The adoption of these more elaborate equations is only likely to improve the estimate of leachate mounding if these assumptions are reasonably applicable for a given situation. In many cases, they are not. If the hydraulic conductivity of the barrier is significantly lower than that of the overlying waste and drainage layers, then it may be appropriate to use the simple equations given above (2.1–2.5), recognizing the simplifications involved. If the hydraulic conductivity of the 'barrier' is of a similar order to that of the waste (e.g. the landfill is constructed directly on a silty sand or silty soil or fractured/weathered till), then the simplified equations are not valid, and it may be necessary to use the finite element technique to model the entire flow system (i.e. waste, barrier and underlying hydrostratigraphy).

Irrespective of the method of analysis or equation used in calculating the leachate head, probably the greatest uncertainty is associated with the hydraulic conductivity of the waste. To date, relatively little research has been conducted into the determination of appropriate values for use in design. A hydraulic conductivity commonly used in design is 10^{-6} m/s; however, there is growing evidence to suggest that the hydraulic conductivity could be one or two orders of magnitude lower near the bottom of the waste or when waste has been disposed with sewage sludge. Recognizing that waste is likely to be heterogeneous, it is not surprising that a limited number of field measurements (e.g. Hughes, Landon and Farvolden, 1971; Page, Raila and Woliner, 1982) indicate a wide range (9×10^{-8} to 8.5×10^{-5} m/s) of hydraulic conductivities. Furthermore, if the waste is anisotropic, then these values may not be representative of the parameters controlling the leachate mounding. Consequently, the design of leachate collection systems and the calculation of potential impact on groundwater quality should involve some consideration of the implications of uncertainty regarding the hydraulic conductivity of the waste.

2.4.3 Examples of mounding calculations

To focus discussion on design considerations, it is useful to begin with a number of examples.

(a) French drains

Figure 2.7(a) shows a schematic of a relatively flat landfill base with French drains (in this case pipe surrounded by granular material) at a spacing of 30 m. Based on equations 2.1–2.3, it is possible to estimate both the maximum and average leachate head acting on the barrier.

Assuming an infiltration $q_0 = 0.15$ m/a and conservatively neglecting flow through the barrier, the heads calculated for two different assumptions concerning the hydraulic conductivity of waste, k_w, are:

1. Assume $k_w = 10^{-6}$ m/s $= 31.5$ m/a
   maximum head (using equation 2.2)
   $h_{max} = 0.5 \times 30(0.15/31.5)^{1/2}$
   $= 1.03$ m
   average head (using equation 2.3)
   $\bar{h} = 0.785 h_{max} = 0.81$ m
2. Assume $k_w = 10^{-7}$ m/s ≈ 3.15 m/a
   maximum head $h_{max} = 3.27$ m
   average head $\bar{h} = 2.57$ m

These values, together with a number of other cases are summarized in Table 2.11.

It is evident from these calculations that substantial leachate mounds can develop on the base of the landfill when French drains are used to control the mounding. The height of the mound can be reduced either by decreasing the pipe spacing or increasing the slope of the barrier between the drains (thereby encouraging more flow to the drains). A few calculations will quickly indicate that reducing the pipe spacing alone is not an effective means of reducing the maximum head on the liner to a nominal (say 0.3 m) value. For example, even with a hydraulic conductivity of the waste of $k_w = 10^{-6}$ m/s, a spacing of less than 8.7 m would be required. To allow for the potential reduction in the hydraulic conductivity to $k_w = 10^{-7}$ m/s (e.g. due to consolidation of the waste) the spacing of drains required to maintain a maximum head of 0.3 m would be about 2.75 m.

By increasing the slope between drains, as shown in Figure 2.7(b), the head on the liner can be reduced, as summarized in Table 2.11. Although the slope clearly has some effect, it only becomes significant in terms of substantially reducing the leachate mound if the hydraulic conductivity of the waste is of the order of 10^{-5} m/s or higher. While this may sometimes occur, it cannot be relied upon.

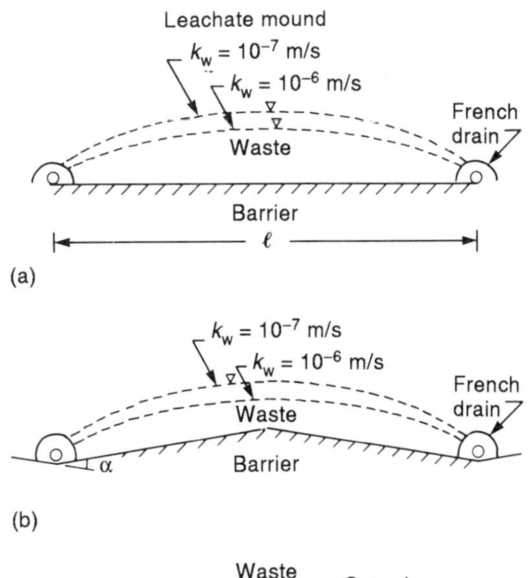

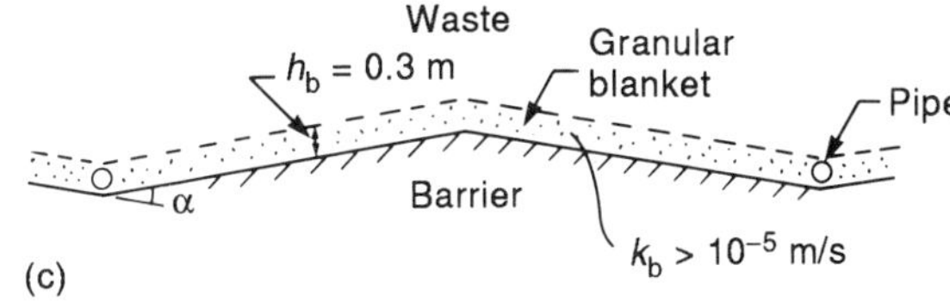

Figure 2.7 Leachate mounding for three cases (not to scale).

(b) Granular blanket drains

An alternative to using French drains to reduce leachate mounding is to use a blanket drain, as shown in Figure 2.7(c). When blanket drains have been used, they have most commonly been constructed from sand with hydraulic conductivities of 10^{-4} to 10^{-5} m/s. For a pipe spacing of 30 m, a maximum leachate head of less than 0.3 m can be readily obtained provided that there is a small slope on the base. With a 2% slope, the spacing between pipes can be increased to about 60 m and still maintain a maximum head of less than 0.3 m on the base.

2.4.4 Clogging of granular blanket drains

While the foregoing works well in theory, experience has indicated that sand blankets

Table 2.11 Calculated maximum head acting on the base of a landfill for a number of assumed conditions: infiltration 0.15 m/a

Spacing of pipes, l (m)	*Hydraulic conductivity of waste, k_w (m/s)*	*Slope of base, s = tan α (Figure 2.6)*	*Maximum head h_{max} (m)*
2.5	10^{-4}	0	0.01
	10^{-5}	0	0.03
	10^{-6}	0	0.09
	10^{-7}	0	0.27
10	10^{-4}	0	0.03
	10^{-5}	0	0.11
	10^{-6}	0	0.34
	10^{-7}	0	1.09
30	10^{-4}	0	0.10
	10^{-5}	0	0.33
	10^{-6}	0	1.03
	10^{-7}	0	3.27
30	10^{-4}	0.01	0.03
	10^{-5}	0.01	0.21
	10^{-6}	0.01	0.90
	10^{-7}	0.01	3.12
30	10^{-4}	0.02	0.02
	10^{-5}	0.02	0.14
	10^{-6}	0.02	0.78
	10^{-7}	0.02	2.98
50	10^{-4}	0.02	0.03
	10^{-5}	0.02	0.24
	10^{-6}	0.02	1.30
100	10^{-4}	0.02	0.06
	10^{-5}	0.02	0.48
	10^{-6}	0.02	2.59

placed directly beneath waste have a tendency to clog (see, for example, the discussion of the performance of the sand layer in the Keele Valley landfill in Chapter 9). This clogging may be a result of a combination of particulate clogging, clogging due to chemical precipitation and clogging due to biofilm growth.

It must be recognized that 'clogging' of a drainage layer is not synonymous with it becoming impermeable. On the contrary, a clogged sand blanket may still be substantially more permeable than, say, an underlying clay liner. 'Clogging' of a drainage layer becomes significant when the hydraulic conductivity of the blanket drops to, or below, the hydraulic conductivity of the overlying waste. For example, consider a leachate collection system similar to that shown in Figure 2.7(c) where a 0.3 m thick granular blanket is constructed over a compacted clay liner. Assume:

1. design maximum head on liner < 0.3 m;
2. drain spacing $l = 30$ m;
3. base slope $s = 0.01$ (1%);
4. hydraulic conductivity of waste $k_w = 10^{-6}$ m/s;
5. initial hydraulic conductivity of blanket $k_s = 10^{-5}$ m/s (315 m/a);
6. infiltration $q_0 = 0.15$ m/a.

Design conditions
Under operating conditions the maximum height of leachate mounding is given by equation 2.4:

$$h_{max} = 0.5l\left[\left(\frac{q_0}{k_s} + s^2\right)^{1/2} - s\right]$$

$$= 0.5 \times 30\left[\left(\frac{0.15}{315} + 0.01^2\right)^{1/2} - 0.01\right]$$

$$= 0.21 \text{ m}$$

Since this is less than the thickness of the 0.3 m thick sand blanket, the design requirement of a maximum head of less than 0.3 m is met.

Moderate clogging
The sand/granular blanket may be deemed to have clogged once its horizontal hydraulic conductivity drops to a value similar to that of the waste. Once this occurs, leachate migration to the drains will no longer be preferential to the drainage blanket, although a significant proportion of the leachate may still be being transmitted horizontally to the drains through the 'clogged' sand blanket.

In this example, assume k_s (moderately clogged) $= k_w = 10^{-6}$ m/s. Thus

$$h_{max} = 0.5l\left[\left(\frac{q_0}{k_s(=k_w)} + s^2\right)^{1/2} - s\right]$$

$$= 0.5 \times 30\left[\left(\frac{0.15}{31.5} + 0.01^2\right)^{1/2} - 0.01\right]$$

$$= 0.90 \text{ m}$$

Here, the head on the base of the liner exceeds the design value by a factor of three, despite the fact that at the location of maximum mound, approximately one-third of the lateral flow is still going through the (moderately clogged) sand blanket, and near the drains almost all the lateral flow is still in the sand blanket.

Severe clogging
Severe clogging of a sand/granular blanket may be regarded to have occurred if the horizontal hydraulic conductivity drops to more than an order of magnitude less than in the waste. Once this occurs, the majority of lateral flow towards the drains will occur in the waste. For example, suppose the hydraulic conductivity of the blanket drops to 10^{-7} m/s. Under these circumstances, the blanket effectively becomes part of the 'barrier' and mounding will occur above the blanket. As a reasonable approximation, the maximum height of mounding can be calculated by adding the thickness of the blanket (0.3 m) to the maximum mound in the waste ($k_w = 10^{-6}$ m/s); therefore

$$h_{max} = h_b + 0.5l\left[\left(\frac{q_0}{k_w} + s^2\right)^{1/2} - s\right]$$

$$= 0.3 + 0.5 \times 30\left[\left(\frac{0.15}{31.5} + 0.01^2\right)^{1/2} - 0.01\right]$$

$$= 1.2 \text{ m}$$

It should be noted that while this level of clogging substantially increased the height of mounding, and hence hydraulic gradient across the barrier, it does not significantly affect the hydraulic conductivity of the barrier because the flow will still be controlled by the clay liner (and/or geomembrane if one is present), whose vertical hydraulic conductivity is likely to be one to three orders of magnitude lower than that of even a severely clogged granular blanket. However, if hydraulic conductivity of the liner is of the order of 10^{-10} m/s (or smaller) then even with mounding as calculated above, diffusion will be a major transport mechanism. In this example, assuming a 1.2 m thick clay liner with a hydraulic conductivity of 10^{-10} m/s and underlain by a second unsaturated drainage layer, it can be shown that the Darcy velocity through the clogged sand and liner is approximately 0.006 m/a (the means of performing this calculation is discussed in Chapter 5). In this context, the reader may wish to refer again to the discussion of Figure 1.22 in section 1.3.7. Under these circumstances, the sand blanket does not contribute to the hydraulic performance of the barrier, but it will become part of the 'diffusion barrier', and the diffusion profile may be expected to begin at the interface between the more permeable waste (where lateral flow dominates) and the less permeable (clogged) granular blanket. An example of this is given with respect to the observed field diffusion profile at the Keele Valley landfill discussed in section 9.3.2.

2.4.5 Minimizing clogging of blanket drains

Clogging of granular blanket drains may be the result of a combination of particulate clogging, clogging due to chemical precipitation and biofilm growth. Thus, the likelihood of clogging occurring can be minimized by

1. maximizing the flow velocity in the drain;
2. maximizing the void size;
3. minimizing the surface area available for biofilm growth.

By maximizing the flow velocity, one reduces the residence time for leachate in the collection system, thereby reducing the amount of sedimentation of particulates and chemical precipitates which can occur. Clearly, the flow velocity can be increased either by increasing the gradient (e.g. by increasing the slope of the landfill base) and/or by increasing the hydraulic conductivity of the granular layer. Ideally both the slope and hydraulic conductivity will be as large as practicable.

Maximizing the void size within the drainage blanket tends to increase the initial hydraulic conductivity and reduces the likelihood of the voids becoming blocked.

Biofilm growth is related to the surface area available on the particles forming the drainage layer. Since the surface area is proportional to the square of a particle's diameter, and the volume is proportional to the cube of a particle's diameter, it is evident that the surface area can be minimized by increasing the diameter of particles used to construct the leachate collection system.

Recent research has demonstrated a clear relationship between the grain size distribution and clogging. For example, Brune *et al.* (1991) obtained samples from clogged leachate collection systems (Figure 2.8) and from laboratory studies. They demonstrated that a graded sandy gravel with grain size in the range 1–32 mm and fine gravel (2–4 mm size) experience severe clogging in laboratory column tests conducted using high strength leachate under anaerobic conditions (landfill 8 listed in Table 2.5). Coarser material (medium gravel 8–16 mm size) experienced clogging of the smaller pores, but the larger pores remained open at the end of the experiment (16 months). Coarse gravel (16–32 mm) in direct contact with waste was covered with a thick film and locally clumped together; however, the large pores remained unclogged throughout the (16-month) test and maintained its hydraulic conductivity.

Similar tests conducted using a nonwoven geotextile filter above the drainage material indicated that the geotextile played a sacrificial role, substantially reducing clogging of the underlying drainage material with the formation of a cake of clogged waste in the upper portion of the geotextile and above the geotextile (Figure 2.9). The beneficial role of the geotextile was most evident for the coarse gravel (16–32 mm), which remained uncemented, with only a very light biofilm around the gravel particles.

Figure 2.8 Clogging of granular material from around a leachate collection pipe in a leachate collection system. (Photo courtesy of Prof. H.J. Collins.)

Analyses of the material which clogged the drainage systems (both in laboratory and field studies) have shown (Brune *et al.*, 1991) that it consists of a network of aggregates of bacteria with deposits of inorganic material on their surface. This process is initiated on the bacterial cells and the slime fibrils which they excrete. Precipitation then centers on these seeds. Chemical analyses indicate that the precipitate consists of calcium, iron, magnesium and manganese combined with carbonate and sulfur.

As indicated by Ramke and Brune (1990) and Brune *et al.* (1991), the precipitation and clogging of drainage material in landfills is caused by the following two processes.

1. Iron-reducing bacteria solubilize Fe(III) by reducing it to Fe(II), while sulfate-reducing bacteria reduce sulfate to sulfide. This bioreduction causes the region immediately around the bacteria to become more alkaline, and leads to the precipitation of sulfur as an insoluble metal sulfide.
2. Calcium is mobilized in the waste as a result of fermentative organisms producing organic acids which lower the pH of the leachate, and hence mobilize calcium (and other metals). As in the case of sulfide, it would appear that calcium carbonate then precipitates in the alkaline region immediately around the bacteria, as discussed above.

Thus, the clogging of drainage systems is the result of a mobilization process involving fermentative bacteria together with iron- and manganese-reducing bacteria, followed by precipitation processes involving primarily methane- and sulfate-reducing bacteria. These latter processes are identical to the precipitation processes described in section 2.3.1(c) on leachate stability. Another example of clogging

Figure 2.9 Clogging of a geotextile in laboratory tests performed by Brune *et al.* (1991). (Photo courtesy of Prof. H.J. Collins.)

of the drainage material around leachate collection pipes in the drain around a landfill has been reported by McBean, Mosher and Rovers (1993), who noted extensive clogging which resulted in excessive leachate mounding and leachate seeps.

Typical specifications require that the hydraulic conductivity of a granular drainage blanket be greater than about 1×10^{-5} m/s. There are many granular materials that would meet this specification, and some are far more susceptible to clogging than others. Based on the foregoing argument, it is evident that the potential for clogging is greatest for well-graded silty sands (which will have a hydraulic conductivity close to 10^{-5} m/s, small voids and a large available surface area), will be less for uniform coarse sand and will be a minimum for uniformly graded coarse stone (e.g. 50 mm clear stone). This would lead one to specify clear stone as the granular blanket; however, if clear stone is placed in direct contact with the waste, one can expect migration of particles (e.g. sand and silt size particles) of the waste into the clear stone. This in turn would reduce the effective void space and the hydraulic conductivity, and would increase the surface area available for biofilm growth. This leaves either or both of the options:

1. placing a select waste above the coarse granular blanket;
2. having a multicomponent collection system involving a filter which will minimize the migration of particles into the granular blanket.

If the stone is to be placed on top of a compacted clay liner, it may also be desirable to install a geotextile separator between the stone and the clay to minimize spoiling of stone and liner. The use of geotextiles as both filter and separator above and below a granular drainage blanket is shown in Figures 2.10 and 2.11 and discussed in the next section.

2.4.6 Geotextiles as filters and separators in leachate collection systems

Geotextiles have been used (as shown in Figure 2.10) to minimize contamination/clogging of granular drainage layers. The upper filter geotextile is intended to minimize the migration of solids from the waste (including a portion of suspended solids in the leachate) while permitting liquid to pass; however, there is a significant risk that this will clog due to biological action and chemical precipitation, as discussed below. The lower separator geotextile shown in Figure 2.10 is intended to minimize

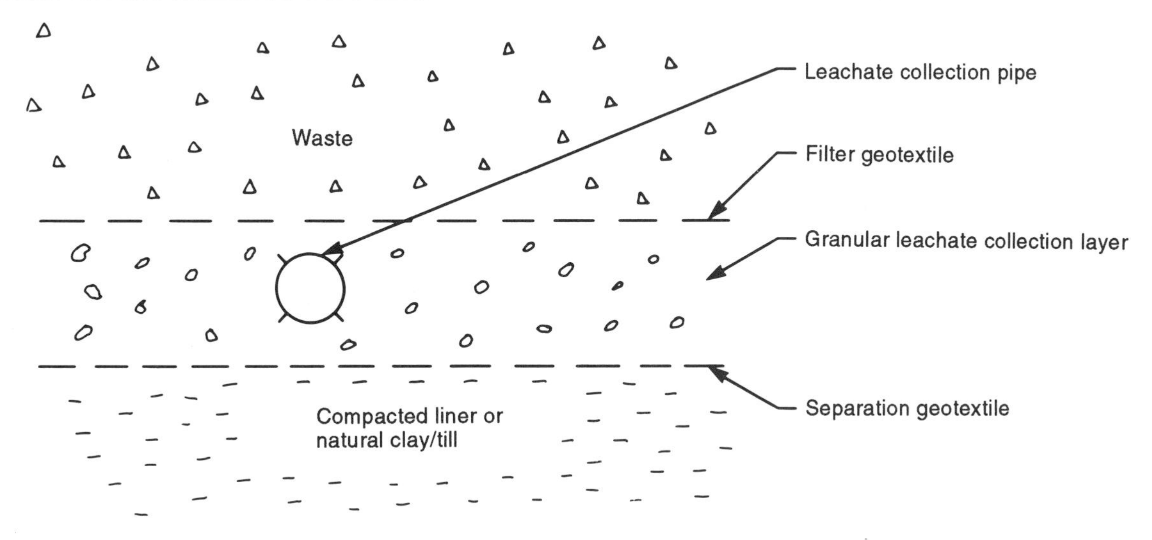

Figure 2.10 Schematic of leachate collection system with geotextile filter, granular drainage layer and separator geotextile over a clay liner.

intermixing of the fine-grained liner material and the coarse-grained drainage stone.

An alternative design, shown in Figure 2.11, makes use of a geotextile filter to separate two components of a granular blanket. This type of detail might be adopted in the design of a landfill as a hydraulic trap (section 1.2.1(d)), where a primary consideration is to avoid the build-up of leachate head directly on the clay liner, since any such build-up in excess of the design value will reduce (and possibly reverse) the inward gradient. With the detail shown in Figure 2.11 there may be clogging of the geotextile and the development of a perched leachate mound (section 2.4.6(b)) in the overlying stone; however, this will not affect the performance of the hydraulic trap. With a system such as that shown in Figure 2.11, it is desirable to provide a direct (pipe) connection between the stone above the geotextile and the manholes (in addition to the normal connection for the pipe in the drainage layer below the geotextile).

It should be noted that the design of leachate collection systems involves consideration of both hydraulic and structural aspects. The structural capacity of the pipe should be considered in the context of the loading conditions and bedding for the pipe (e.g. Moore, 1993).

(a) Geotextile filter design

In filter design, candidate geotextiles must satisfy three requirements: they must have adequate soil retention capability, be sufficiently permeable and have sufficient resistance to clogging. With regard to retention capability, the objective of the geotextile is to filter out particle sizes which would settle out if permitted to migrate into the drain. Thus the design of the geotextile is related to the hydraulic design of the drainage layer – the slower the flow in the granular layer (i.e. the lower the hydraulic conductivity and/or slope of the drainage layer) the slower will be flow, and hence the smaller the particle size which can be permitted to pass through the geotextile. The candidate geotextile must be sufficiently permeable to allow leachate to pass through the waste into the drainage layer; this criterion will be readily satisfied provided that the hydraulic conductivity of the geotextile exceeds that of the waste. The filtration characteristics of the geotextile may be

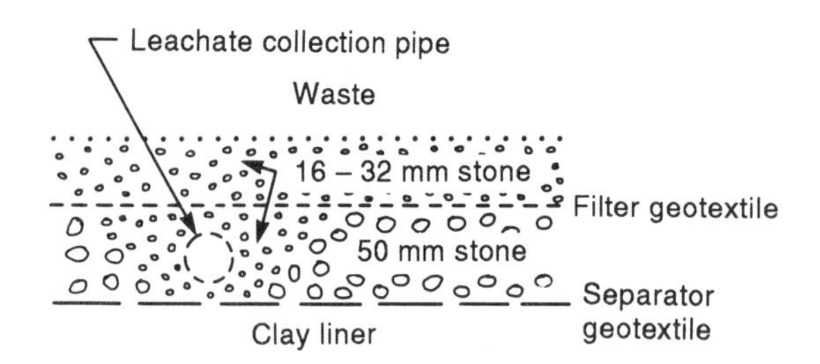

Figure 2.11 Schematic of a two-layered leachate collection system. A geotextile filter is used to separate the upper and lower granular layer.

selected on the basis of a number of techniques. For example, Faure *et al.* (1986) proposed the following guidelines:

1. for uniform soils, FOS $< 1.5d_{85}$;
2. for well-graded soils, $1.5 < \text{FOS}/d_{85} < 3.0$.

d_{85} is based on the grainsize distribution of the soil (or waste) to be filtered, and 85% (by dry weight) of the soil has a particle size smaller than d_{85}; FOS is the filtration opening size of the geotextile as determined by the hydrodynamic sieving technique (e.g. Canadian and General Standards Board Draft Method 148.1-10).

The terms 'uniform' and 'well-graded' relate to the shape of the grainsize curve of the soil to be filtered. This is often quantified in terms of the uniformity coefficient C_u, where

$$C_u = d_{60}/d_{10}$$

where d_{10} and d_{60} correspond to the grainsizes below which 10% and 60% of the soil lies on a grainsize curve (for more details see standard soil mechanics texts). Commonly, soils with $C_u < 3$ to 4 are regarded as being poorly graded; soils with $C_u > 6$ are well-graded provided that the grainsize curve is smooth and reasonably symmetrical.

Giroud (1982, 1994) has proposed an alternative guideline:

	$1 < C_u < 3$	$C_u > 3$
I_D<35%	AOS<$1.0C_u d_{50}$	AOS<$9d_{50}/C_u$
35%<I_D<65%	AOS<$1.5C_u d_{50}$	AOS<$13.5d_{50}/C_u$
I_D>65%	AOS<$2.0C_u d_{50}$	AOS<$18.0d_{50}/C_u$

where d_{50} corresponds to the grainsize below which 50% of the soil lies on a grainsize curve; AOS is the apparent opening size of the geotextile as determined from a dry sieving technique (e.g. ASTM D4751-87); I_D is the density index for the soil which is an indicator of whether the soil is in a loose, compact or dense state of compaction.

It should be noted that the foregoing design criteria were developed for cohesionless soils; compact cohesive soils rarely pose a problem with respect to filtration. It follows from the foregoing that the most difficult soils to filter are fine uniform silts.

As an example of a filter design in a landfill application, Rollin and Denis (1987) have recommended the use of nonwoven geotextiles with FOS < 140 μm for the separator/filter between compacted clay liners and granular (sand) leachate collection systems, and FOS < 75 μm for geotextiles between *in-situ* clay soils and granular (sand) secondary leachate collection/pressure relief systems. For more information concerning filter design, the reader is referred to Rollin and Denis (1987) and Rollin and Lombard (1988).

(b) Geotextile clogging

The high potential for clogging of leachate collection systems is evident from observed clogging of sand layers discussed in the previous section. An indication of the degree of clogging of geotextile filters that can occur due to accumulations of leachate fines and microbial growth may be obtained from recent studies (Cancelli and Cazzuffi, 1987; Koerner and Koerner, 1989, 1990a; Cazzuffi *et al.*, 1991; Brune *et al.*, 1991) and is illustrated in Figure

2.9. These studies suggest that decreases in geotextile permittivity (cross-plane permeability) of many orders of magnitude may occur after exposure to leachate flow. One study (Koerner and Koerner, 1989) indicates a rough correlation between the TS/BOD levels of the leachate and the degree of geotextile clogging. The study by Brune *et al.* (1991) indicated only modest clogging of a geotextile for leachate 7 given in Table 2.5, but considerable clogging for leachate 8 in Table 2.5. As noted in section 2.4.4, even under the latter extreme conditions, the geotextile filter did provide protection of the underlying gravel drainage layer, but did itself experience a substantial decrease in hydraulic conductivity.

While some filter clogging must be expected, there are several strategies which may be employed to decrease the degree of clogging and the rate of fines accumulation and bioslime growth. Two such strategies are suggested by test results reported by Koerner and Koerner (1990a). These texts indicate that inclusion of a granular layer over the geotextile reduces the rate of permeability decrease; however, if sand is used then there is the increased risk that the sand itself will clog. This problem could be addressed by using a clear gravel with particle size in the range of 16–32 mm as shown in Figure 2.11. Decreases in the degree and rate of filter clogging may be achieved by the addition of antimicrobial agents to the geotextile material (Hamilton and Dylingowski, 1989) and by selection of a high percent open area geotextile (Giroud, 1987; Koerner and Koerner, 1989), although this latter criterion must be balanced by the need to provide adequate retention (as discussed above).

The use of one or more of the above strategies may result in a decreased degree and rate of clogging and a more efficient collection of leachate, but in any case, clogging of a filter does not mean that the filter becomes impermeable – it simply means that its permeability has decreased. Typically the geotextile will have had an initial permeability (hydraulic conductivity) several orders of magnitude larger than typical values for compacted waste. The 'clogging' of the geotextile would not be expected to have any significant effect so long as the permeability of the clogged geotextile is above that of the waste. Once the permeability drops to below that of the waste, then the effects of the geotextile clogging will be highly dependent on where the geotextile is used and on what measures are taken to allow leachate drainage once clogging occurs. For example, Rowe (1993) indicated that the use of French drains with a geotextile around the stone would be particularly undesirable, because clogging of the geotextile would result in an increase in the height of the leachate mound which is acting directly on the barrier (e.g. Figure 2.7(a) or (b)). This increased leachate mound would have immediate implications with respect to contaminant migration from the landfill. On the other hand, clogging that decreases the permeability below that of the waste may not be a major concern if the filter is continuous across the site and is underlain by a continuous drainage blanket. A badly clogged filter will result in some perching of leachate above it. This is not problematic with respect to contaminant impact on underlying groundwater/aquifers since the leachate is separated from the barrier by the drainage layer and hence the perched mounding does not represent mounding on the barrier. Clearly, as the perched mounding increases, so too does the gradient across the clogged geotextile, until the gradient is sufficient to provide transmission of infiltration to the drainage layer. For example, if the clogging of a 3 mm thick geotextile causes a decrease in permeability from say 1×10^{-3} m/s to 1×10^{-8} m/s (i.e. by five orders of magnitude), the permittivity of the geotextile would still be high enough to transmit the arriving infiltration of up to about 0.3 m/a without any significant long-term perching of leachate (short-term perching may occur during periods of high infiltration).

To illustrate the implications arising from the use of a sacrificial geotextile and clear stone above a clear stone drainage blanket, consider the cross-section shown in Figure 2.11 applied to a geometry shown in Figure 2.7(c) for the following parameters:

1. design maximum head on liner < 0.3 m;
2. drain spacing $l = 30$ m;
3. base slope $s = 0.01$;
4. hydraulic conductivity of waste $k_w = 10^{-6}$ m/s;
5. hydraulic conductivity of blanket $k_s = 10^{-1}$ m/s;
6. infiltration $q_0 = 0.15$ m/a;
7. initial hydraulic conductivity of geotextile above drainage layer $k = 10^{-3}$ m/s.

Design conditions
Under operating conditions under a unit gradient ($i = 1$) the upper gravel layer and geotextile can transmit a vertical flow

$$q = ki = 10^{-3} \text{ m/s } (31\,540 \text{ m/a})$$

which well exceeds the infiltration of 0.15 m/a, and hence there should be no leachate perching above the drain. Once in the drain the maximum height of leachate mounding is calculated using equation 2.4:

$$h_{max} = 0.0004 \text{ m}$$

This is negligible and the maximum height of mounding would be controlled by irregularities in the surface of the liner, and ponding, which may occur at and below the invert of the collection pipe.

Moderate clogging
Assume that the hydraulic conductivity of the geotextile and overlying clogged granular material drops to a value similar to that of the waste (viz. $k = 10^{-6}$ m/s) under a unit gradient ($i = 1$) across the clogged geotextile; the geotextile can transmit a vertical flow (to the blanket drain) of

$$q = ki = 10^{-6} \text{ m/s } (31 \text{ m/a})$$

which still greatly exceeds the infiltration of 0.15 m/a; hence there should be no long-term leachate perching above the drain. The maximum height of leachate mounding in the drain would be the same as calculated under 'design conditions' above.

Severe clogging
Suppose that the hydraulic conductivity of the geotextile drops to the lowest value found by Cazzuffi *et al.* (1991) of 10^{-9} m/s, the clogged geotextile could transmit a vertical flow

$$q = ki = 10^{-9} \times i \text{ m/s} = (0.031 \times i \text{ m/a})$$

thus in order to transmit the average annual infiltration of 0.15 m/a, a gradient

$$i = 0.15/0.031 = 4.76$$

would be required. This would be achieved with perched mounding of about 0.47 m above the geotextile assuming a 0.1 m thick clogged zone above the geotextile. Once the leachate reaches the granular drainage layer, it would readily drain. Furthermore, even if the hydraulic conductivity of the drainage layer was also reduced by three orders of magnitude, the design criterion of less than 0.3 m head on the liner would still be met.

The foregoing example serves to illustrate the potentially beneficial role of combining a coarse granular drainage blanket with a geotextile filter in the design of a leachate collection system. The extreme case of complete clogging of the geotextile can be addressed by installing the 16–32 mm thick layer of stone over the geotextile, as indicated in Figure 2.11. This zone should be hydraulically connected to the drains; it provides a means of minimizing perched mounding of leachate.

(c) Construction damage

Defects arising from construction damage, poor seaming technique, and rupture from excessive subgrade settlement may cause an intact geotextile to become discontinuous. The survivability of geotextiles during installation has been the subject of some study (e.g. Koerner and Koerner, 1990b; Bonaparte *et al.*, 1988). Damage arising from construction is possible from the puncture of coarse granular materials above and below the geotextile and from tearing due to the action of heavy machinery. The geotextile should have sufficient strength and puncture resistance to survive normal construction. The construction specification should clearly emphasize the importance of maintaining the integrity of the geotextile. Rupture from subgrade settlements can be minimized by the use of low modulus geotextiles or by appropriate subgrade preparation.

The available recommendation for selection of geotextiles with sufficient survivability (based on a classification given by Christopher and Holtz, 1984) is summarized in Table 2.12(a). The corresponding recommendations to the index properties of geotextiles with gradings of 'medium' and 'high' survivability are given in Table 2.12(b).

In a recent study (Rowe, Caers and Chan, 1993), samples of geotextiles meeting the medium and high classifications given in Table 2.12(b) were removed from the interface between:

1. approximately 200 mm of angular 50 mm diameter crushed stone and a compacted clay liner;
2. a compacted clay liner and a rounded 10–30 mm diameter gravel layer.

No significant damage of the high survivability geotextile was observed either in terms of observed holes, or a change in tensile strength as obtained from the wide width tension test. The medium survivability geotextile was indented but generally did not contain holes when placed between the liner and rounded 10–30 mm gravel; when placed between the 50 mm crushed

Table 2.12(a) Categories of required survivability of geotextiles (as presented by Christopher and Holtz, 1984)

Subgrade preparation conditions	*Required survivability for equipment pressure indicated*[a]		
	<27 kPa	*≥27 kPa ≤55 kPa*	*>55 kPa*
Subgrade is smooth and level	Low	Moderate	High
Subgrade has been cleared of large obstacles	Moderate	High	Very high
Minimal site preparation is provided	High	Very high	Not recommended

[a]Recommendations are for 15–30 cm initial lift thickness. For other lift thickness and specific details see above reference.

Table 2.12(b) Geotextile property requirements for geotextiles used as separators[a]

Survivabilility	*Grab strength (N)*	*Puncture strength (N)*	*Tear strength (N)*
Medium	800/510	310/180	310/180
High	1200/800	445/335	445/335

[a]The first value of each set is for geotextiles which fail at less than 50% elongation; the second value is for those that fail at greater than 50% elongation.

stone and a compacted clay (having been subjected to many passes of a loaded water truck) some holes were observed. The reduction in wide width tensile strength was not significantly different for the two exposure conditions. The average reduction in tensile strength of the medium geotextile, due to construction damage, was about 15%.

It is notable that the strain to failure decreased significantly (and the modulus increased) for both geotextiles. This finding is similar to that by Bonaparte *et al.* (1988).

(d) Separator design

The design of the separator layer between the drainage layer and the underlying fine-grained natural or compacted soil is more straightforward than the design of the filter layer discussed above, and follows conventional practice (e.g. Koerner, 1990c). Selection of the geotextile for this layer also requires consideration of ways of minimizing the potential defects as discussed above.

2.4.7 Geosynthetic drainage layers

Geosynthetic drainage layers are typically net materials (geonets) which are structured to include open channels for the in-plane conveyance of relatively large quantities of fluids. The use of these geosynthetic drainage layers for leachate collection and detection at waste disposal facilities is becoming increasingly common (Koerner and Hwu, 1989). For this application, the geosynthetic drainage layer is often used in conjunction with an upstream geotextile filter and downstream geomembrane barrier, to form a composite system (Figure 2.12). The geotextile and the geomembrane restrict the movement of soil particles into the open channels. Figure 2.12 shows a typical leachate collection and leak detection system using geonets, with geotextiles to minimize soil intrusion into the geonets. Figure 2.13 shows a conventional primary leachate collection system and a geosynthetic leak detection system. A primary design consideration for these drainage layers is their *in-situ* flow capacity and the leachate residence time within the drainage layer.

(a) Flow capacity and compressibility

The flow regime within the geosynthetic drains is usually turbulent, and so flow calculations based on the assumption of an intrinsic material permeability or transmissivity are not possible, since flow is not linearly proportional to the hydraulic gradient. Thus the design must be based on direct comparisons of the required and actual material flow rates. In addition to the hydraulic gradient, several other factors have been identified which will influence the flow behavior and capacity of the geosynthetic drainage (Williams, Giroud and Bonaparte, 1984; Koerner *et al.*, 1986; Lundell and Menoff, 1989). These include compression due to overburden loads, compression-induced creep, thickness (or number of layers), boundary conditions, the temperature and viscosity of the flowing fluids, and the directional drainage characteristics.

In order to select suitable geosynthetic drainage material to meet a design flow, it is necessary to have results from laboratory simulation tests on candidate materials, using the actual loads and boundary conditions of the field situation. The ASTM D-4716 test protocol provides guidance for these simulation tests. The test program should be conducted to provide relations between the flow rate, hydraulic gradient and applied pressure. The compressive strength of the drainage layer must be determined to be adequate under expected overburden loads. Consideration must also be given to compressive creep behavior. Koerner *et al.* (1986) and Cancelli, Cazzuffi and Rimoldi (1987) have used a three-element model to estimate creep behavior. This approach shows promise; however, caution is required in extrapolating long-term behavior. Three time-dependent material behaviors are involved: the

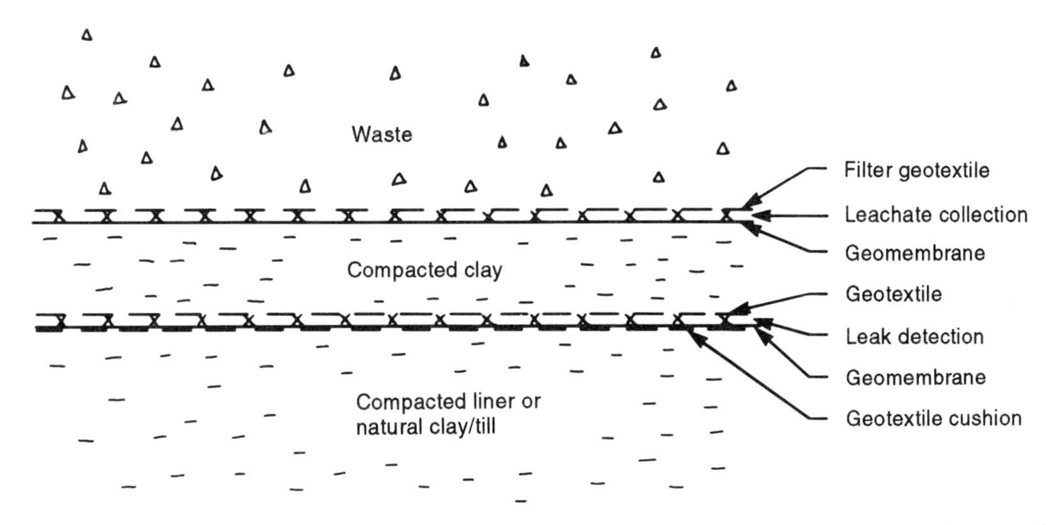

Figure 2.12 Schematic of a geosynthetic primary leachate collection system, composite (geomembrane and clay) primary liner, geosynthetic leak detection system and secondary liner.

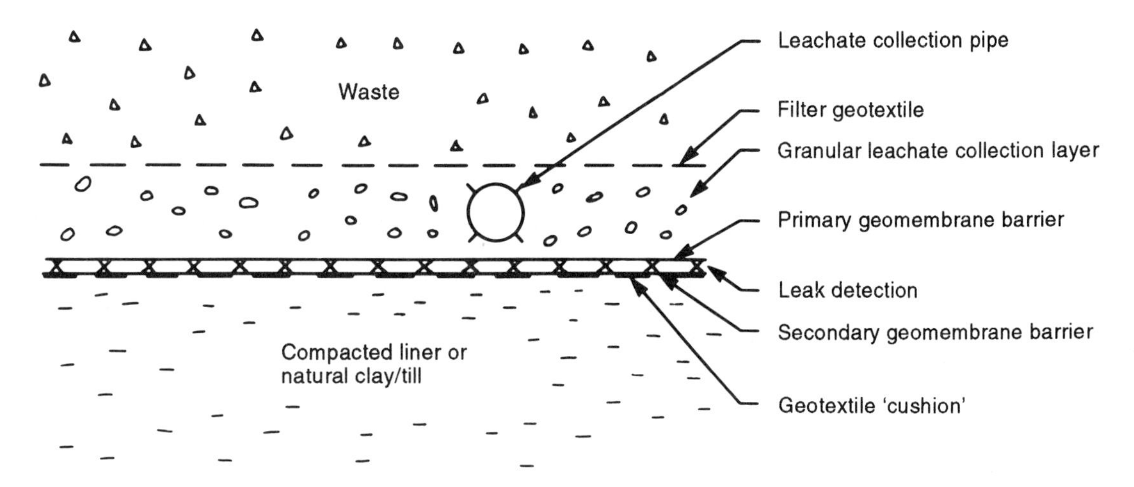

Figure 2.13 Schematic of a conventional primary leachate collection system, primary (geomembrane) liner, geosynthetic leak detection system and secondary liner.

waste, the geosynthetic and the soil. It is difficult enough to predict the longer-term deformation of any one of these components; it is even more difficult to predict the long-term interaction of all three.

As with granular systems, geosynthetic drainage layers have the potential to clog significantly. This should be carefully considered in the design.

The level of confidence in the long-term behavior of these systems decreases with the length of time that the system is required to be operative. In this respect, careful consideration should be given to the contaminating lifespan of the landfill. Generally (Chapter 12), the larger the landfill thickness and the lower the infiltration, the greater will be the concern regarding long-term performance. The longer the contaminating lifespan the greater is the need for backup systems in the event that the leachate collection system fails.

(b) Leak detection

When geosynthetic drains are being designed to detect leakage through a primary liner, the leak detection rates are governed by the transmissivity of the geonet at the design slope and the length of the flow path. Geonet flow capacity may be evaluated for a range of hydraulic gradients and compressive loads, as detailed by Williams, Giroud and Bonaparte (1984).

(c) Reinforcement function of adjacent geosynthetics

The use of geonets for detection and leachate collection imposes a reinforcement function on the adjacent materials (usually geotextiles or geomembranes), which must span openings on the surface of the mat. Tests by Hwu, Koerner and Sprague (1990) indicate that a flow rate reduction of one order of magnitude is possible due to geotextile intrusion into the apertures of geonets. Modest decreases in intrusion (and increase in flow) were observed for geotextiles stiffened by resin treating, burnishing and scrim reinforcing, and a large decrease in intrusion results with the use of a composite fabric (a needle-punched nonwoven over a woven slit-film geotextile). Some mats have been designed with limited opening size adjacent to the geotextile, or with a three-layer system, to minimize the effect of geotextile/soil intrusion. In some cases there may be advantages to using a geocomposite drainage layer with the geotextile heat bonded to the geonet, although care is required to ensure that the permittivity (and hence permeability) of the geotextile is not significantly decreased by this process. A further consideration in the design of geotextiles and geomembranes over geonet openings is long-term tensile creep. Limited studies of this design aspect have been performed (Hwu, Koerner and Sprague, 1990). The earlier comments regarding long-term behavior also apply to this aspect of drain performance.

(d) Selection and acceptability

Published data (e.g. Koerner *et al.*, 1986) show wide variations in flow rate of various geosynthetic drainage products; some have flow capacities of the same order as a 0.3 m thickness of clear stone while others are similar to the flow capacity of 0.3 m of sand. The wide variations in flow rate illustrate the importance of laboratory simulation of the actual field situation for selection purposes. It should be noted that some products may be placed in multiple layers to increase the flow capacity. Test results (Koerner and Hwu, 1989) show that the flow rate increases from the use of additional layers cannot be reliably estimated from tests with a single layer, but must also be determined from simulation testing. An alternative to the use of multiple layers is to use thicker geonets; however, the greater the three-dimensionality of the grid the greater the need to examine carefully the time-dependent response of the system.

With respect to the seaming of synthetic drainage mats, only limited research (notably Zagorski and Wayne, 1990) has been performed, and this research has concerned geonet seams only. Of particular interest is the considerable reduction in geonet transmissivity when overlapping is used, due to intrusion of one geonet into the other.

The acceptability of drainage mats for leachate collection will depend on the regulatory environment, the nature of the leachate (i.e. the potential for clogging) and the length of time that the drainage mat must operate to protect the environment (i.e. the contaminating lifespan). It is a relatively simple matter to design a mat that is only required to function for a 30-

year post-closure period for a landfill that will produce low strength leachate. However, many modern landfills have contaminating lifespans of several hundred years (Ehrig and Scheelhaase, 1993; Rowe and Fraser, 1993b). The long-term performance of any leachate collection system is a major consideration in these cases, particularly when regulations require negligible impact on groundwater in perpetuity. One means of increasing the probability of the long-term operation is to design a redundant system.

2.5 Leakage through liners

Although frequently used, the term 'leakage' is both pejorative and misleading. For example, both a high quality geomembrane liner and a very low permeability clay liner will allow some passage of fluid under an outward hydraulic gradient, and thus some would say that they leak, even though the volume of fluid moving through the liner may be substantially less than that considered in the design, and may be so low that the net impact is negligible. On the other hand, even if there is no actual fluid flow across the liner (i.e. zero leakage), there could be significant outward diffusive mass transport through the liner which could, in some circumstances, have a significant net impact on the environment. Thus the actual magnitude of the flow (leakage) through a liner is not in itself of primary significance. What is of significance is the magnitude of contaminant transport through the liner and the manner in which the barrier system has been designed to accommodate this, and hence provide protection to the environment. However, having said this, it is still necessary to be able to estimate the advective transport through a liner to provide input to the design of the barrier system and, subsequently, the contaminant impact assessment.

2.5.1 Soil liners

Advective transport through the soil is typically represented in terms of the Darcy or discharge velocity, v_a. This, in turn, depends on the hydraulic conductivity (permeability) of the soil and the hydraulic gradient.

(a) Gradients

The gradient may depend upon the initial hydrogeologic conditions, or may be totally controlled by the engineering design. The gradient will often depend on the design of the leachate collection system and the height of leachate mounding as discussed in previous sections. When calculating gradients based on natural hydrogeologic conditions, it is essential to carry out the following.

1. Ensure that there is an adequate data base concerning natural water levels over both short- and long-term periods.
2. Consider the effects construction of the facility itself may have on water levels beneath the site. This is particularly critical for landfills designed as hydraulic traps where the landfill extracts water from the underlying groundwater system (section 1.2.1(d)). It may also be of significance when the landfill simply reduces the natural recharge over a large area. Where necessary, flow modeling (Chapter 5) may be useful for assessing the interaction between the landfill and the hydrogeologic system.
3. Consider the implications of potential long-term changes in climate and/or industrial or municipal development on groundwater recharge groundwater levels. The significance of this factor increases with increasing contaminating lifespan. Systems should be designed such that they are not sensitive to these changes.

Some common pitfalls in the calculation of gradients are discussed in Chapter 5.

(b) Hydraulic conductivity of compacted clay liners

Chapter 3 considers the hydraulic conductivity of compacted clayey liners in some detail and discusses the use of both laboratory and field tests for assessing the hydraulic conductivity of a liner. A major factor affecting the long-term durability of clayey liners is clay–leachate compatibility. This is discussed in Chapter 4. Finally, Chapter 9 discusses the findings from a number of field studies which have demonstrated that low hydraulic conductivity (10^{-10} m/s) clayey liners can be successfully constructed.

Having noted that low hydraulic conductivity clayey liners have been successfully constructed, it should be recognized that some variability in hydraulic conductivity is to be expected and should be considered in both the design and specifications. One of the greatest potential dangers to a well-constructed liner is cracking of the liner if it is not properly maintained following compaction. For example, the liner must be protected from drying (with consequent shrinkage cracks) and freezing (with consequent cracks from ice lensing). If cracking of the liner occurs then the primary path for contaminant transport will obviously be through the cracks/fissures.

It should also be noted that it is a much simpler matter to construct a compacted clay liner on an existing natural soil (e.g. stiff clay, till or dense sand – Figure 1.4) than to construct it on top of other engineered components of a system such as a granular hydraulic control layer (e.g. Figures 1.5–1.8), geosynthetic leak detection system and/or geomembrane liner (e.g. Figure 2.12). Invariably, the need to minimize damage to underlying engineering systems has some effect on the hydraulic conductivity of at least the first lift of an overlying compacted liner. When designing these systems, a test pad which emulates the proposed cross-section is highly desirable in order to identify potential problems and refine the construction specifications (e.g. Rowe, Caers and Chan, 1993).

(c) Hydraulic conductivity of natural barriers

The hydraulic conductivity of undisturbed or natural barriers (e.g. undisturbed clayey deposits) may be assessed by a combination of laboratory and field hydraulic conductivity tests, as discussed in section 3.2. In performing these studies it must be recognized that natural deposits can be variable. This variability should be considered in the design. One issue requiring particular attention when dealing with natural clayey deposits is the potential existence of fractures. It is well-recognized that the weathered crust of a clay or clay till will be fractured. What is not well-recognized is the possibility that the unweathered till or clay could also be fractured. There is some evidence (D'Astous *et al.*, 1989; Herzog *et al.*, 1989) to suggest that many tills in North America may be fractured to depths well below the weathered crust. The findings from a number of cases are summarized in Table 2.13. Where fractures are present, the bulk hydraulic conductivity may be considerably higher than values obtained from laboratory tests on intact samples. Unfortunately, in many cases (e.g. Table 2.13) the presence of fractures was not evident either from borehole samples or hydraulic conductivity tests conducted in boreholes. In most cases the fractures were only readily observed by construction of deep test pits or from the results of a pumping test conducted on an aquifer below the fractured unit.

The design of landfills on fractured clayey soils is discussed in more detail in Chapters 11 and 12. If present, the effect of hydraulically significant fractures can be minimized by

1. operating the landfill as a hydraulic trap;
2. reworking some of the soil as a compacted clayey liner; and/or
3. constructing a composite liner system.

(d) Diffusion through soil liners

As previously noted, the phenomenon of molecular diffusion may be the primary contaminant transport mechanism through a well-designed and constructed liner. This phenomenon is

Table 2.13 Summary of a number of cases where fracturing was encountered below the obviously weathered zone in clayey till deposits

Site	*Approximate depth of weathered zone (m)*	*Approximate observed depth of fractures (m)*	*Was deep fracturing evident in borehole samples?*	*Were deep fractures evident from borehole tests?*	*Was fracturing implied by field pumping tests?*
Southern Ontario 1	4–5	10.5–11	No	No	Yes
2	4–5	7	No	No	Yes
3	4–5	10	No	No	NA
4	4	7	No	Maybe	Yes
5	4–5	>13	No	No	NA
6	7	>12	Yes	NA	NA
7	4–6	9	No	No	NA
8	4–6	8–8.5	No	Maybe	Yes
9	4–6	10	No	No	NA
Illinois	5–6	~15	No	Maybe	NA

NA: not applicable (either not performed or not performed at an elevation to detect fractures).

described in detail in Chapter 6. Chapter 8 discusses the evaluation of diffusion parameters. Natural field diffusion profiles developed over a period of about 10 000 years are examined in Chapter 9 together with results obtained for diffusion from landfills through both natural and recompacted clayey liners. Finally, Chapters 10, 11 and 12 examine the implications of diffusion in the analysis and design of landfills for a range of conditions.

2.5.2 Geomembrane liners

Under certain hydrogeologic and geotechnical conditions, natural soil or compacted clay barriers may be unable to provide the required level of environmental protection from landfill contaminants. Often, the use of geomembranes as a complementary barrier material can provide an economical means of obtaining the additional protection necessary. As with clay liners, there are two primary mechanisms for contaminant transport through geomembranes: advection and diffusion.

(a) Advective transport through geomembranes

Intact geomembranes have a very low hydraulic conductivity to water (values of the order of 10^{-13}–10^{-15} m/s have been reported (e.g. Giroud and Bonaparte, 1989), although, as noted by these authors, the actual mechanism of water migration across the membrane is different from pure advection). The majority of advective flow through a geomembrane will be localized to defects (e.g. tears, inadequate seams) in the geomembrane.

An assessment of leakage through holes in the geomembrane requires an estimate of the size and frequency of such holes. Such an assessment based on a combination of theory and an analysis of geomembrane case histories has been provided by Giroud and Bonaparte (1989) who indicate that even with intensive quality assurances, it is reasonable to expect three to five geomembrane defects per hectare. Equations relating the leakage rate through these holes to the hydraulic head acting for the case where the geomembrane is underlain by a clay layer (composite liner) (e.g. Figure 2.12), and for the case where the liner is a geomembrane is underlain by a pervious material (e.g. a drainage net as shown in Figure 2.13), have been presented by Giroud and Bonaparte (1989), and their findings in terms of estimates of leakage rates are summarized in Table 2.14. This work has been refined and extended by Giroud, Badu-Tweneboah and Bonaparte (1992). This paper considers long cracks in the geomembrane, and presents a number of equations that can be used to estimate the rate of leakage through a composite liner. These equations can be combined with a contaminant transport model as described by Rowe, Booker and Fraser (1994).

The occurrence of defects in the geomembrane liner can be reduced by the use of a geotextile 'cushion' between the geomembrane and adjacent soil layers. When the geotextile is used between the geomembrane and a clay barrier, the geotextile may increase leakage rates by providing a lateral flow plane, increasing the area over which the hydraulic head acts, as discussed in the reference cited above.

The subject matter associated with appropriate design and construction of geomembrane lined landfills is sufficiently extensive to warrant a book on its own. At the date of writing, no such comprehensive study has been published. A reference dealing with the testing of geomembranes has been edited by Rollin and Rigo (1991). Additional information is available from numerous sources, including the US EPA (1988, 1989).

(b) Diffusive transport through geomembranes

The process of molecular diffusion under a chemical potential gradient is well understood

Table 2.14 Estimates of leakage through geomembrane liners (after Giroud and Bonaparte, 1989)

Type of liner	*Leakage mechanism*	*Leakage rate (l/ha per day) for a liquid depth on top of the geomembrane, h (m) of*				
		0.003	*0.03*	*0.3*	*3*	*30*
Geomembrane alone (between two pervious media)	Permeation	0.0001	0.01	1	100	300
	Small hole	100	300	1 000	3 000	10 000
	Large hole	3000	10 000	30 000	100 000	300 000
Composite liner (good field conditions)	Permeation	0.0001	0.01	1	100	300
	Small hole	0.02	0.15	1	9	75
	Large hole	0.02	0.2	1.5	11	85
Composite liner (poor field conditions)	Permeation	0.0001	0.01	1	100	300
	Small hole	0.1	0.8	6	50	400
	Large hole	0.1	1	7	60	500

The large hole has a surface area of 1 cm^2. The frequency of holes is 1 per 4000 m^2. The thickness of the soil layer is 0.9 m and its hydraulic conductivity is 10^{-9} m/s. The HDPE geomembrane thickness is 1 mm. Leakage rates in the case of a composite liner do not significantly depend on the material overlying the geomembrane. In the case of a geomembrane alone, leakage rates were calculated assuming that the geomembrane is overlain and underlain by an infinitely pervious membrane. This assumption is valid for coarse gravel or geonet. Leakage rates through holes would be significantly less if the geomembrane were overlain and/or underlain by sand or a less permeable material.

(e.g. Park, 1986) and applies to geomembranes just as it applies to clay soils. Despite its potential significance, relatively little research has been undertaken with respect to this mode of contaminant transport. The limited available data suggest that diffusion through geomembranes does occur, and that it depends on both the chemical species being examined and the geomembrane. Table 2.15 summarizes the results from a limited number of investigations

Table 2.15 Summary of diffusion data for geomembranes

Geomembrane[a]	*Leachate*	*Bulk diffusion coefficient, nD* (m^2/s)	*Bulk diffusion coefficient, nD* (m^2/a)
Data from Lord, Koerner and Swan (1988)			
PVC	Phenol	0.66×10^{-10}	2.1×10^{-3}
	NaOH	1.3×10^{-10}	4.0×10^{-3}
	H_2SO_4	1.3×10^{-10}	4.0×10^{-3}
	Xylene	0.15×10^{-10}	4.7×10^{-4}
CPE	Water	6.0×10^{-11}	1.9×10^{-3}
	Phenol	8.8×10^{-11}	2.8×10^{-3}
	NaOH	6.0×10^{-11}	1.9×10^{-3}
	H_2SO_4	6.3×10^{-11}	2.0×10^{-3}
HDPE	Water	1.6×10^{-12}	5.0×10^{-5}
	Phenol	1.6×10^{-12}	5.0×10^{-5}
	NaOH	1.6×10^{-12}	5.0×10^{-5}
	H_2SO_4	1.6×10^{-12}	5.0×10^{-5}
	Xylene	1.8×10^{-12}	5.9×10^{-5}
Data from Hughes and Monteleone (1987)			
CPE	Water	4.0×10^{-11}	1.3×10^{-3}
	Nonsolvent	3.3×10^{-11}	1.0×10^{-3}
	Solvent	1.3×10^{-10}	4.1×10^{-3}
HDPE	Water	2.6×10^{-12}	8.2×10^{-5}
	Nonsolvent	2.6×10^{-12}	8.2×10^{-5}
	Solvent	3.8×10^{-12}	1.2×10^{-4}
Data from Luber (1992)			
HDPE	Dissolved		
	$CHCl_3$	1.7×10^{-11}	5.4×10^{-4}
	CCl_4	4.8×10^{-11}	1.5×10^{-3}
	C_2HCl_3	7.3×10^{-11}	2.3×10^{-3}
	C_2Cl_4	7.8×10^{-11}	2.4×10^{-3}
MLDPE	$CHCl_3$	2.5×10^{-11}	7.9×10^{-4}
	CCl_4	5.7×10^{-11}	1.8×10^{-3}
	C_2HCl_3	8.5×10^{-11}	2.7×10^{-3}
	C_2Cl_4	8.7×10^{-11}	2.7×10^{-3}
PVC	$CHCl_3$	11×10^{-11}	3.5×10^{-3}
	CCl_4	9.7×10^{-11}	3.1×10^{-3}

[a]PVC: polyvinyl chloride; CPE: chlorinated polyethylene; HDPE: high density polyethylene; MLDPE: medium–low density polyethylene.

that have reported the results of diffusion experiments for HDPE liners. It appears that the value of nD which would be used in modeling contaminant transport through HDPE geomembranes for organic compounds is of the order of 8×10^{-11} to 1×10^{-12} m^2/s.

Considerable additional work is required to evaluate the significance of diffusion through geomembranes. Until such time as more definitive studies are completed, it would be prudent to consider diffusion when examining the potential impacts of landfills. This can be readily done using the theoretical models described in Chapter 7. For example, a geomembrane overlying a compacted clay liner can be modeled as separate layers. The finite layer technique (Chapter 7) readily allows modeling of a 1–2 mm thick geomembrane in contact with a clay layer which may range in thickness from a few centimeters to tens of meters. In modeling the geomembrane, it would appear that the value of nD (equation 1.14b) used would be as indicated in Table 2.15, and the value of nv would represent the average Darcy flux (leakage) through the geomembrane. The implications of performing modeling based on these assumptions are discussed in Chapter 12.

2.6 Leak detection systems

Just as the concept of leakage through a liner is subject to a number of interpretations, so too is the role (and consequent design) of leak detection systems. Considerable work has been conducted in the USA with respect to leak detection (e.g. Geoservices, 1984b) and the performance of a primary liner is judged in terms of the leakage (i.e. flow) monitored in leak detection systems such as those shown in Figures 2.12 and 2.13. The effectiveness of these leak detection systems will depend on the detection sensitivity and the time between when leakage occurs and its detection.

The implementation of contingency plans in case of failure of all or part of the waste containment system will occur when unacceptable levels of leachate are collected from the leak detection layer. However, there are other sources of flow from leak detection layers, as identified by Gross, Bonaparte and Giroud (1990). If leakage through the primary liner into the leak detection layer occurs simultaneously with flow from these other sources, the presence of the leachate may be masked. Chemical analysis of the flows may identify the presence of leachate, but there may be considerable dilution. The leak detection system should be designed such that in case of a failure (i.e. a large rate of leakage) the time between detection and implementation of contingency measures will not be so long that environmental damage will occur.

A secondary leachate collection system (such as that shown in Figure 2.14) provides an additional means of monitoring for failure of the primary leachate collection system. A significant increase in leachate collected in this system (i.e. over and above that which would be reasonably predicted for a properly operating system) would provide a good indication of the failure of the liner and/or primary leachate collection system. However, the converse is not necessarily true, as is explained below.

It may be possible to have a failure of part of the primary leachate collection system which is not clearly evident in the flows in the secondary leachate collection system. This can arise because the change is masked by the variability of infiltration and flows to the secondary system with time, combined with the gradual nature of the failure of the leachate collection system. An additional factor is the escape of leachate which passes through the first barrier into the secondary barrier. The sensitivity of the volume of leachate collected in the secondary leachate collection system to migration of leachate through the primary liner is least when the area of the landfill is large, when the slopes on the lower leachate collection system are small, and when the material beneath the secondary

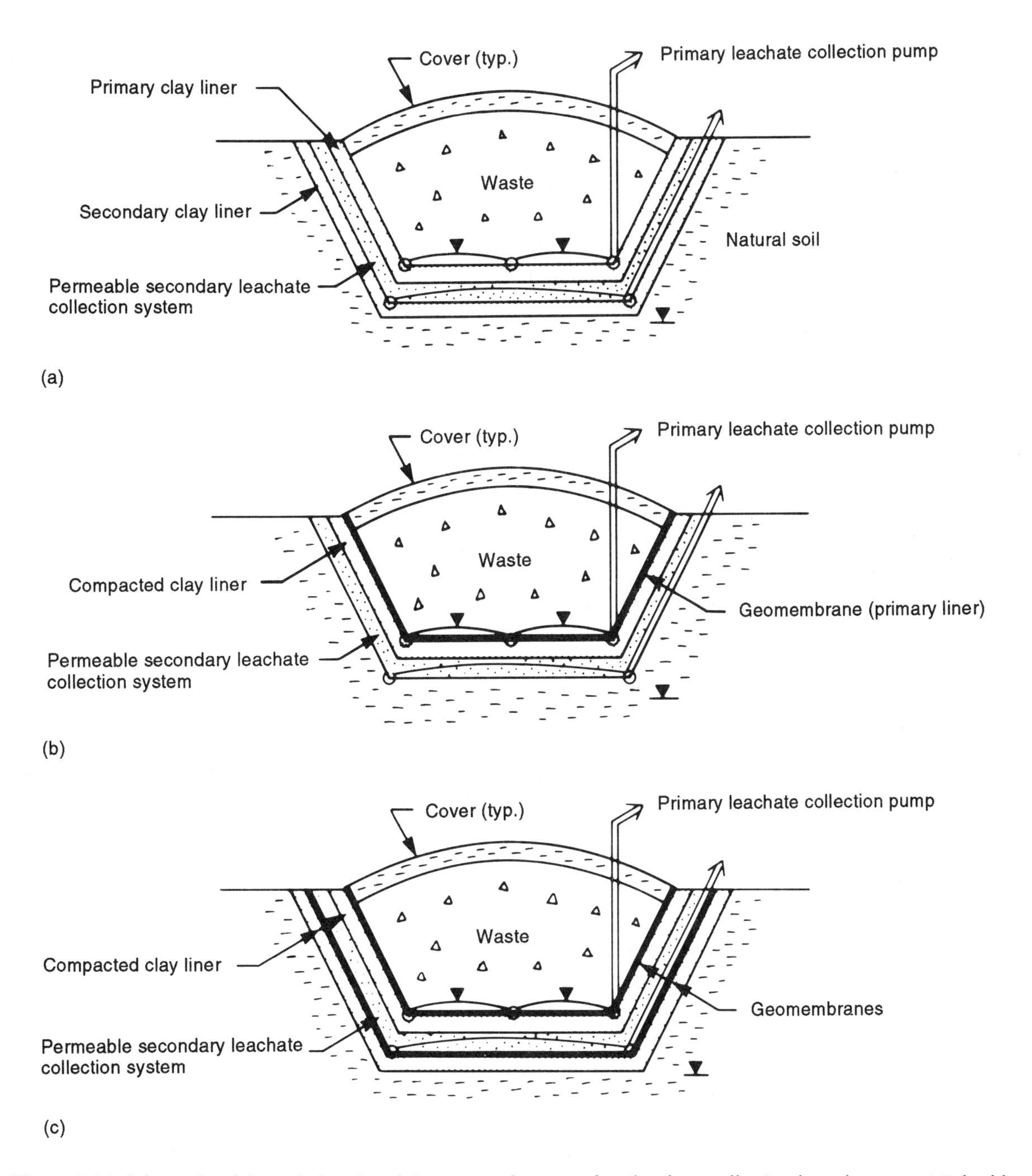

Figure 2.14 Schematic of three designs involving a granular secondary leachate collection layer between: (a) double clay liners; (b) a composite primary liner and natural clay; (c) double composite liners.

leachate collection system has a permeability of the order of 10^{-9} m/s or greater. The effectiveness can be increased by one or more of the following:

1. reducing the distance contaminant may have to travel before it reaches a collection point;
2. increasing the slopes to provide for faster transmission of leachate;
3. ensuring that the underlying material has as low a hydraulic conductivity as is practical (e.g. 10^{-10} m/s liner and/or the use of a geomembrane over the soil/clay liner).

2.7 Landfill capping and the control of infiltration

2.7.1 Geosynthetic and clay covers

A schematic of a design which uses geosynthetics as part of the leachate collection, leak detection and leachate containment systems are shown in Figure 2.15. This schematic also shows the geosynthetic components of the landfill capping system used to control the amounts of leachate generated by rainfall infiltration. The geosynthetic capping system will typically comprise a geomembrane barrier, a geosynthetic drainage layer for rainfall collection, a geotextile filter between the drainage layer and the final soil cover, and a geotextile drainage layer below the geomembrane for the transmission of landfill gases to a higher vent or collection location. The use of geotextiles for gas transmission has been discussed elsewhere (Koerner, Bove and Martin, 1984). These systems can be quite effective in constructing a relatively tight cap which will place a control on the infiltration into the landfill facility. However, it is important to recognize the interrelationship between the cover design (which controls infiltration), the contaminating lifespan of the landfill and the service life of the components of the leachate collection, leak detection and leachate containment systems.

Two key factors affecting the contaminating lifespan are the mass of waste per unit area of landfill (the height of waste) and the infiltration through the landfill cover (i.e. long-term leachate generation per unit area). Increasing the thickness of a landfill or reducing the infiltration both tend to increase contaminating lifespan, as discussed in Chapter 12. It is also important to recognize the significance of these two variables when attempting to extrapolate data from existing landfills to infer contaminant decay curves for future landfills.

2.7.2 Controlling infiltration

There are various philosophies to controlling the generation of leachate. One is to construct as impermeable a cover as possible, as soon as possible, so as to minimize the generation of leachate. This approach has the benefits of minimizing both the amount of leachate that must be collected and treated, and the mounding of leachate within the landfill; it also has the disadvantage of extending the contaminating lifespan.

Because of the heterogeneous nature of waste, some leachate will be generated almost immediately, even for landfills designed with low infiltration covers. With low infiltration, it may take from decades to centuries before the field capacity of the waste is reached and full leachate generation occurs. This means that the full capacity of the leachate collection system may not be required for many decades after construction. However, during this period of time, degradation and biological clogging of the leachate collection system can be occurring unless the waste has been pre-treated to convert it to an essentially inert form. Furthermore, due to the variable nature of leachate generation, it may be difficult to assess whether there has been a failure of the leachate collection system, by monitoring the volumes of leachate extracted.

An alternative philosophy is to allow as much infiltration as would practically occur. This

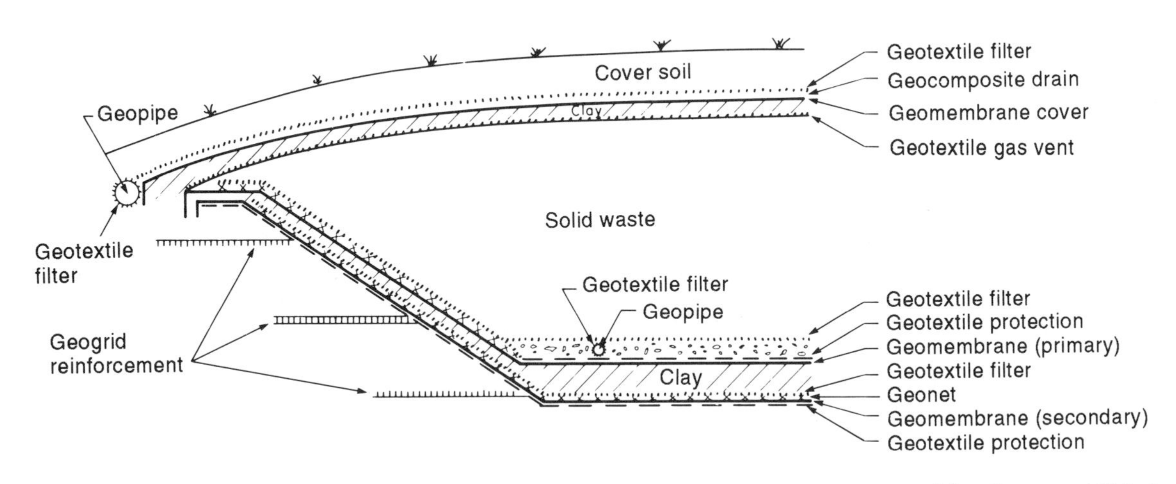

Figure 2.15 Schematic of a complete landfill system showing geosynthetic components. (After Koerner, 1990a.)

would bring the landfill to field capacity quickly and allow the removal of a large proportion of contaminants (by the leachate collection system) during the period when the leachate collection system is most effective and is being carefully monitored (e.g. during landfill construction and, say, 30 years after closure). The disadvantages of this approach are two-fold. Firstly, larger volumes of leachate must be treated; this has economic consequences for the proponent. Secondly, if the leachate collection system fails, a high infiltration will result in significant leachate mounding.

The low infiltration philosophy is readily suited to meet environmental regulations that only require a limited period during which the landfill must not cause an unacceptable impact (be it 30 years or 100 years). However, engineers have a moral responsibility also to consider the longer-term consequences. In some areas (e.g. Ontario, Canada), there is also a regulatory responsibility to consider environmental protection in perpetuity.

When long-term protection is considered, it may be desirable to find a balance between the low/high infiltration philosophies. Although it is not always practical (e.g. because of after-use requirements), one option is to allow high infiltration during construction of the entire landfill (not just the cell) and for a period after landfilling ceases. Once the landfill has been largely stabilized or the leachate collection system starts to degrade noticeably, then a low infiltration cover would be constructed. This approach rapidly brings the landfill to field capacity and removes a substantial portion of the potential contaminants early in the life of the landfill when the engineered components are at their most reliable, thereby reducing the contaminating lifespan. This approach also reduces the infiltration and potential mounding of leachate after the low infiltration cover has been constructed, and hence minimizes potential problems once the performance of key components of the engineered system (like the primary leachate collection system) have degraded.

These factors should be considered in selecting the barrier system, in developing the monitoring and contingency plans, and in assessing the service life and contaminating lifespan of the facility.

It should also be recognized that while it may be possible to construct relatively tight landfill covers that permit very little infiltration into the landfill, these covers will degrade with time due to factors such as:

1. differential settlement of the waste (subsidence);
2. freezing and drying;
3. damage by insects and rodents;
4. degradation of synthetic material with time (due to physical, chemical and biological causes);
5. vegetation;
6. end-use of the finished landfill;
7. erosion.

Thus, in order to maintain low infiltration it may be necessary to repair or indeed replace landfill covers on a regular basis over the contaminating lifespan of the landfill. For a tight cover, this may be a very long time.

2.8 Choice of barrier system and service life considerations

The choice of barrier system should always be site specific and should take account of:

1. local hydrogeology;
2. nature of the waste;
3. size of the landfill (in both vertical and areal extent);
4. climatic conditions;
5. availability of suitable materials;
6. potential impact of failure of the system on health and safety;
7. local social and regulatory systems.

It must be acknowledged that there is always more to learn, and it is not possible to guarantee that a facility thought to be safe today will last, say, 500 years. On the other hand, professionals responsible for environmental management should, in the authors' opinion, design facilities to ensure that, based on what we know today, and what we know that we do not know, a waste disposal facility will not cause an unacceptable environmental impact or risk to human health and safety, irrespective of when this might occur. This design philosophy recognizes that disposal sites will have a number of phases.

Firstly, during the active life of the landfill, careful quality control and monitoring will be required to ensure that the engineered features are working as intended – if not, problems should be rectified. This will involve careful inspection of existing clayey deposits, and careful control and monitoring of compacted liners, geosynthetic liners, drainage systems and leachate collection systems. Double liner systems (e.g. Figures 2.12, 2.13 and 2.14) have an engineered backup which can contain much of the leachate in the event of a failure of the primary system. In other situations, the hydrogeologic setting must be selected such that it provides a backup in the event of a failure of the primary engineering. The facility should be designed such that it will not impact on the reasonable use of the groundwater or the environment during its operational life.

Secondly, it must be recognized that eventually some engineering components of the landfill may fail. The effective life of drainage and leachate collection systems and, to a lesser extent, geosynthetic liners, is the subject of much debate.

Reviews of the potential for clogging of leachate collection systems (e.g. Bass, Ehrenfeld and Valentine, 1984) have indicated that clogging may in fact occur due to physical, chemical, biochemical and biological mechanisms. Based on their study, Bass *et al.* concluded, *inter alia*, that 'it is reasonable to expect clogging to occur in a probabilistic manner during the active and post-closure operational lifetime'. Section 2.4.5 discussed means of prolonging the life of these systems. Notwithstanding this, the potential for clogging of the leachate collection system, and in particular the clogging of the granular component and pipes which cannot be cleaned, should be considered when performing environmental impact assessments. One cannot justifiably assume that, because the pipes are clean, the leachate system is necessarily functioning correctly. Similarly, it is difficult to justify arguments that failure of the leachate collection system can be detected by monitoring the volume of leachate extracted. Monitoring of head in the granular component of the leachate collection system is one means of monitoring its performance. However, consideration should then be given to the potential impact of mounding, which is not detected by these monitors, and the potential impact should be quantified. Depending on the results of these analyses, additional levels of redundancy may be required as part of the engineered system (e.g. in some cases it may be necessary to use a double liner with either secondary leachate collection capabilities or hydraulic control).

In addition to consideration of clogging of the leachate collection system by a combination of biological and particulate clogging, reduction in transmissivity of geonet drainage systems due to time-dependent intrusion of soil should be considered. Similarly, when dealing with geomembranes, time-dependent degradation of the geomembrane should also be considered. As noted by Koerner (1990a), 'degradation of any geosynthetic material must come from molecular chain scission of the polymeric resin itself or bond breaking and reactions with the compounding materials. ... The process is too slow to follow in the laboratory and experience in the field is still too short to make long-term degradation predictions.' This issue was discussed in some detail at the Symposium on Durability and Aging (Koerner, 1990b).

Degradation mechanisms include ultraviolet (UV) exposure (during construction), chemical reaction with leachate, swelling due to chemical absorption, extraction, oxidation and biological attack (Koerner, Halse and Lord, 1990).

The severity of the degradation mechanisms listed above will govern the service life of the various geosynthetic components of the facility. Installation damage, the magnitude of the imposed mechanical stresses, and synergism between the various mechanisms may increase the rate of degradation (Koerner, Halse and Lord, 1990). Accelerated tests have been developed which have been used to predict service lifetimes; however, these tests are not particularly realistic. With respect to geomembranes, Koerner (1990a) has noted that 'the service lifetimes of the various geomembranes are rarely in excess of 30 years'. This should not be interpreted as meaning that geomembranes will only last 30–40 years; they are likely to last much much longer with current estimates being about 200 years. However, the limited experience should be carefully considered in design.

2.9 Geotechnical considerations

Many aspects of geotechnical and geosynthetic engineering, beyond that directly associated with leachate collection, detection and contaminant migration, need to be considered in the design of barrier systems. Most of these issues are dealt with in standard textbooks on geotechnical engineering. Landva and Clark (1990) provide a good general discussion of the geotechnics of waste, and Landva and Knowles (1990) provide numerous articles dealing with different aspects of waste geotechnics. The objective of this section is to highlight some important considerations.

2.9.1 Settlement and bearing capacity

Differential settlement of the soil beneath a landfill, and of compacted clay liners, may reduce the effectiveness of leachate collection systems. Similarly, settlement of the cover (due to settlement of underlying waste) will reduce the effectiveness of the cover. Leachate (or water) will pond in areas where settlement occurs, increasing the hydraulic head and consequent flow through the underlying soil.

With the use of geomembranes, an additional consideration is that the straining of the geomembrane over the settlement may initiate holes or ruptures. The formation of tension cracks in the underlying compacted or natural clay barrier is also a possibility under some extreme circumstances. An expected increase in the use of vertical landfill expansions to satisfy demand for landfill space may be expected to increase the occurrence of problems related to differential settlements. The use of geogrid reinforcement over areas of potential differential settlements may reduce such impacts.

Bearing capacity and settlement is a concern when designing manholes to be constructed within the waste areas. Bearing failure or excessive settlement of the manhole may reduce the efficiency of the system and/or damage the barrier. Factors to be considered in assessing the settlement of manholes include not only the weight of the manhole itself, but also the downdrag forces that may develop due to settlement/consolidation of adjacent waste (Gartung, Pruhs and Nowack, 1993; Bierner and Sasse, 1993). There is a direct analogy with the design of piles for downdrag (e.g. Poulos and Davis, 1980). One technique of avoiding large shafts in landfills is the use of tunnels beneath the waste (Geusebroek and Luning, 1993).

2.9.2 General stability

A landfill can represent a major loading on the underlying soil. Just as with the design of any structure on soil, care should be taken to ensure the overall stability of the landfill under both static and, where appropriate, seismic conditions. This involves an evaluation of the geotechnical conditions and appropriate stability analyses to identify the most critical conditions in the development of the landfill. These analyses should consider:

1. stability of the excavated trenches for the landfill (where appropriate);
2. stability during development of the landfill;
3. stability of the completed landfill.

Particular care is required not to underestimate the load that can be applied by partially saturated waste, and a conservative estimate of the unit weight of waste should be considered when evaluating landfill stability. This is particularly critical for landfills being constructed in soft soils, sensitive soil, near escarpments or on slopes. Landfill failures have occurred due to inadequate assessment of overall stability! A number of potential failure mechanisms are shown in Figure 2.16.

2.9.3 Blowout or basal heave

In addition to considering general stability due to bearing capacity or slope failure (section 2.9.2), consideration should also be given to the potential for blowout (basal heave) of the bottom of the excavation. This is particularly critical for landfills being designed as hydraulic traps (section 1.2.1) where the water pressure in an underlying aquifer is to be used to minimize contaminant transport from the landfill. This

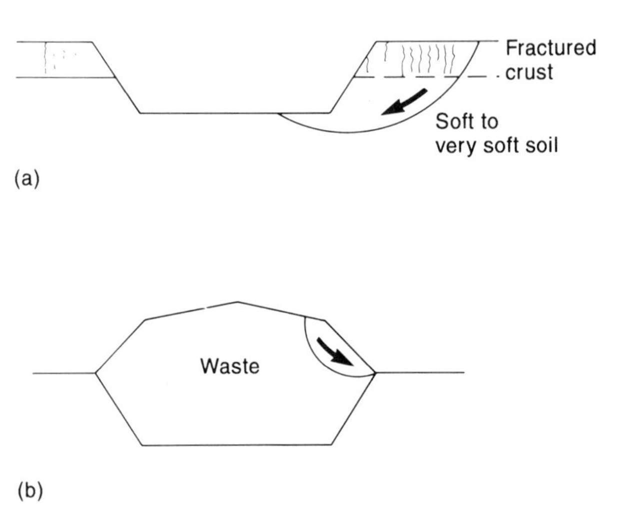

Figure 2.16 Potential failure mechanisms. (a) Rotational type failure after excavation of cell; (b) landfill slope instability.

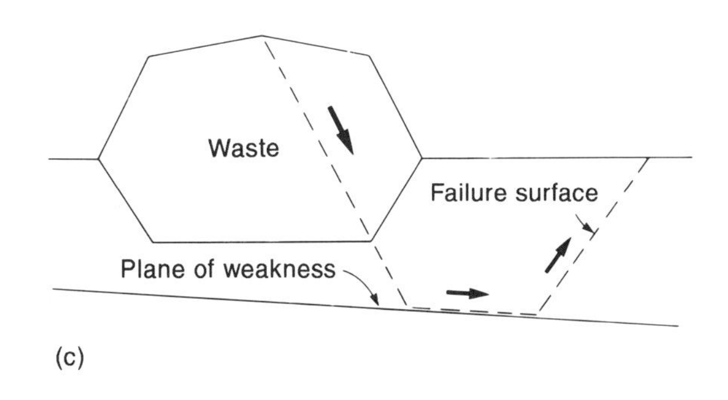

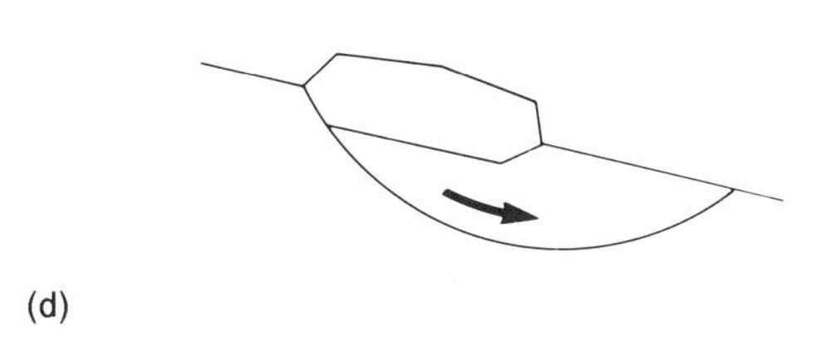

Figure 2.16 (c) Failure along plane of weakness; (d) general slope failure initiated by presence of landfill.

same water pressure (if not controlled) can cause blowout of the base of the landfill. Blowout considerations may necessitate pumping of underlying aquifers to reduce water pressure (and hence to ensure an adequate factor of safety against blowout) during excavation of a cell, placement of the barrier systems and the waste.

For example, consider the system shown in Figure 2.17. Assuming a unit weight of the clay till of $\gamma_t = 21$ kN/m^3, of the engineered barrier system of $\gamma_e = 20$ kN/m^3 and of the waste of $\gamma_{ws} = 6$ kN/m^3, the maximum head, h_p, in the aquifer for a factor of safety against blowout of FS = 1.4 may be calculated as follows (where $\gamma_w = 9.8$ kN/m^3 is the unit weight of water):

$$\text{Factor of safety} = \frac{\text{Weight of material above aquifer}}{\text{Water pressure acting on overlying aquitard}}$$

1. After excavation to base of landfill and before placing engineered system,

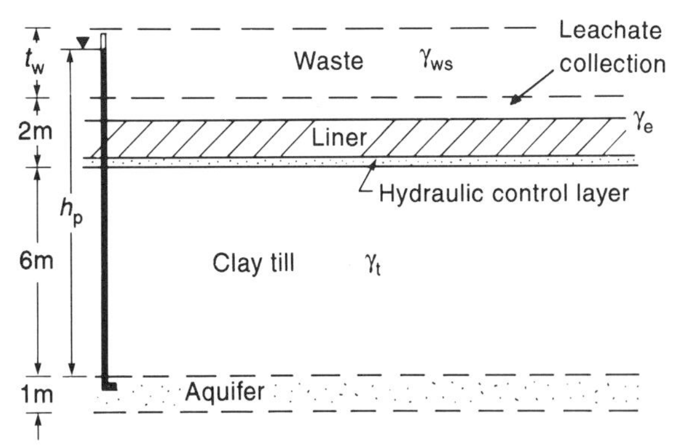

Figure 2.17 Schematic of natural soil, liner system and waste used for example blowout calculation.

$$FS = \frac{6\gamma_t}{h_p\gamma_w} \text{(referring to Figure 2.17)}$$

Rearranging terms gives

$$h_p = \frac{1}{\gamma_w}\left(\frac{6\gamma_t}{FS}\right) = \frac{1}{9.8}\left(\frac{6 \times 21}{1.4}\right)$$
$$= 9.2 \text{ m}$$

2. After construction of the (2 m thick) engineered barrier system,

$$h_p = \frac{1}{\gamma_w FS}(6\gamma_t + 2\gamma_e)$$
$$= \frac{1}{9.8 \times 1.4}(6 \times 21 + 2 \times 20)$$
$$= 12.1 \text{ m}$$

3. Assuming that the natural head in the aquifer, h, is 14 m, one could then estimate how much waste must be placed before pumping could be terminated and the aquifer allowed to return to natural conditions:

$$FS = \frac{6\gamma_t + 2\gamma_e + t_w\gamma_{ws}}{14\gamma_w}$$
$$\therefore \gamma_{ws}t_w = 14\gamma_w FS - 6\gamma_t - 2\gamma_e$$
$$= (14 \times 9.8 \times 1.4) - (6 \times 21) - (2 \times 20) = 26$$
$$\therefore t_w = \frac{26}{6} = 4.35 \text{ m}$$

That is, 4.35 m of waste would need to be placed in order to have an adequate factor of safety against blowout with the natural hydraulic conditions in the underlying aquifer.

2.9.4 Stability of trenches

A landfill may be designed on a crusted soil such that the base of the landfill lies with the crust and overall stability is satisfied (section 2.9.2). However, problems may still occur if careful consideration is not given to stability of pipe trenches cut into any underlying soft, or very soft, soil. It is important that consideration be given not only to the global stability of the landfill, but also to the stability of individual components.

2.9.5 Stability of engineered systems on side slopes

Figure 2.15 shows an engineered system involving geomembranes, geonet drains, geotextiles and compacted clay all extending up a side slope. Tensile forces will be mobilized in the geosynthetic components lining the side slopes of the waste containment facility due to the waste overburden loads, waste settlement and consolidation (downdrag forces), and from the self-weight of the geosynthetic components themselves. The geosynthetics in the landfill capping system will also develop tensile forces due to the sliding potential of the final soil cover. The current way to evaluate these tensile forces is by static equilibrium methods (Richardson and Koerner, 1988). This evaluation requires knowledge of the friction characteristics between the various components of the lining system. These characteristics may be estimated on the basis of direct shear friction testing (Bove, 1990). An example of a failure attributed to sliding along the interfaces within a composite, multilayered geosynthetic–clay liner system has been reported by Mitchell, Seed and Seed (1990a,b) and Seed, Mitchell and Seed (1990). When interface friction alone is not sufficient to prevent sliding, either slopes must be reduced or tensile elements must be introduced to carry loads that would otherwise be expressed as shear along a potential failure surface. The geosynthetics in tension must be anchored at the top of the slope, and descriptions of anchoring methods and determination of the anchorage capacity are available (Richardson and Koerner, 1988).

Figure 2.15 shows the inclusion of geogrids to steepen the side slopes, which will increase the available landfill airspace. Jewell (1991) has published design charts for geogrid reinforced slopes. Note that an increase in slope angle will be accompanied by an increase in the tensile forces mobilized in the geosynthetics lining the slope and an increased risk of sliding, unless appropriate measures are taken.

The geosynthetics in tension must have sufficient strength to withstand the tensile forces to which they are subjected, and high safety factors should be used since strength losses due to installation damage, long-term creep, and degradation mechanisms must be expected. Possible degradation mechanisms include UV exposure during construction and on unprotected slopes, chemical reactions with leachate, swelling due to chemical absorption, extraction, oxidation and biological attack (Koerner, Halse and Lord, 1990). The possibility of chemical and swelling degradation can be minimized by the selection of geosynthetics chemically resistant to the leachate. The US EPA has developed Method 9090 (US EPA, 1985) in which geomembrane specimens are exposed to leachate in a controlled environment, followed by physical and mechanical testing. Changes in mechanical behavior after exposure are evaluated, and the compatibility of the geomembrane with the leachate is assessed. This method may be readily extended to include all geosynthetic components which may be exposed to the landfill leachate (Koerner,

1990a). Similarly, compatibility of the geosynthetics above the waste with landfill gases should be assured.

2.9.6 Post-construction desiccation of liners

When compacted clay liners are isolated from the groundwater (e.g. Figures 2.12 and 2.13), consideration should be given to the potential for the liner cracking due to desiccation. The risk of cracking is lowest when the liner is constructed of clays of low activity with a high applied vertical stress (e.g. due to the weight of the waste). Conversely, the risk of cracking is greatest for highly active clays under conditions of low confining stress.

2.10 Summary

This chapter has discussed some of the factors to be considered in environmental impact assessment of a number of barrier systems.

Barriers should not be designed in isolation. Careful consideration must be given to the site hydrogeology, geotechnical conditions and climate as well as the size of the footprint and capacity of the landfill. Depending on these conditions, landfills which will not significantly affect groundwater quality may range from very simple designs involving natural clayey barriers to highly engineered systems involving multiple liners and leachate collection systems. In general, the larger the landfill, the longer the contaminant lifespan and the more elaborate the barrier system required to provide adequate environmental protection.

A barrier that is perfectly adequate for a given landfill at one site may be totally inadequate for a larger landfill at the same site or for the same landfill at a second site. In particular, it should be emphasized that recommendations such as those by the US EPA represent minimum technological requirements for US conditions. While these documents provide a very useful resource, their recommendations should not be adopted without careful evaluation on a site-specific basis.

Both natural and synthetic materials have advantages and disadvantages. A good design will recognize this and make appropriate use of both types of material. For example, if designed appropriately, giving due consideration to the function and service life of the various components in the context of the contaminating lifespan of the landfill, geosynthetics can provide a means of considerably enhancing the performance of these facilities. However, appropriate design is only the first step. It remains to construct the facility. Although it is beyond the scope of this book, it should be emphasized that construction quality control and assurance (CQC/CQA) are critical for both the natural and synthetic components of the engineered system. The CQC/CQA procedure should be such as to ensure that the designer's intentions are met (Rowe and Giroud, 1994).

It is relatively straightforward to design landfills to meet environmental requirements for a 30-year post-closure period. The longer the period of environmental protection\required, the more difficult is the design. The design of landfills which, given what we know today, can be reasonably expected not to cause unacceptable environmental impact at any time in the future, often requires the balancing of a number of conflicting criteria. The challenge is to use effectively the engineered components of the barrier system such that eventual failure of key components, like the leachate collection system or a geomembrane liner, combined with the relatively long contaminant travel times through suitable clayey liners does not create environmental problems for future generations.

Chapter three

Clayey barriers: compaction, hydraulic conductivity and clay mineralogy

3.1 Introduction

Soil barriers, containing enough clay minerals to provide low permeability (hydraulic conductivity), are used extensively to prevent the rapid advective migration of various leachates from waste disposal sites. The clayey barriers vary from thin bentonite liners (1–3 cm thick), to compacted clayey liners (0.9–3 m thick), to natural undisturbed clayey barriers up to 30 or 40 m thick.

Undisturbed clayey barriers have a long record of performance with respect to the containment of chemical species often found in a waste disposal site. For example, the 10 000-year salt profile through a freshwater clay deposit overlying a naturally occurring high-salt bedrock is discussed in Chapter 9. The study of this migration profile provides excellent data on the long-term behavior of clayey barriers. Chapter 9 also discusses a 20-year investigation of migration of contaminants from a municipal waste disposal site which indicates long-term performance consistent with what one would expect on the basis of shorter-term laboratory and field data. However, good performance of clay as a barrier to municipal, industrial and hazardous wastes cannot be assumed *a priori*. The hydraulic performance will depend on a number of important factors to be discussed in this chapter. These include the method of placement and compaction of the clay and the clay mineralogy. The potential interaction between clay minerals and leachate, and its effect on hydraulic conductivity, will be examined in Chapter 4.

3.1.1 Undisturbed clayey deposits

The hydraulic conductivity of undisturbed clayey deposits will depend on the mineralogy, the manner of deposition and the stress history of the deposits. The *in-situ* hydraulic conductivity of these deposits may be assessed by a number of means, including:

1. laboratory tests such as triaxial and fixed ring hydraulic conductivity apparatus;
2. rising head, falling head and constant head tests conducted in piezometers in the field (slug tests);
3. pumping tests conducted on aquifers underlying (or overlying) the undisturbed clayey deposit.

These techniques for assessing the hydraulic conductivity of undisturbed clay are discussed in section 3.2. Attention will then be focused on clay mineralogy and potential changes in the hydraulic conductivity of clayey soils due to clay–leachate interaction (sections 3.4 and 3.5).

3.1.2 Compacted clayey liners

As with undisturbed clayey deposits, the hydraulic conductivity of compacted clayey liners will also depend on the mineralogy, the manner of placement and the stress history of the liner. The primary difference between natural and compacted barriers is that some control over the manner of placement and the stress history is possible for compacted clay liners. The performance of a clay liner will depend on both the choice of material (i.e. the geologic origin, grainsize distribution and mineralogy) and the manner in which this material is broken up and re-compacted. The methods of assessing hydraulic conductivity for compacted clay liners include:

1. laboratory tests on liner samples by triaxial and fixed ring hydraulic conductivity apparatus;
2. large fixed ring infiltrometers;
3. large scale field lysimeters;
4. falling head tests in short boreholes into liners etc.

These techniques are discussed in section 3.2.

3.1.3 Liner specifications

Although the detailed requirements for natural and compacted clayey barriers vary greatly from one jurisdiction to another, the following criteria seem commonly to apply.

1. The liner shall have a hydraulic conductivity, k, of 10^{-9} m/s or less and be free of natural or compaction-induced fractures. From a chemical flux point of view, 10^{-10} m/s is preferable, since diffusion often becomes the dominant migration mechanism.
2. Because low hydraulic conductivity is normally associated with the presence of clay minerals, a minimum of 15–20% of the soil with a particle size smaller than 2 μm (<2 μm), a plasticity index greater than 7% and an activity of 0.3 or greater may be specified. Alternatively, a minimum cation exchange capacity of about 10 milliequivalents (meq) per 100 g of soil may be specified.
3. The clayey barrier shall be compatible with the leachates to be retained. In other words, the hydraulic conductivity, k, shall not increase significantly on exposure to leachate similar to that to be contained by the barrier.
4. The minimum thickness of a compacted clayey liner for a domestic waste facility will generally be about 0.9–1 m; however, this is highly variable from jurisdiction to jurisdiction and in some cases is reduced to 0.6 m if the clayey liner is used together with a geomembrane to form a composite liner.
5. The minimum thickness of a clayey liner for an industrial/toxic facility shall generally be 3–4 m, although some jurisdictions require up to 15 m or multiple composite liner systems, such as those discussed in Chapter 1.

It should be emphasized that while the foregoing represent common characteristics of clayey liners, the design of a liner is site-specific and its characteristics must be selected such that the barrier system (of which the liner is usually only part) will provide adequate control on contaminant migration. This can only be assessed by appropriate site-specific contaminant impact modeling of the landfill–barrier system. The reader is referred to sections 2.2 and 2.5 and Chapters 10 to 12 for more details.

Compaction–hydraulic conductivity (permeability) relationships are discussed in section 3.3, and clay mineralogy and clay colloid chemistry are discussed in sections 3.4 and 3.5.

Clay leachate compatibility and its effect on hydraulic conductivity is discussed in Chapter 4.

3.2 Methods of assessing hydraulic conductivity

3.2.1 Laboratory tests

Laboratory hydraulic conductivity, or *k*-testing, involves application of a total head drop across a soil specimen and calculation of *k* from Darcy's law. The many methods available are discussed in most standard texts and in specialty volumes devoted to clay barriers like *Pollution Technology Review* No. 178 (Goldman *et al.*, 1990).

The objective of this section is not to review all of the test procedures, but to highlight problems that develop with those most commonly used, namely, fixed wall and flexible wall testing.

(a) Fixed wall testing

This form of testing normally involves application of either a constant or falling head across a soil sample compacted into a fixed wall cell. Since a good seal is required between the soil and the cell walls, application of a static vertical effective stress is highly recommended so that lateral yield guarantees a seal. Such a system is often called a consolidation permeameter. Special variations involving constant flow rates, *q*, have been successfully used by Fernandez and Quigley (1991) (Chapter 4).

Although small gradients close to field values are preferred, high gradients give correct values for *k* provided samples are saturated and seepage stresses do not cause consolidation and thus reduction in void ratio and *k*-values.

If set up properly, fixed ring tests can generally provide a good simulation of field conditions and should provide reliable *k*-test results. However, problems do develop, as itemized below.

1. Tests at zero vertical static stress are very prone to failure by side wall leakage, especially for soils compacted at a water content less than the Standard Proctor optimum water content (referred to as being compacted 'dry of optimum') or for fractured natural soils.
2. Very high vertical stresses may be required for stiff fractured soils to obtain a side seal.
3. Most soils compacted at a water content about 2% greater than the Standard Proctor optimum water content (referred to as being compacted 'wet of optimum') will have a saturation of about 95%. Since most fixed wall devices do not have back pressure systems, testing is done on unsaturated specimens. This may or may not be bad, since the field soil would also be in this condition initially.
4. Clay–leachate compatibility tests on soils that shrink normally result in abrupt failure by side wall leakage. Thus, fixed wall compatibility *k*-testing must be done at various stress levels such that the relationship between leachate interaction (shrinkage) and applied stress can be evaluated.

(b) Flexible wall testing

Triaxial or flexible wall testing involves wrapping the soil specimen in a membrane and running the test in a triaxial cell at the appropriate cell pressure. A major benefit is that the degree of saturation may be increased by applying a high back pressure to dissolve air in the voids. Normally, full saturation produces the highest possible *k*-values.

Although triaxial *k*-testing tends to be favored by industry (using gradients less than 50) it does have both advantages and disadvantages, as follows.

1. Cell pressure limitations prevent the use of high gradients, so that even after several weeks of flow, rarely has more than 10% of the total pore volume of the sample been replaced. Thus, compatibility testing requiring replacement of several pore

volumes is not normally feasible in a flexible wall device, even if bladders are part of the system.
2. If a soil shrinks in a compatibility test, the membrane contracts with the specimen, preventing any abrupt increases in k, which is easily observed in a fixed wall device.
3. Very low stress environments are much easier to simulate in a flexible wall permeameter.
4. Flexible wall testing is probably the only option for rock specimens and hard fractured clay specimens.

In summary, the authors consider that both types of testing are viable, and in some cases necessary, for the same project.

If the test soils are soft enough, k-values calculated from consolidation tests provide a very useful third check on the magnitude of the results of fixed and flexible wall testing. However, k-values from consolidation tests are usually not as reliable as k-values from fixed and flexible wall testing.

(c) Stress levels

Of more importance than the actual test method employed are the stress levels at which the soils are tested: Shelby tube sampling of a barrier clay, subsequent cutting and trimming etc. all cause disturbance, stress release, gas release, fracture separation etc. Testing at very low effective stresses may yield k-values 10 to 100 times higher than those run at a final field stress of say 100–150 kPa. A major consideration in developing proper protocol is establishing the appropriate confining stress. In a comparative study presented by King *et al.* (1993), flexible wall k-tests run at a confining stress of 145 kPa yielded k-values ranging from 1.2×10^{-10} to 1.2×10^{-11} m/s and averaging 8.5×10^{-11} m/s (Figure 3.1). The values were quite similar to field values of k deduced from lysimeter data after field loading to similar stress levels.

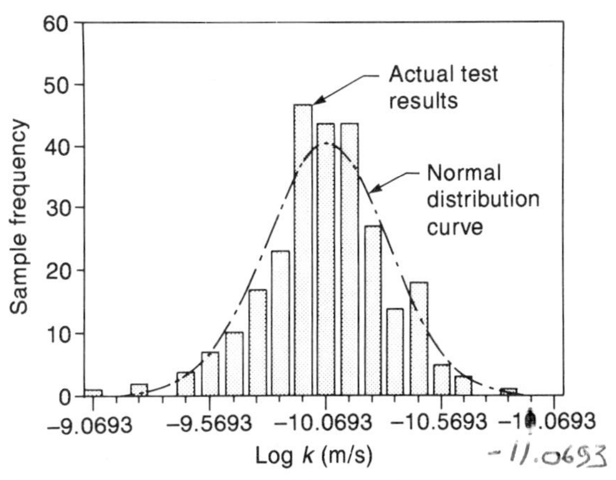

Figure 3.1 Log of normal distribution of k-test values obtained on samples of compacted clayey silt from a high quality landfill barrier. (After King *et al.*, 1993; reproduced with permission of the *Canadian Geotechnical Journal.*)

3.2.2 Borehole–piezometer field tests

Piezometers installed in natural clay, till, silt, sand or gravel deposits are frequently used to assess the hydraulic conductivity of the deposit. The properties often play an important role in modeling the potential impact of a disposal facility. Figure 3.2 shows an idealized hydrostratigraphy into which a series of piezometers have been placed. Typically, the piezometer installation will involve a screen and/or a sand pack (of diameter D) in the intake portion of the piezometer (e.g. Figures 3.2 and 3.3) and a tube (of diameter d) for controlling flow or monitoring water levels (Figure 3.3). Two basic types of test can be performed using these piezometers:

1. variable head tests (sometimes called falling head, rising head, slug or Hvorslev tests) in which the change in water level with time is monitored following some perturbation of the water level in the piezometer;
2. constant head tests in which the head in the piezometer is maintained at some specified level (relative to static conditions) and the flow is monitored.

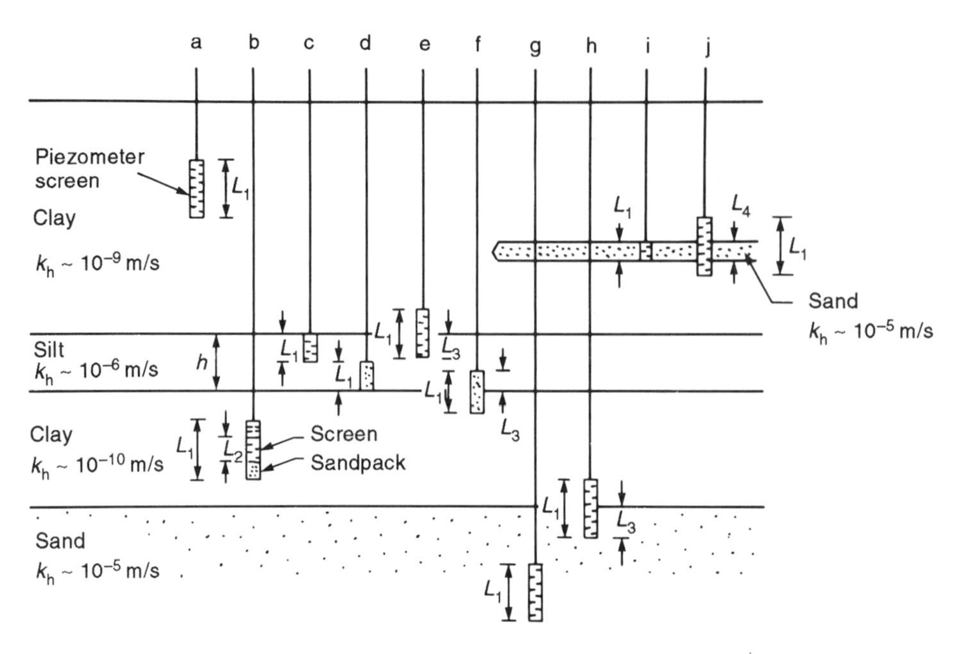

Figure 3.2 Schematic of various piezometer configurations that have been used in performing 'falling head', 'rising head' or 'constant head' tests to estimate the hydraulic conductivity of different strata. (Not to scale.)

In principle, either test can be performed on all types of deposits; however, in practice variable head tests are most commonly used for low permeability units (e.g. silts, clays, tills) and constant head tests are most commonly used for highly permeable deposits (e.g. sands, gravels). The reason for this is related to the ease of measurement. For low permeability deposits it is relatively easy to monitor the gradual rise or fall in water level with time after some perturbation in water level, but it is somewhat more difficult and/or expensive to monitor accurately the flow corresponding to a constant head difference between the piezometer and the equilibrium water level for these soils. For permeable deposits ($k > 10^{-6}$ m/s) the change in head in the piezometer resulting from some perturbation may be so quick that it is difficult and/or expensive to record accurately. In these cases, however, it is usually quite easy to measure the flow corresponding to a constant head difference.

The following subsections will discuss the interpretation of results from these types of test, both for simple cases such as that shown for piezometers 'a' to 'd', 'g' and 'i' in Figure 3.2, and for more complex situations as shown for piezometers 'e', 'f', 'h' and 'j'. In all cases, piezometer tests such as those discussed in this section:

1. primarily give an estimate of horizontal hydraulic conductivity;
2. provide an estimate of the hydraulic conductivity within a relatively small region of soil close to the piezometer.

For larger scale estimates of horizontal hydraulic conductivity in aquifers, or vertical hydraulic conductivities in aquitards adjacent to aquifers, a pumping test (as discussed in section 3.2.3) may be required.

(a) Variable head tests

Variable head tests are performed by perturbing

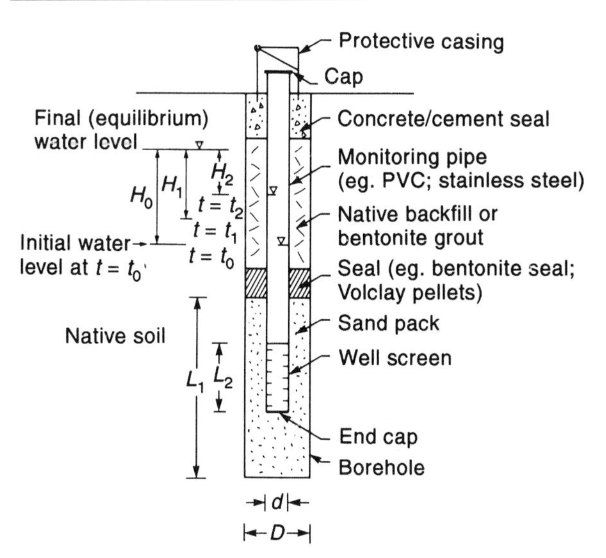

Figure 3.3 Schematic of a piezometer installation.

the water level in the piezometer and then monitoring the change in piezometer water level (head) with time. This perturbation may be generated by either of the following.

1. The water level can be raised above the equilibrium value by, for example, putting a metal slug down the hole which, by displacing the water, causes a water level rise. With time, this water will flow from the piezometer into the surrounding soil and the water level in the piezometer will fall; hence this is called a falling head test.
2. Water is removed from the piezometer by pumping or bailing, and the increase in water level to the static level is monitored as water flows into the piezometer from the surrounding soil. This is sometimes called a bail down or rising head test.

The analysis of both forms of the variable head tests is similar and is based on the classic paper by Hvorslev (1951); for this reason, variable head tests are sometimes generically referred to as Hvorslev tests. They are also sometimes generically referred to as slug tests.

Piezometers in a thick uniform soil ('a', 'b' and 'g' in Figure 3.2)
For this case the horizontal hydraulic conductivity is approximately given by

$$k_h = \frac{d^2 \ln\left\{\dfrac{mL}{D} + \left[1 + \left(\dfrac{mL}{D}\right)^2\right]^{1/2}\right\}}{8L(t_2 - t_1)} \ln\left(\frac{H_1}{H_2}\right) \tag{3.1a}$$

where k_h is the horizontal hydraulic conductivity [LT^{-1}]; d is the diameter of the monitoring tube (Figure 3.3) [L]; $m = (k_h/k_v)^{1/2}$ ($m = 1$ for isotropic soil $k_h = k_v$) [–]; D is the diameter of the piezometer intake (Figure 3.3) [L]; H_1 is the piezometer head at time $t = t_1$ (relative to static water level) [L]; H_2 is the piezometer head at time $t = t_2$ (relative to static water level) [L]; L is the intake length of the piezometer [L]. The selection of this length will be discussed below for three situations 'a', 'b' and 'g' shown in Figure 3.2.

For long piezometers such that $mL/D > 4$, equation 3.1a reduces to

$$k_h = \frac{d^2 \ln\left(\dfrac{2mL}{D}\right)}{8L(t_2 - t_1)} \ln\left(\frac{H_1}{H_2}\right) \tag{3.1b}$$

The horizontal hydraulic conductivity can also be calculated based on the basic time lag T. This time lag can be established graphically by plotting the piezometric head H (on a log scale) versus the time t (on a linear scale), as shown in Figure 3.4. The basic time lag T_0 is the time corresponding to $H = 0.37H_0$ (i.e. $\ln(H_0/H) = 1$), where H_0 is the difference in maximum head from the final static level (i.e. it is the perturbation head at $t = 0$; Figure 3.3). As noted by Hvorslev, the advantage of plotting this diagram is that it reveals irregularities in the data caused by volume changes or stress adjustment time lag, and permits easy advance adjustment of the results of the test.

$K_h = \frac{d^2 \ln(2mL/D)}{8LT_0}$

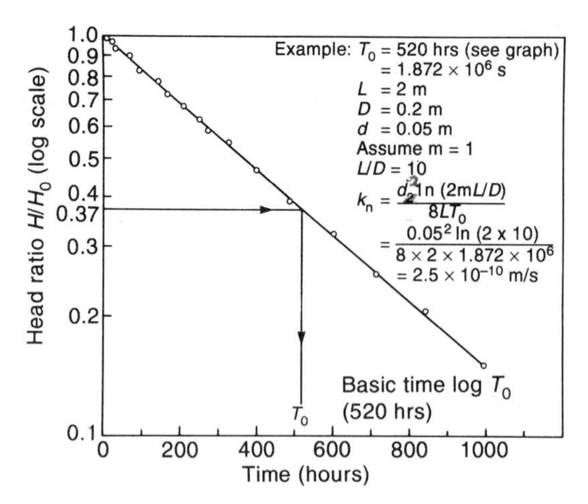

Figure 3.4 Determination of Hvorslev's basic time lag and example calculation of hydraulic conductivity for piezometer 'b' shown in Figure 3.2.

Once the basic time lag has been evaluated, the hydraulic conductivity can be calculated from

$$k_h = \frac{d^2 \ln\left\{\frac{mL}{D} + \left[1 + \left(\frac{mL}{D}\right)^2\right]^{1/2}\right\}}{8LT_0} \tag{3.2a}$$

which, for $mL/D > 4$, reduces to

$$k_h = \frac{d^2 \ln\left(\frac{2mL}{D}\right)}{8LT_0} \tag{3.2b}$$

and all terms are as previously defined.

When the final equilibrium water level is not known, the basic time lag, T_0, may be estimated prior to full stabilization by observing successive changes in piezometer level for equal time intervals. Suppose that the change in head over some time interval Δt since the beginning of the test is Δh_1 and that over the subsequent time interval Δt the **change** in head is Δh_2 then

$$T_0 = \frac{\Delta t}{\ln(\Delta h_1/\Delta h_2)} \tag{3.3}$$

The evaluation of the piezometer intake length L warrants some discussion. Consider piezometers 'a', 'b' and 'g' shown in Figure 3.2.

Piezometers 'a' and 'g' involve a screen length which is the entire length of the borehole below the seal. In this case the intake length is equal to the screen length ($L = L_1$); this is the case irrespective of whether the soil in which the piezometer is placed is a uniform clay, silt or sand.

Piezometer 'b' involves a screen length L_2 and a total sandpack length L_1. If the sandpack is substantially more permeable than the surrounding soil (e.g. the clay in Figure 3.2), then the hydraulic conductivity is best estimated by taking the intake length L equal to the entire length L_1 (i.e. $L = L_1$) which includes the sandpack.

Piezometers at the boundary of a confined aquifer ('c'–'f' and 'h' in Figure 3.2)

Consider the situation indicated by piezometers 'c' and 'd' which monitor a portion of a more permeable layer (e.g. silt) overlain (or underlain) by a much lower permeability layer (e.g. clay). Provided that the penetration length L_1 is small compared to the thickness of the layer being monitored (i.e. $L_1 \ll h$), then the hydraulic conductivity of the monitored layer may be estimated from

$$k_h = \frac{d^2 \ln\left\{\frac{2mL}{D} + \left[1 + \left(\frac{2mL}{D}\right)^2\right]^{1/2}\right\}}{8L(t_2 - t_1)} \ln\left(\frac{H_1}{H_2}\right) \tag{3.4a}$$

which, for $mL/D > 2$, reduces to

$$k_h = \frac{d^2 \ln\left(\frac{4mL}{D}\right)}{8L(t_2 - t_1)} \ln\left(\frac{H_1}{H_2}\right) \tag{3.4b}$$

where all the variables are as previously defined.

In terms of the basic time lag T_0, the hydraulic conductivity is given by

$$k_h = \frac{d^2 \ln\left\{\frac{2mL}{D} + \left[1 + \left(\frac{2mL}{D}\right)^2\right]^{1/2}\right\}}{8LT_0} \quad (3.5a)$$

which, for $mL/D > 2$, reduces to

$$k_h = \frac{d^2 \ln\left(\frac{4mL}{D}\right)}{8LT_0} \quad (3.5b)$$

Provided that the sandpack is substantially more permeable than the monitored layer (an order of magnitude or more), the length L is equal to the full length of the screen/sandpack in the monitored layer (i.e. $L = L_1$).

Consider the situation shown for piezometers 'e', 'f' and 'h', where the screen/sandpack is of length L_1, but only part of the length (L_3) penetrates the monitored layer (i.e. $L_3 < L_1$). If the hydraulic conductivity of the aquitard is more than an order of magnitude lower than that of the monitored layer then the length, L, may be reasonably estimated as being equal to the portion of the intake interval actually in the more permeable layer (i.e. $L = L_3$).

Factors affecting piezometer performance

Equations 3.1–3.5 may be used to estimate hydraulic conductivity from the observed response of piezometers to a perturbation in piezometer water level. However, it should be recognized that these equations are approximate, and that the response of the piezometer may be affected by:

1. leakage through the piezometer seal;
2. clogging of the intake;
3. removal of fine grained particles from the surrounding soil;
4. storage in the piezometer casing;
5. accumulation of gases (air bubbles) near the intake.

(b) Constant head tests

There are a number of advantages to the use of the constant head test in preference to the more commonly used variable head tests discussed above. These include:

1. constant head tests tend to be less affected by storage effects in the sandpack and are unaffected by storage within the piezometer casing;
2. for permeable deposits, the constant head test is conducted over a much longer time period than variable head tests and hence can test a much larger volume of soil around the piezometer/well;
3. for permeable deposits, the test is easier to perform than the variable head test and involves less potential error.

The constant head test may involve either adding or removing water from the piezometer as required to maintain the water level to some value H_c relative to the equilibrium (static) water level, and then observing the flow required to maintain this constant water level.

Piezometers in a thick uniform soil ('a', 'b' and 'g' in Figure 3.2)

The most commonly used expression for assessing the hydraulic conductivity k_h in a constant head test is that developed by Hvorslev (1951):

$$k_h = \frac{q \ln\left\{\frac{mL}{2R} + \left[1 + \left(\frac{mL}{2R}\right)^2\right]^{1/2}\right\}}{2\pi L H_c} \quad (3.6a)$$

where q is the steady-state flow required to maintain a constant head H_c, [L^3T^{-1}]; H_c is the change in water level (relative to the static/equilibrium water level) [L]; R is the radius of the piezometer/well as discussed below [L]; L is the length of the intake as discussed below [L]; $m = (k_h/k_v)^{1/2}$. For $mL/R > 8$ this reduces to

$$k_h = \frac{q \ln\left(\frac{mL}{R}\right)}{2\pi L H_c} \quad (3.6b)$$

If the test is performed in a unit whose hydraulic conductivity is substantially less than that of the sandpack (e.g. piezometers 'a' and 'b') then the length $L = L_1$ and the radius $R = D/2$ (Figure 3.3). If the piezometer is installed in a granular unit with a hydraulic conductivity similar to that of the sandpack (e.g. if the sandpack is really material from the same unit) then the length L is best approximated by the screen internal length $L = L_2$ and the radius is the radius of the well/screen, $R = d/2$ (Figure 3.3).

Although not commonly used (at the time of writing), constant head tests can be reliably performed even in low permeability materials. In the past, problems with performing tests in clay have been related to remolding of the soil (e.g. by borehole construction) and clogging of the permeameter during installation. Self-boring permeameters have now been developed which minimize these effects (e.g. Tavenas, Tremblay and Leroueil, 1983, 1986). These permeameters tend to be shorter than conventional piezometers and the approximation involved in Hvorslev's equation needs to be carefully evaluated. Accordingly, various investigators have also refined Hvorslev's equation (e.g. Brand and Premchitt, 1982; Randolph and Booker, 1982; Tavenas *et al.*, 1986; Tavenas, Diene and Leroueil, 1990). Based on these studies, a simplified (approximate) equation can be developed (assuming isotropy; $m \approx 1$) for $2 \leqslant L/D \leqslant 10$:

$$k_h = \frac{q}{(5D + 2.1L)H_c} \tag{3.7}$$

where q is the steady-state flow $[L^3T^{-1}]$; H_c is the change in water level giving rise to the flow q [L]; D is the diameter of the intake [L]; L is the length of the intake [L].

Equation 3.7 will give lower estimates of hydraulic conductivity than equation 3.6. There are data (e.g. Tavenas *et al.*, 1986) to support the lower estimates based on equation 3.7.

Piezometers at the boundary of a confined aquifer ('c'–'f' and 'h' in Figure 3.2)

Based on Hvorslev (1951), the hydraulic conductivity k_h (m/s) can be estimated from a constant head test using

$$k_h = \frac{q \ln\left\{\dfrac{mL}{R} + \left[1 + \left(\dfrac{mL}{R}\right)^2\right]^{1/2}\right\}}{2\pi L H_c} \tag{3.8a}$$

and for $mL/R > 8$, this reduces to

$$k_h = \frac{q \ln\left(\dfrac{2mL}{R}\right)}{2\pi L H_c} \tag{3.8b}$$

where all terms are as defined in the discussion of equation 3.6, including the choice of L and R.

Piezometers screened across a confined aquifer ('i' and 'j' in Figure 3.2)

Again, based on Hvorslev (1951), the hydraulic conductivity k_h (m/s) for a constant head test can be estimated from

$$k_h = \frac{q \ln\left(\dfrac{R_0}{R}\right)}{2\pi L H_c} \tag{3.9}$$

where R_0 is the effective radius to the source of supply (i.e. the estimated zone of influence of the test) [L]; R is the radius of the screen/well [L]; q is the steady-state flow $[L^3T^{-1}]$; H_c is the change in water level (head) giving rise to the flow q [L]; L is the length of the intake [L]. For piezometer 'i' in Figure 3.2, $L = L_1$. For piezometer 'j', the length of the screen exceeds the thickness of the deposit, L_4 (i.e. where $L_1 > L_4$), and so the length L is taken as the thickness of the permeable deposit ($L = L_4$).

Factors affecting interpretation

Hvorslev's equations only consider the hydraulics of the permeable layer being monitored. They ignore effects such as leakage from adjacent aquitards, and boundary conditions such as that shown in Figure 3.2, where piezometers 'i' and 'j' are monitoring a lens of finite extent. These effects may make it difficult to obtain a true steady-state flow q. Judgement is required to assess the reasonableness of using Hvorslev's equations for a given field situation. Where serious doubt exists, more elaborate methods of analysis may be adopted as discussed in the following section.

3.2.3 Pumping tests

Much has been written on the subject of pumping tests and space does not permit a detailed review here. Rather, this section is intended to:

1. summarize some commonly adopted methods for estimating the transmissivity and storativity of an aquifer using a pumping test;
2. discuss methods of estimating the bulk hydraulic conductivity of an adjacent aquitard based on its response to the pumping of an adjacent aquifer.

Typically, the performance of a pumping test involves construction of a pumping well in the aquifer to be pumped. The well is then pumped at a constant flow rate and the drawdown of the water level is observed with time and is plotted on semi-log paper (drawdown versus log(time)). The primary differences between this test and the constant head test discussed in section 3.2.2 are as follows.

1. The pumping test is a transient test (i.e. steady-state conditions are not required for interpretation).
2. The pumping well is usually larger than the piezometer used in a constant head test.

In many cases, piezometers will be installed at different distances from the pumping well, and the response of these piezometers to the pumping test can be used to assess the hydraulic characteristics of the aquifer. In some cases, monitors are also placed in an adjacent aquitard, as shown in Figure 3.5. The response of these piezometers may be used to estimate aquitard properties.

(a) Cooper and Jacob's (1946) interpretation of aquifer properties

Assuming:

1. a uniform aquifer with the same hydraulic conductivity in all directions;
2. uniform thickness h of the aquifer and infinite extent;
3. a fully penetrating pumping well (e.g. Figure 3.5);
4. a well with 100% efficiency;
5. laminar flow in the well and aquifer;
6. all water removed comes from aquifer storage (i.e. no recharge or leakage);
7. the potentiometric surface has no significant slope.

The drawdown response of a well (e.g. Figure 3.6) can be used to estimate the transmissivity T ($= kh$, i.e. hydraulic conductivity times aquifer thickness) and hence the hydraulic conductivity from the equation

$$T = \frac{0.183q}{\Delta s} \tag{3.10}$$

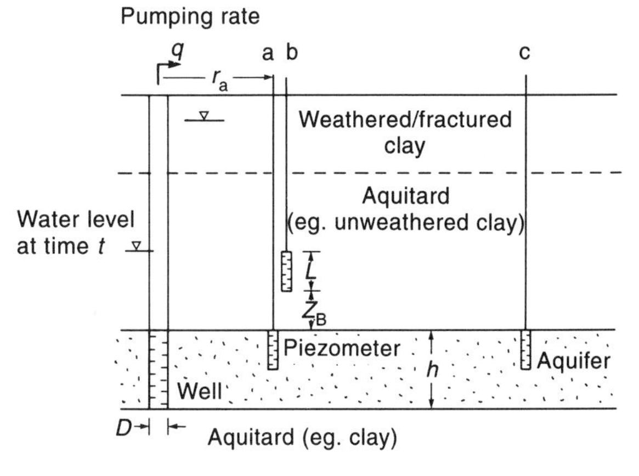

Figure 3.5 Schematic of a pumping test. (Not to scale.)

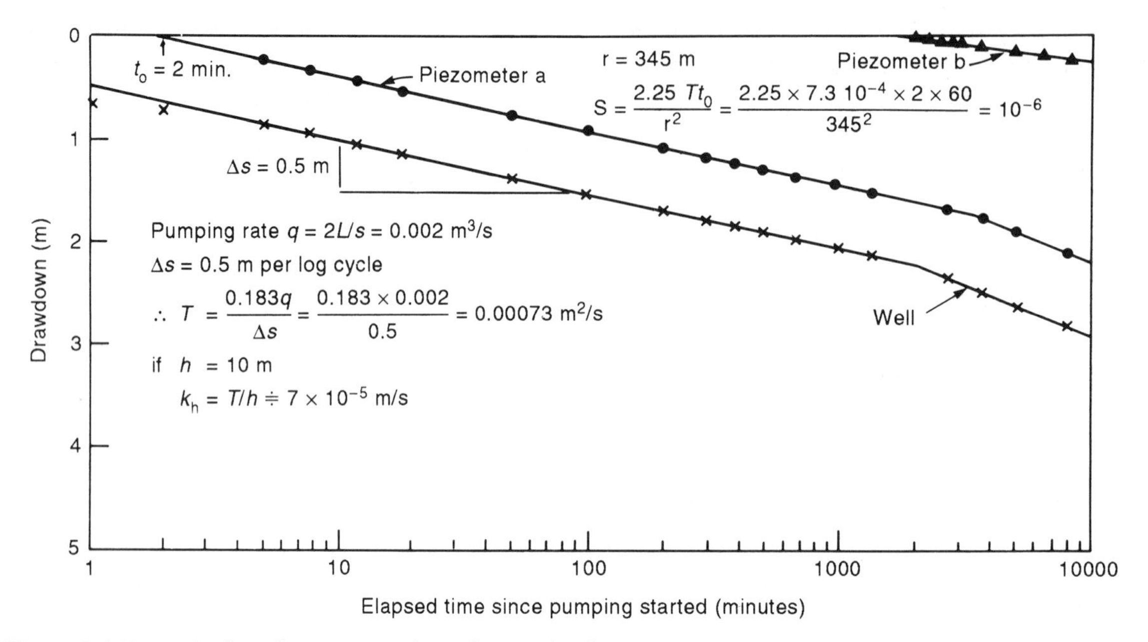

Figure 3.6 Example drawdown versus elapsed time plot for a pumping test.

where T is the transmissivity $[L^2T^{-1}]$; q is the flow $[L^3T^{-1}]$; Δs is the slope of the drawdown graph (expressed as the change in drawdown per log cycle time) [L].

The hydraulic conductivity is then given by

$$k_h = \frac{T}{h} \tag{3.11}$$

where k_h is the horizontal hydraulic conductivity $[LT^{-1}]$; T is the transmissivity (from equation 3.10) $[L^2T^{-1}]$; h is the thickness of the aquifer [L].

An example calculation is shown in Figure 3.6. The drawdown curve shown in Figure 3.6 exhibits a significant change in slope at about 2000 minutes. This suggests that some boundary effect, such as a change in aquifer transmissivity, occurs at some distance away from the pumping well.

The storativity, S, of the aquifer can also be calculated from the time–drawdown plot of a piezometer by extrapolating the drawdown curve back until it intercepts the zero drawdown axis. The corresponding time t_0 is noted and the storativity is calculated from

$$S = \frac{2.25Tt_0}{r^2} \tag{3.12}$$

where S is the storativity [–]; T is the transmissivity $[L^2T^{-1}]$; t_0 is the zero drawdown intercept [T]; r is the distance from the pumped well to the piezometer [L]. An example calculation is shown in Figure 3.6.

The method of interpretation presented above assumes no leakage into the aquifer from above or below. In reality, there is likely to be leakage but, as noted by Neuman and Witherspoon (1972), the errors associated with the use of equations 3.11 and 3.12 which neglect leakage will be small if the data are collected close to the pumping well. The errors increase with the distance of the monitoring point from the piezometer. As a general rule, early drawdown data are less affected by leakage than later time data and hence, provided the data are reliable, most emphasis should be given to early time data.

If leakage is considered important, techniques such as that proposed by Hantush (1956, 1960) can be used for estimating transmissivity and storage coefficient.

(b) Neuman and Witherspoon's (1972) interpretation of aquitard hydraulic conductivity properties

Neuman and Witherspoon (1972) proposed a simple method of estimating the bulk hydraulic conductivity of an aquitard by monitoring the response of the aquitard to a pumping test conducted on an adjacent aquifer. For example, consider the schematic shown in Figure 3.5. Suppose that a piezometer 'a' is installed in an aquifer at some distance r from the pumping well. Suppose that a second piezometer 'b' is installed in the aquitard at some distance z_B above the aquifer and that the length of the piezometer is L. The drawdown of the aquifer, s, and of the aquitard, s', in piezometers 'a' and 'b' can be monitored with time. From the early time response of piezometer 'a', the aquifer transmissivity T and storativity S may be calculated using equations 3.11 and 3.12 respectively. At some time t after the aquitard piezometer begins to respond (e.g. piezometer 'b' in Figure 3.6), the ratio s'/s of the drawdown of the aquitard piezometer, s', to that of a nearby aquifer piezometer, s, can then be calculated. Also, a dimensionless time factor t_D can be calculated for this time:

$$t_D = \frac{Tt}{Sr^2} \tag{3.13}$$

where T is the transmissivity of the aquifer [L^2T^{-1}]; S is the storativity [–]; r is the distance from the pumping well to piezometer 'a' [L]; t is the time of interest [T].

Using t_D and s'/s, Figure 3.7 can be used to obtain a second dimensionless time factor t'_D, where

$$t'_D = \frac{k_v t}{S'_s z^2} \tag{3.14}$$

where k_v is the bulk vertical hydraulic conductivity [LT^{-1}]; t is the time of interest [T]; S'_s is the specific storage ($S'_s = m_v\gamma_w$, where m_v is the coefficient at volume change [$M^{-1}LT^2$] and γ_w is the unit weight of water [$ML^{-2}T^{-2}$]) [L^{-1}]; z is the distance from the aquifer to the monitoring point [L].

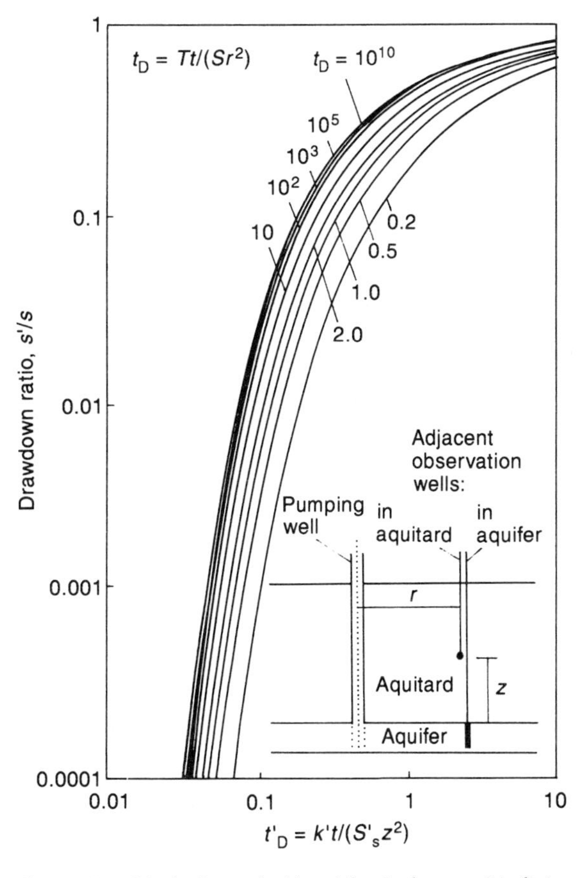

Figure 3.7 Variation of s'/s with t'_D for semi-infinite aquitard. (After Neuman and Witherspoon, 1972; reproduced with permission of *Water Resources Research*. Copyright: American Geophysical Union)

Knowing t'_D, t, z and S'_s (e.g. from a consolidation test), the hydraulic conductivity can be calculated from equation 3.14. Neuman and Witherspoon's charts were developed assuming that the monitoring piezometer is a point. In practice, the monitor length L is usually significant compared with the distance z_B from the aquifer to the bottom of the piezometer; this complicates the interpretation as illustrated in the following example.

Example
Consider the data presented in Figure 3.6. Based on the response of piezometer 'a',

$$T = \frac{0.183 \times 0.002}{0.5} = 0.00073 \text{ m}^2/\text{s}$$

$$S = \frac{2.25 T t_0}{r^2}$$

$$= \frac{2.25 \times 0.00073 \times 2 \times 60}{345^2}$$

$$= 1.66 \times 10^{-6}$$

At about 2000 minutes, piezometer 'b' begins to respond. At $t = 2500$ minutes the drawdowns are

$s = 1.65$ m (for 'a')
$s' = 0.05$ m (for 'b')

Therefore the dimensionless time factor

$$t_D = \frac{Tt}{Sr^2} = \frac{0.00073 \times (2500 \times 60)}{1.66 \times 10^{-6} \times 345^2} = 555$$

and the drawdown ratio $s'/s = 0.05/1.65 = 0.03$.

From Neuman and Witherspoon's charts (Figure 3.7) for $t_D = 555$, $s'/s = 0.03$,

$$t'_D \approx 0.11$$

Assuming that the specific storage $S'_s = 10^{-3}$ m^{-1} (e.g. as obtained from consolidation tests on the aquitard; $S'_s = \gamma_w m_v$, where γ_w is the unit weight of water and m_v is the coefficient at volume change), the bulk vertical hydraulic k_v conductivity may be estimated as

$$k_v \approx \frac{S'_s z^2 t'_D}{t} = \frac{10^{-3} \times z^2 \times 0.11}{2500 \times 60}$$

$$= 7.3 \times 10^{-10} z^2 \text{ m/s}$$

The major problem is selecting the distance z. Suppose the piezometer is installed such that $z_B = 0.9$ m and $L = 1.6$ m. The value of z would be a minimum for $z = z_B = 0.9$ m, giving $k_v = 5.9 \times 10^{-9}$ m/s, whereas for the average $z = z_B + L/2 = 1.7$ m, $k_v = 2 \times 10^{-9}$ m/s. In this case the choice of z makes a three-fold difference to the estimated hydraulic conductivity. Furthermore, Neuman and Witherspoon (1972):

1. used diffusion theory which fails to consider the effect of stress changes in the aquitard due to pumping;
2. did not consider time lag in the piezometer.

It may be concluded that Neuman and Witherspoon's method is most applicable for short piezometers with a small time lag (i.e. where a pressure transducer is used to monitor pressure changes in the piezometer).

Rowe and Nadarajah (1993) have examined the effect of stress change in the aquifer and the finite size of the piezometer as discussed in the following subsection.

(c) Rowe and Nadarajah's (1993) interpretation of aquitard hydraulic conductivity

Using Biot theory (Biot, 1941), Rowe and Nadarajah (1993) examined the effect of the finite size of piezometers, the proximity of the piezometer to the aquifer, time lag and effective stress changes within the aquitard on the interpretation of pumping test results. They developed simple correction factors β_1, for the effect of time lag, and β_2, for the effects of size of the piezometer and its proximity to the aquifer, which can be used in association with the conventional Neuman and Witherspoon (1972) chart given in Figure 3.7.

The correction factors β_1 and β_2 can be expressed in terms of a dimensionless parameter λ:

$$\lambda = 1.5lS'_s \left(\frac{1-\nu}{1+\nu}\right) \frac{k_h}{k_v} \frac{\pi D^2}{A} \tag{3.15}$$

where l is the length of the piezometer screen plus the length of any sand/gravel packed between the screen and the aquifer being pumped [L]; D is the diameter of the borehole [L]; A is the cross-sectional area of riser pipe used for monitoring the change in water level in the piezometer ($\pi d^2/4$) [L^2]; ν is Poisson's ratio for the drained aquitard soil [–] (typically 0.3–0.35 for silt; 0.3–0.4 for soft clay; 0.2–0.3 for stiff clay); S'_s is the specific storage [L^{-1}] ($= m_v\gamma_w$, where m_v is the coefficient of volume change [$M^{-1}LT^2$] and γ_w is the unit weight of water [$ML^{-2}T^{-2}$]); k_h is the expected horizontal hydraulic conductivity of aquitard [LT^{-1}]; k_v is the expected vertical hydraulic conductivity of aquitard [LT^{-1}]; z_B is the thickness of aquitard from aquifer being pumped to the bottom of the borehole [L]; $z = z_B + 0.5l$.

Figure 3.8(a) shows the relationship between the correction factor β_1 and the parameter λ for a number of values of s'/s, the ratio of the drawdown in the aquitard, s', to the drawdown in the aquifer, s. Curves are shown for different values of z/D.

The horizontal hydraulic conductivity k_h can be estimated from a falling head test in the aquitard piezometer, as described in section 3.2.2(a).

Figure 3.8(b) shows the relationship between the correction factor β_2 and parameter λ for a number of values of s'/s. In this case, curves are shown for different values of z/l.

The effects of time lag and piezometer location can then be incorporated into the assessment of hydraulic conductivity, k, as follows:

$$k_v = \frac{t'_D S'_s z^2}{t} \frac{\beta_2^2}{\beta_1} \tag{3.16}$$

where t'_D is the time factor determined from Neuman and Witherspoon's chart (Figure 3.7) [–]; t is time; β_1 is the correction factor for time lag determined from Rowe and Nadarajah's chart (Figure 3.8(a)) [–]; β_2 is the correction factor for piezometer location determined from Rowe and Nadarajah's chart (Figure 3.8(b)) [–]. The other parameters are as defined for equation 3.15.

The time lag of a piezometer can also be affected by entrapped air. The potential effect of entrapped air increases with the size of the sand/gravel pack. Thus, the size of the sand/gravel pack should be kept as small as is practical.

Changes in atmospheric pressure can give rise to apparent changes in head in the piezometers. The atmospheric pressure should be monitored during the test, and this should be used in the interpretation of the field data.

The specific storage may be estimated from the results of one-dimensional consolidation tests. However, these test results may overestimate S'_s due to:

1. sample disturbance;
2. entrapped air;
3. larger stress–strain changes than expected in the field.

In some cases, the specific storage can be bracketed by the values obtained in compression and expansion over the relevant stress range. The effect of uncertainty regarding the specific storage can be assessed by evaluating the hydraulic conductivity for the upper and lower values over the reasonable range of uncertainty.

Example

Figure 3.9 shows the results from a pumping test conducted to help define the hydraulic conductivity of an aquitard. The drawdowns

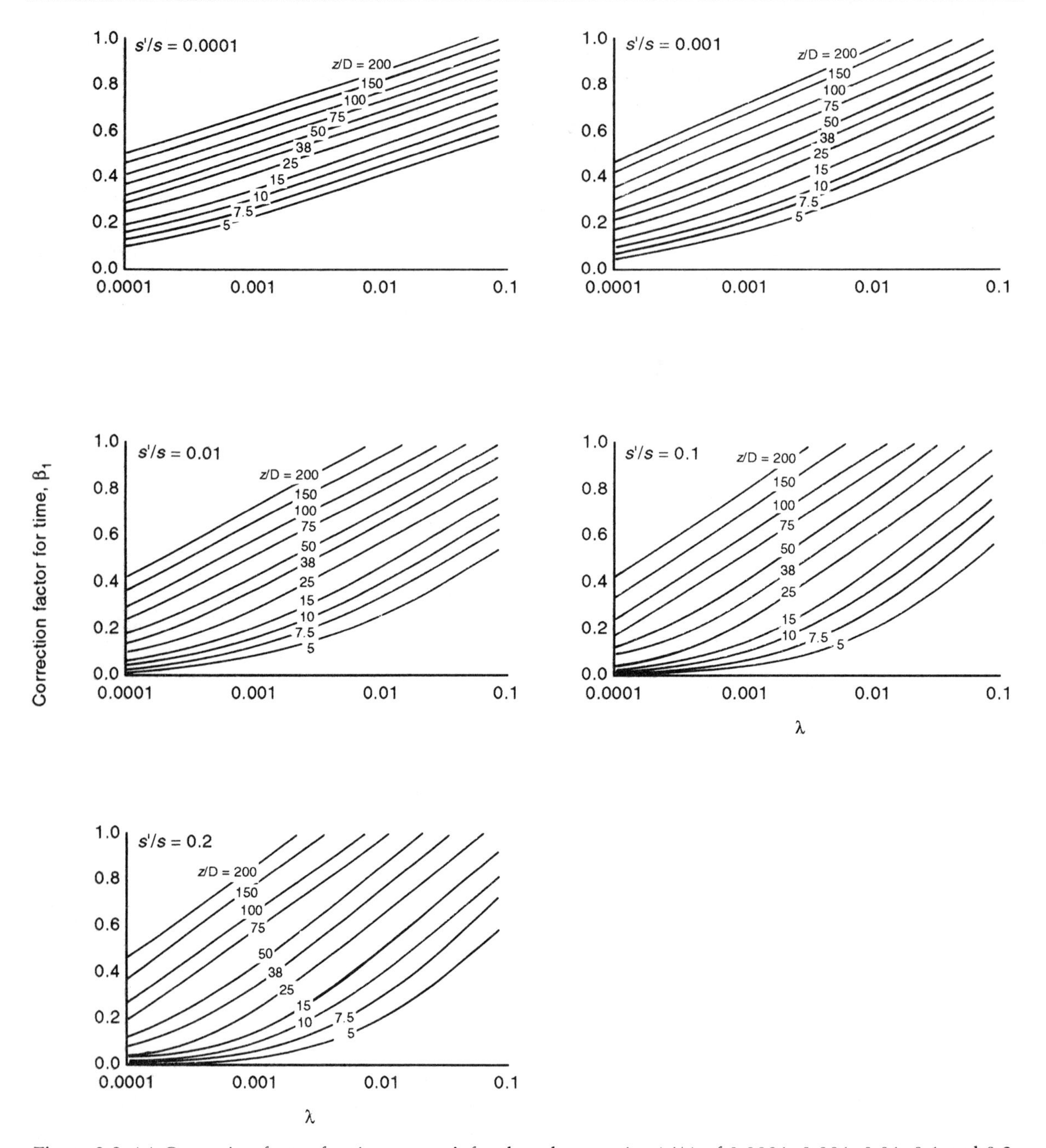

Figure 3.8 (a) Correction factor for time versus λ for drawdown ratios (s'/s) of 0.0001, 0.001, 0.01, 0.1 and 0.2. (From Rowe and Nadarajah, 1993; reproduced with permission of the *Canadian Geotechnical Journal.*)

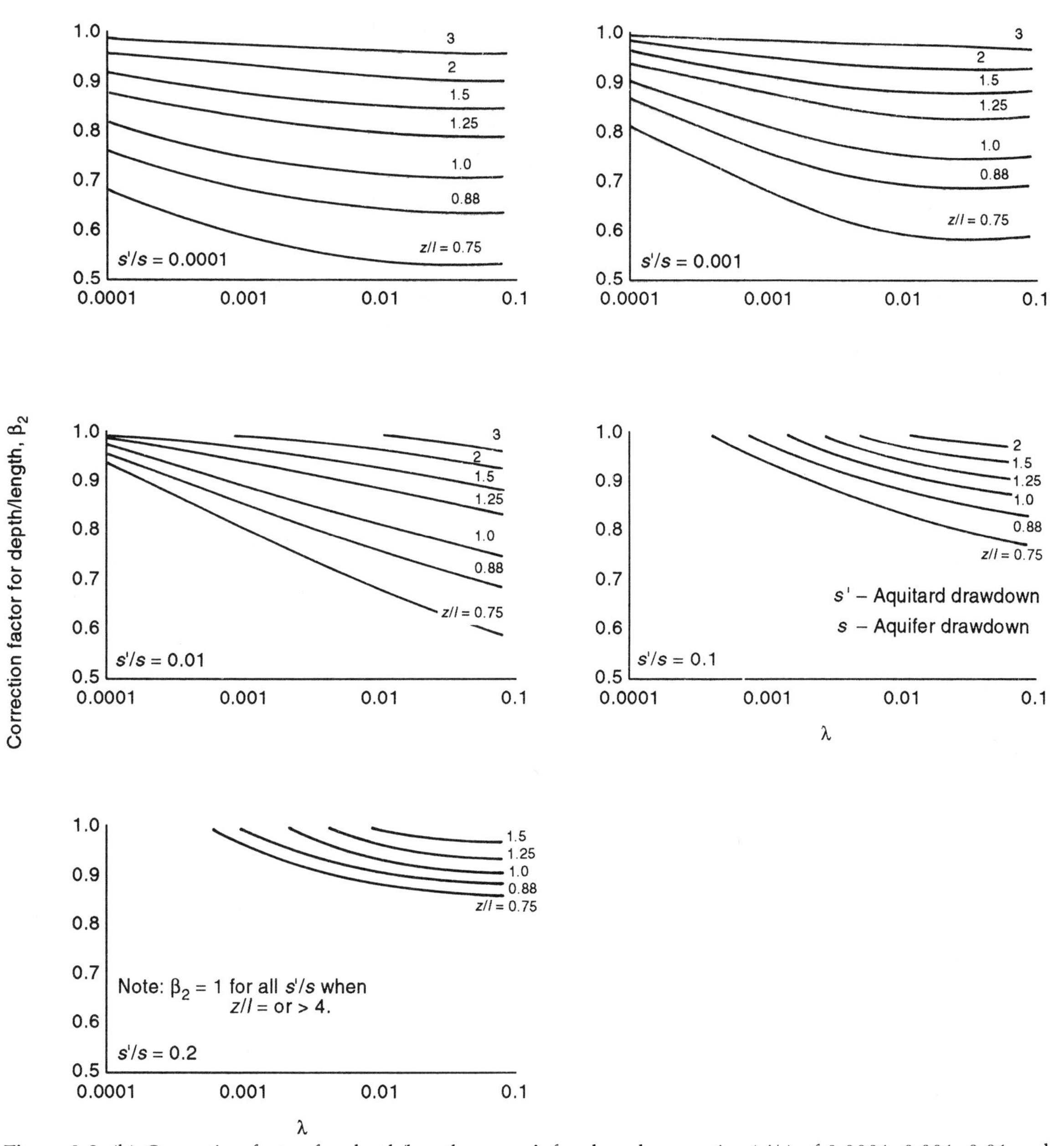

Figure 3.8 (b) Correction factor for depth/length versus λ for drawdown ratios (s'/s) of 0.0001, 0.001, 0.01 and 0.1. (From Rowe and Nadarajah, 1993; reproduced with permission of the *Canadian Geotechnical Journal*.)

were monitored using observation wells which consist of a 100 mm diameter flush-joint PVC pipe, installed with a 0.31 m machine slot screen in a 0.23 m diameter borehole, with a gravel pack extending 0.3 m above the top level of the screen (no gravel pack was placed below the bottom of the screen).

From Figure 3.9, when t = 400 min, s' = 0.029 m and s = 3.66 m, so the drawdown ratio becomes

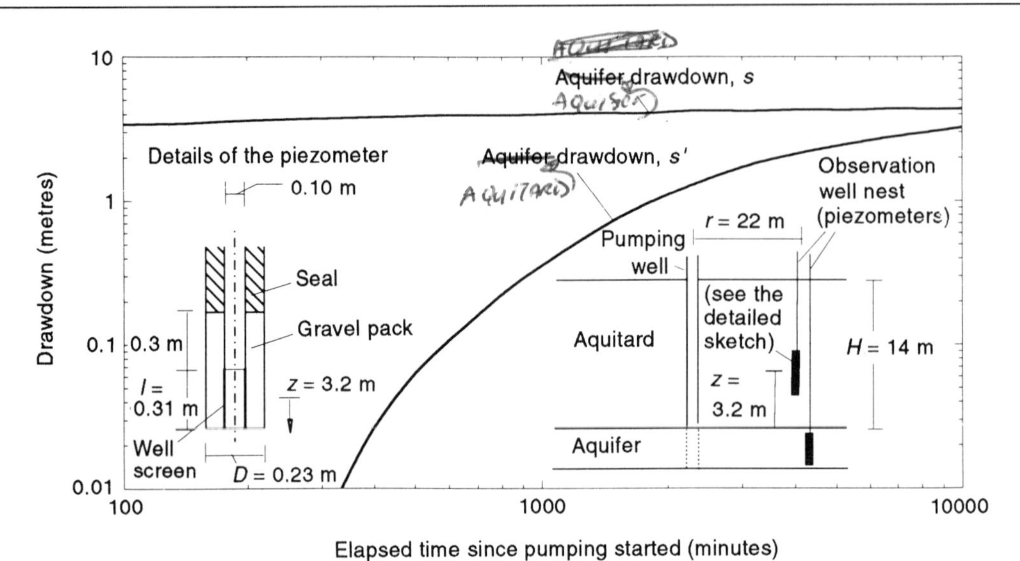

Figure 3.9 Drawdown versus elapsed time in a pumping test: example. (After Rowe and Nadarajah, 1993; reproduced with permission of the *Canadian Geotechnical Journal.*)

$$\frac{s'}{s} = \frac{0.029}{3.66} = 0.0079$$

The transmissivity $T = 0.0184\ \mathrm{m^2/s}$ and the storativity $S = 1.1 \times 10^{-4}$ were obtained on the basis of a conventional analysis of the aquifer response using the initial straight line portion of the drawdown curve and Jacob's method (Cooper and Jacob, 1946). For a monitoring nest located 22 m from the pumping well:

$$t_D = \frac{Tt}{Sr^2} = \frac{0.0184 \times (400 \times 60)}{1.1 \times 10^{-4} \times 22^2} = 8.3 \times 10^3$$

For the *in-situ* stress range, an average specific storage of the aquitard obtained from consolidation tests is taken to be $7.9 \times 10^{-4}\ \mathrm{m^{-1}}$. Assuming $\nu = 0.3$ and $k_h/k_v = 1$, λ can be calculated as

$$\lambda = 1.5 l S_s' \frac{(1-\nu)}{(1+\nu)} \frac{\pi D^2}{A}$$

Therefore

$$\lambda = 1.5 \times 0.31(7.9 \times 10^{-4}) \times \frac{1-0.3}{1+0.3} \times \frac{\pi(0.23)^2}{\pi(0.05)^2} = 0.004$$

$z = 3.2$ m, $D = 0.23$ m, $l = 0.31$ m, therefore

$$\frac{z}{D} = \frac{3.2}{0.23} = 13.9 \qquad \frac{z}{D} = \frac{3.2}{0.31} \approx 10.3$$

From Figure 3.8(a) the time correction factor β_1 is 0.33, and from Figure 3.8(b) the depth/length correction factor β_2 is 1, since $z/l > 4$. So the gross correction factor is

$$\frac{\beta_2^2}{\beta_1} = \frac{(1.0)^2}{0.33} = 3$$

Thus, in this particular case the errors that would arise from neglecting the time lag are such that the hydraulic conductivity obtained from the normal application of the ratio method would underestimate the bulk hydraulic conductivity by a factor of 3.

From Figure 3.7, $t'_D = 0.075$ for $s'/s = 0.0079$ and $t_D \approx 8.3 \times 10^3$. The hydraulic conductivity of the aquitard is therefore given by

$$k_v = \frac{t'_D S'_s z^2}{t} \frac{\beta_2^2}{\beta_1}$$
$$= \frac{0.075 \times (7.9 \times 10^{-4}) \times 3.2^2}{(400 \times 60)} \times 3$$
$$= 7.6 \times 10^{-8} \text{ m/s}$$

(d) Estimating hydraulic conductivity for complex aquitard systems

As discussed in the preceding subsections, one common means of estimating the bulk vertical hydraulic conductivity is to perform a pumping test on the underlying aquifer and then monitor the pore pressure response in the overlying aquitard. The techniques discussed so far were developed for a uniform soil layer above (or below) the aquifer (e.g. Neuman and Witherspoon, 1972; Rowe and Nadarajah, 1993). These techniques can be readily applied to situations where the layer adjacent to the aquifer is the primary barrier to contaminant migration. However, in many practical situations, the clayey deposit is layered (e.g. Figure 3.10) and the primary barrier (Layer 3 in Figure 3.10) may be separated from the aquifer by one or more layers of clayey material with different hydraulic conductivities (and coefficient of consolidation) from that of the primary layer. Under these circumstances, the techniques of Neuman and Witherspoon (1972) and Rowe and Nadarajah (1993) cannot be used to provide a suitable estimate of the hydraulic properties of the aquitard. Usher and Cherry (1988) have modified a two-layer heat flow solution to allow the analysis of a two-layer aquitard. However, even this approach is not adequate in some practical situations.

Finite layer techniques are ideally suited to the analysis of multilayer problems such as that indicated in Figure 3.10. For example, the finite layer contaminant transport model proposed by

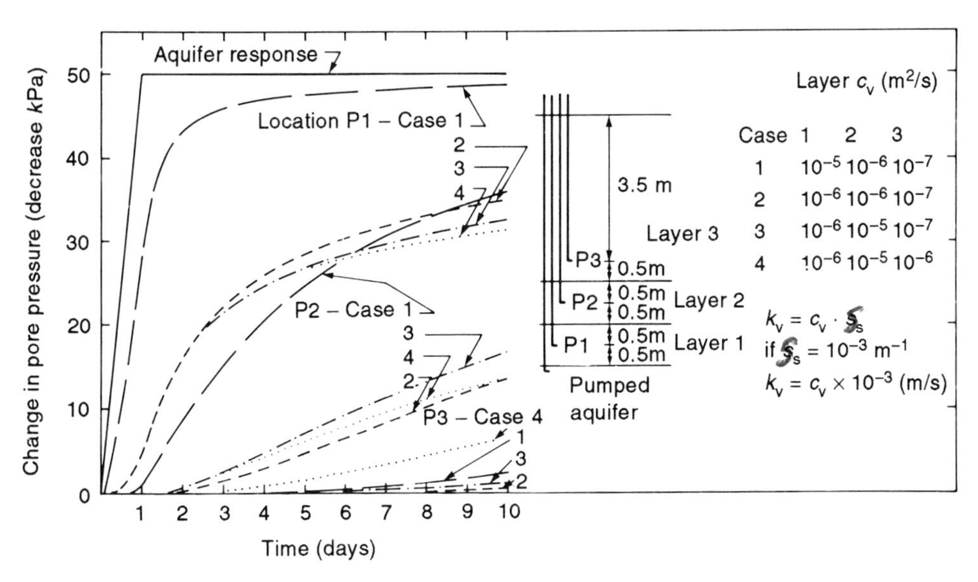

Figure 3.10 Response of a three-layered aquitard to a pumping test. (After Rowe, 1991d; reproduced with permission, Balkema.)

Rowe and Booker (1986, 1987) and described in Chapter 7 can be readily used to analyze any number of layers, and since the solution is analytically based, large contrasts in hydraulic properties between layers do not cause any numerical error or numerical problems. Recognizing that in the absence of advection and sorption, the equations governing contaminant diffusion are the same as those governing pore pressure diffusion (where contaminant concentration is analogous to excess pore pressure), one can model the aquifer loading and pore pressure response using the theory presented in Chapter 7.

To illustrate the application of this approach, the hydrogeologic conditions shown in Figure 3.10 were modeled for a 10-day pumping test. The aquifer was drawn down approximately 5 m beneath the monitoring points P1, P2 and P3 over a one-day period. The pumping rate was then adjusted to maintain a 5 m drawdown for the remaining nine days. Figure 3.10 shows the response of a piezometer located in each of the three layers, as calculated using program POLLUTE (Rowe and Booker, 1994), for four different combinations of hydraulic properties. An inspection of the relative response of piezometers P1, P2 and P3 for the four cases clearly shows the sensitivity of the analysis (and expected system response) to changes in hydraulic properties. Using small volume piezometer sampling points and pressure transducers to monitor the response, the effects evident in Figure 3.10 could be readily detected, and by adjusting the hydraulic diffusivity (i.e. coefficient of consolidation, c_v) for each layer to obtain a fit to the observed response, the hydraulic properties of the layered system can be readily estimated from $k_v = c_v S'_s$.

3.2.4 Ring infiltrometers

Large diameter (0.5–2 m) ring infiltrometers have been recommended (Day and Daniel, 1985) for field use to establish the k-values of liners prior to waste disposal. In some government jurisdictions they are actually mandatory. Such testing is expensive, time consuming and fraught with so many problems that mandatory use is most unfortunate.

A good discussion of large infiltrometers was presented by Daniel (1989) and the schemes are shown in Figure 3.11. These devices do a good job of measuring fairly high k-values on inferior liners ($k > 10^{-8}$ m/s) which are poorly compacted, slickensided, fractured and/or fissured by desiccation or frost, etc. Laboratory tests on such barrier soils tend to be conducted on intact interfissure clods or after closure of fractures by application of triaxial cell pressures. Therefore, as presented in Table 3.1(a), the resulting ratio k_{field}/k_{lab} can be as high as 2000 to 100 000 in the extreme case.

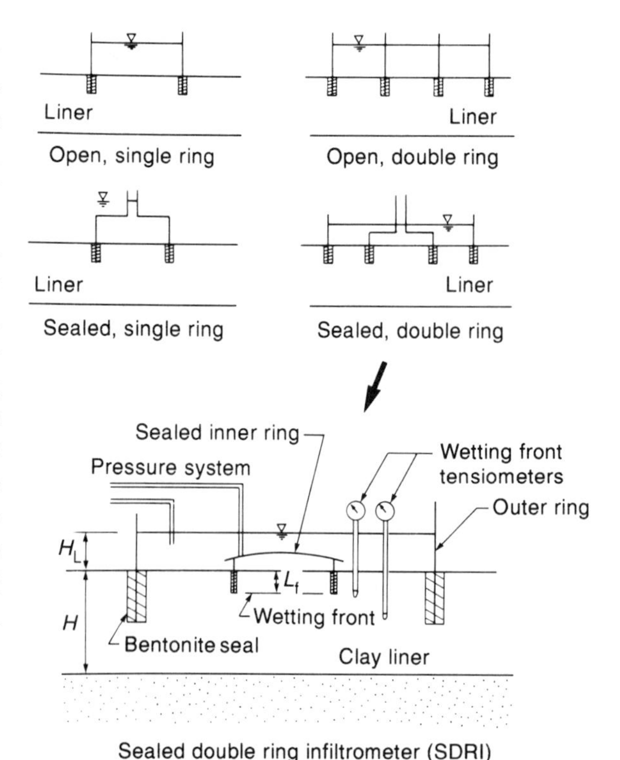

Figure 3.11 Open and sealed, single and double ring infiltrometers. (Adapted from Daniel, 1989; Daniel and Trautwein, 1986.)

Table 3.1 Hydraulic conductivity of clay liners

(a) Ring infiltrometer measurements	
	k_{field}/k_{lab}
Project (Daniel, 1984)	
Central Texas	25–100 000
Northern Texas	10–2 000
Southern Mexico	25–200
Northern Mexico	5–200
	k-value (m/s)
Bracci, Giardi and Paci (1991)	
k_{field}	10^{-6}–10^{-8}
k_{lab}	10^{-8}–10^{-11}
(b) Lysimeter measurements	
	k-value (m/s)
Three Wisconsin landfill liners (specified $k = 10^{-9}$ m/s) (Gordon, Huebner and Miazga, 1989)	
'Marathon'	$0.5–2 \times 10^{-10}$
'Portage'	$1–9 \times 10^{-11}$
'Sauk'	$0.6–4 \times 10^{-10}$
Keele Valley liner (specified $k = 10^{-10}$ m/s) (King *et al.*, 1993)	
1984 (stabilizing)	$\sim 5 \times 10^{-10}$
1988 (stabilized)	$\sim 5 \times 10^{-11}$

Use of ring infiltrometers (be they single, double, open or sealed) on high quality barriers ($k < 10^{-9}$ m/s) may involve many weeks of testing, and for this reason alone may be quite impractical, especially open systems where evaporation losses may far exceed infiltration into the barrier. Another major problem is determining the length of the flow path, in calculations of gradient. Normally this is taken as the distance from the water reservoir to the wetting front, and suction head at the front is ignored in the estimation of Δh_t. Special potentiometers or post-test excavation are required to estimate the wetting front location.

Since inactive clayey barriers compacted ~2% wet of Standard Proctor optimum have very positive self-healing characteristics, due to consolidation during waste disposal, combinations of flexible wall lab testing and field lysimeter monitoring are probably adequate controls. Again, stringent construction quality control and construction quality assurance is required during barrier compaction to obtain a low hydraulic conductivity liner. In addition, a suitable post-placement protocol is required to prevent drying or swelling since this can be critical to final performance at the compacted clay liner.

3.2.5 Lysimeters

Lysimeters are large scale seepage collection devices installed within or below clayey liners in the field to measure bulk hydraulic conductivity. A typical lysimeter might be 15 m × 15 m in horizontal area and lined on the bottom and sides with an HDPE geomembrane (Figure 3.12). The lysimeter would be filled with sand possibly filtered upwards to prevent migration of clay from the liner. Water passing through the liner is collected in the lysimeter and transferred through drainage tubes to a collection or sampling point at the margin of the liner. Using Darcy's law, $Q = kiA$, the hydraulic conductivity, k, may be calculated if flow, Q, gradient, i, and area, A, can be determined.

Like most field instrumentation, measurement is complicated by problems such as those discussed by King *et al.* (1993). Difficulties include the following.

1. Intermittent flow, Q, from a lysimeter, resulting from barometric fluctuations, means that long-term averaging is required.
2. The true area of the lysimeter and the true thickness of the overlying clay barrier (typically 0.3–0.5 m) are subject to construction errors.
3. The true pressure heads on the liner above a lysimeter require installation of some kind of

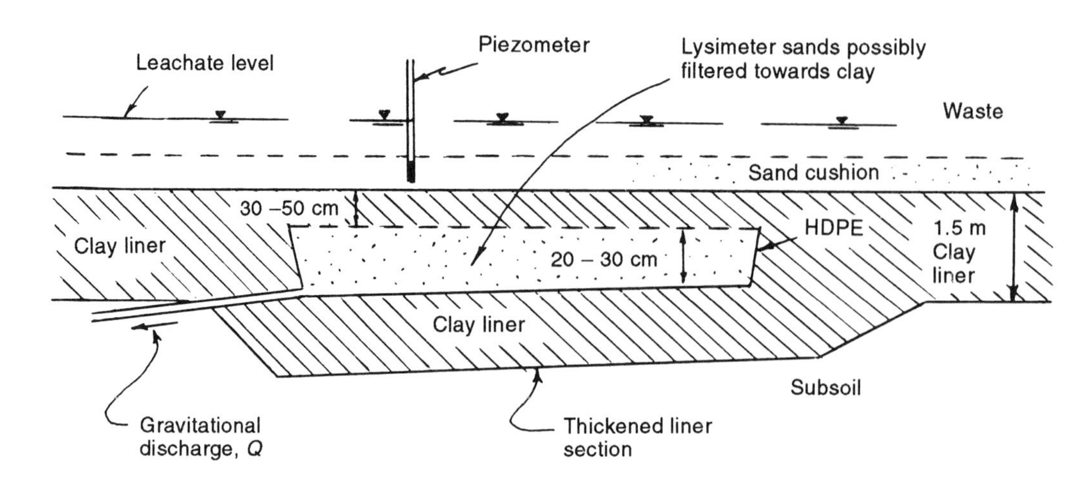

Figure 3.12 Schematic of a lysimeter installation.

strategically located piezometer for determination of *i*.

4. Possible suction effects created by negative pore pressures associated with capillary rise within the lysimeter sands to the bottom surface of the clay may influence *i*.
5. The competing effects of decreasing void ratio and increasing saturation with time as waste is placed in the landfill above the lysimeter make the hydraulic conductivity *k* time dependent.
6. The effects of double layer expansion as Na^+ is adsorbed and the potential effects of mineralogical changes due to potassium or ammonium fixation may influence *k* with time.

As discussed by King *et al.* (1993), it is probably good policy to collect liner samples from an adjacent dummy lysimeter for laboratory *k*-testing. The lysimeter *k*-test results presented by King *et al.* for the Keele Valley clayey silt till liner showed a significant (one order of magnitude) decrease in hydraulc conductivity over three years as the lysimeter was loaded with waste. This particular liner was compacted ~2% wet of Standard Proctor optimum and was amenable to self-healing on load application above ~100 kPA.

Chemical analysis on the lysimeter discharge *Q* may also be effectively used to monitor the time rate of arrival of effluent chemicals and to check any advection–diffusion modeling done prior to construction.

3.3 Compaction–permeability relationships

3.3.1 Laboratory experiments

The compaction characteristics of soils are well understood (Lambe, 1958, 1960; Mitchell, Hooper and Campanella, 1965). As shown in Figure 3.13(b), the dry unit weight of a soil, γ_d (and hence the dry density) varies with the molding water content, ω, at which it is compacted. By definition, the maximum dry density is obtained at what is referred to as the optimum water content, ω_{opt}. Figure 3.13(a) shows the corresponding relationship between the hydraulic conductivity (permeability) *k* and

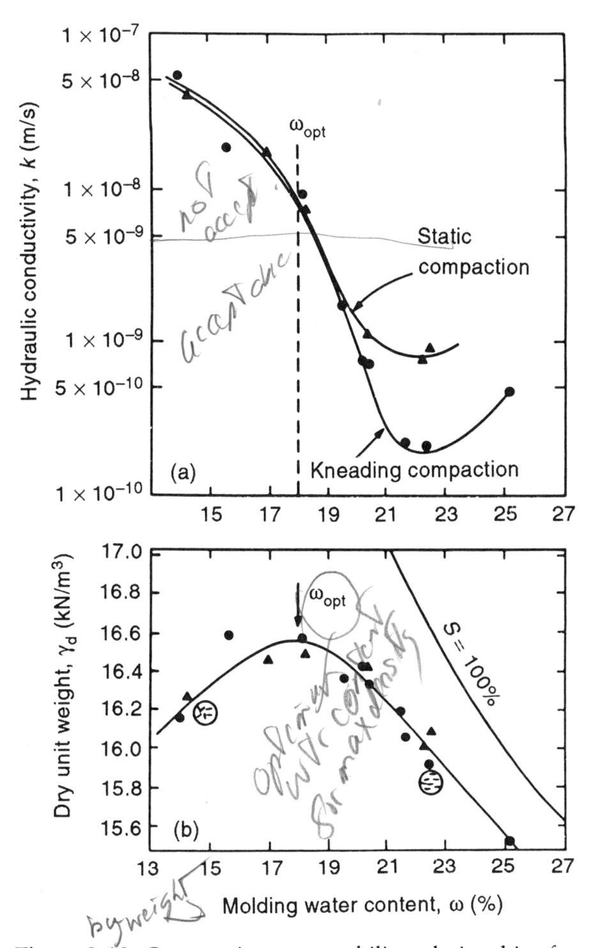

Figure 3.13 Compaction–permeability relationships for a silty clay. (Adapted from Mitchell, Hooper and Campanella, 1965; Lambe, 1960.)

molding water content. The hydraulic conductivity normally decreases from high values when compacted at water contents less than the optimum water content for the soil (this is referred to as being dry of optimum water content) to minimum values at water contents 2% to 4% above the optimum water content (this is referred to as being wet of optimum). In Figure 3.13(a), the decrease amounts to about two orders of magnitude, from 10^{-8} m/s at optimum to 10^{-10} m/s for kneading compaction 4% wet of optimum for this particular soil.

These decreases are normally attributed to the presence of a more dispersed soil structure when the soil is compacted wet of optimum. This really means that the clay platelets are largely horizontal and parallel to one another, thereby providing a more tortuous flow path for water to follow, and hence a lower value of hydraulic conductivity k. Since the lowest values of k are obtained usually with kneading compaction, it is common to make a great effort in the field to knead the soils repetitively with many passes by pad-foot or wedge-foot rollers that destroy soil clods and interclod macropores.

Generally speaking, liner compaction specifications are based on Standard Proctor or Standard AASHO compaction data. As shown in Figure 3.14, somewhat lower hydraulic conductivity can be achieved by laboratory compaction using more energy (i.e. heavier equipment) at lower water contents. However, care is required when specifying field compaction at water contents drier than Standard Proctor. The soil clods become hard and difficult to compact, resulting in interclod macropores. Also, the soil is brittle, increasing the likelihood of compaction-induced fractures along the sides of the compaction feet. This combination of interclod flow channels and fracture flaws makes representative sampling of drier liners for control k-testing more difficult because of the extra scatter in the k-test results.

This type of scatter is illustrated in Figure 3.15, which shows that a silty clay soil compacted by kneading to a unit weight of 17 kN/m^3 at moisture contents between 19.0 and 20.3% can exhibit laboratory-measured k values varying between 10^{-7} and 10^{-10} m/s. Obviously, a 1000-fold range in k caused by lack of a proper laboratory compaction protocol could completely sabotage a clay–leachate compatibility testing program.

Papers emphasizing clod destruction have been published by Benson and Daniel (1990) and Elsbury *et al.* (1990). Figure 3.16, adapted from these authors, illustrates how reduction of

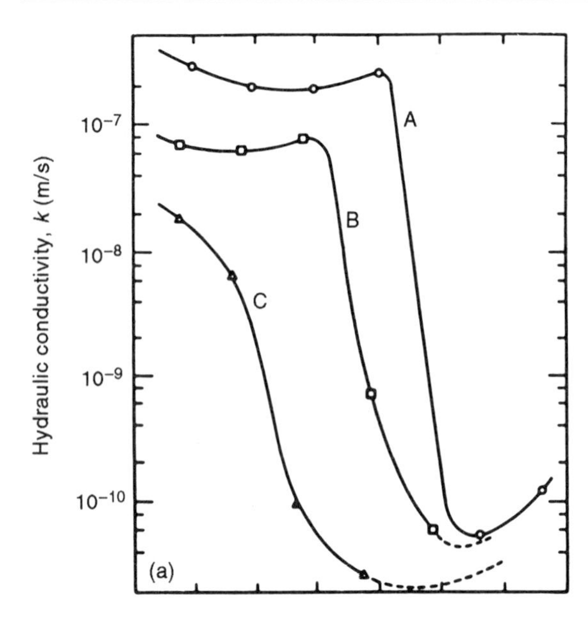

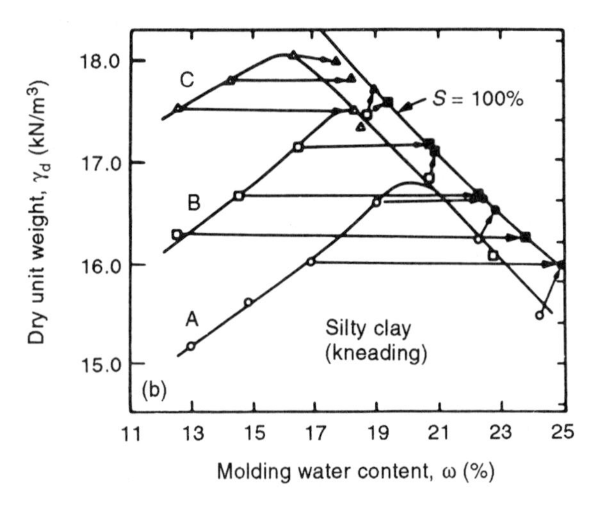

Figure 3.14 Compaction–permeability relationships for low energy (curves A) to high energy (curves C) compaction. (Adapted from Mitchell, Hooper and Campanella, 1965.)

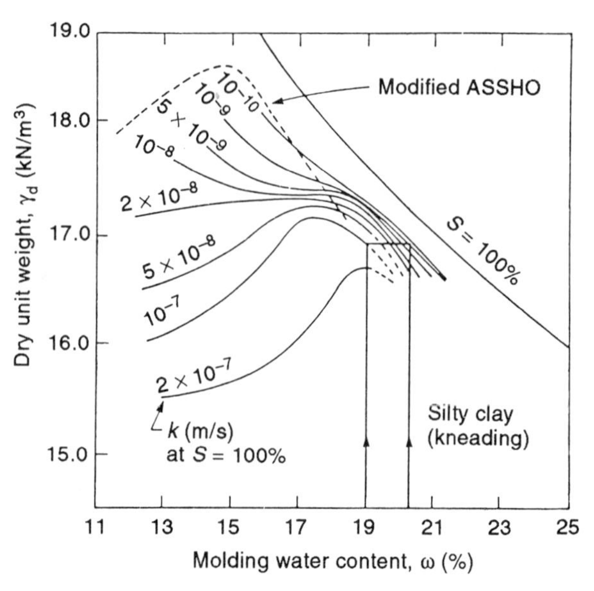

Figure 3.15 Fracture-induced variability in laboratory-measured hydraulic conductivity versus dry unit weight and molding water content. (Adapted from Mitchell, Hooper and Campanella, 1965.)

clod size from 19 mm down to 4.8 mm before Standard Proctor compaction produces a reduction of four to five orders of magnitude in laboratory measured *k*-values. Such extreme variations in *k* imply open voids between hard clods that must be destroyed by field compaction procedures (pre-crushing, extended wetting, heavier equipment, extra passes, etc.).

Even if one decides to compact wet of Standard Proctor using extra passes to achieve extra kneading and destruction of soil clods and the contacts between them, a few risks may still remain. As shown in Figure 3.17, all such soils shrink on drying. The kneaded soil demonstrates the greatest axial (vertical) shrinkage, probably because of the horizontal preferred orientation of the clay platelets. Since horizontal shrinkage would cause immediate cracking rather than shrinkage, it is important that the clay liner not be allowed to dry out, since this can rapidly induce vertical desiccation cracking.

Many clay barriers are covered with a drainage layer of sand and/or a geosynthetic. This permits time for quality control testing for the hydraulic conductivity of the clay liner, which may require a month or so before waste is applied. With good field control, the liner

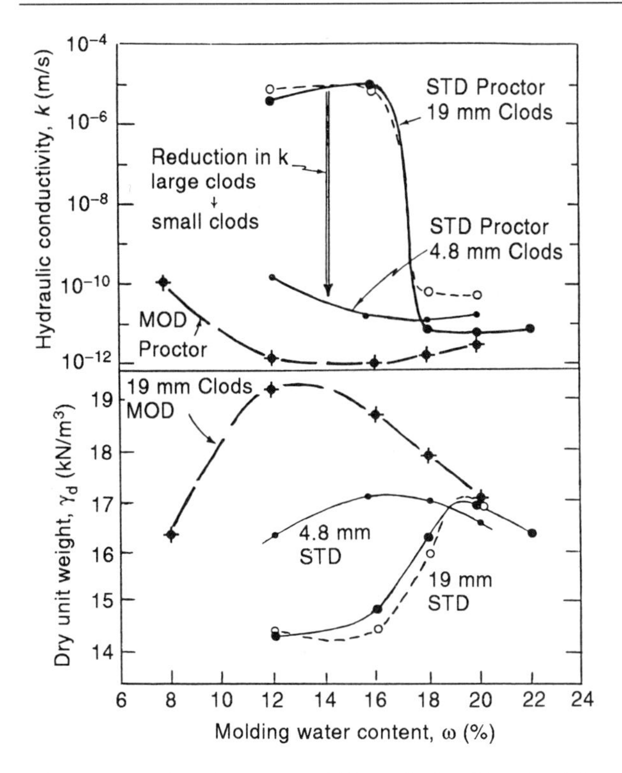

Figure 3.16 Density–permeability relationships for lab compaction of a soil of variable initial clod size. (Adapted from Benson and Daniel, 1990; Elsbury *et al.*, 1990.)

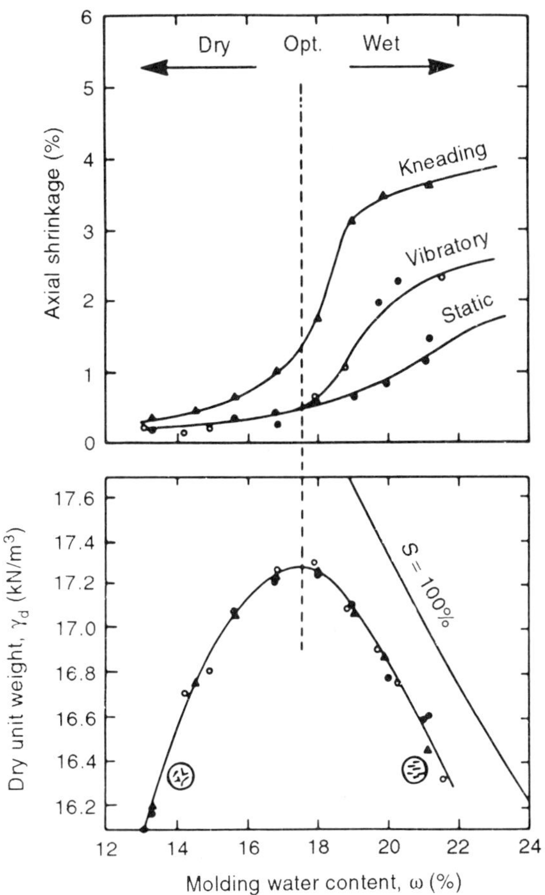

Figure 3.17 Axial shrinkage as a function of compaction method and molding water content. (Adapted from Mitchell, Hooper and Campanella, 1965; see also Lambe, 1960.)

should stay damp, and the risk of shrinkage can be minimized. For example, three exhumations of liners in Ontario showed no evidence of vertical desiccation cracking. This is partly attributable to wet-of-optimum compaction which produced horizontal clay platelet parallelism and one-directional shrinkage (vertical) rather than horizontal shrinkage and cracking.

Another factor to be considered is thixotropy. Figure 3.18 (adapted from Mitchell, Hooper and Campanella, 1965) shows increases in hydraulic conductivity *k* for certain soils if allowed to 'age' after compaction. This figure is not a simple one, since all soils are noted as having $\gamma_d = 17$ kN/m^3. For the range of water contents and saturations shown, a wide variety of compaction energies must have been employed. Thixotropy as an influence on *k* was also mentioned specifically by Dunn and Mitchell (1984).

Two of the test soils show increases in *k* by factors of two to three as a result of aging at constant volume. This 'thixotropic' increase in *k* probably results from re-orientation of the clay platelets into a more flocculated or aggregated state with larger interped pores and less tortuosity. Such re-orientation would certainly be

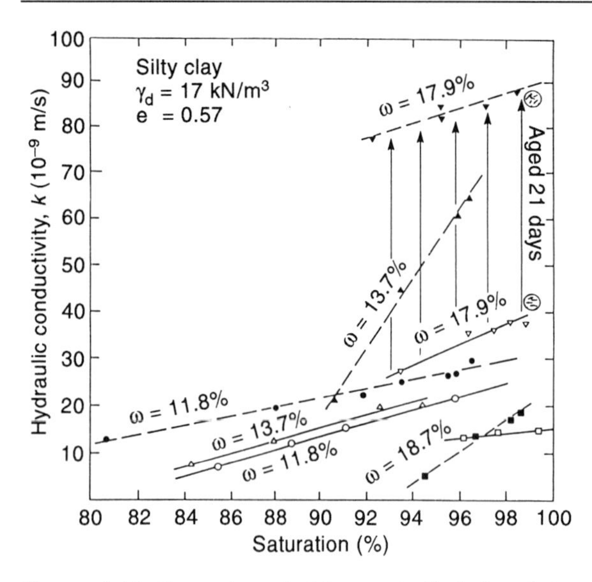

Figure 3.18 Examples of thixotropically-induced increases in *k* caused by aging. Solid line: tested immediately; dashed line: tested after 21 days. (Adapted from Mitchell, Hooper and Campanella, 1965.)

more likely to occur wet of Standard Proctor optimum and under conditions of low confining stress, as in a new liner awaiting waste after protocol testing. Normally active soils (smectites) are considered more thixotropic than inactive soils. While the increases are not large, they are certainly significant enough to complicate greatly interpretation of long-term clay–leachate compatibility test results obtained on soils compacted wet of optimum.

Thixotropic strength gain is very common in geotechnical practice wherever very soft or active clays are involved. Examples are unwanted strength gain in liquid limit tests and increased remolded strength measurement for calculation of sensitivity. It is not commonly considered in the context of *k*-testing. The authors have noted that freshly compacted, inactive clayey soils from southern Ontario, Canada, exhibit an apparent pre-consolidation pressure of ~100 kPa, that may be somewhat 'age dependent' or thixotropically controlled. This set-up in soil structure tends to resist compression early in the loading history of a liner, but will be eliminated once field stresses exceed ~100 kPa.

3.3.2 Field behavior of compacted liners

The reliability of clayey liners has been a contentious issue. A number of failures related to very poor control has led to great suspicion and to prescriptive regulations involving the use of geomembrane liners along with clay. A few words are therefore in order about field performance, although this is discussed in more detail in Chapter 9.

Daniel (1984) published a comparative study of four liners in the southern United States and demonstrated the presence of field *k*-values five to 100 000 times greater than laboratory values (Table 3.1(a)). This led to the recommended use of large diameter, double ring permeameters to measure field *k* before placement of waste. Daniel noted that post-construction desiccation cracking may have played a role in the failure of the four liners. A similar comparison implying a ratio of field to laboratory *k*-values of between two and 1000 times was published by Bracci, Giardi and Paci (1991).

A series of problems related to under-compaction of two prototype clay liners (Day and Daniel, 1985) and field *k*-testing at low effective stresses led to a long sequence of discussions in ASCE regarding the question of proper control and protocol.

Articles by Gordon, Huebner and Miazga (1989) and King *et al.* (1993) have demonstrated marked reductions in field permeability a few years after construction, as measured by lysimeter installations. Their *k*-test results are summarized in Table 3.1(b), and show field *k*-values generally varying from 5×10^{-10} to 5×10^{-11} m/s. These data are discussed again in Chapter 9.

It would appear that the inconsistency between the k-test results for various barriers may be related to as many as four factors. Firstly, it would appear that the high field values obtained in some cases may have been related to under-compaction by equipment that was too light. Secondly, cracking by desiccation on dry hot days and fissures from ice lenses in winter appear to be major problems. When cracking occurs it will obviously result in a liner having much higher hydraulic conductivity than is desired, and the cracked material should be excavated and recompacted or replaced. Thirdly, undamaged liners may become damaged during installation of liner k-test systems such as ring infiltrometers or shallow wells, and in these cases the results are not representative. Fourthly, field tests are often conducted at stress levels which are not representative of later loaded field conditions or the corresponding laboratory test conditions. For example, the data presented by King *et al.* (1993), as summarized in Table 3.1(b), show a significant decrease in hydraulic conductivity with time which appears to be related to consolidation under the applied stress caused by the overlying waste and cover materials.

In summary, it would appear that for well-designed and constructed clay liners, there is a good correlation between laboratory and field values under true stress conditions when field lysimeter data are compared with laboratory data obtained with significant applied stresses. However, laboratory tests may not detect macrostructures such as large uncompacted clods or desiccation cracking. These must be controlled by good field supervision. Field tests such as the ring infiltrometer may detect faulty construction, but typically only in the upper portion of the liner. The results obtained from this test represent low stress k-values not characteristic of higher stress field conditions.

3.3.3 Summary comments concerning compaction and hydraulic conductivity of inactive soils

1. Field compaction should normally be wet of Standard Proctor optimum, and probably close to the plastic limit. This should produce a clayey barrier soft enough to self-heal on stress application.
2. Compaction should normally be by excessive passes of pad-foot or club-foot rollers so that the interclod voids are destroyed.
3. Most borrow pits contain stratified materials, so careful control may be required to obtain a uniform field distribution of minimum hydraulic conductivity (k) values.
4. Even homogeneous clays are often variably weathered so that successive lifts have different ω_{opt}, $\gamma_d(\max)$ and k-values.
5. Liners should be immediately covered and kept moist enough to prevent shrinkage until control k-testing permits waste disposal. Similarly, smectitic liner components must not be allowed to free-swell.
6. Desiccated liner sections (dry weather compaction) should be removed, pulverized, rewetted and recompacted.
7. Liner sections must be protected from frost action and ice lensing, and if damaged should be replaced.
8. Wet-of-optimum compaction that produces a minimum k-value, produces a clay wet enough to experience chemical shrinkage on exposure to certain leachates. Clay–leachate compatibility testing is therefore necessary for soils for which no data exist or whose clay mineralogy indicates reactive mineral constituents.
9. A major benefit of wet-of-optimum compaction is that the pre-consolidation pressure of inactive clayey soils seems to be 100–150 kPa. Thus, if the amount of effective stress applied by the mounded waste exceeds about 200 kPa, the wet clay liner

will consolidate. The resultant decrease in void ratio should enable the wet clay to self-heal even in the presence of deleterious chemicals, as indicated by the lysimeter *k*-values of King *et al.* (1993) in Table 3.1(b), which is discussed further in Chapter 9.

10. The above comments apply to inactive clayey barriers and possibly to calcium smectites. The problems associated with high swell Na^+ bentonites require special discussion and more study.

3.4 Clay mineralogy

3.4.1 Introduction

Only a brief overview of clay mineralogy and colloid chemistry is provided in this text, with emphasis being placed on those factors which are responsible for critical interactions with leachate. Good texts on clay mineralogy include: Mineralogical Society (1980), Van Olphen (1977) and Grim (1953). Good applied texts with useful abbreviated discussions include: Mitchell (1993), Yong and Warkentin (1975), Lambe and Whitman (1979) and Grim (1962).

3.4.2 Abbreviated classification of the clay minerals

Most clay minerals are sheet silicates (phyllosilicates), and a useful abbreviated classification adapted from Bailey (1980) is presented in Table 3.2. The single most important characteristic of the clay minerals is that they carry a net negative charge which is expressed as the number of deficient electrons/unit cell of ten oxygens with respect to charge neutrality, as emphasized in the table. If there are no fixed cations between the sheets, the negative charge deficiency correlates directly with the cation exchange capacity (CEC) which is the measure of the number of positive cations in milliequivalents to neutralize 100 g of clay. Since the layer combinations and both the size and the location of the charge deficiency influence all aspects of engineering behavior, including the cation exchange capacity, it is important to understand these phenomena, at least conceptually, as presented in the following sections.

3.4.3 Structural components of clay minerals

The two most important elementary structural elements of the clay minerals are the silicon tetrahedron and the aluminum or magnesium octahedron. The charges present on a given clay mineral are located in either or both of these units, due to isomorphous substitution of cations for one another.

(a) The silicon (and aluminum) tetrahedron

The silicon tetrahedron, illustrated in Figure 3.19, comprises four oxygens (O^{2-}) surrounding a central Si^{4+} which fits snugly into a central space ~0.50 Å (0.050 nm) in diameter. The Si–O bond is the strongest in the silicate mineral system and is the basis of silicon technology. The net charge on an individual tetrahedron is minus four electrons ($-4e$).

During formation of tetrahedra, probably in the high temperature/pressure atmosphere of a molten rock magma, aluminum (Al^{3+}) may substitute for silicon, Si^{4+}. The net charge on an aluminum tetrahedron is $-5e$ compared to $-4e$ for silicon.

(b) The aluminum and magnesium octahedron

The individual octahedron illustrated in Figure 3.20 consists of six hydroxyls (OH^-) surrounding one Al^{3+} or Mg^{2+}, both of which are frequently present in clay minerals.

The charge on an individual Al octahedron is $-3e$, compared to $-4e$ for an Mg octahedron. Many other elements fit conveniently into the 0.65 Å space between the hydroxyls, and the basis of detailed clay mineral classification is

Table 3.2 Abbreviated classification of the clay minerals[a]

Layer type	*Negative[b] charge per unit cell of ten oxygens*	*Mineral group[c]*	*Selected mineral species*
Amorphous	N/A	Allophane*	
1:1	0–0.015	Kaolin*	Kaolinite, dickite, halloysite, amesite
		Serpentine	
2:1	0	Talc/pyrophyllite	Talc, pyrophyllite
	0.2–0.6	Smectite*	Montmorillonite, beidellite, saponite, hectorite, sauconite, etc. (bentonite is a generic name)
	0.6–0.9	Vermiculite*	Vermiculites, often weathering products
	0.6–1.0	Mica/illite*	Muscovite, biotite, phlogopite, illite etc.
	0.6–1.0	Chlorite*	Chamosite etc.

[a]Roughly adapted from Bailey (1980).
[b]Number of deficient electrons/unit cell of ten oxygens.
[c]Asterisks indicate the important soil minerals and the name normally used in geotechnical practice.

Figure 3.19 Illustration of a tetrahedron. (10 Å = 1.0 nm.)

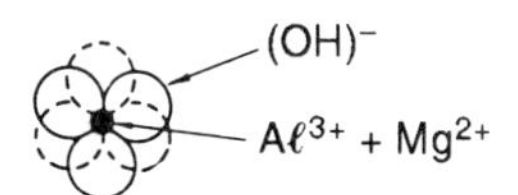

Figure 3.20 Illustration of an octahedron. (10 Å = 1.0 nm.)

often the dominant cationic species in the octahedral positions.

(c) Tetrahedral sheets

The silica tetrahedra are linked together by sharing oxygens in a wide variety of nets, chains and sheets to form many of the minerals in the silicate system. Except for a few species, the majority of the clay minerals are sheet silicates or phyllosilicates.

The structure of the silica sheet is illustrated in both cross-section and plan in Figure 3.21, complete with a repeating unit cell of dimensions a_0 and b_0. The individual tetrahedra are tied together by sharing oxygens in such a way that the top of the sheet consists of a network of hexagonal holes surrounded by oxygens. These holes are about the same size as the potassium cation, K^+, which occupies them in the structure of mica and illite.

If all of the tetrahedral positions are filled with Si^{4+}, the tetrahedral sheet is neutral. If Al^{3+} is present, the sheet becomes negative.

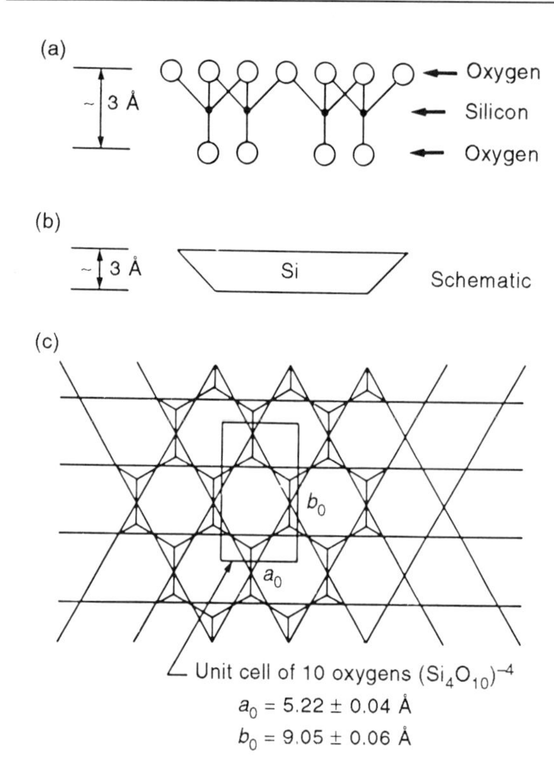

Figure 3.21 Structure of the tetrahedral sheet: (a) and (b) cross-sections; (c) top view showing hexagonal holes. (10 Å = 1.0 nm.)

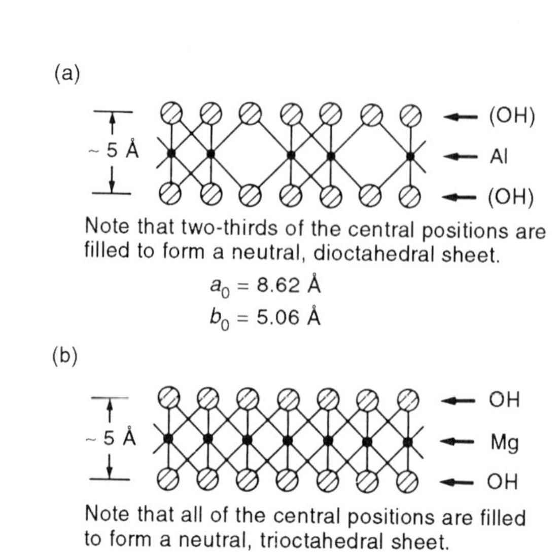

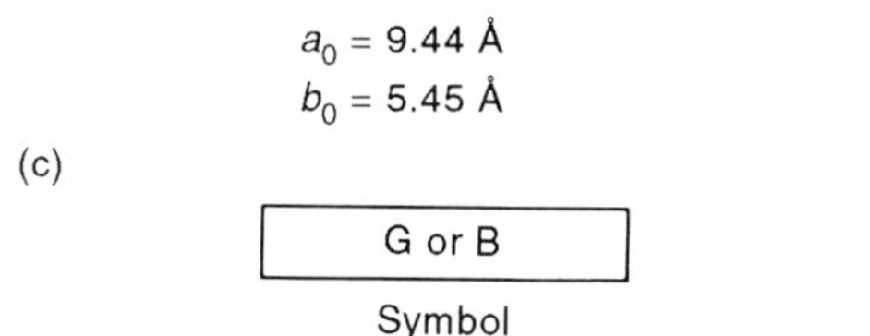

Figure 3.22 Structure of the octahedral sheet: (a) gibbsite sheet; (b) brucite sheet; (c) schematic. (10 Å = 1.0 nm.)

Both potassium, K^+, and ammonium, NH_4^+, are abundant in domestic waste leachate. Both tend to adsorb onto certain high-charge clay minerals, fixing themselves into the holes in the surface of the tetrahedral sheets. If this causes the *c*-axis of the mineral to contract and/or causes the CEC to decrease, a potential incompatibility problem may exist.

(d) Octahedral sheets

The individual octahedra are linked together by sharing hydroxyls to form sheets as shown in Figure 3.22. If only Al^{3+} is present, two-thirds of the positions are filled to form a neutral, dioctahedral sheet known as gibbsite. Gibbsite, a mineral in its own right, is a common product of tropical weathering and is found in bauxite deposits.

If only magnesium, Mg^{2+}, is present, all of the octahedral positions are filled, and a trioctahedral brucite sheet is formed. Brucite is also a mineral in its own right, which is mined for magnesium.

The sheets are normally represented by rectangles labeled G (for gibbsite) and B (for brucite), as shown in Figure 3.22(c).

Isomorphous substitution of divalent cations such as Fe^{2+} or Mg^{2+} for Al^{3+} in the octahedral gibbsite sheet will render it negative. Alternatively, if Mg^{2+} or Fe^{2+} fill some of the empty holes in a gibbsite sheet it could become positive.

(e) Two-layer and three-layer units

When clay minerals form during weathering, the silica sheets grow simultaneously with the octahedral sheets to form two-layer and three-layer units (Figure 3.23).

Under slightly acidic tropical conditions (pH 5–6), aluminum is retained and silica is gradually depleted from the soil so that two-layer units or 1:1 clays tend to form. As shown in Figure 3.23(a), these clays consist of one silica sheet merged with one octahedral sheet. The kaolins are typical of these clays.

Under conditions of incomplete leaching (arid, flooded etc.) three-layer units or 2:1 layer clays tend to form. As shown in Figure 3.23(b), these clays consist of two silica sheets sandwiching one octahedral sheet. The smectites and illites are typical of these clays.

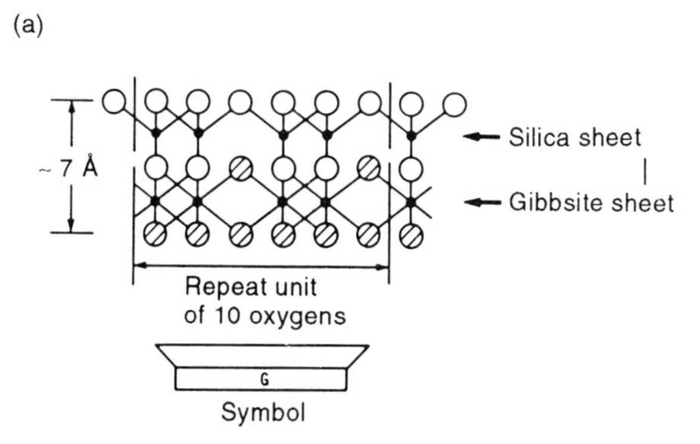

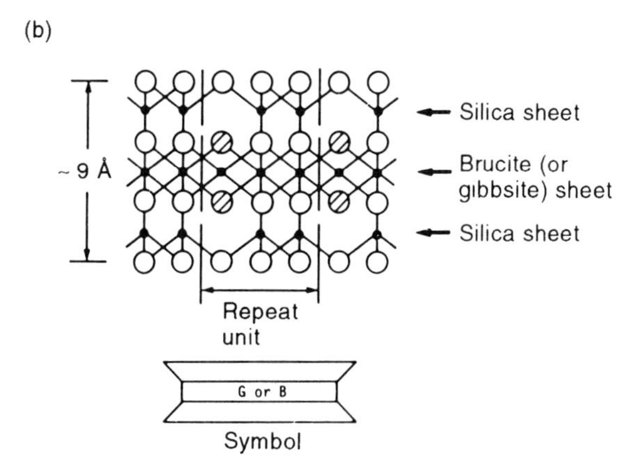

Figure 3.23 Structure of the principal phyllosilicate units: (a) 1:1 layer clays; (b) 2:1 layer clays. (10 Å = 1.0 nm.)

The clay minerals form very small crystals rarely exceeding 2–3 μm in diameter. The reason for this is distortion of the layer structure because of the poor match between the unit cell dimensions (a_0 and b_0) of the tetrahedral and octahedral sheets (compare the a_0 and b_0 values in Figures 3.21 and 3.22). Halloysite, a form of tubular kaolin, is the most spectacular example of distortion, since it forms in curved sheets with the gibbsite sheet inside the tetrahedral sheet (Figure 3.24). Montmorillonite forms very small curved particles, since there is no interlayer bonding between the 2:1 layer units, whereas muscovite may form large, table-sized slabs, due to a very strong, interlayer K^+ bond which resists the distortion.

The formulae of the clay minerals are all conveniently written in terms of a unit cell of ten oxygens. For clarity, repeat units comprising ten oxygens are shown schematically for both the 1:1 and 2:1 clays in cross-section on Figure 3.23. A count of the elemental species present in

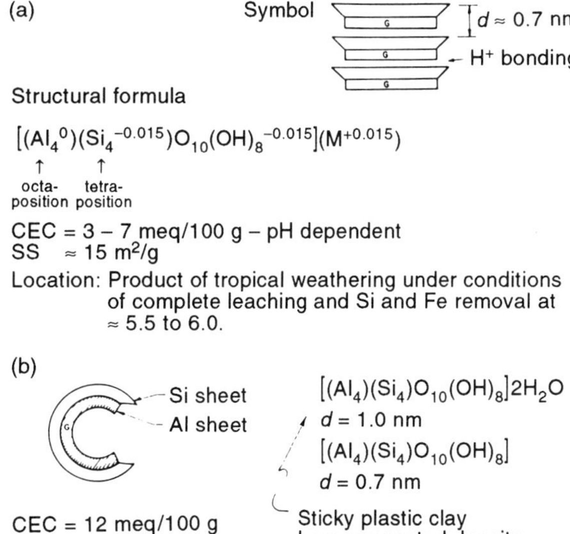

Figure 3.24 The 1:1 layer clays: (a) kaolinite; (b) halloysite.

the repeat unit actually yields the correct formula for the phyllosilicate part of the clay minerals shown.

3.4.4 Important clay minerals found in soil

The clay mineral species normally encountered in engineering practice are the kaolins, illites, smectites, vermiculites and chlorites. Emphasis is placed on the importance of the cation exchange capacity, CEC, which is the measured abundance of exchangeable cations required to be adsorbed onto the negative clay platelets to render them neutral (measured in milli-equivalents of cations/100 g of clay). These five mineral types are now described in some detail.

(a) Kaolinite (and halloysite)

A sketch showing the structure of kaolinite is presented in Figure 3.24(a) along with its structural formula. This clay mineral consists of stacked 1:1 layer units which have a *c*-axis lattice spacing, *d*, of 0.7 nm (7 Å).

As shown by the structural formula, there is very little charge on this mineral. The formula shows neutral octahedral layers and very slightly charged tetrahedral layers, probably by isomorphous substitution of Si^{4+} by Al^{3+}. As a result, very few cations are adsorbed onto this clay, and the cation exchange capacity (CEC) is very low at ~5 meq/100 g.

The bonding between the sheets is a fairly strong hydrogen ion bond. The H^+ is supplied by the $(OH)^-$ forming the gibbsite sheet, and tends to co-bind with O^{2-} on the surface of the tetrahedral sheet. The strength of this bond enables kaolinite to form fairly large crystals so that kaolinite has a fairly low surface area (specific surface) of only 15 m^2/g. The specific surface of a mineral is its measured surface area per unit mass or volume as desired.

Although the double layers around kaolinite may be fairly thick, this mineral is generally very inactive, has little adsorptive capacity and a hydraulic conductivity *k* that is rarely less than 10^{-8} m/s, unless very pure and fine-grained. Fairly rapid advective transport and little retarding capacity are characteristics. Kaolinite, therefore, is not an ideal clay mineral for a liner, even though it is relatively immune to damage when exposed to many chemicals.

The clay mineral halloysite is actually a tubular form of kaolinite, as shown in Figure 3.23(b). This tropical mineral crystallizes with water between the sheets. Since the unit cell of the Al sheet is slightly smaller than that of the Si sheet, the sheets curl with the gibbsite sheet on the inside. When dried, the interlayer water is irreversibly lost, and *d* changes from 1.0 nm to a permanent 0.7 nm, similar to kaolinite.

This mineral is difficult to compact because of its tubular structure, so that *k*-values tend to be high and erratic (see Lambe and Martin, 1955, for a discussion of properties of halloysite).

(b) Illite

A sketch showing the structure of illite is presented in Figure 3.25(a). Illite is really a form of mica, and much soil illite appears to be fine-grained muscovite, and occasionally biotite, derived by mechanical weathering from igneous and metamorphic rocks (glacial grinding).

This mineral consists of stacks of 2:1 layer units held together by a very strong potassium bond to form a *c*-axis repeat unit, $d = 1.0$ nm (10 Å). The high strength of this bond is related to two main factors:

1. a very high negative charge within the tetrahedral sheets (-0.6 to $-1.0e$/unit cell of ten oxygens);
2. the holes in the tetrahedral sheets into which the K^+ is partially countersunk.

As shown by the structural formula, the cause of this negative charge is isomorphous substitution of one in four Si^{4+} by Al^{3+}.

Most of the large negative charge on illite is satisfied by fixed potassium shown in square brackets in the structural formula. A small

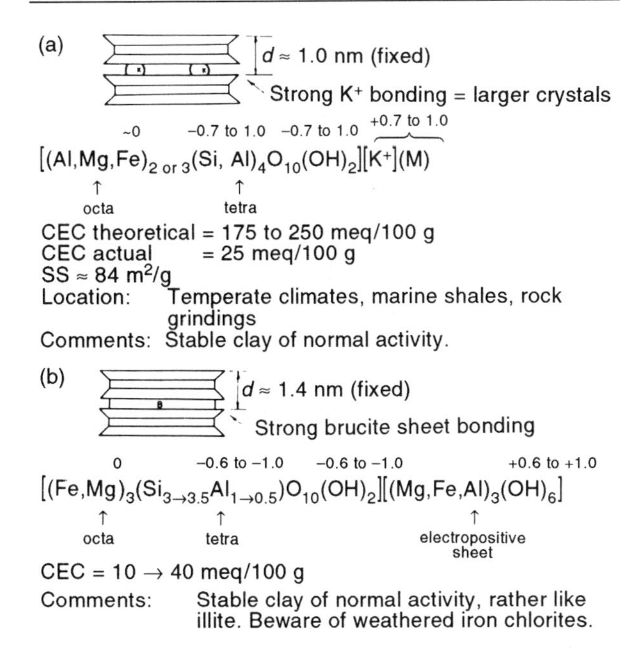

Figure 3.25 The nonswelling, 2:1 layer clay minerals: (a) illite (hydrous or soil mica); (b) chlorite.

unbalanced charge corresponding to a CEC of 25 meq/100 g remains and these cations are what make illite a clay mineral of normal activity.

Illite is efficiently compacted or consolidated to form clayey soils having a hydraulic conductivity, k, of 10^{-9} to 10^{-11} m/s depending on the void ratio. Also, the CEC of 25 meq/100 g is adequate to permit abundant adsorption of undesirable species such as heavy metals. Finally, there is no interlayer, c-axis expansion or contraction possible, so illite is often considered to be one of the most desirable clay minerals for use in engineered clay liners for municipal solid waste.

In temperate areas of acid weathering, the interlayer K^+ is sometimes leached out of illite, enabling c-axis expansion and formation of degraded illite or soil vermiculite. Such degradation would also occur if illite was leached by acid mine or tailings waters with a pH of about 2. Further discussion of c-axis expansion/contraction problems are presented later.

(c) Chlorite

Chlorite is a 2:1 layer clay bonded together by positively charged octahedral sheets having Mg, Al or Fe in the central position (Figure 3.25(b)). The resulting mineral has a d-value of 1.4 nm and the bond is normally strong enough to prevent c-axis swelling.

In many respects, chlorite has properties similar to illite, and is often favored as an effective, nonreactive barrier clay for MSW.

Some iron chlorites, however, are highly susceptible to oxidation weathering. Fe^{2+} in the octahedral position oxidizes to Fe^{3+}, reducing the charge deficiency enough to permit swelling and formation of vermiculites or smectites. These latter two minerals can be very troublesome and are discussed next.

(d) Smectites

Smectite is a group name for the 2:1 layer **swelling** clays which have a range in charge deficiency from $-0.2e$ to $-0.6e$ per unit cell (Table 3.2). Normally, this charge is present in the octahedral layer, and is too low to fix cations and bind the layers together. This mineral, therefore, experiences interlayer, c-axis swelling from 1.7 nm to over 10 nm depending on the chemical state of the clay. These features are illustrated in Figure 3.26.

It is important to realize that a smectite with Ca^{2+} cations surrounding and neutralizing its negative charge will have a d-value of ~1.8 nm, and actually consists volumetrically of 0.9 nm of water for each 0.9 nm of solid (i.e. a void ratio of unity for the solid phase!). When the double-layer water and free pore water are added in, it is not hard to understand why smectites have such high void ratios.

Smectites consist of very tiny particles which produce a specific surface of up to 800 m^2/g. Thus, smectites may have very low hydraulic conductivities ($k \approx 10^{-11}$ to 10^{-15} m/s) even at high void ratios. For this reason they are marketed (as bentonites) for use as additives to

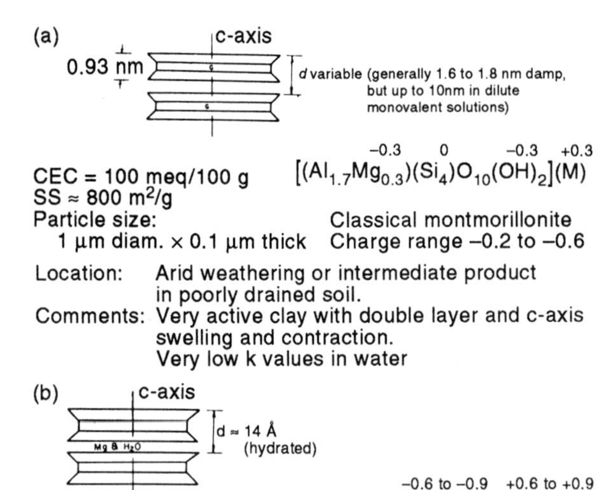

CEC (theoretical and actual) = 150 meq/100 g

Location: Acid weathering product of illite degradation. Oxidation weathering product of chlorite.

Comments: Limited swelling properties but highly susceptible to K^+ fixation, c-axis contraction, shrinkage and reduction in CEC

Figure 3.26 The swelling, 2:1 layer clays: (a) smectites (montmorillonite); (b) vermiculite.

sand to make clayey liners (or in combination with geotextiles to form geosynthetic clay liners). Great care must be taken, however, to ensure that they are compatible with the leachate to be retained, since they may be prone to chemically-induced *c*-axis contraction, double-layer shrinkage and cracking. High-charge smectites (−0.6*e*/unit cell, Table 3.2) require particular attention since they fix K^+, making them even more prone to shrinkage and reduction in CEC.

(e) Soil vermiculites

Vermiculite is a common soil additive used by gardeners as a modifier. A fine-grained equivalent occurs in many soils where it can be troublesome if abundant enough.

The illustration for vermiculite in Figure 3.26(b) shows a high charge, 2:1 layer clay. The charge of −0.6*e* to −0.9*e* per unit cell is satisfied by hydrated interlayer Mg^{2+}. Vermiculite rarely expands beyond 1.4 nm, but if it comes into contact with K^+, it promptly fixes it, and collapses to 1.0 nm illite. This *c*-axis contraction causes shrinkage of the soil phase and an abrupt decrease in CEC. This converts an active, adsorbent clay into a much less active illite. If sufficient quantities of vermiculite are present, shrinkage, and possibly cracking, seems probable in the absence of high effective stresses.

Smectites and vermiculites ‘overlap’ at a charge deficiency of −0.6*e*/unit cell (Table 3.2), and it is very difficult to decide which mineral is present and how it will behave in a barrier contacted by leachate. The amount of *c*-axis contraction to be expected can be assessed by x-ray diffraction analysis, and examples are presented later.

(f) Summary

A summary of values of cation exchange capacity and specific surface minerals of the clay is presented in Table 3.3. Generally speaking, inert clay minerals of modest CEC, such as chlorite and illite, probably produce the most reliable clayey liners.

Vermiculites tend to fix K^+ and NH_4^+ causing reductions in CEC and thus require compatibility testing if present in significant amounts.

The kaolins are the most inert, but suffer from very low CEC values and problems in obtaining low values of hydraulic conductivity.

Smectites may, in the end, be the most difficult clay mineral to control (especially in the Na^+ state), due to possible *c*-axis contraction caused by cation exchange and changes in electrolyte concentration. Another factor to be considered is that smectite barriers tend to be very thin (in some cases, as little as a few centimeters), and consequently a fairly large diffusion flux may be expected to pass through them, unless they are used in conjunction with another soil which will act as a diffusion barrier.

Most natural soils contain two or three different clay minerals, often mixed with quartz,

Table 3.3 Specific surface and cation exchange capacity (CEC) of common clay minerals (Adapted from Yong and Warkentin, 1975)

Mineral (edge view)	*Surface area* m^2/g	*CEC meq/100 g*
Montmorillonite	800	100
Clay mica (illite)	80	25
Chlorite	80	25
Vermiculite	±250	150
Kaolinite	15	5

feldspar and carbonates. To make matters even more complex, some clay minerals are interstratified with one another. For example, illite–smectite, illite–vermiculite and illite–chlorite are quite common in southern Ontario, Canada.

Quantitative mineralogical analyses, complete with CEC measurements and an assessment of the adsorbed cation regime are now normally run as part of each site appraisal. Assessment of diffusion coefficients, and occasionally clay–leachate compatibility assessment by hydraulic conductivity testing may also be required. Chapters 4 and 8 deal with these topics.

3.5 Clay colloid chemistry

3.5.1 The nature of the clay–water micelle

A single, tubular-shaped clay particle is shown in Figure 3.27. The negatively charged particle is surrounded in nature by a swarm of hydrated cations which balance the charge. The particle and its surrounding double layer of water together compose a 'micelle'.

Even though most clay particles are far larger than the typical colloidal size of 1 μm to 1 nm, they behave like colloids because they carry a negative charge. Their electrical properties, therefore, have a far greater effect on their physical properties than do gravitational factors, and under the right chemical conditions, clays (especially smectites) will form a stable suspension.

The concentration of hydrated cations is highest next to the charged particle, decreasing with distance to the same value as the free pore water. Similarly, the concentration of anions is virtually zero next to the clay particle (except the edges) and increases to that of the free pore water with distance. Cation concentrations of

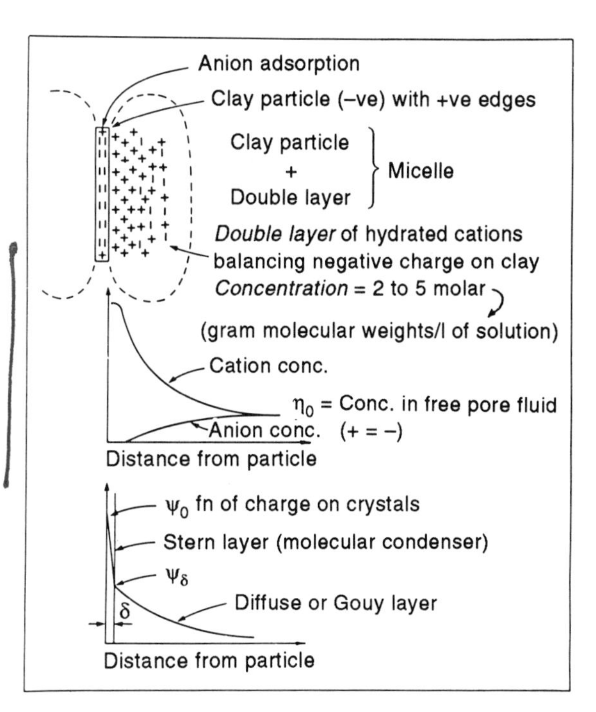

Figure 3.27 Nature of the clay–water micelle including the Stern layer concept.

2–5 M (molar: gram molecular weights per liter) are typical of the double layer.

The electric potential illustrated in Figure 3.27 incorporates the Stern layer concept in which the proximity of the first layer of firmly attached cations results in an abrupt drop in potential through the Stern layer, as shown. The distance δ corresponds to the center-line of the first layer of cations, and ψ_δ corresponds to the reduced potential which actually controls the size of the double layer. ψ_δ is somewhat variable, since an increase in concentration will add more cations to the Stern layer, reducing ψ_δ. Also, changes in the cation species can greatly alter δ.

Perhaps the most important factor in clay–leachate interaction is expansion and contraction of the double layers. As illustrated in Figure 3.28, a large double-layer contraction at constant void ratio, sometimes referred to as flocculation, creates a large increase in free void space. This may cause increases in hydraulic conductivity, and possibly in the diffusion coefficient as well. Conversely, a chemical change that peptizes, disperses or expands the double layers may eliminate most of the free pore space and reduce hydraulic conductivity.

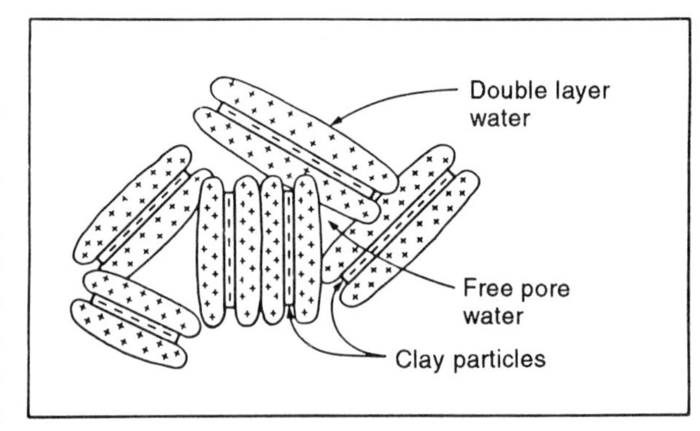

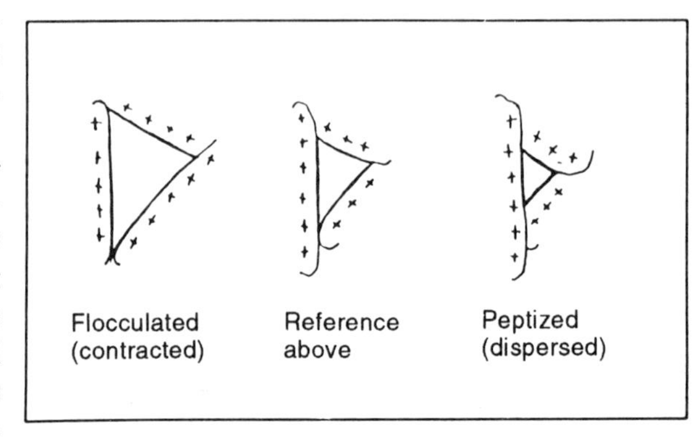

Figure 3.28 Naure of a soil–water system, showing effects of 'flocculation' or 'dispersion' on free pore space at constant void ratio.

For most conceptual needs in engineering practice, the Gouy–Chapman theory of potential adequately explains double-layer behavior. The Gouy–Chapman equation is presented in Figure 3.29 along with a sketch of the double-layer potential. The center of gravity of the area under the potential curve occurs at a value of $x = 1/\varkappa$, and many discussions now refer the $1/\varkappa$ as the double-layer thickness (Mitchell, 1993).

The main messages to be derived from the equation for potential are as follows.

1. If the pore water concentration is increased, the double layers contract.
2. If the cations are changed from monovalent to divalent or trivalent the double layers contract (e.g. from Na^+ to Ca^{2+}).
3. If the dielectric constant, ε, is reduced, the double layers contract (e.g. from 80 for water to 35 for ethanol).
4. If the temperature is increased, the dielectric constant ε decreases and the double layers contract (this is not indicated by the equation).
5. All four of the above factors will cause the hydraulic conductivity to increase (sometimes dramatically) if the void ratio remains constant.

Two plots of potential versus distance from the clay particle are presented in Figure 3.30 to illustrate concentration, valency and dielectric effects. The upper plot (Figure 3.30(b)) demonstrates how monovalent to divalent cation exchange can reduce the double-layer thickness from 100 nm to 50 nm at dilute concentrations

$$\psi = \psi_0 e^{-\kappa x}$$

$$\kappa = \sqrt{\frac{4\pi e^2 \Sigma n_{i0} z_i^2}{\varepsilon kT}}$$

n_{i0} = conc. of ions (i) in bulk suspension
z_i = valence of ions
k = Boltzmann's constant (1.38×10^{-16} erg/°K)
T = Absolute temperature (°K)
kT = 0.4×10^{-13} erg at room temperature
ε = dielectric constant of pore fluid
e = elementary charge = 4.77×10^{-10} esu
= 16.0×10^{-20} Coulomb
e = 2.71828

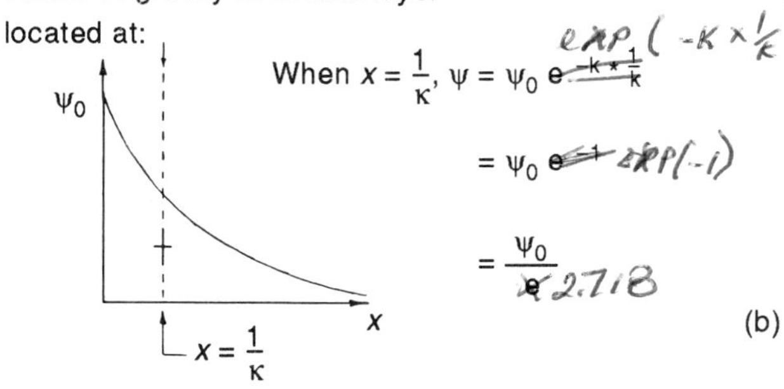

Figure 3.29 The Gouy–Chapman theory of potential.

(0.01 M). At **very** high concentrations, double-layer thicknesses of only 0.5–1.0 nm are indicated (Van Olphen, 1977, p. 35).

The second plot for changes in dielectric constant (Figure 3.30(c)) is equally dramatic and will form the basis of the discussion of clay–liquid hydrocarbon compatibility in Chapter 4.

3.5.2 Adsorbed cation regimes and cation exchange

Cations normally found in nature are summarized as follows.

1. Ca^{2+} and Mg^{2+} dominate most systems.
2. Na^+ and K^+ are common in marine soils and certain volcanic ash deposits.
3. Al and Fe hydroxide cations having 'partial valences' of ~0.5^+ exist in low pH podsolic environments (pH < 4).

(a)

Ion conc. (moles/l) (ψ_0 const)	Double layer thickness	
	Monovalent	Divalent
0.01	100 nm	50 nm
1.0	10 nm	5 nm
100.	1 nm	.5 nm

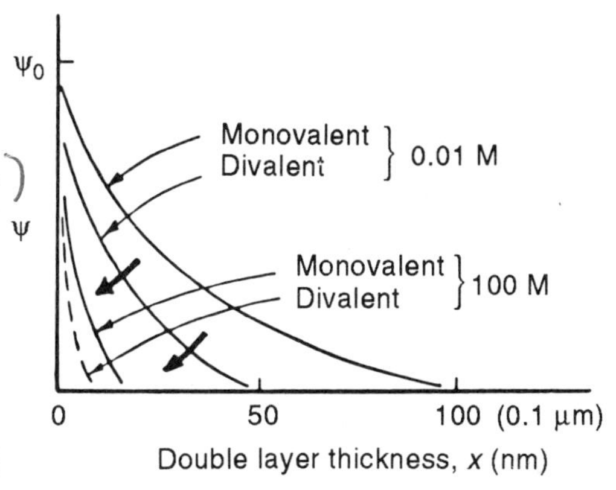

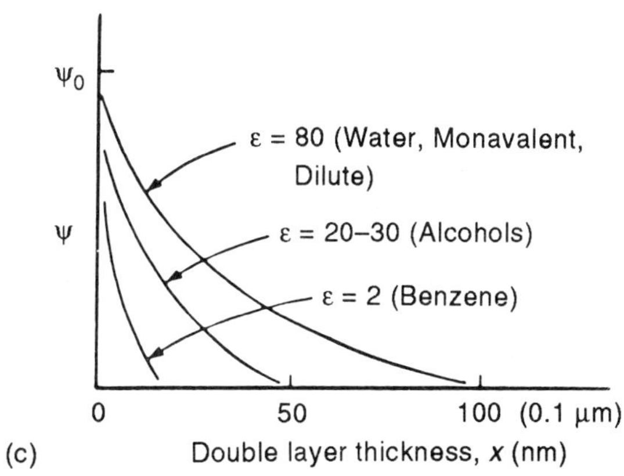

Figure 3.30 Double layer thickness variations: (a) data from Van Olphen (1977); (b) concentration and valency effects; (c) dielectric effects.

4. H^+ and H_3O^+ are common in acid soils.
5. Metallics are rare, and are rapidly adsorbed or precipitated, except in very low pH environments.

The replacing power of the normally occurring cations can be arranged in the following sequence (Yong and Warkentin, 1975):

$$Al^{3+} \gg Ca^{2+} > Mg^{2+} \gg NH_4^+ > K^+ > H^+ > Na^+ > Li^+$$

The Gapon exchange equation can be used to calculate roughly the relative abundance of the various cations on a clay, but it is normally more useful to do an NH_4Cl exchange or a combined Ag–thiourea/KCl exchange and actually measure the adsorbed cations.

Since Na^+ is the most common and troublesome monovalent cation found on natural soils, measurement of the sodium adsorption ratio (SAR) is useful, and often required during geotechnical assessment.

$$SAR = \frac{Na^+}{[0.5(Ca^{2+} + Mg^{2+})]^{1/2}}(meq/l)^{1/2}$$

Analyses are usually run on soils wetted to their saturation water content (i.e. the water content close to the liquid limit at which a shine appears on the wet soil) and all concentrations are in meq/l.

The SAR (which is actually a form of Gapon's equation) is particularly useful because it often correlates with the amount of exchangeable sodium (ESP) (Mitchell, 1993; Sherard, Decker and Ryker, 1972). Dispersed, low-salt soils containing abundant adsorbed Na^+ may initially have low hydraulic conductivities but they may also be susceptible to chemical shrinkage and cracking if chemically altered, especially at low confining stresses.

Cation exchange

The levels of Na^+, K^+ and NH_4^+ in MSW leachate are sufficiently high that they effectively exchange some of the Ca^{2+} and Mg^{2+} present on natural clays during advection and diffusion. Since it takes two Na^+ to exchange one Ca^{2+}, this reaction should expand the double layers and decrease k.

The role of the similarly sized cations, K^+ and NH_4^+, is quite different. These two species tend to fix onto vermiculites and high charge smectites and are highly retarded. Worse than this, however, they may contract the *c*-axis spacing, causing a reduction in the mineral or solids volume which should increase hydraulic

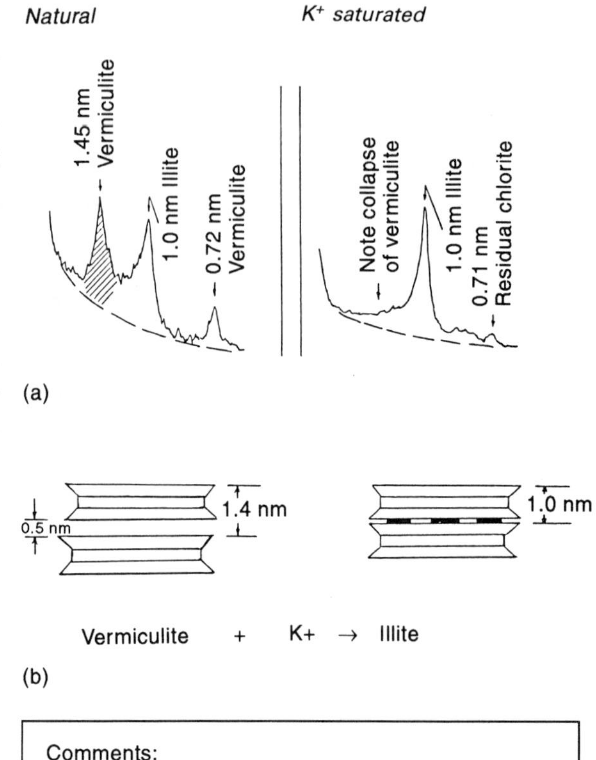

Figure 3.31 Potassium fixation by soil vermiculites and high charge smectites. (a) X-ray diffraction traces showing vermiculite in natural clay collapsing to illite on KCl treatment and K^+ fixation; (b) schematic of vermiculite to illite conversion; (c) significance of K^+ fixation.

conductivity, k. If they do fix into the holes in the tetrahedral sheets they also may reduce the CEC. This may contract the double layers, owing to charge reduction, causing large increases in free pore space at constant void ratio. In turn, this may cause increases in k, and possibly even liner cracking. These phenomena were discussed previously in sections 3.4 and 3.5 and are illustrated by the x-ray traces and sketches shown in Figure 3.31.

Chapter four

Clay–leachate compatibility by measurement of hydraulic conductivity

4.1 Introduction

Extensive clay–leachate compatibility testing is now being conducted as part of many clay liner designs for landfill sites. This chapter discusses this form of testing, with examples cited with respect to clays from southern Ontario, Canada.

Major concerns exist regarding potential pollution problems related to contamination by toxic liquid hydrocarbons (both soluble and insoluble), and so a major portion of this chapter will be devoted to organic liquids. Public pressure has also made it necessary to assess the effects of domestic waste leachate on proposed clayey liners for domestic waste landfills. This subject will be considered first.

The hydraulic conductivity equipment employed for the work presented in this chapter is a constant flow rate system described by Fernandez and Quigley (1985, 1991). The system (Figure 4.1) generates a constant flow, q, through the test specimens, and the induced total head drop across the specimen is used to calculate the hydraulic conductivity, k, using Darcy's law. This procedure, extensively described by Olsen (1966), is particularly useful for the volatile, toxic liquid hydrocarbons which are sealed in the reservoir cylinders (syringes). The single circuit shown for one permeameter is actually expanded to eight circuits, with the reservoir syringes mounted in a triaxial frame. The upward displacement generates constant flow that can be varied from 6×10^{-6} to 1×10^{-2} ml/s. A dial gage mounted on the compression frame allows continuous monitoring of the flow rates. The pressure head at the inlet end of the specimen is measured using a pressure transducer, and the outlet end is kept at atmospheric pressure. The total head drop is essentially equal to the pressure head drop for high pressure heads, the elevation head correction becoming more important at lower inlet pressures.

The permeameter cell, illustrated in Figure 4.2, consists of either an aluminum or stainless steel cylinder (A). Both ends of the cylinder are machined to contain viton O-rings (B) sealing the contact between the cylinder and the aluminum or steel plates (C). The fluid outlet (D) allows collection of the effluent permeant for constant flow assessment or chemical analysis. The two fittings in the upper plate are the fluid

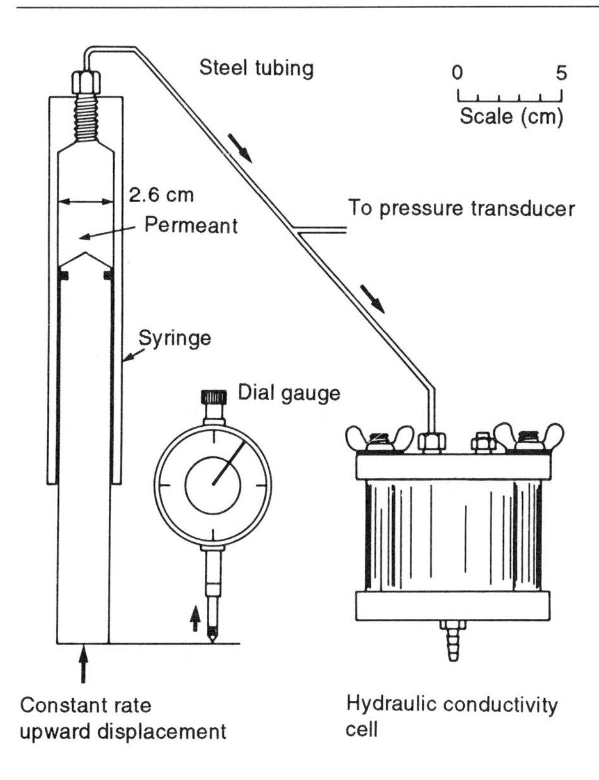

Figure 4.1 Schematic diagram of constant flow permeameter. (Fernandez and Quigley, 1985; reproduced with permission of the *Canadian Geotechnical Journal.*)

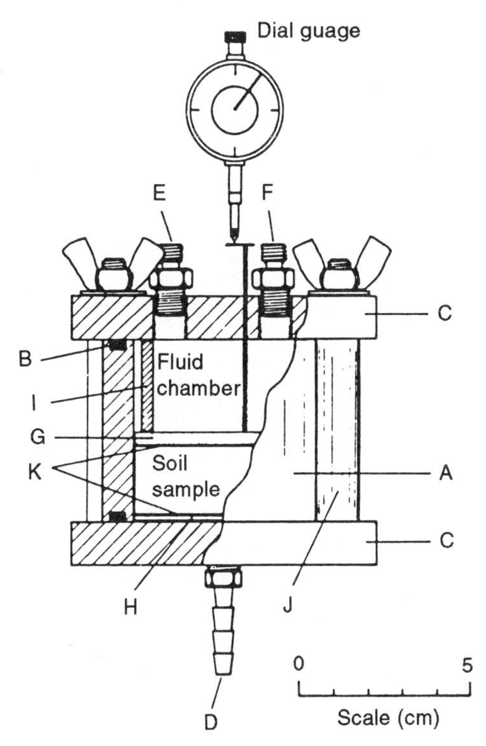

Figure 4.2 Schematic of one-permeameter cell. (Adapted from Fernandez and Quigley, 1985.)

inlet (E) and a valve for escape of air during filling of the fluid chamber (F). Port F can also be used as the pressure transducer mount. A brass porous disk (G) approximately 3 mm thick and a polyethylene filter (H) 1.5 mm thick are placed on top and below the soil sample respectively. (I) is either a rigid spacer on the porous disk to prevent swelling at zero static effective stress ($\sigma'_v \approx 0$) or a set of springs capable of applying static effective stress σ'_v up to 320 kPa before and during permeation. The assembled cell is held together by four threaded and sleeved rods (J) fixed to the lower plate. Filter paper (K) is normally placed between the soil specimens and the filter disks (G and H). Finally, a settlement measuring rod extends through the cap to the top of the sample and, by means of another dial gage, chemically induced consolidation may be measured. This apparatus might be regarded as an elaborate constant flow rate consolidation permeameter.

The test soils to be described consist of brown and gray clays from Sarnia, Ontario. A brief description of the mineralogy of samples from 0.3 m and 11 m depth is presented on Table 4.1. The gray soils at depth, which are parent materials to the brown weathered surface soils, contain more carbonate, more chlorite and very little (<1%) smectite. Near the soil surface, smectite constitutes about 15% of the soil. This smectite is derived from the iron-rich chlorite by oxidation weathering.

These soils provide a range of potential responses to leachate since, at depth, they are fairly inactive yet good liner clays, and at surface they contain enough smectite to demonstrate the high activity effects of montmorillonite.

Table 4.1 Silty clay composition (Sarnia)

	Brown (0.3 m)	*Gray (11.0 m)*
Quartz and feldspar	~20%	~17%
Carbonates	~10%	~35%
Illite	~50%	~25%
Chlorite	~8%	~22%
Smectite	~15%	~1%
CEC	~25 meq/100 g	10 meq/100 g
<2 μm	~60%	~42%

The effluent chemistry is carefully monitored during hydraulic conductivity (k) testing for clay–leachate compatibility assessment. Normally the criterion to stop a test is not a constant value for k, but chemical equilibrium between the clay and the influent leachate. Once the effluent concentrations, c, reach influent concentrations, c_0, this equilibrium is normally considered to have been reached.

Typical c/c_0 curves are presented in Figure 4.3 to illustrate a wide spectrum of possible interactions between soil (either undisturbed or compacted) and the chemical species used as a test permeant. These curves will be considered individually.

1. Uniform, undispersed advective movement (piston flow) would arrive as a 100% concentration front (i.e. $c/c_0 = 1$) at exactly one pore volume (PV) of flow. At this point all fluid previously in the soil would be displaced by the permeant; curve (1).
2. Random dispersion tends to spread the front so that some chemicals arrive before 1 PV and c/c_0 does not reach unity until after 1 PV. Curve (2) therefore has a $c/c_0 = 0.5$ at 1 PV, and is unretarded but somewhat dispersed.
3. Curve (3) represents a strongly retarded species which is totally retained by the soil for 2 PV, and then slowly approaches equilibrium at more than 7 PV.
4. Curve (4) is described as retarded and tailed, yet still demonstrates some arrival before 1.0 PV. This type of behavior normally indicates channel or fracture flow of a highly reactive species with equilibrium obtained by diffusion into adjacent soil peds.
5. In comparison, curve (5) represents early arrival and tailing of a nonreactive species, $c/c_0 = 0.5$, arriving at only 0.5 PV. Again, the only explanation is rapid channel flow, with equilibrium reached by a long process of tailing caused by diffusion from the fractures (or macropores) into the adjacent micropores (sections 1.4.2 and 7.6 and Chapter 11 give more information on this phenomenon, which is referred to as matrix diffusion). It is conceivable that with an even higher flow rate, c/c_0 would approach unity, and one would never know if equilibrium was reached.
6. Curve (6) is a typical desorption curve, often reflecting calcium (Ca^{2+}) and magnesium (Mg^{2+}) displaced by sodium (Na^+) and potassium (K^+) in a permeating leachate. Occasionally the Ca^{2+} curve does not return to unity, in which case long-term dissolution of carbonate or gypsum might be inferred. It is often important not to run such tests too long, in case loss of solids causes a large increase in laboratory hydraulic conductivity (breakthrough), something that could never happen in a field situation.

With the above summary of typical effluent curves it is now appropriate to look at some actual clay–leachate compatibility tests.

4.2 Soil–MSW leachate compatibility

Most inactive soils whose clay minerals consist of illites and chlorites are relatively insensitive to leachate from municipal solid waste. In fact, the hydraulic conductivity will normally decrease (Griffin *et al.*, 1976) probably because of Na^+ adsorption and double-layer expansion, and possibly because of bacterial clogging.

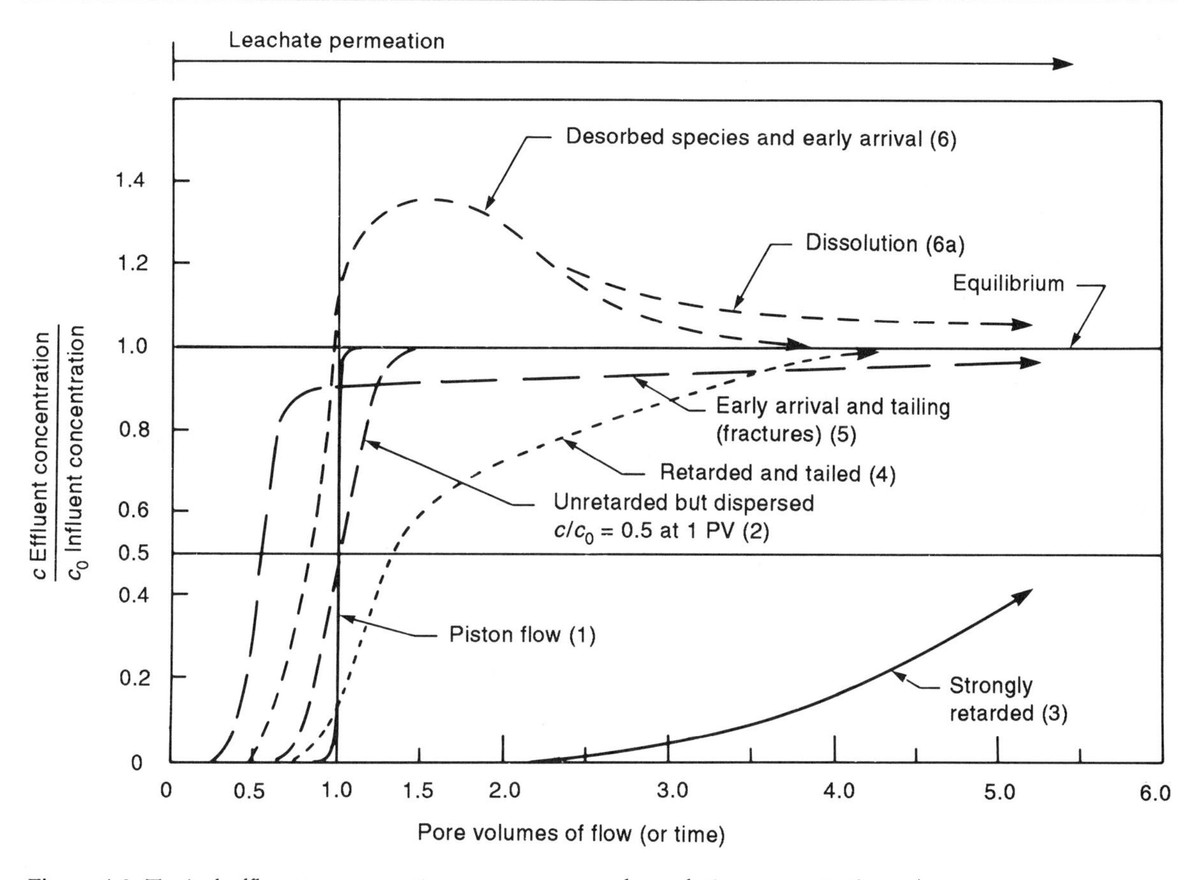

Figure 4.3 Typical effluent concentration curves expressed as relative concentration, c/c_0.

If the soils contain significant amounts of swelling clay (here defined as vermiculite, montmorillonite or interlayered illite–smectite (section 3.4)), *c*-axis contraction or expansion is another complicating factor than can respectively increase or decrease hydraulic conductivity. For example, K^+ fixation by vermiculite causes a 28% decrease in crystal volume plus a contraction of the double-layer thickness, because the charge deficiency, and hence the cation exchange capacity, is reduced equivalent to the amount of K^+ fixed. Such phenomena should contract the soil peds, open further the voids or fractures between them and thus cause increases in the hydraulic conductivity. If a high effective stress is present on the clay liner, chemically-induced consolidation should help compensate for any chemically-induced increase in the size and frequency of the macropores. An illustration of these phenomena was presented in Figures 3.28 and 3.31.

4.2.1 Hydraulic conductivity

Hydraulic conductivity tests run on three samples of Sarnia clay from depths of 0.3 and 11 m are presented in Figure 4.4. These results will be used to demonstrate both the effect of clay mineralogy and the effect of compaction at a water content too close to Standard Proctor optimum, which may result in a dense, strong liner, but can also give rise to microfractures in

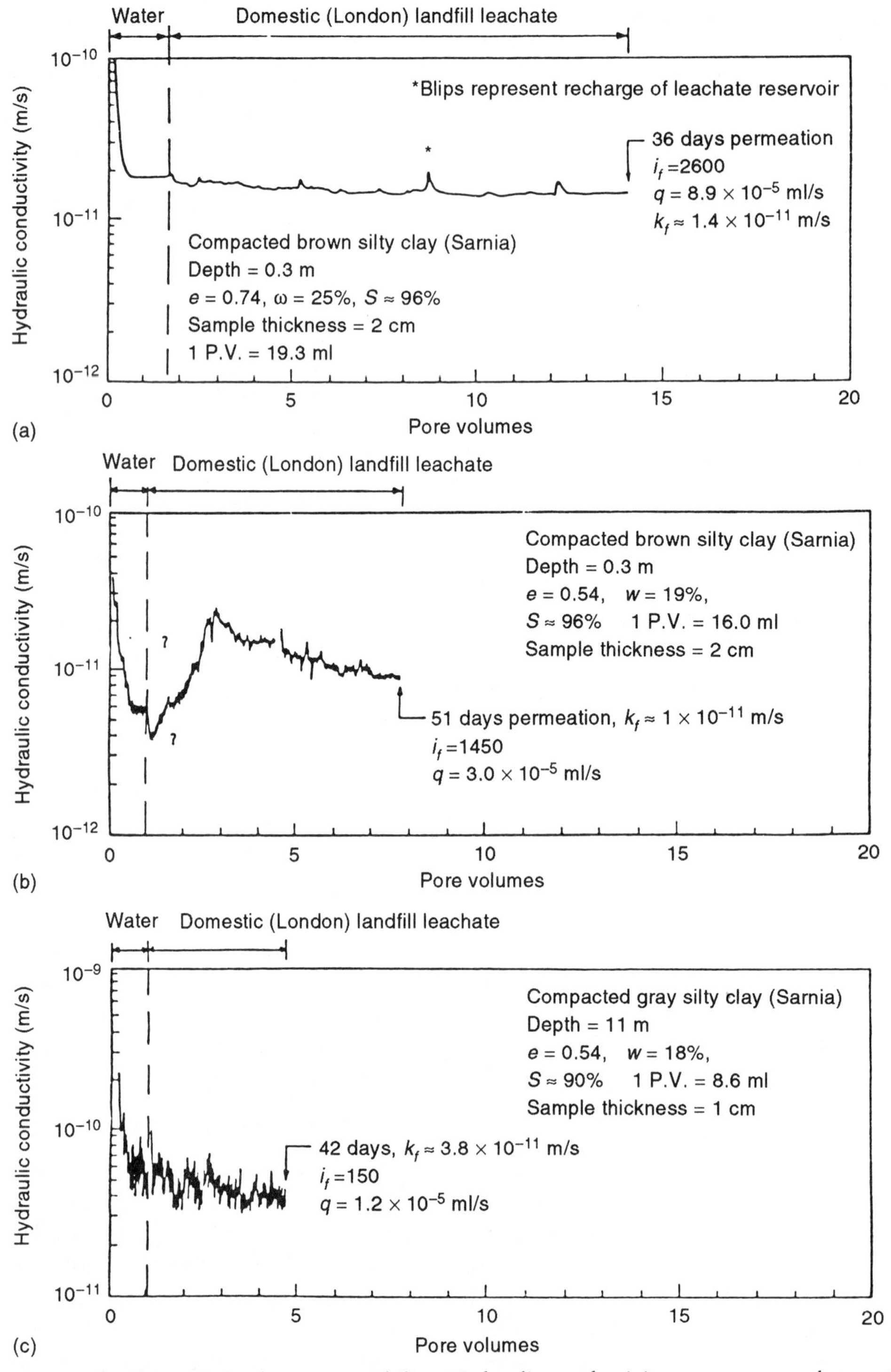

Figure 4.4 Sarnia clay/domestic leachate compatibility. Hydraulic conductivity versus pore volumes: (a) brown, compacted, $e = 0.74$, $\omega = 25\%$; (b) brown, compacted, $e = 0.54$, $\omega = 19\%$; (c) gray, compacted, $e = 0.54$, $\omega = 18\%$.

the clay, which will influence both the hydraulic conductivity and chemical characteristics of the movement of leachate through the barrier. The tests were run for the most conservative case of zero vertical static stress on the soil samples. The gradients developed by the constant flow rate test system are dependent on hydraulic conductivity, and for the three tests shown in Figure 4.4 were 2600, 1450 and 150 for the two brown clays and one gray clay, respectively. The corresponding values of the maximum seepage stresses, J_{max}, at the base of the test specimens, were approximately 510, 284 and 15 kPa, respectively.

The following observations can be made.

1. Domestice waste (MSW) leachate causes a slight reduction in hydraulic conductivity for both the brown and gray test clays (Figures 4.4(a) and (c)).
2. The gray clay has a higher hydraulic conductivity than the brown clay (3.8×10^{-11} compared with 1.4×10^{-11} m/s), even though it is at a lower void ratio ($e = 0.54$ compared with $e = 0.74$).
3. The greater smectite content of the brown clay is partly responsible for its lower *k*-value. Also, as will be shown later, the denser, gray clay (sample (a)) contains compaction-induced fractures that allow channel flow.
4. The brown clay compacted to $e = 0.54$ (Figure 4.4(b)) produced an erratic hydraulic conductivity versus pore volume curve that suggests rapid clogging of fractures followed by stabilization at a reasonable hydraulic conductivity of 1×10^{-11} m/s.

4.2.2 Chemical controls

Companion curves for the concentration ratio (c/c_0) are shown in Figure 4.5 for the three hydraulic conductivities versus pore volume (PV) curves shown in Figure 4.4. The upper set of curves (Figure 4.5(a)) for the looser sample compacted to a void ratio of 0.74 indicates homogeneous flow through equal-sized voids, and arrival of the 50% chloride front (c/c_0 = 0.5) at close to 1 PV of leachate flow.

The set of curves in Figure 4.5(b) for the denser brown clay compacted to a void ratio of 0.54 demonstrates early arrival and tailing, both indicative of channel flow through compaction-induced fractures. This is quite certain since the presence of more clay at $e = 0.54$ should cause greater K^+ retardation, not less!

The final set of curves in Figure 4.5(c) for the gray soil represents a problem from a chemical point of view, even though the hydraulic conductivity test results in Figure 4.4(c) appear to be quite reliable. The presence of fractures in this sample has allowed early arrival of all chemical species, at least at the gradient of 150 employed.

Considering Figures 4.5(a) and (b) for the brown clay only, the following aspects merit detailed comment.

1. Both sets of curves show Na^+ and K^+ retardation due to adsorption onto the clays, and resultant displacement of Ca^{2+} and Mg^{2+} to form a hardness halo effect which arrives early.
2. Conservative chloride reaches $c/c_0 = 0.5$ at ~0.8 PV of leachate flow for the looser soil (Figure 4.5(a)), compared with 0.1 PV for the dense soil (Figure 4.5(b)), indicating almost homogeneous flow through the former, and channel (fracture) flow through the latter.
3. In the absence of macropores, the denser soil should retard more K^+ than the looser soil, whereas the opposite appears to occur. The higher (early) K^+ values indicate channel flow along compaction-induced fractures.
4. For sodium, c/c_0 reaches 0.5 at 1.4 PV of leachate flow, indicating the expected retardation (by adsorption) for the looser soil (Figure 4.5(a)). For the denser soil, c/c_0 reaches 0.5 at only 0.8 PV, again indicating

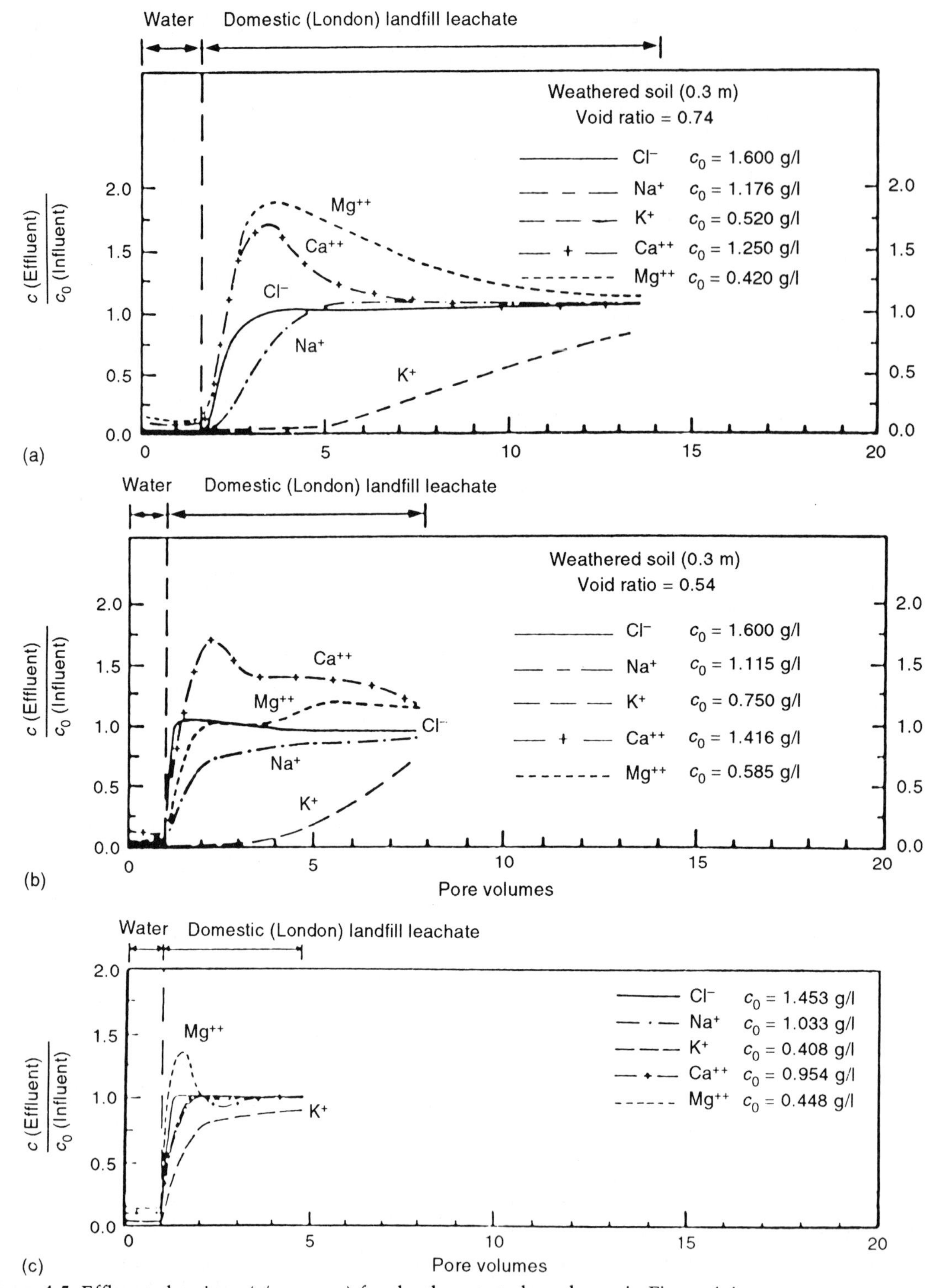

Figure 4.5 Effluent chemistry (c/c_0 curves) for the three test clays shown in Figure 4.4.

early arrival due to channel flow through this sample. This occurs in spite of the much lower gradient on the denser sample compared with the looser sample ($i_f = 1450$ compared with 2600).

5. The abrupt flattening of the sodium curve (Figure 4.5(b)) and the long slope towards $c/c_0 = 1$ (tailing) indicates fracture or macropore flow with gradual diffusion of Na^+ into the pore fluid of the adjacent soil peds, and ultimately cation exchange onto the clay and desorption of Ca^{2+} and Mg^{2+}.

The c/c_0 curves in Figure 4.5(c), while technically a problem (as noted previously), do yield some useful information.

1. Mg^{2+} appears to be the major exchange cation, since it yields a desorption peak rather than Ca^{2+} observed for the brown soils.
2. Some mineral component is also retarding K^+ in the gray clay, possibly an interlayered species.
3. The test should be repeated at a lower gradient than the 150 used so that retardation and equilibrium may be properly observed.

(a) The importance of K^+ retardation

The extent of K^+ adsorption from domestic leachate by these soils increases with increasing amounts of vermiculitic smectite in the specific test specimens. Since the smectite increases from about 1% in the gray clays to about 15% near the surface, as a result of oxidation weathering of chlorite, much more retardation should occur with the passage of leachate through near-surface soils.

Many borrow pits used for construction of clayey barriers demonstrate weathering changes similar to the Sarnia site. The mineralogy of a clayey soil barrier thus may vary significantly from one stage of construction to the next as the borrow depth varies.

The significance of this weathering phenomenon is illustrated in Figure 4.6, which shows c/c_0 curves for K^+ generated by MSW leachate influent passing through clayey samples taken from 0.3, 1.7 and 11.0 m depths. The three samples were essentially free of fractures and macropores, and thus the various degrees of K^+ retardation reflect increasing vermiculite–smectite towards the surface.

The c/c_0 curves for K^+ are also useful to indicate the role of fracture flow. Figure 4.7 compares curves for fractured (dashed) and unfractured (solid) samples from 0.3 and 11.0 m depths. The early arrival of K^+, evident for both soils with a void ratio $e = 0.54$, occurs as a result of fracture flow. For the two unfractured soils, from 0.3 and 11.0 m, passage of ~15 and ~6 PV of leachate was required for the effluent and influent liquids to contain equal amounts of K^+. If flow through the soil was homogeneously through equal pore sizes, it could be reasonably assumed that equilibrium had been reached, at least with respect to soluble chemical species. Testing times for these two tests (0.3 and 11.0 m) were five and seven weeks at gradients of 2600 and 2000, respectively.

For the two fractured samples, equilibrium might be difficult to establish because K^+ would have to diffuse from the flow channels into the pore fluid in the adjacent soil peds (by matrix diffusion, as discussed in section 1.4.2) and then sorb onto the clay minerals before equilibrium is reached. Arrival of c/c_0 values equal to 1 after a long period of tailing would probably indicate equilibrium. If, however, the fractures opened further, owing to soil shrinkage, the hydraulic conductivity would increase and c/c_0 would equal unity after a short testing time without equilibrium being reached.

Like K^+, NH_4^+ is also abundant in raw domestic waste leachate, and it will also fix onto vermiculitic minerals in borrow soils.

(b) Post-testing assessment

Post-testing assessment is quite an elaborate process, and the following procedures are often required.

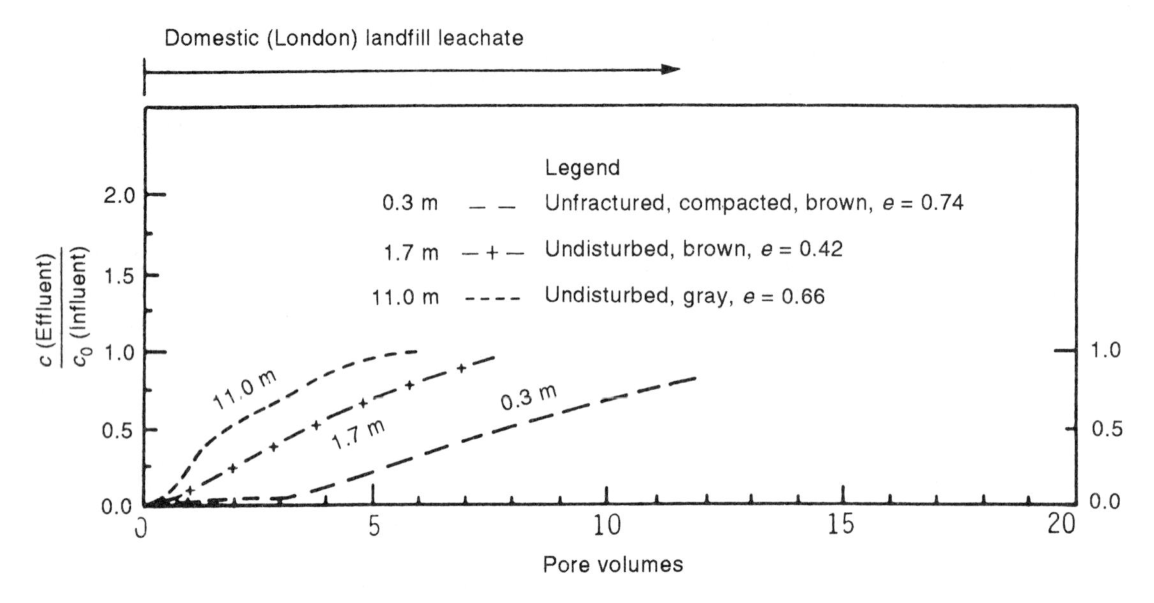

Figure 4.6 Potassium retardation as a function of sample depth (or smectite content), undisturbed or unfractured samples.

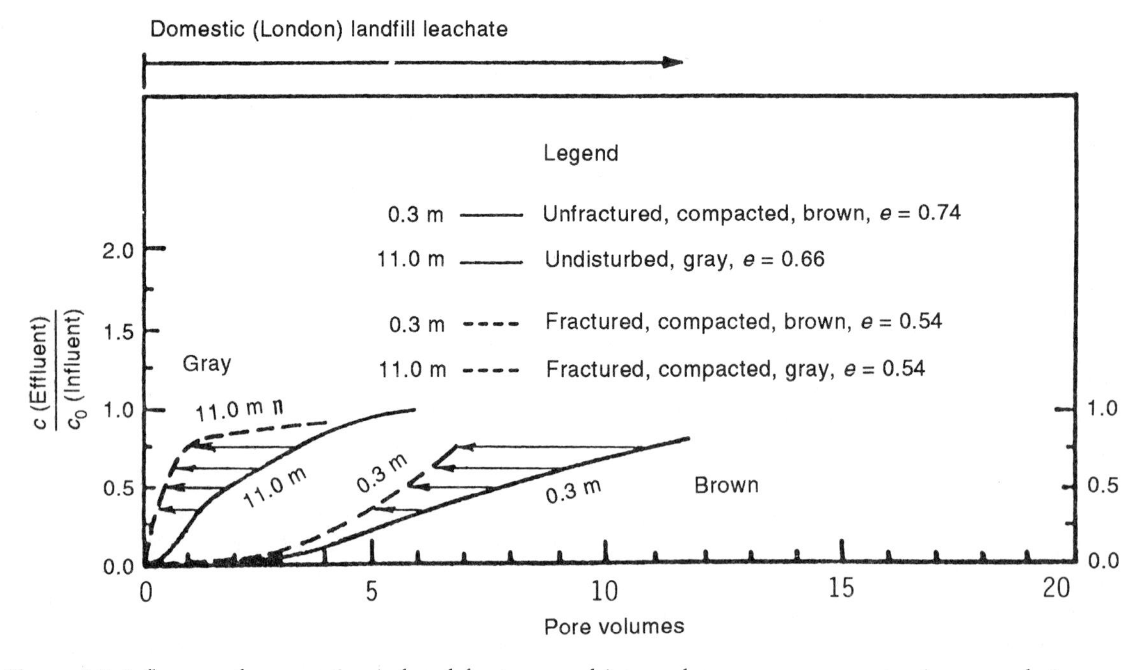

Figure 4.7 Influence of compaction-induced fractures and interped macropores on potassium retardation curves (comp: kneading compaction).

1. A pore water squeeze and chemical analysis is required to ensure constant pore fluid composition at the end of testing equal to the effluent, and to compare with pre-testing conditions.
2. For thick specimens, pore water compositions at the top, middle and bottom might be required to demonstrate identical composition throughout.
3. The adsorbed cation regime and CEC should be assessed before and after testing. This is particularly important if significant K^+ retardation has been observed and *c*-axis contraction of vermiculite is suspected. A drop in CEC would certainly signal the possibility of increased *k*-values unless consolidation has occurred.
4. X-ray diffraction traces for the test soil before and after permeation should be compared, again looking for *c*-axis changes.
5. A very careful inspection should be made of shrinkage cracks which might have developed near the top of the samples, complete with a photographic record.

For the Sarnia soil example described to this point, a series of x-ray traces were obtained, as illustrated in Figure 4.8. Considering the gray unweathered clay first (bottom traces on (a) and (b)), it can be seen that the clay minerals are dominated by illite and iron chlorite. KCl treatment and air drying both have relatively little effect on the x-ray traces, indicating an absence of smectite–vermiculite, and a soil probably relatively immune to changes caused by MSW leachate, which the *k*-tests indicated to be the case.

The weathered brown surface clay in the air-dry state (dashed trace, Figure 4.8(b)) shows significant collapse of smectite due to 0.5 N KCl treatment if air-dried (solid trace). The 1.4 nm peak is greatly reduced and the 1.0 nm peak greatly strengthened. In the water-wet state (Figure 4.8(a)), however, little collapse occurs, except in the interlayered illite–smectite phase indicated by the high background between 1.0 and 1.4 nm that develops after KCl treatment.

As noted in the figure, water-wet and leachate-wet samples may have to be x-rayed to define properly the behavior of a clayey barrier which will normally remain wet in the field. For example, in Ontario, Canada, many of the apparent vermiculites in the brown weathered surface soils do not collapse on exposure to MSW leachate, but stay expanded and perform satisfactorily as compacted landfill barriers.

In other cases, higher charge vermiculites may well collapse in the wet state, possibly causing shrinkage cracks and loss of barrier integrity.

4.2.3 Clay–leachate compatibility testing costs

Hydraulic conductivity testing performed as an assessment tool in clay–leachate compatibility studies (when performed with appropriate chemical controls) is expensive. Once experience has been gained with local soils, it is probable that a mineralogical report including an x-ray assessment of the effects of K^+ saturation by both KCl and MSW leachate will suffice. A few hydraulic conductivity tests using leachate as permeant are always a useful confirmation of compatibility.

4.2.4 Weathered Leda clay

The high moisture content, and often sensitive, Champlain Sea clays of eastern Canada are generally not suitable for compaction, unless they have been desiccated, as is commonly the case in stiff near-surface crusts. The desiccated soils are also weathered brown, and original soil chlorites may be weathered to vermiculite. A recent study of such a weathered Leda clay from Ottawa, Canada, found that this soil was susceptible to K^+ fixation when contacted by MSW leachate. After permeation with nine pore volumes of MSW leachate, the x-ray traces on

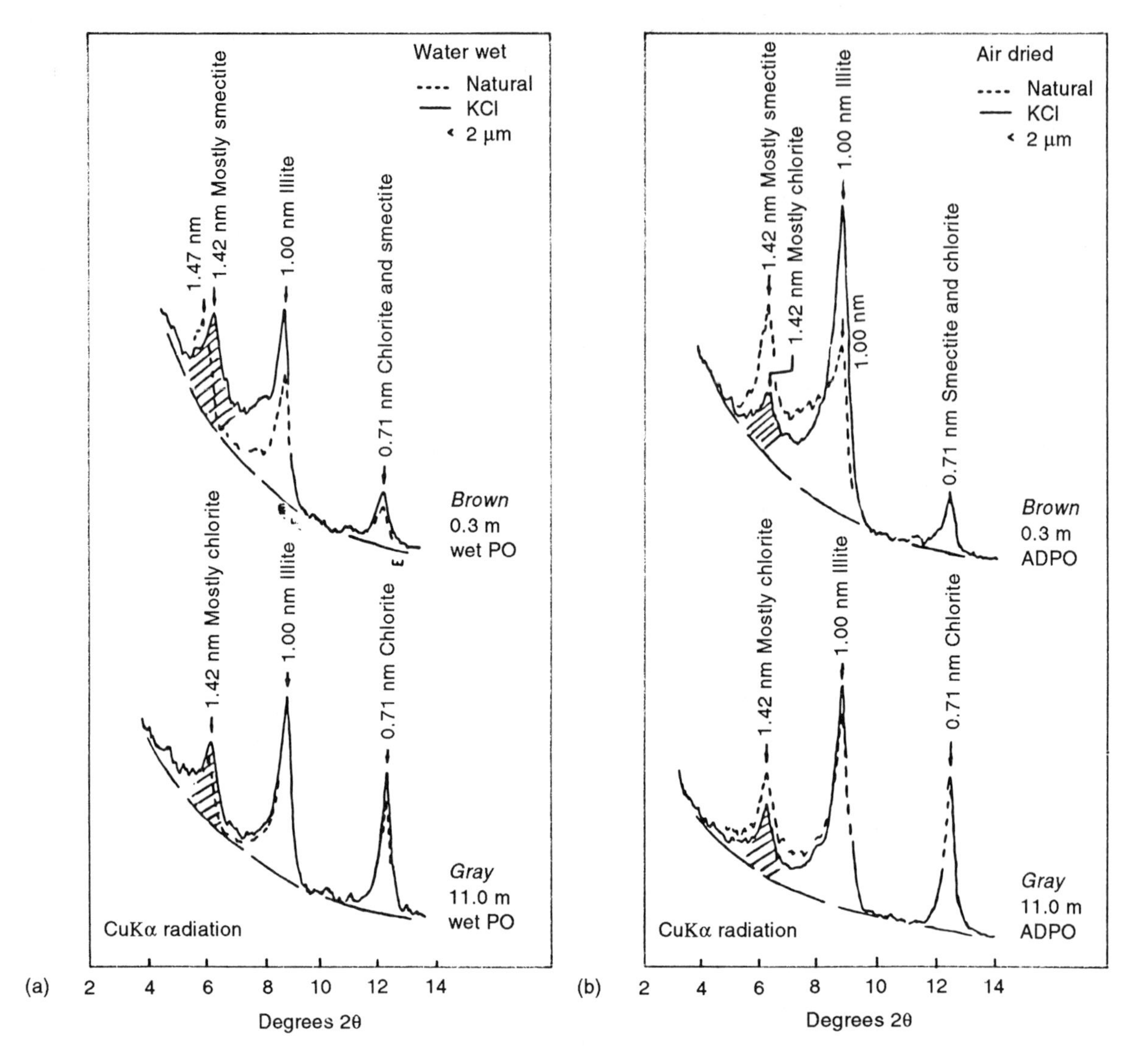

Figure 4.8 X-ray diffraction traces on oriented, <2 μm fines of brown and gray Sarnia clays, natural and KCl-treated. (a) Water-wet and leachate-wet x-ray traces did not demonstrate *c*-axis contraction, explaining the absence of increases in *k* on permeation with leachate in spite of extensive K^+ retardation. (b) Standard air-drying procedures strongly suggest vermiculite, which should incur *c*-axis contraction, reduction in CEC and **increases** in *k* on exposure to leachate.

pressure-oriented post-test samples altered, as illustrated in Figure 4.9. The natural clay contained abundant vermiculite–smectite and interlayered illite–vermiculite that produced a CEC value of 32 meq/100 g. The traces for the wet, post-testing sample show complete collapse of the vermiculite to illite and a drop in CEC to 17 meq/100 g.

The sample was compacted for hydraulic conductivity testing at its natural water content of 34.5%, and subjected to a static effective stress of 40 kPa. During leachate permeation,

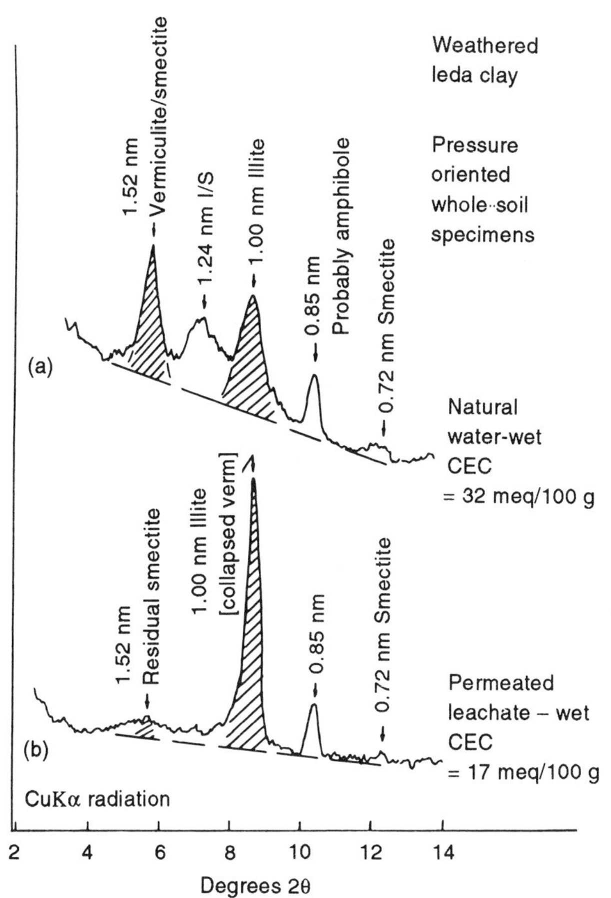

Figure 4.9 X-ray traces obtained on pressure oriented bulk Leda clay: (a) natural water-wet; (b) permeated with 9 PV of MSW leachate.

the soil creep-consolidated throughout the test period, as illustrated in Figure 4.10(b). One might expect a steady decrease in hydraulic conductivity as the void ratio decreased, but this did not happen (Figure 4.10(a)). Potassium fixation and double-layer contraction appear to have compensated for the void ratio decrease.

This example shows that if compaction is carried out on a soil that is wet enough (for example, 2% wet of Standard Proctor optimum for the clays discussed here), consolidation in the presence of effective vertical stresses may compensate for deleterious chemical and mineralogical alterations.

4.3 Compatibility of clays with liquid hydrocarbon permeants

The following discussion of hydrocarbon liquids is presented under six major headings in an attempt to organize and rationalize a very complex physico-chemical problem. Although much has been written, this section will dwell primarily on work done at the University of Western Ontario. Selected references to the work of others are noted in the text.

4.3.1 Dry clay–organic mixtures

The only way to obtain complete clay mineral interaction with many liquid hydrocarbons is to mix dry clay with the test liquid under consideration. In Figure 4.11, a plot of hydraulic conductivity for smectitic Sarnia soil versus void ratio is

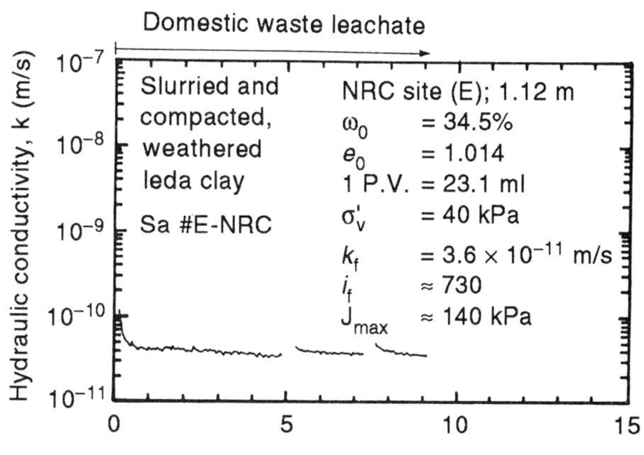

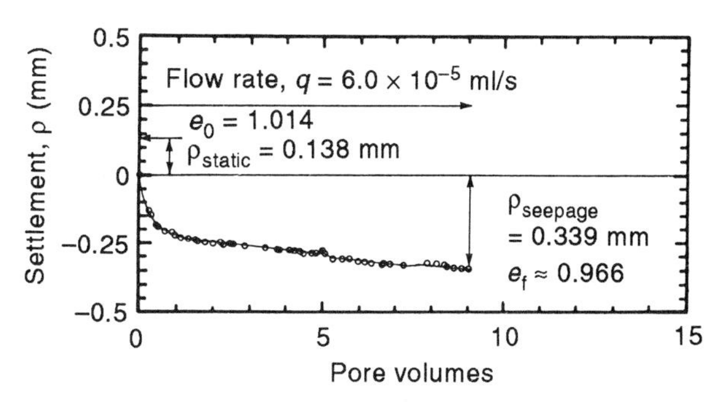

Figure 4.10 Hydraulic conductivity and settlement versus pore volumes; NRC Site (E), 1.12 m depth, slurried and compacted. Permeant: domestic waste leachate.

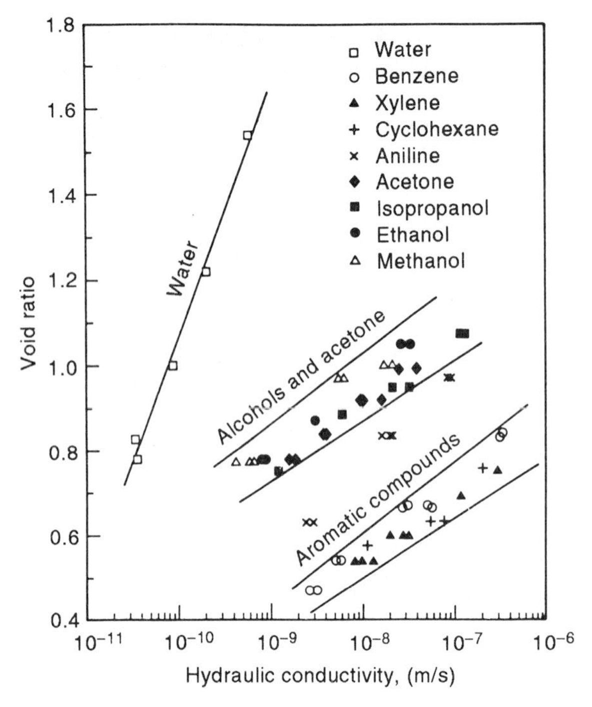

Figure 4.11 Directly measured hydraulic conductivity versus void ratio, all samples molded and permeated with the fluid indicated. (Modified from Fernandez and Quigley, 1985.)

presented for a variety of liquid organics having dielectric constants varying from ~80 (polar water) to ~25 (ethanol) to ~2 (aromatics). At a void ratio of 0.8, for example, an increase in hydraulic conductivity from 3×10^{-11} m/s for water ($\varepsilon = 80$) to 1×10^{-6} m/s for simple aromatics ($\varepsilon = 2$) is observed.

These trends are well-known and were especially well-described by Mesri and Olson (1971) on dry clay–organic mixtures subjected to oedometer testing (Figure 4.12). Smectite (CEC = 100) is particularly susceptible to organics, demonstrating up to million-fold increases in hydraulic conductivity as the dielectric constant of the molding fluid decreases from 80 to 2. Illite (CEC = 25) is much less susceptible, demonstrating increases in hydraulic conductivity of up to 1000 times. Finally, kaolin (CEC = 5) seems relatively inert to large changes. Similar work dates back to the 1950s (e.g. Michaels and Lin, 1954). Since dry clays flocculate in the presence of liquid organics, due to their low ε values (Figure 3.30(c)), soil fabric or structure is directly responsible for the range in observed hydraulic conductivities. To illustrate this, scanning electron photomicrographs taken of vertical fracture surfaces after freeze-drying are presented in Figure 4.13. Much larger interfloc pores occur in the benzene-molded clay (Figure 4.13(d)) compared with those observed for ethanol-molded or water-molded clays (Figures 4.13(a)–(c)). When visually inspected, the benzene-flocculated clay actually looks like a loose assortment of sand particles.

A further correlation with dielectric constant is presented in Figure 4.14 (Acar and Seals, 1984). Two plots of liquid limit versus dielectric constant of the molding fluid are presented for

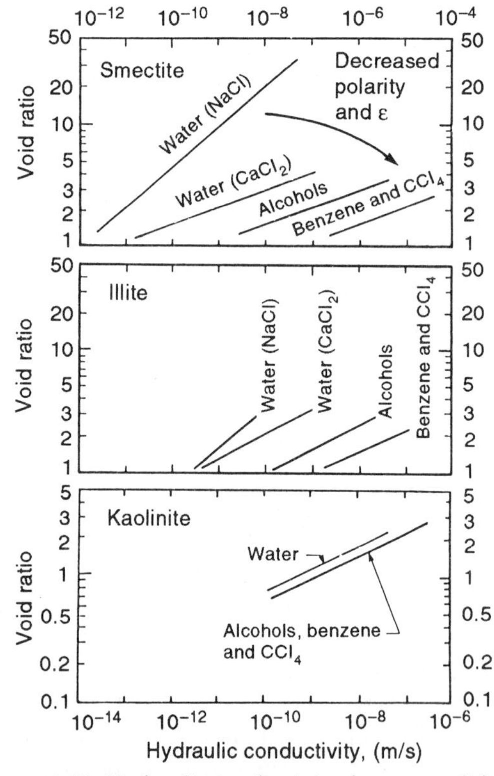

Figure 4.12 Hydraulic conductivity from consolidation tests versus void ratio for smectite, illite and kaolinite. (Adapted from Mesri and Olson, 1971.)

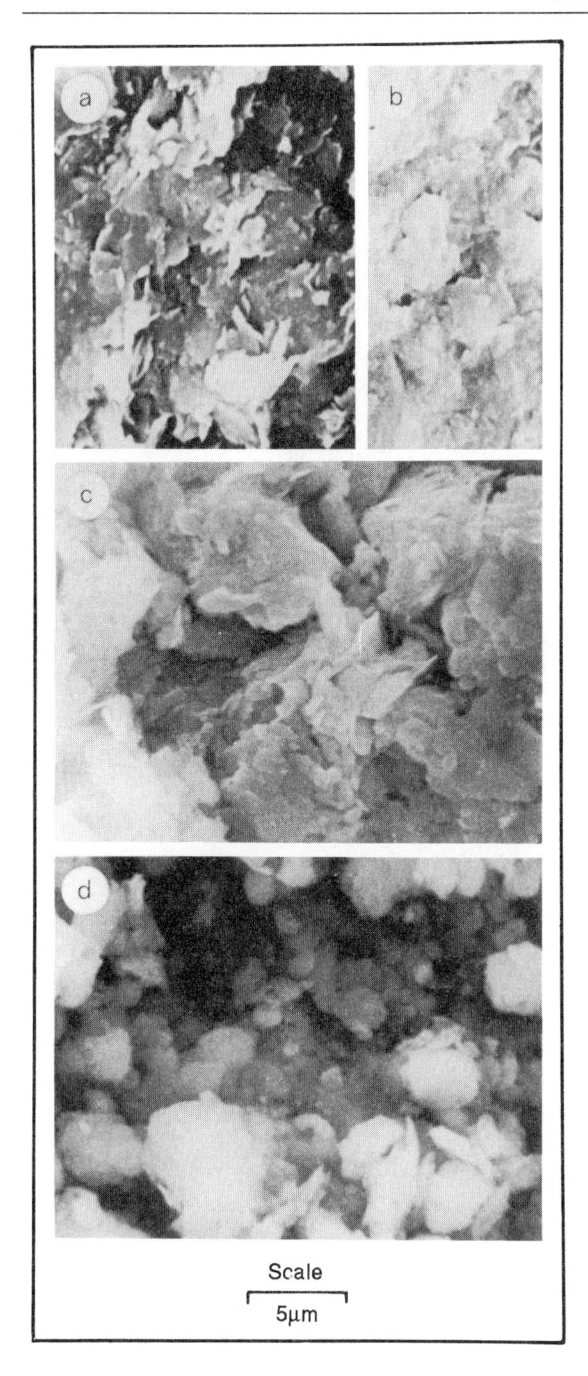

Figure 4.13 Scanning electron photomicrographs of vertical surface test samples: (a) fractured, water-molded clay; (b) cut and smeared water-molded; (c) alcohol-molded; (d) benzene-molded. (Fernandez and Quigley, 1985; reproduced with permission of the *Canadian Geotechnical Journal*.)

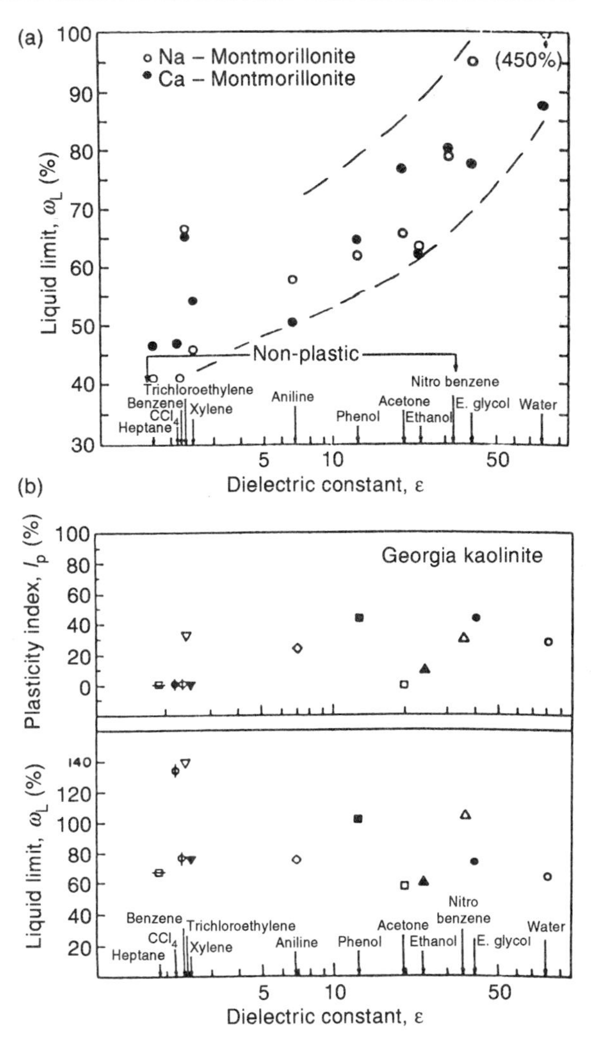

Figure 4.14 Liquid limit versus dielectric constant for a variety of molding fluids: (a) smectite; (b) kaolinite. (Acar and Seals, 1984; reproduced with permission, *Hazardous Waste*.)

smectite and kaolinite. Again it appears that smectites are markedly influenced, whereas the kaolins are fairly inert.

4.3.2 Compacted, water-wet clays and insoluble organics

Flow through clayey soils is generally governed by Darcy's law, which can be written as

$$q = kiA \tag{4.1}$$

which relates the flow, q [L^3T^{-1}], through a cross-sectional area of soil, A [L^2], to the hydraulic gradient, i [–] and hydraulic conductivity, k [LT^{-1}]. A plot of q versus i for water-saturated clays yields a series of straight lines having a slope proportional to the hydraulic conductivity k, as illustrated in Figure 4.15. Inactive kaolins and relatively inactive illites and chlorites compacted dry of optimum with compaction-induced fractures have relatively high k-values, and thus yield a high q at low gradients. Illites and chlorites compacted wet of optimum without the fractures have a much lower k-value, and thus yield low q and high gradients. High quality Na^+ bentonites have such low k-values that low flows are generated even at extraordinary gradients. This is one factor that makes their study very difficult.

The most important aspect of Figure 4.15 with respect to the current discussion is the linearity of the q versus i lines, provided no changes in chemistry, void ratio etc. occur during testing of the water-saturated clays.

If an insoluble lyophobic organic liquid (i.e. a nonaqueous-phase liquid, NAPL) is placed on top of a water-wet clay, the interparticle surface

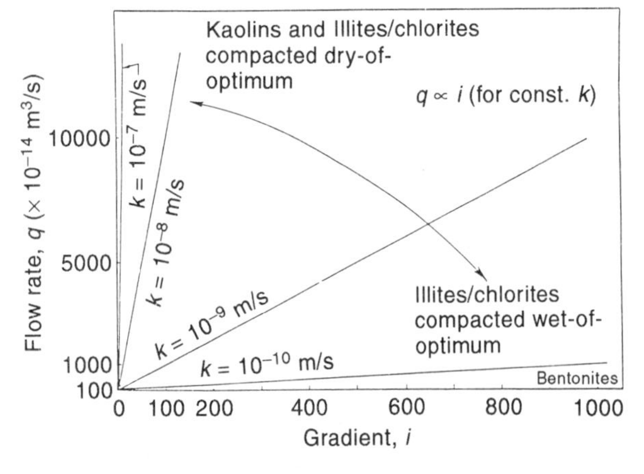

Figure 4.15 Darcy's law plotted as q versus i lines with the slope of each line representing a different k-value (unit cross-sectional area A assumed).

tension of the highly structured water creates an extremely strong membrane which resists penetration and flow. For this reason, DNAPLs perch on top of clayey layers in field situations.

If constant flow rate hydraulic conductivity equipment is used to study NAPL flow through water-wet clays, an abrupt experimental nonlinearity of the q versus i plots is produced, and this can be used to infer breakthrough pressures and gradients.

To start this discussion, Figures 4.16(a) and (b) are presented to illustrate the resistance to lyophobic cyclohexane flow caused by the surface tension of water at the face of an unfractured, water-compacted clay. At a constant head of 0.26 m above a 20 mm thick clay specimen, total penetration amounted to 2.4 ml, increasing in two steps to 3.7 ml at 3.46 m head without breakthrough for this soil at a void ratio of 0.913 (Figure 4.16(a)). Although the initial static effective stress on the clay specimens was zero, the nonpenetrating liquid head would operate like an effective stress in this case.

The sketch in Figure 4.16(b) illustrates schematically this increasing penetration which has almost reached breakthrough. At this point, the penetration amounts to ~17% of the total pore volume of 21.9 ml in this compacted fine-grained sample. The presence of flaws such as cracks and fractures in a soil would probably enable much lower breakthrough values.

Because of the lyophobic behavior of cyclohexane and water, the constant flow rate hydraulic conductivity test system can also be used to measure approximate values of breakthrough pressure or gradient. This is illustrated in Figure 4.17 which combines both constant head and constant flow tests to illustrate the breakthrough point. Breakthrough is inferred to have occurred at a gradient of ~175 corresponding to a head of ~3.5 m across the 20 mm thick test samples. A corresponding linear curve for water is also shown for reference.

Once breakthrough occurs, the insoluble cyclohexane is forced along macropore channels

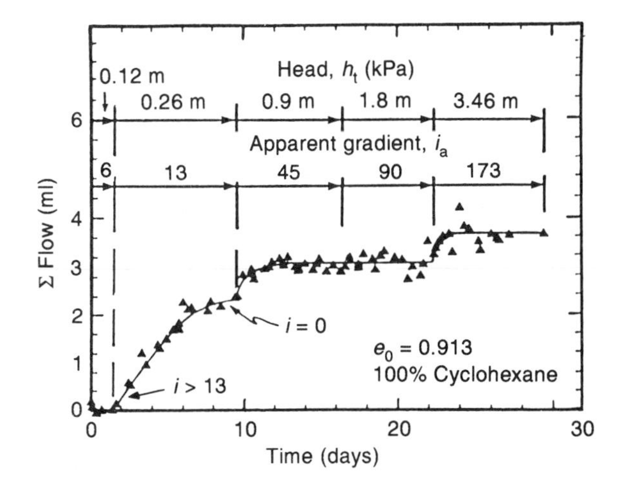

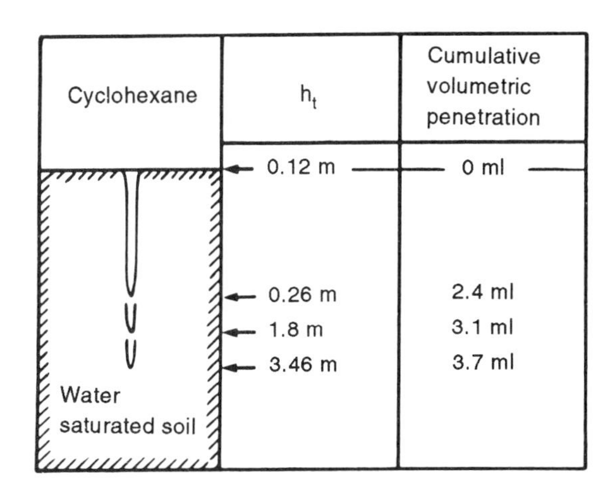

Figure 4.16 (a) Constant head tests for insoluble cyclohexane penetration into water-compacted clay (all data before breakthrough). (b) Schematic illustrating volumetric penetration of cyclohexane without breakthrough at three values of constant total head (one pore volume = 21.9 ml).

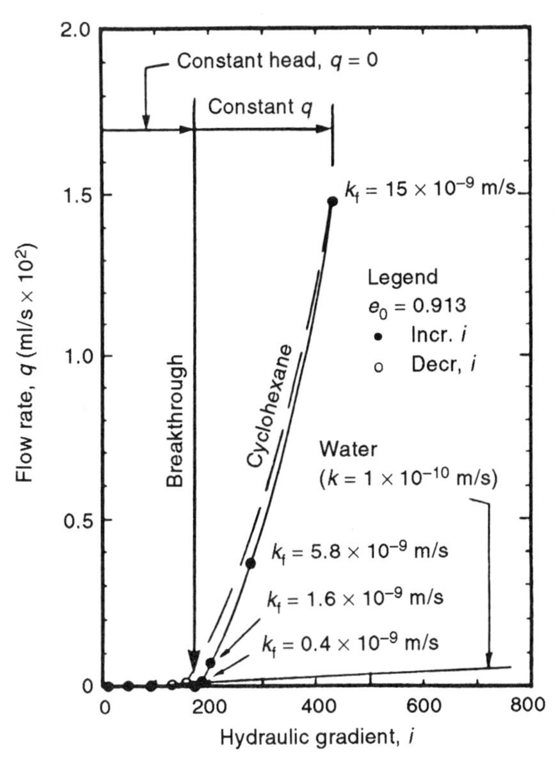

Figure 4.17 Flow rate versus gradient for insoluble cyclohexane flow through water-compacted brown silty clay. (Note that five constant head and four constant flow rate values are plotted.)

or microfissures which expand and contract (in this relatively loose sample) in proportion to the head. This causes the measured hydraulic conductivity to be gradient-dependent, as shown by the curved plot of flow rate q versus hydraulic gradient i. A corresponding linear curve of q versus i for water is also shown for reference.

Provided the soil has no fractures, the breakthrough head or gradient should increase with decreasing void ratio. This is demonstrated in Figure 4.18 for cyclohexane at two soil void ratios of 0.707 and 0.913. The breakthrough gradients were ~175, ~410 and ~920, for void ratios of 0.913, 0.707 and 0.58 respectively.

A three-point summary plot of breakthrough gradient versus void ratio is presented in Figure 4.19. The breakthrough gradient (based on constant flow rate tests) increases with decreasing void ratio on a hyperbolic curve that plots as a straight line on a semi-log plot. Since the capillary strength of a water film increases with decreasing pore size, it is inferred that this is a major factor in defining breakthrough heads or gradients in the case of lyophobic liquids.

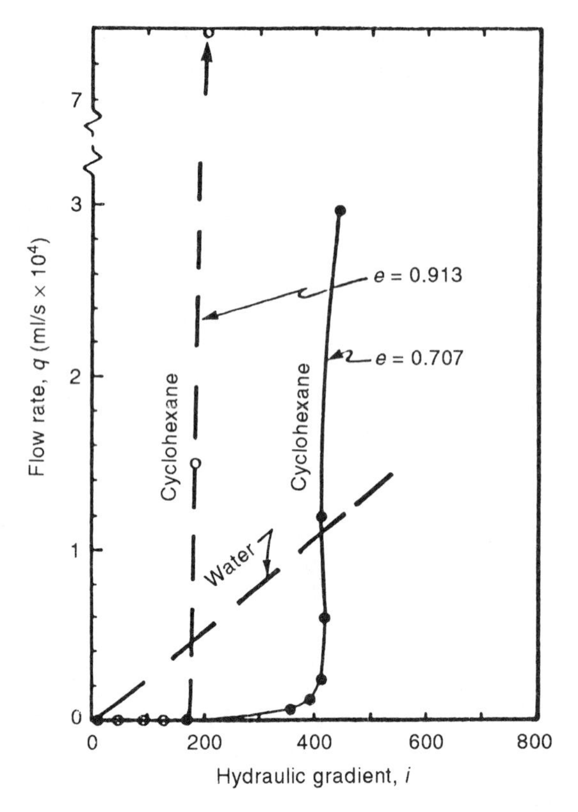

Figure 4.18 Flow rate versus gradient for insoluble cyclohexane flow through water-compacted silty clay. (Note increase in breakthrough gradient at lower void ratio.)

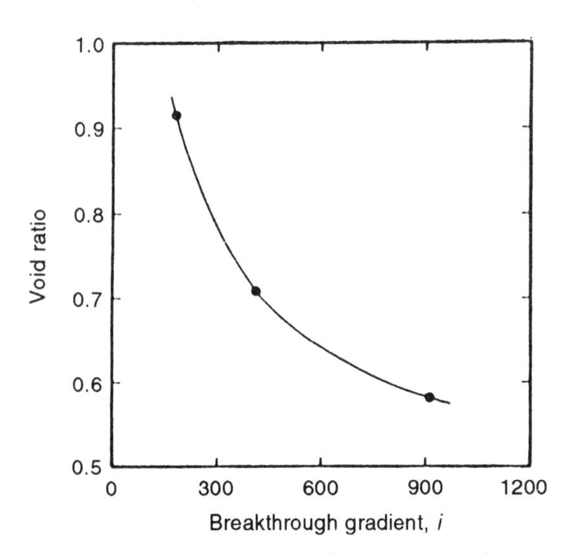

Figure 4.19 Breakthrough gradient versus void ratio.

Finally, it is noted that cyclohexane and other insoluble liquid hydrocarbons rarely occupy more than 6–10% of the soil pore space at the end of testing, a factor indicating macropore or channel flow. Also, previous studies and reviews by Fernandez and Quigley (1985), Mitchell and Madsen (1987), Foreman and Daniel (1986) and Acar *et al.* (1985) all indicate that forced permeation of insoluble, lyophobic organics does not increase hydraulic conductivity beyond that of water. This is illustrated in Figure 4.20, where almost 2 PV of cyclohexane flow left only 7% cyclohexane in the pore fluid and yielded the same hydraulic conductivity as did water.

On Figure 4.17, however, it was clearly shown that the constant flow rate system used measures an apparent hydraulic conductivity which may be highly gradient- or head-dependent. If enough different flow rates are

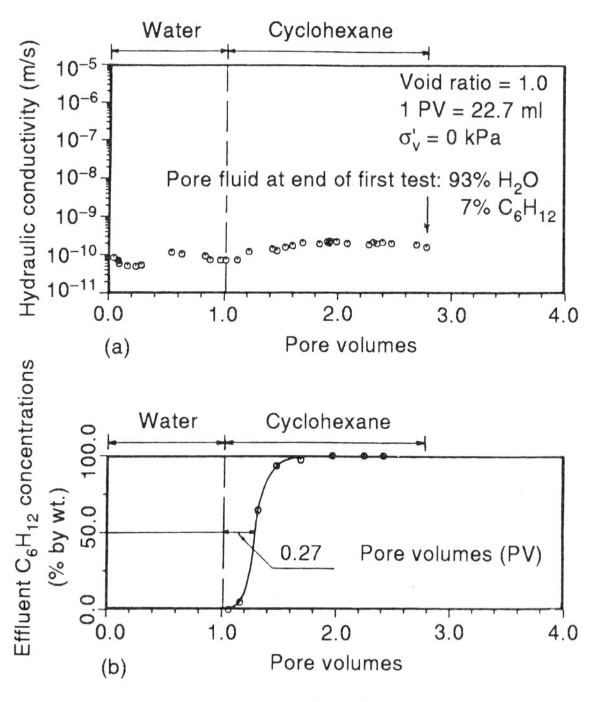

Figure 4.20 Permeation of cyclohexane through water-saturated clayey soil: (a) hydraulic conductivity; (b) relative concentration versus pore volumes.

employed to establish gradients both below and above the breakthrough gradient, the resulting curves may be extrapolated back to a zero flow condition, thereby identifying a gradient which is believed to be close to the breakthrough gradient measured by constant head testing.

Quigley and Fernandez (1992) have shown that an increasing static effective stress on a soil also caused a large increase in breakthrough gradient, possibly related to the greater difficulty in expanding the flow channels associated with fingering once the water meniscus is penetrated.

The principles just described have been known for a long time, and were effectively used by Lambe (1956) to store oil in a water-compacted clayey reservoir in Venezuela.

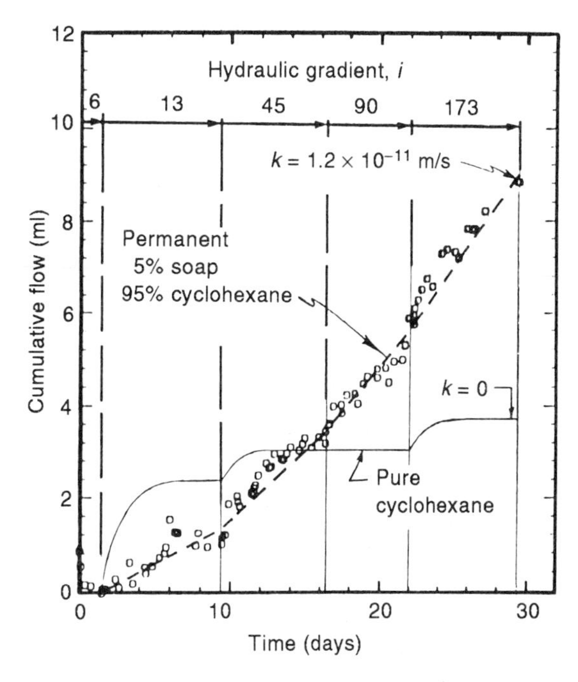

Figure 4.21 Constant head tests for insoluble cyclohexane penetration into water-compacted clay in the presence of 5% surfactant (soap). (Quigley and Fernandez, 1989; reproduced with permission of Balkema.)

4.3.3 The role of surfactants

A surfactant is a surface-active agent which reduces the interfacial surface tension of liquids, and should greatly reduce the strength of the water menisci of a water-compacted clay in contact with insoluble liquid hydrocarbons.

The presence of 5% surfactants (such as soap) in the cyclohexane permeant discussed in the previous section has the important effect illustrated in Figure 4.21. The constant head testing conditions are exactly the same as those used to produce Figure 4.16(a) for pure cyclohexane (replotted in Figure 4.21 as a solid curve). Surface tension in water at the cyclohexane–soil interface was apparently completely destroyed, and there appears to have been little resistance to cyclohexane penetration and flow, even at low gradients. Also, the cumulative flow versus time plot is only slightly nonlinear.

Comparison of the cumulative flow at 29 days at a gradient $i = 173$ shows nonbreakthrough penetration of 3.7 ml for pure cyclohexane compared with a total flow volume of ~9 ml for the 5% soap–95% cyclohexane mixture. It is important to note that a hydraulic conductivity of zero ($k = 0$) was obtained for the pure cyclohexane system, since there was no continuing flow, whereas a hydraulic conductivity, $k = 1.2 \times 10^{-11}$ m/s, was obtained for the cyclohexane–surfactant system in which there was slow continuous flow through the sample. Again, since many types of surfactants are probably present in domestic and industrial leachates, it is suggested that one should not rely on the high breakthrough pressures of a good clay barrier to retain insoluble hydrocarbon liquids.

4.3.4 Permeation with water-soluble ethanol and dioxane mixed with domestic leachate

Mixing water-soluble liquid hydrocarbons with water or municipal solid waste (MSW) leachate produces some important changes in their

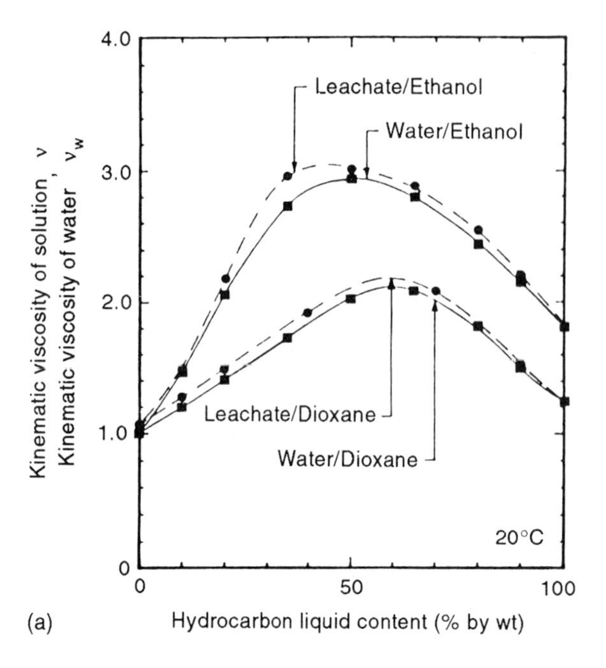

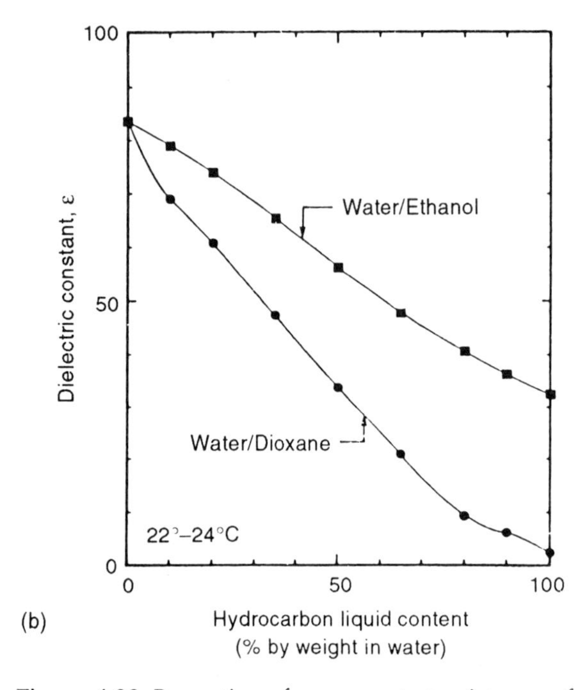

Figure 4.22 Properties of aqueous test mixtures of water-soluble liquid hydrocarbons: (a) relative kinematic viscosity; (b) dielectric constant. (Adapted from Fernandez and Quigley, 1988a.)

physical and electrochemical properties, as shown in Figure 4.22. Both ethanol and dioxane, when mixed with water, produce two- to three-fold increases in kinematic viscosity at concentrations of 50 to 60% (Figure 4.22(a)). The same two organics cause the dielectric constant of the mixture to decrease as their percentage increases (Figure 4.22(b)).

The effect of these two factors on the hydraulic conductivity k of a water-compacted clay is shown in Figure 4.23(a) for ethanol–leachate mixtures used as permeant. At all concentrations up to ~70% ethanol in MSW leachate, the hydraulic conductivity decreases in a manner which mirrors the increases in viscosity shown in Figure 4.22(a). Above 70%, ethanol apparently enters the double layers around the clay particles in sufficient quantity to cause double-layer contraction in accordance with Gouy–Chapman theory (Figure 3.30(c)). Since the tests shown in Figure 4.23(a) are for zero static stresses (only seepage stresses influence the sample) large increases in free pore space develop at a constant void ratio (Figure 3.28) resulting in 100-fold increases in hydraulic conductivity k. This work was discussed in considerable detail by Fernandez and Quigley (1988a).

The intrinsic permeability, K [L^2], is defined by

$$K = k\mu/\gamma \qquad (4.2)$$

where k is the hydraulic conductivity [LT^{-1}]; μ is the dynamic viscosity [$ML^{-1}T^{-1}$]; γ is the unit weight of permeant [$ML^{-2}T^{-2}$].

Figure 4.23(b) shows a plot of intrinsic permeability versus the percentage of ethanol. By considering the intrinsic permeability, one removes the complicating effects of viscosity and density. This plot shows nearly constant K at ~6×10^{-18} m^2 up to about 50% ethanol at which point K increases by about 100-fold for 100% ethanol. This suggests that ethanol does not significantly access the double layers at concentrations below 50–60%. This is to be

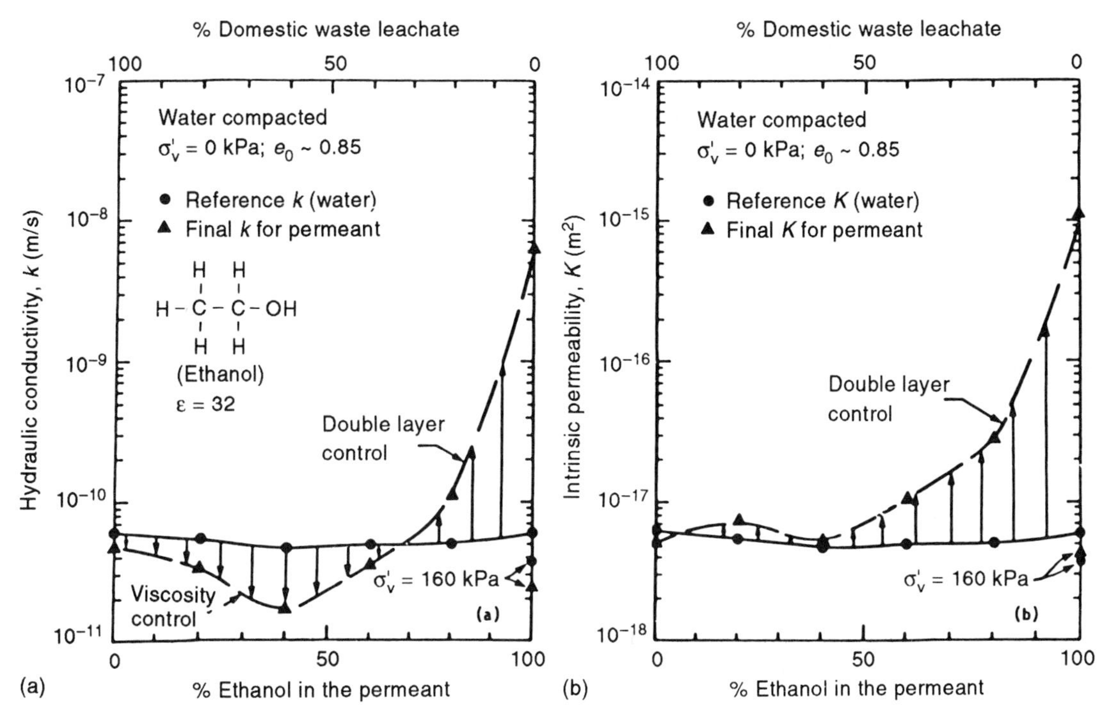

Figure 4.23 (a) Hydraulic conductivity and (b) intrinsic permeability of water-compacted brown Sarnia clay permeated with leachate–ethanol mixtures. (Modified from Fernandez and Quigley, 1988a.)

expected since the double-layer cations are hydrated with more-polar water ($\varepsilon = 80$) and should reject the less-polar ethanol ($\varepsilon = 32$).

Similar plots are presented in Figure 4.24 for water-soluble dioxane which has a dielectric constant ε of only 2.3. Dioxane is nonpolar, yet completely soluble in water, owing to the presence of two oxygens replacing carbon in the benzene ring (Figure 4.24(a)). The oxygen apparently associates readily with the H–O system of polar water. The plots for k and K are very similar to those for ethanol. Decreases in k for less than 75% dioxane relate to the viscosity increases and exclusion of dioxane from the double layers. Once the concentration reaches 75%, however, high increases are observed in both k and K as the double layers contract.

The data shown in Figures 4.23 and 4.24 were obtained on samples with a static applied vertical effective stress of zero ($\sigma'_v \approx 0$ kPa). Two important data points for a static effective stress of 160 kPa applied to the samples prior to pure hydrocarbon permeation, are also shown. These two points indicate a decrease in hydraulic conductivity k for ethanol and only a slight increase for dioxane. Chemically-induced consolidation in the presence of stresses was shown to inhibit greatly the large increases in k observed at low stress levels (Fernandez and Quigley, 1988a, b).

By way of summary, the intrinsic permeability ratio K_f/K_w is plotted in Figure 4.25 for the water-compacted brown smectitic Sarnia clay permeated with both ethanol–leachate and dioxane–leachate mixtures. Since the effects of viscosity and density have been removed, the values of K_f/K_w reflect the effects of dielectric constant ε on the double layers, and hence on both intrinsic permeability K and hydraulic conductivity k.

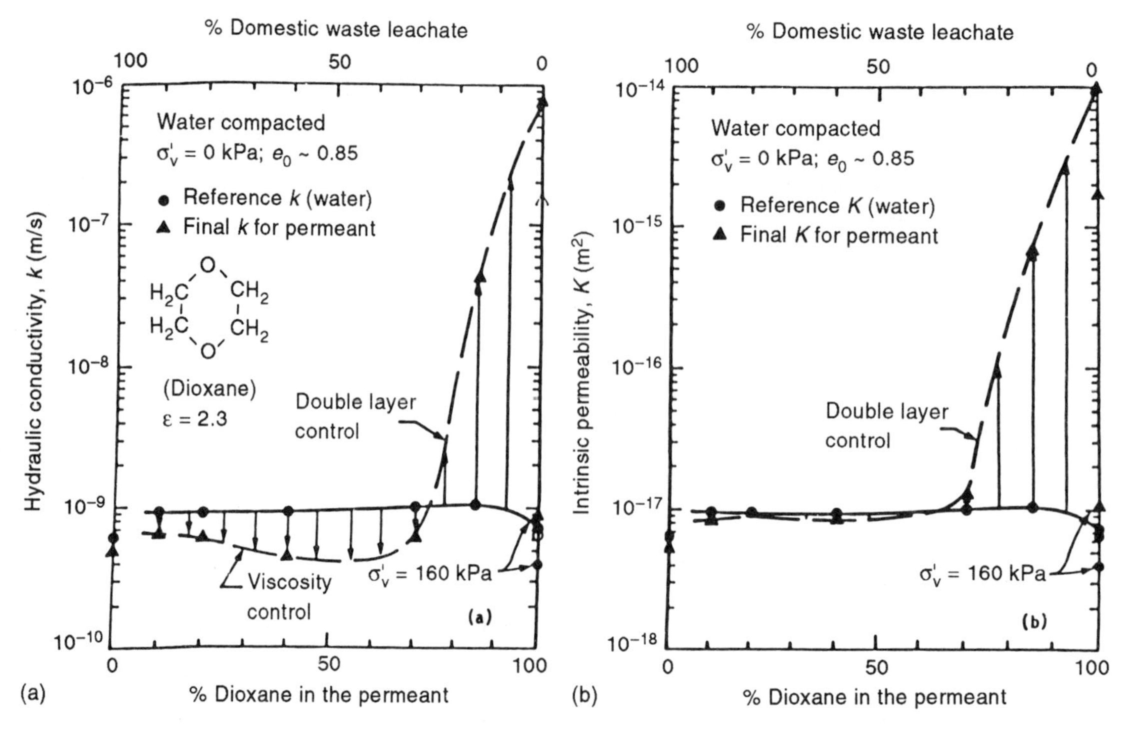

Figure 4.24 (a) Hydraulic conductivity and (b) intrinsic permeability of water-compacted brown Sarnia clay permeated with leachate–dioxane mixtures. (Modified from Fernandez and Quigley, 1988a.)

For dioxane concentrations greater than 70%, intrinsic permeability *K* increases, rapidly reaching values 1000 times greater than water for pure dioxane. Remarkably, no effects on *K* are observed below 70%. Although dioxane is completely soluble in water, because of oxygen replacing two carbons in the benzene ring, it is also nonpolar. This means that cations present in the double layers around the clay particles must retain their affinity for polar water and reject dioxane, preventing double-layer collapse until concentrations in the range of 70% are reached.

In the case of ethanol, which is more polar than dioxane, the trends are somewhat different. Small increases in *K* are apparent at concentrations possibly as low as 25%, suggesting partial entry into the double layers.

Chemical confirmation of these hypotheses is quite difficult, since the double layers occupy only a small portion of the pore space in these high void ratio soils. Also, salt dumping (precipitation) by addition of these organics may also create large increases in salinity or free salt crystals.

4.3.5 The role of effective stress

One of the most significant conclusions of the review of hydraulic conductivity by Mitchell and Madsen (1987) was: 'Permeation with pure hydrocarbon leads to a decrease in hydraulic conductivity when flexible wall permeameters are used.' The reason for this is that the triaxial cell pressures consolidate the test specimens if any chemical reactions occur, effectively masking any increases in hydraulic conductivity *k*. For this reason, the above authors recommended the use of 'consolidometer permeameters'.

In Figures 4.23, 4.24 and 4.25, data points

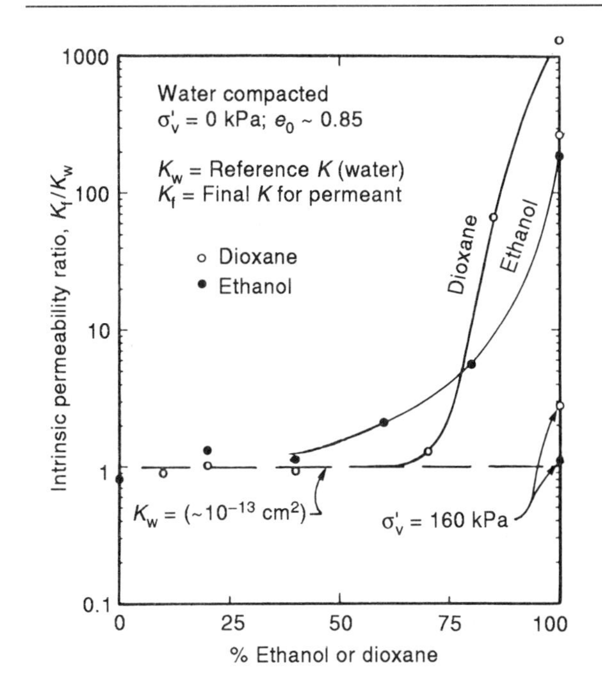

Figure 4.25 Ratio of final to initial values of intrinsic permeability, K, for water-compacted clay permeated with mixtures of ethanol–leachate and dioxane–leachate. (Quigley and Fernandez, 1989; reproduced with permission of Balkema.)

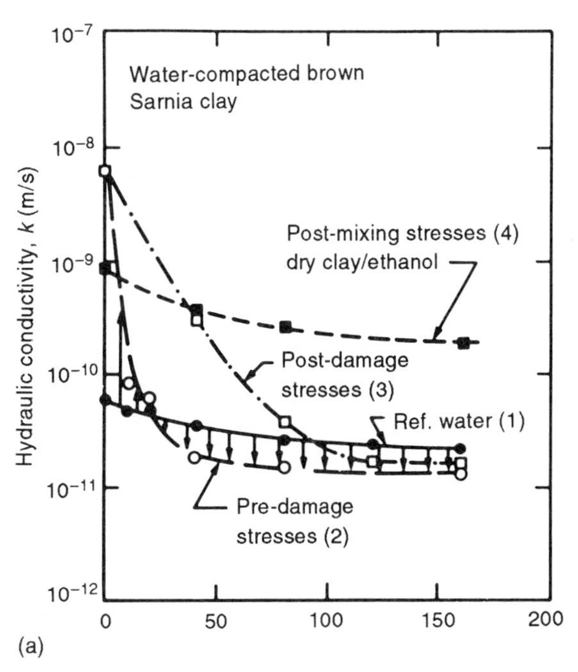

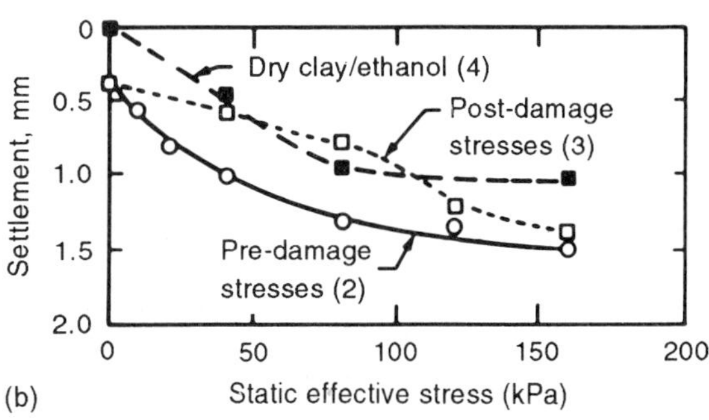

Figure 4.26 Hydraulic conductivity and settlement for pure ethanol permeation: (a) k versus stress level for four stress scenarios described in the text; (b) settlement versus stress level. (Modified from Fernandez and Quigley, 1991.)

were plotted for hydraulic conductivity tests run with pure ethanol and dioxane at an applied effective stress $\sigma'_v = 160$ kPa to compare with those run at $\sigma'_v = 0$ kPa. In each plot, the presence of the stresses eliminated the increases in k-value for ethanol and reduced them to small increases for dioxane.

The effective stress work of Fernandez and Quigley (1991) is presented in Figures 4.26 and 4.27 to illustrate the importance of stress level during k-testing. For ethanol (Figure 4.26(a)), each water-compacted specimen was consolidated to the appropriate stress level and the hydraulic conductivity k was measured for water (solid circles on curve (1)). Pure ethanol was then passed through each sample, and k measured at equilibrium (open circles on curve (2)). At $\sigma'_v = 0$, k increased by up to 100 times, as noted in Figure 4.23, whereas at $\sigma'_v = 20$ kPa there was no change and at all higher stresses up to 160 kPa, k actually decreased.

Curve (3) was obtained from tests on water-wet soil damaged by pure ethanol permeation at $\sigma'_v = 0$ kPa (open squares). Stress was then applied in an attempt to heal the damaged barrier. The figure shows that up to 100 kPa of static vertical stress is required, but that healing

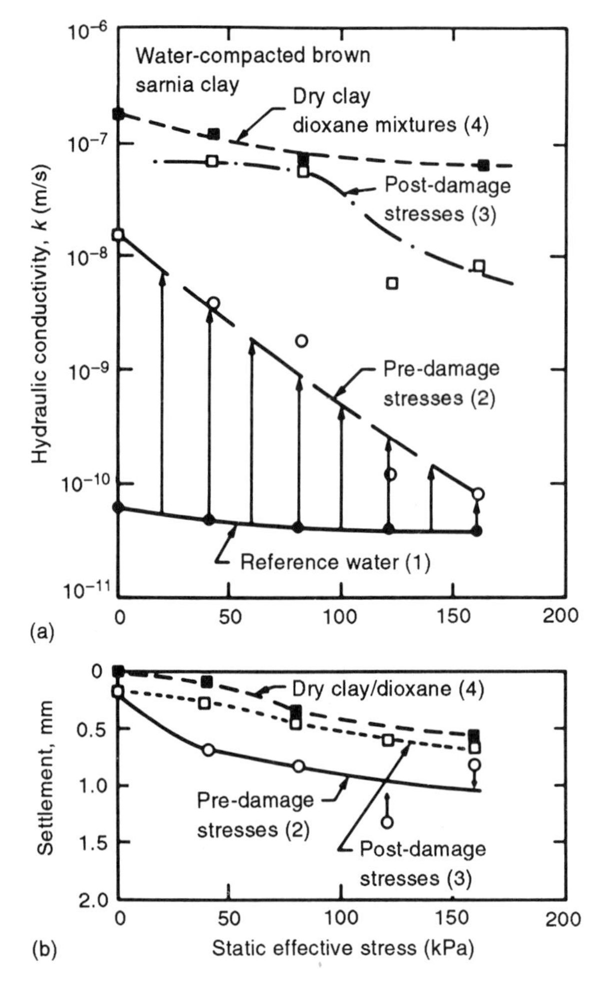

Figure 4.27 Hydraulic conductivity and settlement for pure dioxane permeation: (a) k versus stress level for four stress scenarios described in the text; (b) settlement versus stress level. (Modified from Fernandez and Quigley, 1991.)

can be effected in the case of pure ethanol damage.

Curve (4) in Figure 4.26 (solid squares) represents soil molded and compacted with ethanol, then permeated with ethanol at the indicated stress level. These samples were flocculated, as discussed in section 4.3.1, and the strength and structure derived from this pre-treatment apparently resisted efficient soil consolidation and recovery of low k-values. In spite of the settlement shown in Figure 4.26(b), the flow channels apparently remained open.

For dioxane (Figure 4.27) the trends are quite different. Permeation by pure dioxane resulted in large increases in hydraulic conductivity k that seemed inversely proportional to the stress level on the water-compacted sample, and even at 160 kPa there was a small increase (open circles on curve (2)).

Curve (3) in Figure 4.27 represents post-damage application of stress, and shows only minor healing, even at 160 kPa. This was in spite of significant consolidation which apparently was not accompanied by flow channel closure.

Curve (4) in this figure represents dioxane-molded and compacted specimens which were then consolidated at the σ'_v shown and subsequently permeated with dioxane. Again the highly flocculated structure of this pre-treated soil enabled it to resist consolidation and closure of the flow channels, so the k-values remained very high, even at $\sigma'_v = 160$ kPa.

On the basis of this preliminary work, it is suggested that water-compacted clayey barriers which are heavily stressed before a spill of water-soluble organics may self-heal during exposure by consolidation. If the spill occurs on an unloaded or unstressed clay barrier, however, large increases in hydraulic conductivity k may occur with little chance of subsequent healing by stress application.

4.3.6 The role of association liquids

An association liquid (for the purposes of this book) is one that is mutually soluble in two immiscible lyophobic liquids, and which if added produces a single-phase liquid over a certain concentration range. In complex mixtures of waste containing hundreds of compounds, it is probable that such liquids and processes exist.

The effects of these complex mixtures on the

integrity of clay barriers has received little study. A single, complex plot is presented in Figure 4.28 to illustrate the probable effects. In this figure, ethanol is used as the association liquid and cyclohexane as the insoluble light non-aqueous phase liquid (LNAPL).

Considering zero effective stresses first: If pure cyclohexane is passed through the sample, the hydraulic conductivity is close to that of water but is gradient-dependent, as illustrated by the four solid circles on the 100% cyclohexane line. If mixtures of alcohol and cyclohexane are passed through, k increases by a factor of 10 to 100 depending on the ratio of cyclohexane to ethanol. In all cases k was greater than for water. This feature is quite the opposite to the trends for alcohol–MSW leachate mixtures shown in Figure 4.23 until ethanol concentrations exceed 60–70%.

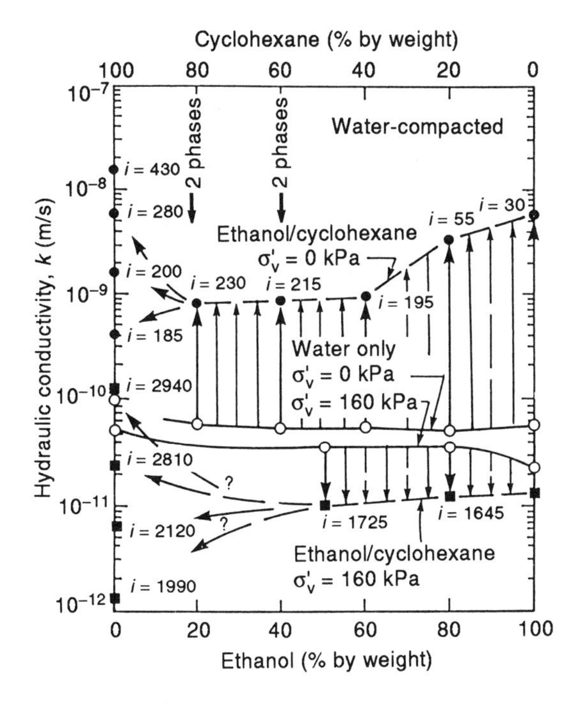

Figure 4.28 Influence of mixed organics on the hydraulic conductivity of water-compacted clay at σ'_v = 0 and σ'_v = 160 kPa. (Modified from Fernandez and Quigley, 1991.)

At a static vertical effective stress σ'_v = 160 kPa, all samples yielded k-values below those for water, except the pure cyclohexane at one high gradient of 2940 kPa. Even at the slowest flow rate, the low hydraulic conductivity of the clay ($\sim 10^{-11}$ m/s) gives rise to high values of gradient i from 1645 to 2810.

In summary, the curves in Figure 4.28 indicate that the application of an effective stress σ'_v = 160 kPa prevents the increases in hydraulic conductivity k observed at zero effective stress σ'_v = 0 kPa for cyclohexane–ethanol mixtures. The gradients indicated next to each point are generally highly variable and often very high. In addition, two of the mixtures split into two phases, making interpretation difficult. Figure 4.28, therefore, well illustrates the complex problems associated with the study of mixed organic liquids containing association liquids and surfactants.

4.4 Summary and conclusions

Engineered clayey liners are often used as barriers for domestic waste leachate. Based on the discussion presented in this chapter, the following concluding comments may be made.

1. Municipal solid waste (MSW) leachate will normally cause a slight decrease in the hydraulic conductivity of inactive clays with divalent cations on the exchange sites.
2. High concentrations of K^+ and NH_4^+ cause c-axis contraction of high charge smectites and vermiculites, resulting in decreases in CEC, double-layer contraction and increases in the free pore space at constant void ratio. The clay mineralogy should carefully identify the clay types and their collapse characteristics.
3. The amount of vermiculite–smectite necessary to cause significant damage remains to be established, especially in relation to the stress level on the barrier clays.
4. Three field studies of hydraulic conductivity below landfills are presented in Chapter 9.

The effects of liquid hydrocarbons on the integrity of clayey liners range from negligible to large. Although there is the need for further research, the following conclusions may be drawn from the work presented herein.

1. Water-soluble liquid hydrocarbons may cause large increases in hydraulic conductivity at concentrations above 70% in low stress environments. In higher stress environments the increases may be largely eliminated by chemical consolidation. Contraction of the double layers related to low values of dielectric constant appear to be the cause.
2. Water-soluble liquid hydrocarbons present at concentrations below 70% often induce decreases in hydraulic conductivity due to increases in viscosity of the aqueous mixtures.
3. Water-compacted clayey barriers are resistant to penetration by insoluble organic liquids due to surface tension effects. Even after breakthrough, the hydraulic conductivity, k, remains low, due to lyophobicity effects and flow restricted to macropores. The presence of fractures, however, creates higher velocity flow channels and much higher hydraulic conductivity values.
4. Surfactants and mutually soluble liquid organics destroy surface tension effects, permitting easy entry of the insolubles, which suggests that little reliance should be placed on the presence of breakthrough pressures in the field.
5. Dry clay–organic mixtures are a useful guide to behavior if carefully interpreted with respect to soil fabric and pore size. The latter play an adverse role not necessarily encountered in a water-compacted clay liner system.
6. The presence of high effective stresses on a water-wet clayey liner is beneficial since consolidation occurs and compensates for chemical damage. Since post-damage application of stresses is much less efficient in healing a barrier, it is suggested that a suitable HDPE geomembrane on top of a clay barrier would represent a good design. The geomembrane would protect the clay against organic contact long enough for establishment of high effective stresses on the clayey portion of the composite system.
7. Hydration of a dry clay using uncontaminated water results in a substantial improvement in performance compared to clays that are brought into contact with contaminant (e.g. organic chemicals) before full hydration by uncontaminated water.

Chapter five

Flow modeling

5.1 Introduction

The design of barrier systems typically requires some associated contaminant impact assessment. As discussed in Chapters 1 and 2, this should involve consideration of both advective and diffusive transport. The calculation of the advective components requires that an assessment be made of the flow in the barrier system. In some cases, this may be isolated from the general groundwater system (e.g. in an arid region where there is a thick unsaturated zone above the aquifer and negligible on-site recharge to the aquifer either before or after landfill construction). However, in most cases, the construction of a landfill will involve some interaction between the engineered system and the local hydrogeology. Thus the objective of this chapter is to summarize a number of approaches for calculating flows in barrier systems.

In section 5.2, consideration is given to one-dimensional steady state flow systems where simple hand calculations can be used to estimate flows. These simple calculations are then used to provide background for the interpretation of field data.

In some cases, the flow system may be too complicated to use one-dimensional hand calculations. For these situations, it is usual to use an analytical model (e.g. Rowe and Nadarajah, 1994a) or perform numerical (finite element) analyses. Section 5.3 provides the theoretical basis for two- and three-dimensional flow modeling. Sections 5.4, 5.5 and 5.6 provide background regarding finite difference, finite element and boundary element techniques, respectively, for solving the governing differential equations (given in section 5.3) subject to appropriate boundary conditions.

5.2 One-dimensional flow models

5.2.1 Basic concepts

Flow modeling is based on Darcy's law

$$v_a = ki \tag{5.1a}$$

$$q = v_a A = kiA \tag{5.1b}$$

$$q = ki \quad \text{for } A = 1 \tag{5.1c}$$

where v_a is the Darcy velocity (flux) $[LT^{-1}]$; k is the hydraulic conductivity at the location where flow is being evaluated $[LT^{-1}]$; i is the hydraulic gradient at the same location [–]; q is the flow at the location of interest $[L^3T^{-1}]$; and when $A = 1$, q is flow per unit area; A is the cross-sectional area through which flow is occurring at the location of interest $[L^2]$; $A = 1$ in this case.

Hydraulic gradient [–] may be defined as a change in (total) head $\Delta\phi$ (L) over a given distance, H (L).

$$i = \frac{\Delta\phi}{H} \tag{5.1d}$$

Total head is the sum of two significant components – the pressure head h_p, and the position head, z. To illustrate this, consider the situation shown in Figure 5.1 where a 0.3 m deep pond of water is above a 1.2 m thick clayey liner of low permeability soil. For simplicity it is assumed that the clayey liner is underlain by a free draining layer and that atmospheric pressure exists at the bottom of this liner. This layer might correspond to a stone secondary leachate collection system which is kept pumped, as shown in Figures 1.5 and 1.12.

Total head is always measured relative to some datum. Since flow only depends on the change in total head (over a given distance) the choice of datum is not important. For purposes of illustration, in Figure 5.1, the datum is taken to be at the bottom of the liner. If the position head is defined as the vertical distance, z, above the datum, then the total head, ϕ, at any point of interest, is given by

$$\phi = h_p + z \tag{5.2}$$

where ϕ is the total head [L]; h_p is the pressure head [L] ($= u/\gamma_w$, where u is the pore water pressure [$ML^{-1}T^{-2}$] and γ_w is the unit weight of water [$ML^{-2}T^{-2}$]); z is the position head [L].

For example, in Figure 5.1, the total head at point 'a' at the bottom of the liner, ϕ_a, is equal to the pressure head, which is zero for atmospheric pressure ($h_p = 0$), plus the position head (which is also zero since the datum is at point 'a'):

$$\phi_a = h_p + z = 0 + 0 = 0 \tag{5.3a}$$

At point 'b', the pressure head $h_p = 0$ (atmospheric conditions above the water) and the position head is 1.5 m. Therefore the total head, ϕ_c, is given by

$$\phi_b = h_p + z = 0 + 1.5 = 1.5 \text{ m} \tag{5.3b}$$

At the top of the liner (point 'c'), the pressure head h_p is equal to the height of water above this point, $h_p = u/\gamma_w = 0.3$ m and the pressure head is $z = 1.2$ m. Therefore the total head, ϕ_c, is given by

$$\phi_c = 0.3 + 1.2 = 1.5 \text{ m} \tag{5.3c}$$

The gradient across the clayey liner is then given by

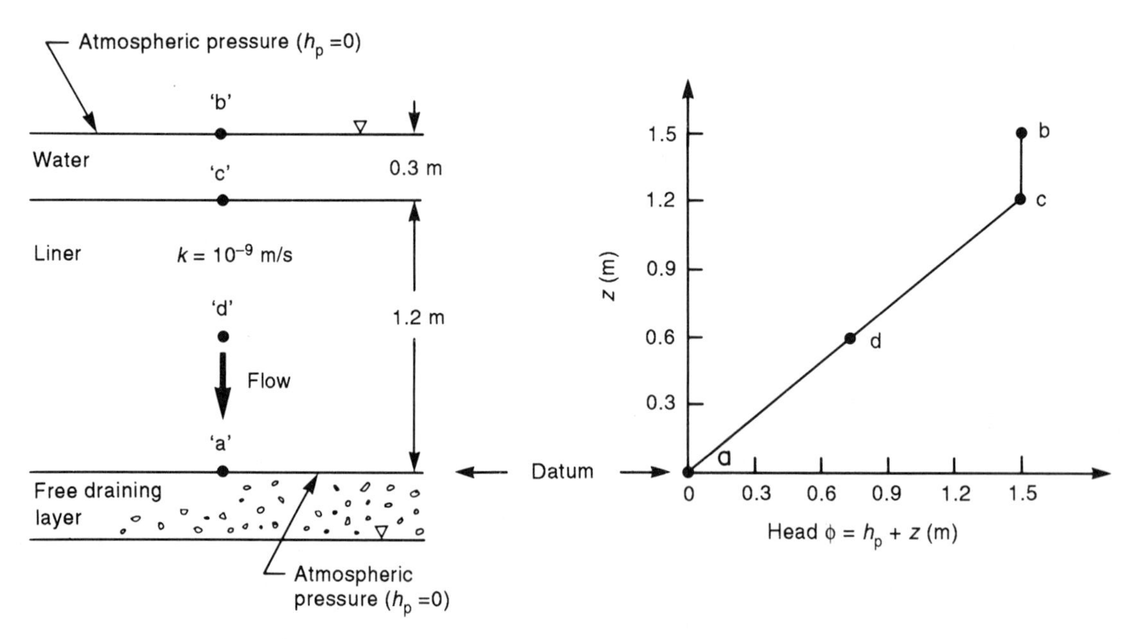

Figure 5.1 Simple flow model – head distribution through a liner from a pond when the liner is underlain by a free draining layer.

$$i = \frac{\Delta\phi}{H} = \frac{\phi_c - \phi_a}{H} = \frac{1.5 - 0}{1.2} = 1.25 \quad (5.4)$$

and the flow across the liner (per unit area)

$$q = ki = 1 \times 10^{-9} \times 1.25 = 1.25 \times 10^{-9} \text{ m/s} = 0.0394 \text{ m/a} \quad (5.5)$$

Consider a point 'd' at the middle of the liner. The flow between point 'c' and point 'd'

$$q_1 = ki = k\left(\frac{\phi_c - \phi_d}{0.5H}\right) \quad (5.6a)$$

Similarly, the flow between point 'd' and point 'a'

$$q_2 = ki = k\left(\frac{\phi_d - \phi_a}{0.5H}\right) \quad (5.6b)$$

Now for one-dimensional flow conditions, the flow between points 'c' and 'd' must be the same as the flow between points 'd' and 'a', since there is nowhere else for the water to go for one-dimensional flow conditions.

Thus, continuity of flow requires that

$$q_1 = q_2$$

and from equations 5.6a and b,

$$k\left(\frac{\phi_c - \phi_d}{0.5H}\right) = k\left(\frac{\phi_d - \phi_a}{0.5H}\right) \quad (5.6c)$$

For a uniform layer the hydraulic conductivity k is a constant throughout and it follows (from equation 5.6c) that

$$\phi_c - \phi_d = \phi_d - \phi_a \quad (5.6d)$$

hence

$$\phi_d = \frac{\phi_a + \phi_c}{2} = \frac{0 + 1.5}{2} = 0.75 \quad (5.6e)$$

This implies that the change in total head is linear across a uniform layer (as shown graphically in Figure 5.1).

5.2.2 Multilayered systems

Consider the one-dimensional flow situation, shown in Figure 5.2, in which one liner of thickness $H_1 = 0.3$ m and hydraulic conductivity $k_1 = 2 \times 10^{-9}$ m/s overlies a second liner of thickness $H_2 = 1.2$ m and hydraulic conductivity

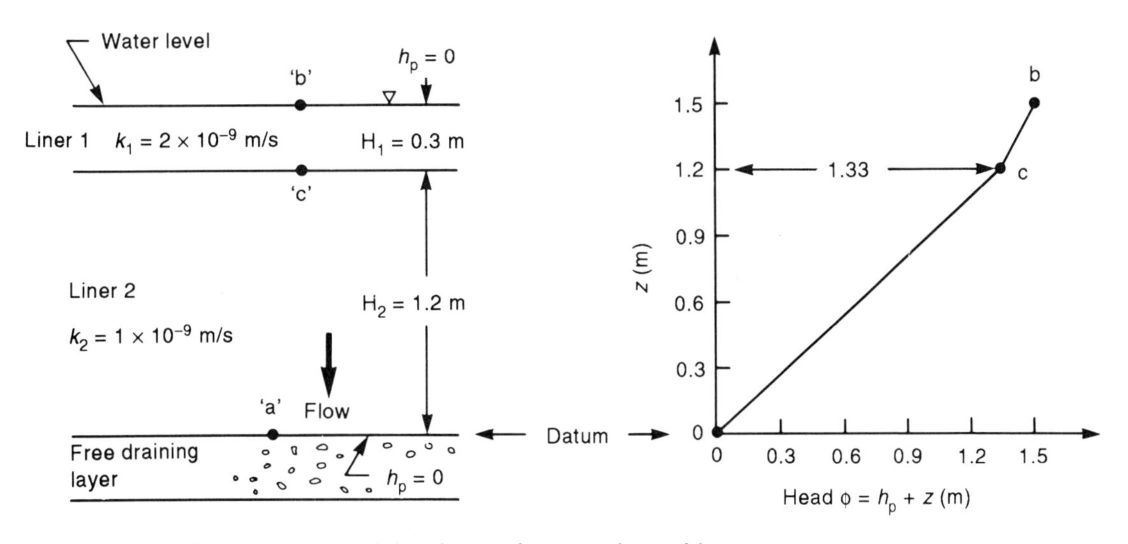

Figure 5.2 Simple flow model – head distribution for a two-layered liner system.

$k_2 = 1 \times 10^{-9}$ m/s, assuming that the water table is at the surface of liner 1, and that liner 2 is underlain by a drain, such that the pressure head at the bottom of liner 2 is zero. Taking account of datum at the bottom of liner 2, the total head, ϕ_a, at the bottom of liner 2 ($h_p = 0$, $z = 0$) is given by

$$\phi_a = h_p + z = 0 \text{ m} \tag{5.7a}$$

and is shown as point 'a' in Figure 5.2. The total head, ϕ_b, at the top of liner 1 ($h_p = 0$, $z = 1.5$ m) is given by

$$\phi_b = h_p + z = 1.5 \text{ m} \tag{5.7b}$$

and is shown as point 'b' in Figure 5.2.

In order to plot the head distribution, it is necessary to calculate the total head, ϕ_c, at point 'c' which is at the boundary of liner 1 and liner 2. This can be obtained from consideration of continuity of flow per unit area as follows.

$$\text{Flow in liner 1, } q_1 = k_1 i_1 = k_1\left(\frac{\phi_b - \phi_c}{H_1}\right) \tag{5.8a}$$

$$\text{Flow in liner 2, } q_2 = k_2 i_2 = k_2\left(\frac{\phi_c - \phi_a}{H_2}\right) \tag{5.8b}$$

Continuity of flow requires that $q_1 = q_2$, thus

$$k_1 i_1 = k_2 i_2 \tag{5.9a}$$

or

$$k_1\left(\frac{\phi_b - \phi_c}{H_1}\right) = k_2\left(\frac{\phi_c - \phi_a}{H_2}\right) \tag{5.9b}$$

where k_1, k_2, ϕ_a, ϕ_b, H_1, H_2 are known, and ϕ_c is to be determined. Rearranging equation 5.9b gives

$$\phi_c = \frac{1}{\left(\frac{k_1}{H_1} + \frac{k_2}{H_2}\right)}\left(\frac{k_1}{H_1}\phi_b + \frac{k_2}{H_2}\phi_a\right)$$

and substituting $\phi_a = 0$, $\phi_b = 1.5$, $H_1 = 0.3$, $H_2 = 1.2$, $k_1 = 2 \times 10^{-9}$ m/s, $k_2 = 1 \times 10^{-9}$ m/s,

$$\phi_c = \frac{1}{\left(\frac{2 \times 10^{-9}}{0.3} + \frac{1 \times 10^{-9}}{1.2}\right)} \times\left(\frac{2 \times 10^{-9}}{0.3} \times 1.5 + 0\right)$$

Multiplying the numerator and denominator by 10^{-9},

$$\phi_c = \frac{1}{\left(\frac{2}{0.3} + \frac{1}{1.2}\right)}\left(\frac{2 \times 1.5}{0.3}\right) = \frac{10}{7.5} = 1.33$$

The flow can then be calculated from either equations 5.8a or 5.8b:

$$\begin{aligned} q_1 &= 2 \times 10^{-9} \times \left(\frac{1.5 - 1.33}{0.3}\right) \\ &= 1.11 \times 10^{-9} \text{ m/s} \\ &= 0.035 \text{ m/a} \end{aligned}$$

$$\begin{aligned} q_2 &= 1 \times 10^{-9} \times \left(\frac{1.33 - 0}{1.2}\right) \\ &= 1.11 \times 10^{-9} \text{ m/s} \\ &= 0.035 \text{ m/a} \end{aligned}$$

Alternatively, one can show that the equivalent hydraulic conductivity of the system is given by the harmonic mean $\bar{k}$ where

$$\frac{\Sigma H_i}{\bar{k}} = \Sigma \frac{H_i}{k_i} \tag{5.10}$$

where the summation is performed over all layers. In this case,

$$\frac{H_1 + H_2}{\bar{k}} = \frac{H_1}{k_1} + \frac{H_2}{k_2}$$

$$\therefore \frac{1.5}{\bar{k}} = \frac{0.3}{2 \times 10^{-9}} + \frac{1.2}{1 \times 10^{-9}}$$

$$\therefore \bar{k} = 1.11 \times 10^{-9} \text{ m/s}$$

The flow can then be calculated:

$$\bar{q} = \bar{k}\bar{i} \tag{5.11}$$

where $\bar{i}$ is the hydraulic gradient across the entire system. In this case,

$$\bar{i} = \frac{\phi_b - \phi_a}{H_1 + H_2} = \frac{1.5 - 0}{0.3 + 1.2} = 1$$

$$\therefore \bar{q} = \bar{k}\bar{i} = 1.11 \times 10^{-9} \times 1 \text{ m/s}$$
$$= 0.035 \text{ m/a}$$

Knowing

$$\bar{q} = q_1 = k_1\left(\frac{\phi_b - \phi_c}{H_1}\right) \tag{5.12}$$

one could then calculate ϕ_c by rearranging terms to give

$$\phi_c = \phi_b - \frac{H_1}{k_1}\bar{q}$$
$$= 1.5 - \frac{0.3}{2 \times 10^{-9}} \times 1.11 \times 10^{-9}$$
$$= 1.33 \text{ m}$$

This latter approach (equations 5.10, 5.11 and 5.12) often represents the most convenient method of calculating a head distribution.

It also follows from equation 5.9a that

$$\frac{i_1}{i_2} = \frac{k_2}{k_1} \tag{5.13}$$

Thus, the change of slope in the total head distribution is directly related to the change in hydraulic conductivity between layers.

5.2.3 Layered systems with high hydraulic conductivity contrasts

Figure 5.3 shows a layer system identical to that shown in Figure 5.2, except that in this case it is assumed that the upper 0.3 m is waste with a hydraulic conductivity of 1×10^{-6} m/s. Based on equation 5.13 (which is based on continuity of flow),

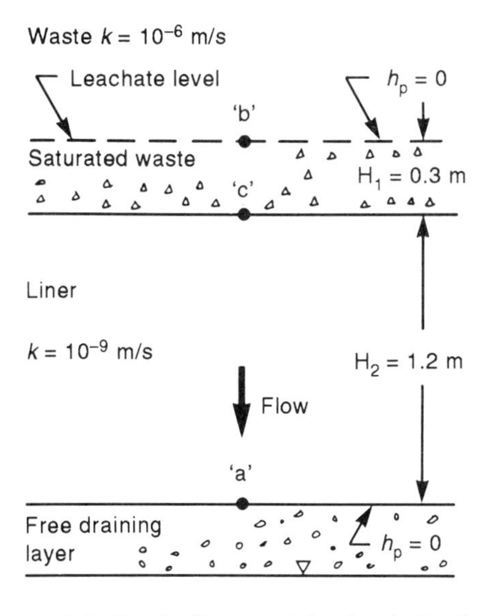

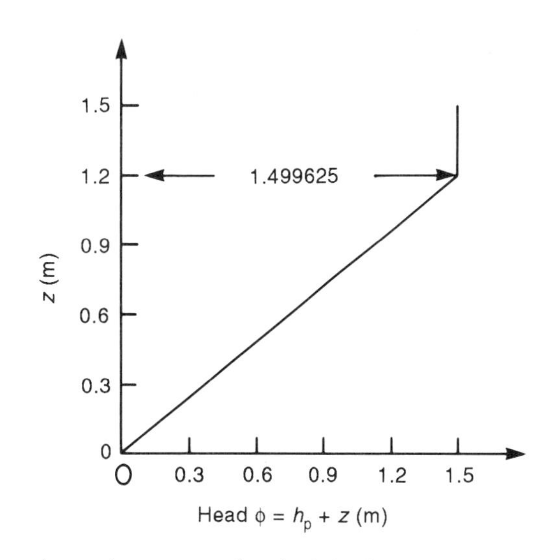

Figure 5.3 Simple flow model – head distribution for a two-layered system with a high hydraulic conductivity contrast between layers.

$$\frac{i_1}{i_2} = \frac{[(\phi_b - \phi_c)/H_1]}{[(\phi_c - \phi_a)/H_2]} = \frac{k_2}{k_1} = \frac{10^{-9}}{10^{-6}} = 10^{-3} \tag{5.14}$$

Since $\phi_b = 1.5$ m and $\phi_a = 0$, as already demonstrated in the previous section, and $H_1 = 0.3$ m, $H_2 = 1.2$ m, it follows that

$$\frac{[(\phi_b - \phi_c)/0.3]}{[(\phi_c - \phi_a)/1.2]} = 10^{-3}$$

$$\therefore \frac{\phi_b - \phi_c}{\phi_c - \phi_a} = 0.25 \times 10^{-3}$$

$$\therefore \phi_b - \phi_c = 0.25 \times 10^{-3} (\phi_c - \phi_a)$$

$$\therefore 1.00025\ \phi_c = (\phi_b + 0.25 \times 10^{-3}\ \phi_a) = 1.5$$

$$\therefore \phi_c = 1.499625 \approx 1.5$$

which implies that the total head change across the waste is negligible. This is because the waste is so much more permeable than the clay. The gradient across the liner is given by

$$i \approx \frac{1.5}{1.2} = 1.25$$

which is essentially the same as that calculated in section 5.2.1,

$$q = ki \approx 1 \times 10^{-9} \times 1.25 = 1.25 \times 10^{-9}\ \text{m/s} \approx 0.0394\ \text{m/a}$$

Thus when there are high hydraulic conductivity contrasts, the flow is controlled by the least permeable layer for one-dimensional conditions.

5.2.4 Interpretation of data

The change in hydraulic gradient which may be associated with significant changes in hydraulic conductivity between different layers, as illustrated in the previous section, should be considered when interpreting data from field installation of piezometers. For example, consider the hydrostratigraphy shown in Figure 5.4. The head distribution from the upper fine sand aquifer to the bottom gravel layer is shown, together with a number of piezometers and their corresponding water levels (relative to some datum).

The important point to be made is that when calculating vertical gradients for systems such as that shown in Figure 5.4, it is essential to consider the hydrostratigraphy, and not simply calculate gradient as the difference in water level in piezometers divided by the distance between piezometers. The calculation of gradients in the various units will be discussed below.

Fine sand unit

Piezometers 'a' and 'b' record essentially the same water level (50 masl – meters above sea level) to the accuracy of measurement, and hence one would calculate essentially zero vertical gradient in this unit. However, this does not mean that there is zero vertical flow – only that the vertical gradient and flow are small.

Clay unit

Piezometer 'b' is partly screened across the fine sand and partly across the clay unit. However, because of the large hydraulic conductivity contrast, piezometer 'b' is recording the head in the fine sand rather than the clay, and for purposes of calculating a gradient, the head for piezometer 'b' may be regarded as corresponding to the top of the clay unit.

Piezometer 'c' is located 9–10 m below the top of the clay (having a 1 m long sand pack). The total head in piezometer 'c' is 41.09 masl. Since this piezometer is embedded completely in the clay, the head is considered to correspond to that at the centroid of the sand pack, therefore

$$i_{b\text{–}c} \approx \frac{50 - 41.09}{9.5} = 0.937 \approx 0.94$$

where the distance of 9.5 m is measured from the centroid of piezometer 'c' to the top of the clay unit.

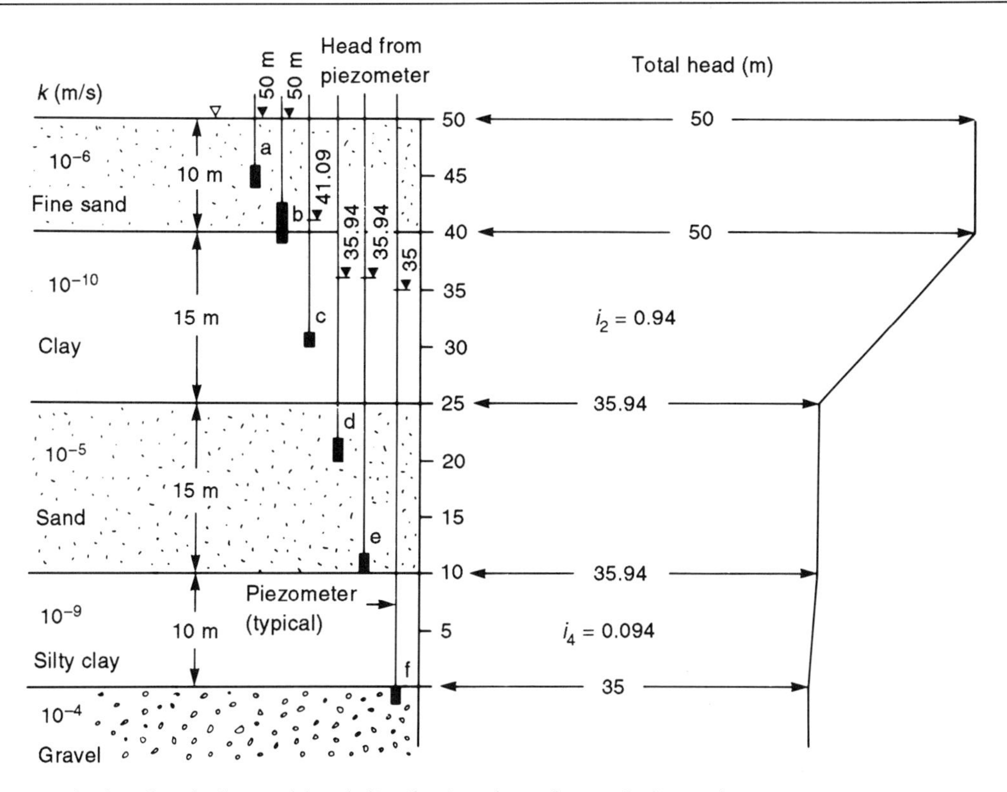

Figure 5.4 Calculated and observed head distribution through a multi-layered system.

A second estimate of the gradient across the clay unit can be obtained using piezometers 'c' and 'd', but recognizing that, because of the hydraulic conductivity contrast, the head of ~35.94 masl recorded by piezometer 'd' is essentially the same as at the bottom of the clay unit:

$$i_{c\text{-}d} \approx \frac{41.09 - 35.94}{5.5} = 0.937 \approx 0.94$$

where 5.5 m is the distance between the centroid of piezometer 'c' and the bottom of the clay unit.

A third estimate of gradient across the clay unit can be obtained using piezometers 'b' and 'd', recognizing that these record the head at the top and bottom of the clay unit respectively. Thus

$$i_{b\text{-}d} = \frac{50 - 35.94}{15} = 0.937 \approx 0.94$$

where the 15 m distance is the thickness of the clay unit.

The agreement between the three calculations of gradient across the clay unit indicates that the unit has a uniform vertical hydraulic conductivity.

It should be noted that the gradient that one would calculate based on the distance between the centroid of piezometers 'b' and 'd',

$$\frac{50 - 35.94}{20} = 0.70$$

when used in conjunction with conventional

interpretations, gives a misleading indication of the vertical flow in the system which is being controlled by the clay layer.

The problems of gradient interpretation can be minimized by placing piezometers close to boundaries between materials having significantly different hydraulic conductivities, for example, as shown for piezometers 'e' and 'f' in Figure 5.4. Based on the head measured in these piezometers, the gradient across the silty clay layer is fairly readily estimated:

$$i = \frac{35.94 - 35}{10} = 0.094$$

based on the 10 m thickness of the silty clay layer.

When dealing with deposits of wide lateral extent, there are often situations where there is a gradual change in grainsize with depth. This will generally give rise to a gradual change in hydraulic conductivity; in situations where there are significant differences in vertical head, this will be manifested by a change in gradient. For example, Figure 5.5 shows a number of piezometers located in a silt (some clay) silty clay layer where there is a change from lower clay content at the top to a high clay content near the bottom. The water levels and corresponding total head distribution are shown in Figure 5.5. These data alone indicate a significant change in hydraulic conductivity with depth, with the bottom 5 m being responsible for 60% of the head loss. The change in gradient for a one-dimensional system such as this indicates the relative changes in hydraulic conductivity. If data are available to allow an estimate of hydraulic conductivity to be made for one layer, then an estimate can also be made for the entire deposit. For example, suppose that the results of a pumping test (e.g. section 3.2.3) conducted on the lower aquifer and monitored at piezometers 'd' and 'c' indicates a bulk (vertical) hydraulic conductivity of the soil between 'c' and 'd' of 1.6×10^{-10} m/s.

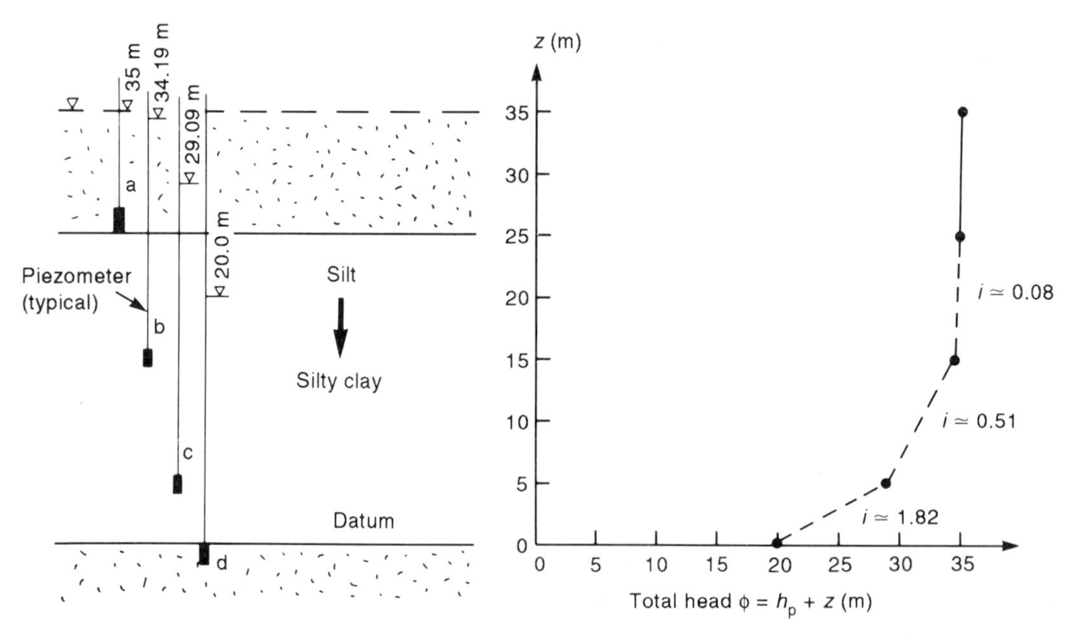

Figure 5.5 Head distribution through a gradually varying silt–silty clay layer and the estimation of hydraulic conductivity distribution based on water level data.

The gradient between each pair of piezometers can be readily calculated:

$$i_{c-d} = \frac{29.09 - 20}{5} = 1.82$$

(where the 5 m distance is from the centroid of the sand pack of piezometer 'c' and the bottom of the clay layer, for the reasons discussed earlier). Also

$$i_{b-c} = \frac{34.19 - 29.09}{10} = 0.51$$

$$i_{a-b} = \frac{35 - 34.19}{10} = 0.08$$

For essentially one-dimensional (vertical) flow through this silt–silty clay layer the ratio of gradient is equal to the inverse ratio of bulk hydraulic conductivity. Thus from equation 5.13,

$$\frac{i_{c-d}}{i_{b-c}} = \frac{k_{b-c}}{k_{c-d}}$$

$$\therefore \frac{1.82}{0.51} = \frac{k_{b-c}}{1.6 \times 10^{-10}}$$

$$\therefore k_{b-c} \approx 5.7 \times 10^{-10} \text{ m/s}$$

and similarly,

$$\frac{i_{c-d}}{i_{a-b}} = \frac{k_{a-b}}{k_{c-d}}$$

$$\therefore \frac{1.82}{0.08} = \frac{k_{a-b}}{1.6 \times 10^{-10}}$$

$$\therefore k_{a-b} \approx 3.6 \times 10^{-9} \text{ m/s}$$

The total vertical flow in this system can be estimated from the hydraulic conductivities and gradients evaluated between any pair of piezometers. Thus

$$q = k_{c-d} i_{c-d} = 1.6 \times 10^{-10} \times 1.82$$
$$\approx 2.9 \times 10^{-10} \text{ m/s} = 0.009 \text{ m/a}$$

This is a small but potentially significant flow with regard to contaminant transport through the silt–silty clay system. Note that it could be quite misleading to use a gradient determined between one pair of piezometers (e.g. a–d) and a hydraulic conductivity between a second pair of piezometers (e.g. c–d). For example, this would give an estimate of flow

$$q = \frac{15}{25} \times 1.6 \times 10^{-10}$$
$$= 9.6 \times 10^{-11} \text{ m/s (0.003 m/a)}$$

which is quite incorrect. If the gradient across the entire unit is to be used, it must be used with the appropriate harmonic mean hydraulic conductivity, $\bar{k}$, which can be calculated using the values of k_{a-b}, k_{b-c}, k_{c-d} estimated above:

$$\frac{H}{\bar{k}} = \frac{H_{a-b}}{k_{a-b}} + \frac{H_{b-c}}{k_{b-c}} + \frac{H_{c-d}}{k_{c-d}}$$

$$\therefore \frac{25}{\bar{k}} = \frac{10}{3.6 \times 10^{-9}} + \frac{10}{5.7 \times 10^{-10}} + \frac{5}{1.6 \times 10^{-10}}$$

$$\therefore \bar{k} = 4.85 \times 10^{-10} \text{ m/s}$$

$$\therefore q = \bar{k}\bar{i} = 4.85 \times 10^{-10} \times \frac{15}{25}$$
$$= 2.9 \times 10^{-10} \text{ m/s (0.009 m/a)}$$

5.2.5 Head distributions in some simple barrier systems

Figure 5.6 shows a barrier system consisting of a primary leachate collection system, a primary clay liner with a hydraulic conductivity $k_1 \approx 3 \times 10^{-10}$ m/s underlain by a leak detection–secondary leachate collection layer which, it is assumed, is kept free-draining without significant build-up of head in this layer. This is then

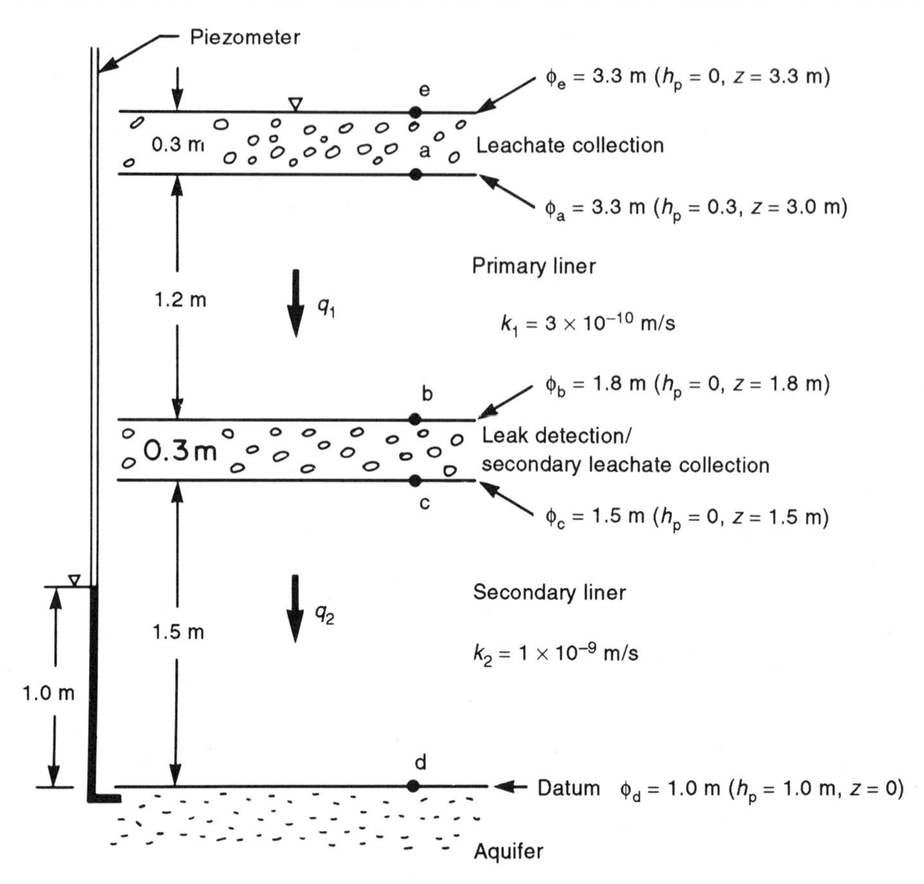

Figure 5.6 Head distribution and schematic for a multilayered system involving a leak detection–secondary leachate collection layer.

underlain by a 1.5 m thick natural clay (secondary liner) ($k_2 = 10^{-9}$ m/s) and an aquifer. It is assumed that leachate mounding in the primary collection system is 0.3 m above the primary liner and that the water level (potentiometric surface) in the aquifer is 1.0 m above the top of the aquifer.

The vertical flow in the two liner systems can be readily calculated using the concepts presented in the previous sections. Since the primary and secondary leachate collection systems are both far more permeable than the liner, the gradient loss across the primary liner is given by

$$i_1 = \frac{\phi_a - \phi_b}{z_a - z_b} = \frac{3.3 - 1.8}{3.0 - 1.8} = 1.25$$

$$q_1 = k_1 i_1 = 3 \times 10^{-10} \times 1.25$$
$$= 3.75 \times 10^{-10} \text{ m/s } (0.012 \text{ m/a})$$

The vertical flow through the secondary liner can be calculated as follows:

$$i_2 = \frac{\phi_c - \phi_d}{z_c - z_d} = \frac{1.5 - 1.0}{1.5 - 0} = 0.33$$

$$q_2 = k_2 i_2 = 3.3 \times 10^{-10} \text{ m/s } (0.0105 \text{ m/a})$$

Since $q_2 < q_1$, some of the flow will be collected by the secondary leachate collector. However, this system has the potential to allow the majority of flow passing through the primary liner also to pass through the secondary liner, and hence the effectiveness of the 'leak detection–secondary leachate collection' for the system must be questioned.

If the head in the aquifer was higher (e.g. if $\phi_d \approx 1.5$ m) or if the secondary liner was less permeable, either because of lower hydraulic conductivity clay or the use of a geomembrane to form a composite liner, then the secondary collection system would be more effective.

Figure 5.7 shows a system with similar physical characteristics to that in Figure 5.6, but with quite a different hydraulic head distribution arising from a much higher head in the aquifer, and not pumping the second granular layer. Under these conditions, the head at points 'b' and 'c' will depend on the hydraulic conductivity contrast between the primary and secondary liner. Since it is assumed that no fluid is added (or removed) from the hydraulic control layer and since this layer is far more permeable than either the primary or secondary liner, continuity of flow requires that $q_1 = k_1 i_1 = q_2 = k_2 i_2$. Therefore

$$\frac{i_1}{i_2} = \frac{k_2}{k_1} \qquad (5.15a)$$

(as already established in equation 5.13).

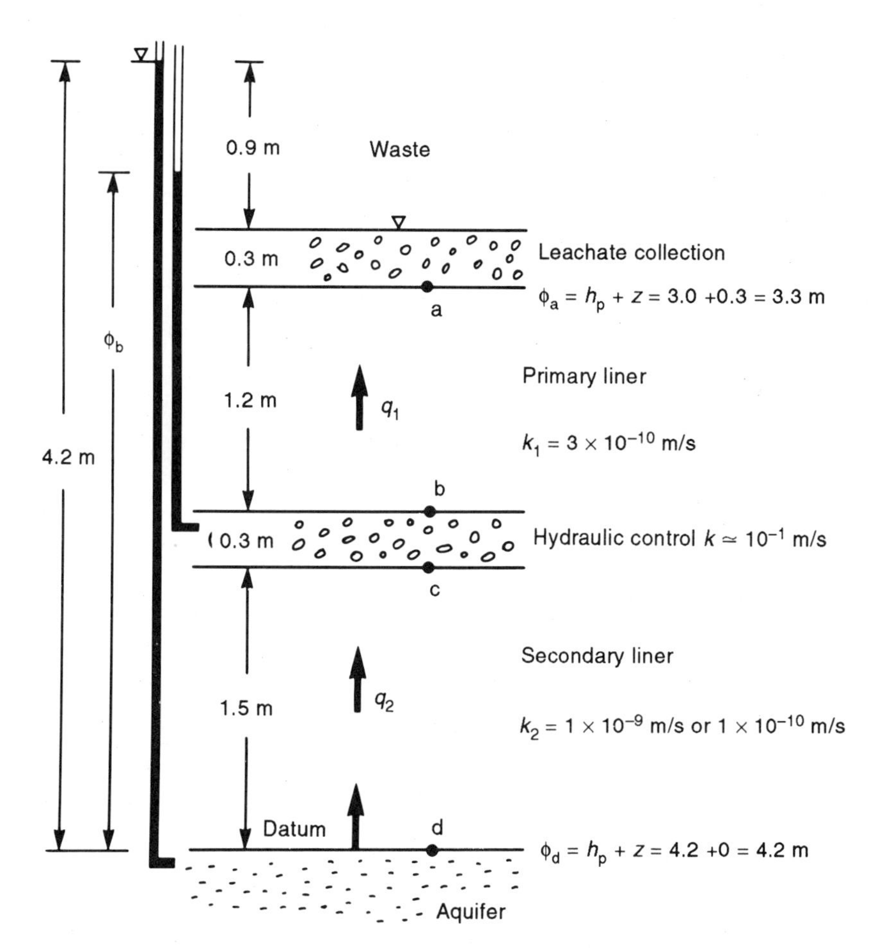

Figure 5.7 Head distribution and schematic for a multilayered system with a hydraulic control layer between two liners.

Thus, taking $k_1 = 3 \times 10^{-10}$ m/s and $k_2 = 1 \times 10^{-9}$ m/s gives

$$\frac{i_1}{i_2} = \frac{1 \times 10^{-9}}{3 \times 10^{-10}} = 3.33 \tag{5.15b}$$

Now

$$i_1 = \frac{\phi_a - \phi_b}{1.2} \tag{5.16a}$$

$$i_2 = \frac{\phi_c - \phi_d}{1.5} \tag{5.16b}$$

where

$$\phi_b \approx \phi_c \tag{5.16c}$$

Then, from equations 5.15 and 5.16,

$$\frac{\phi_a - \phi_b}{\phi_b - \phi_d} = 3.33 \times \frac{1.2}{1.5} = 2.66$$

$$\therefore \phi_b = \frac{\phi_a + 2.66\phi_d}{3.66} = \frac{3.3 + 2.66 \times 4.2}{3.66} = 3.95 \text{ m}$$

$$\therefore q_1 = k_1 i_1 = 3 \times 10^{-10} \times \frac{3.3 - 3.95}{1.2} = -1.64 \times 10^{-10} \text{ m/s } (-0.005 \text{ m/a})$$

$$q_2 = k_2 i_2 = 1 \times 10^{-9} \times \frac{3.95 - 4.2}{1.5} = -1.64 \times 10^{-10} \text{ m/s } (-0.005 \text{ m/a})$$

where the negative sign indicates upward flow from the aquifer to the leachate collection system. This provides a hydraulic trap, as discussed in Chapter 1.

The preceding calculation assumes that the head in the aquifer is not affected by the upward flow. This is only likely to be true if it is a highly transmissive aquifer which did not rely on recharge from the overburden beneath the waste site for a significant proportion of its flow. This will be discussed in more detail in Chapter 12.

The effect of the hydraulic conductivity contrast between the layers can be illustrated by taking $k_2 = 10^{-10}$ m/s and repeating the foregoing calculations. Thus

$$\frac{i_1}{i_2} = \frac{10^{-10}}{3 \times 10^{-10}} = 0.33$$

$$\frac{\phi_a - \phi_b}{\phi_b - \phi_d} = 0.33 \times \frac{1.2}{1.5} = 0.26$$

$$\therefore \phi_b = \frac{\phi_a + 0.26 \times 4.2}{1.26} = 3.49 \text{ m}$$

$$\therefore q_1 = k_1 i_1 = 3 \times 10^{-10} \times \frac{3.3 - 3.49}{1.2} = -4.7 \times 10^{-11} \text{ m/s } (-0.0015 \text{ m/a})$$

$$q_2 = k_2 i_2 = 1 \times 10^{-10} \times \frac{3.49 - 4.2}{1.5} = -4.7 \times 10^{-11} \text{ m/s } (-0.0015 \text{ m/a})$$

Thus the lower hydraulic conductivity of the secondary liner relative to the previous example reduces flow through the entire system. This will reduce the effectiveness of the hydraulic trap, as will be discussed in subsequent chapters.

In these two examples the hydraulic control layer was used in a passive mode, wherein no water was added or subtracted from this layer. By active operation of this layer the head, ϕ_b ($\approx\phi_c$), can be controlled and either increased or decreased relative to the leachate head, ϕ_a, and potentiometric surface in the aquifer, ϕ_d.

5.3 Analysis of two- and three-dimensional flow

While the simple one-dimensional calculations are adequate for some important practical applications, there are situations where the flow regime is more complex, and it is necessary to use more elaborate analysis techniques to model it adequately, although one can often do useful checks using simple one-dimensional hand calculations.

5.3.1 Governing equations

Suppose that the occurrence of some event, such as the construction of a landfill, the installation of cut-off wall or the introduction of a well or drain, disturbs the equilibrium of the pore water in the soil. At first there will be a period during which the pore water pressure and the stress distribution will change, but ultimately the pore water pressure and the stress state will reach a new equilibrium, and steady-state flow condition will be established. It is usually convenient to analyze the steady-state flow in terms of the total head, ϕ, defined by equation 5.2. For two- and three-dimensional cases, the form of Darcy's law becomes

$$v_x = k_{xx} i_x$$
$$v_y = k_{yy} i_y \qquad (5.17)$$
$$v_z = k_{zz} i_z$$

where

$$\boldsymbol{v} = [v_x \quad v_y \quad v_z]^{\mathrm{T}}$$

is the Darcy velocity vector, and

$$\boldsymbol{i} = [i_x \quad i_y \quad i_z]^{\mathrm{T}}$$

is the hydraulic gradient vector, with

$$i_x = -\frac{\partial \phi}{\partial x}$$
$$i_y = -\frac{\partial \phi}{\partial y} \qquad (5.18)$$
$$i_z = -\frac{\partial \phi}{\partial z}$$

The negative sign has been introduced in recognition that water flows from high head to low head, and so $\partial\phi/\partial x$ will be negative for flow in the positive x-direction, and similarly for $\partial\phi/\partial y$ and $\partial\phi/\partial z$ in the y- and z-directions.

Since a steady state has been reached, there will be no change in the effective stress state, and thus an element of soil will undergo no deformation, and in particular will undergo no change in volume. If the pore fluid can be considered as incompressible, then continuity of flow within an element will be satisfied, provided that

$$\frac{\partial v_x}{\partial x} + \frac{\partial v_y}{\partial y} + \frac{\partial v_z}{\partial z} = 0 \qquad (5.19)$$

This is called the continuity equation.

It follows from Darcy's law (5.17) that the total head satisfies

$$\frac{\partial}{\partial x}\left(k_{xx}\frac{\partial \phi}{\partial x}\right) + \frac{\partial}{\partial y}\left(k_{yy}\frac{\partial \phi}{\partial y}\right) + \frac{\partial}{\partial z}\left(k_{zz}\frac{\partial \phi}{\partial z}\right) = 0 \qquad (5.20a)$$

If the soil is homogeneous and isotropic, this reduces to Laplace's equation,

$$\nabla^2\phi = \frac{\partial^2 \phi}{\partial x^2} + \frac{\partial^2 \phi}{\partial y^2} + \frac{\partial^2 \phi}{\partial z^2} = 0 \qquad (5.20b)$$

For two-dimensional plane flow (in the x–z-plane), this becomes

$$\frac{\partial^2 \phi}{\partial x^2} + \frac{\partial^2 \phi}{\partial z^2} = 0 \qquad (5.20c)$$

5.3.2 Boundary conditions

Figure 5.8 illustrates the flow from a leachate pond through a clay cut-off wall overlying an impermeable stratum, and into a drain. This example serves to illustrate four common types of boundary conditions:

1. On AB the water pressure is given by $u =$

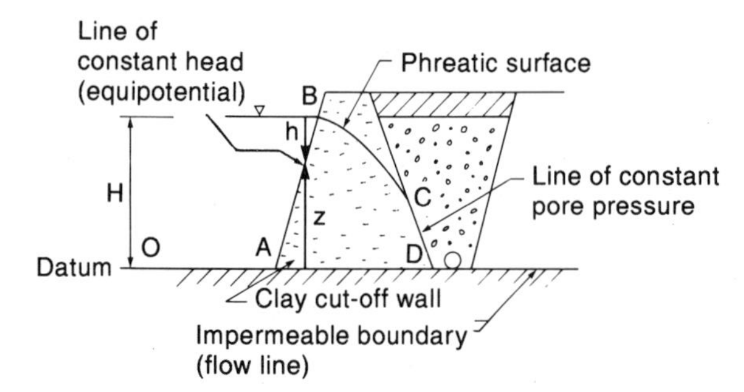

Figure 5.8 Boundary conditions for steady-state seepage.

$\gamma_w h$, and thus for the datum shown in Figure 5.8, $\phi = h + z = H$ at any arbitrary point along AB. Thus the line AB is a line of constant head, or equipotential. This is a special case of what is often called a Dirichlet or first type boundary condition, where the head is specified along a boundary.

2. The line AD is an impermeable surface. There can be no flow across such a surface and thus the direction of flow must be parallel to AD. The line AD is called an impermeable surface or a flow line. This is a special case of the Neumann or second type boundary condition where the flux (flow) normal to the boundary is specified; in this case, the flow across the boundary is zero.
3. On CD the pore water pressure is equal to atmospheric pressure (usually taken as zero) and since $\phi = +z$, the head varies linearly with distance above the datum. Such a line is called a line of constant pressure.
4. When seepage occurs in the clay cut-off wall illustrated in Figure 5.8, the entire soil mass does not remain saturated. The material above BC becomes unsaturated and ultimately the water drains from there to a boundary called the phreatic surface which represents the boundary between unsaturated and saturated materials. The phreatic line BC is both a line of constant pore water pressure, which is equal to atmospheric pressure and taken as zero, and a flow line. The position of this phreatic line is not known initially, and is determined by an iterative algorithm.

5.3.3 Continuity conditions

Natural soil deposits and artificial barriers (e.g. below a waste facility), often consist of a number of distinct layers, as illustrated in Figure 5.9. At the interface of two such layers, the pore water pressure just above the interface must equal the pore water pressure just below the interface, and thus

$$\phi_A = \phi_B \tag{5.21a}$$

Also, there can be no net flow into the interface, and thus

$$v_{nA} = v_{nB} \tag{5.21b}$$

Any solution of equation 5.20 must satisfy these continuity conditions.

5.4 Finite difference approximation

The determination of the variation of head in any particular case depends upon the solution of the governing differential equation (equation 5.20), subject to boundary conditions of the type discussed in section 5.3.2 and continuity equations of the type given by equations 5.21a, b.

The explicit determination of the distribution of head is only feasible for very simple flows, and for more complex cases it is necessary to employ numerical techniques. The earliest

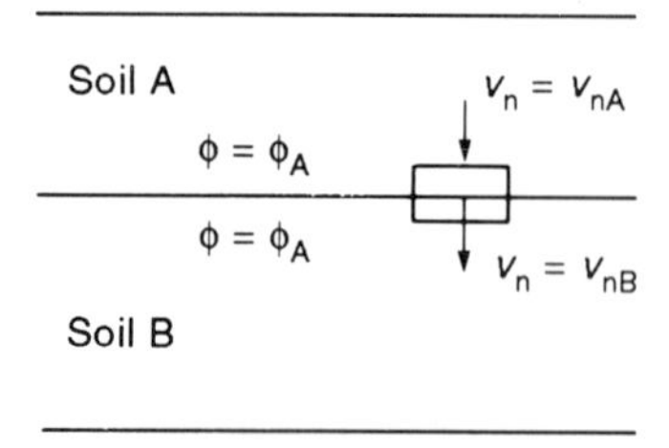

Figure 5.9 Continuity conditions.

approach to this was the finite difference approach. Figure 5.10(a) shows the section of a cut-off wall where it is assumed that the sand layer is far more permeable than the waste and the underlying silt–clayey silt soil. The head in the sand is assumed to be approximately uniform on each side of the cut-off wall, but with a lower head in the sand to the right of the wall. In the finite difference method, there is no attempt to determine the distribution of head at all points; rather, the heads are determined at a finite number of points on a grid, as shown in Figure 5.10(b).

The governing differential equation, in this case

$$\frac{\partial^2 \phi}{\partial x^2} + \frac{\partial^2 \phi}{\partial z^2} = 0 \qquad (5.22)$$

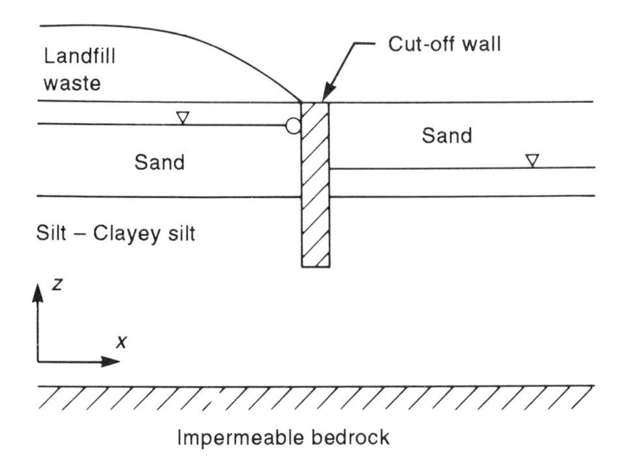

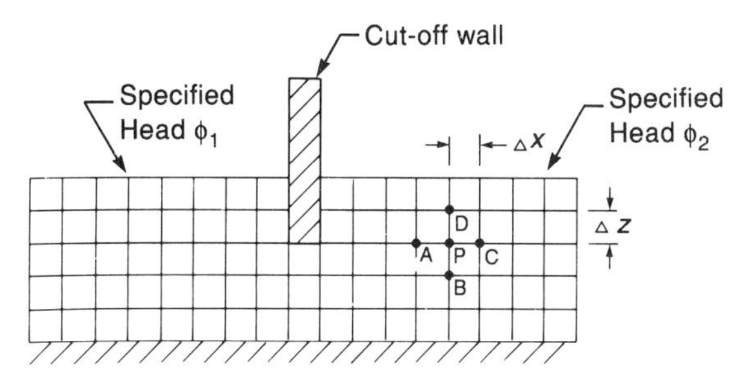

Figure 5.10 (a) Schematic description of cut-off wall. (b) Finite difference grid.

is then approximated in terms of the values of head at these grid points, thus at the point P, equation 5.22 is approximated by

$$\frac{\phi_C - 2\phi_P + \phi_A}{\Delta x^2} + \frac{\phi_D - 2\phi_P + \phi_B}{\Delta z^2} = 0 \qquad (5.23)$$

It is possible to incorporate boundary conditions and to take into account continuity conditions at the junction of different soils, although this is quite difficult for anything other than the simplest rectangular configurations.

Equation 5.23, together with the boundary conditions and continuity conditions, lead to a set of linear equations which can be solved to obtain the head at each grid point. The finer the finite difference mesh, the more accurate the approximation to the governing differential equation.

5.5 Application of the finite element method to the analysis of plane flow

As noted above, there are some difficulties in applying the finite difference method to seepage analyses involving complex geometries and non-homogeneous soils. These difficulties are largely overcome by the finite element method.

The solution of any seepage problem involves satisfying

1. the continuity equation (equation 5.19);
2. the constitutive behavior (Darcy's law), equations 5.17;
3. the boundary conditions;
4. the continuity conditions at the interface of soils having different properties.

Consider the landfill situation shown in Figure 5.11, where there is leachate mounding between a set of drains along the base of the landfill, and in addition there is a known location of the water table outside the landfill, which means that the head is defined along the boundary S_ϕ

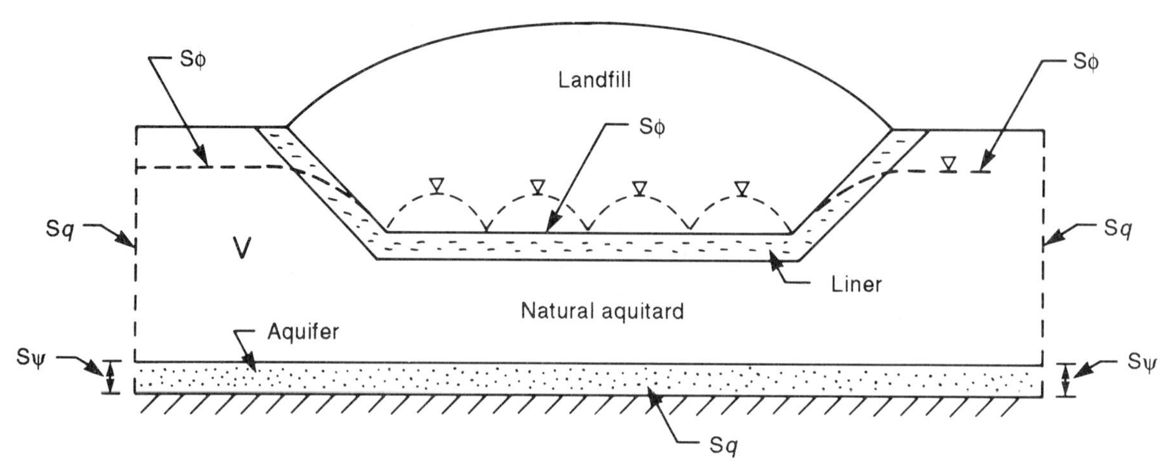

Figure 5.11 Simple seepage problem.

(this includes boundary conditions 1 and 4 discussed in section 5.3.2). At the upgradient and downgradient edge of the aquifer (along boundary S_ψ) either the head may be specified (i.e. boundary condition 1, section 5.3.2) or the flow may be specified (i.e. a Neumann or second type boundary condition). Along the boundary S_q the flow is specified to be zero (i.e. boundary condition 2, section 5.3.2). Thus the problem is defined along the surface $S = S_\phi + S_\psi + S_q$. Within the volume, V, inside this boundary surface, there are three materials with three different sets of properties – the liner, the natural aquitard and the aquifer. Thus we need to solve equations 5.17 and 5.19 within the region V, subject to the boundary conditions along S, while requiring that continuity conditions are satisfied at the material boundaries within the volume V.

The finite element method has a totally different philosophy to the finite difference method. In the finite element method, the continuity equation, Darcy's law and the boundary conditions are combined to derive an equation of 'virtual work'. The finite element method then seeks approximately to satisfy the virtual work equation:

Internal virtual work
= Applied virtual work (5.24)

This method of analysis has been applied extensively to the analysis of a wide variety of physical phenomena (Zienkiewicz, 1977). The procedure is straightforward, and is outlined below.

First, the body is divided into a number of elements, as shown schematically in Figure 5.12. Each element has a number of nodes; for example at the corners of the elements shown in Figure 5.12. The finite element method seeks to determine the value of head at these nodes, and does so by a systematic approximation of the equation of virtual work. The quantities to be determined at the heads at the nodal points 1, 2, . . . , viz. ϕ_1, ϕ_2, ... The essential idea is to approximate the continuous variation of head ϕ within an element in terms of nodal heads.

Similarly, the variation of the continuous 'virtual head' $\delta\phi$ is approximated in terms of the values of 'virtual' nodal heads $\delta\phi_1$, $\delta\phi_2$, ... It is perhaps worth remarking that this process of approximation is essentially a local one, so that the variation of head within a given element is expressed in terms of the heads at the nodes of that element. This approximation technique leads to approximations of the form

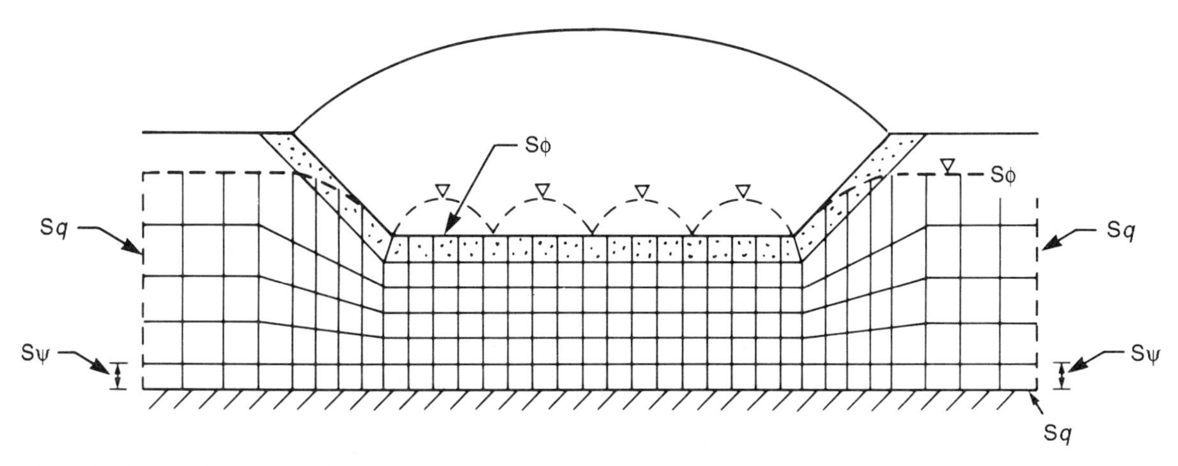

Figure 5.12 Division of body into elements.

$$\text{Internal virtual work} = \int_V \delta i^{\mathrm{T}} v \,\mathrm{d}V$$
$$\approx \delta a^{\mathrm{T}} K a \qquad (5.25)$$

$$\text{Applied virtual work} = \int_S q \,\delta\phi \,\mathrm{d}S$$
$$\approx \delta a^{\mathrm{T}} r \qquad (5.26)$$

In these equations, $\delta\phi$ is a virtual change in head, which is zero at any point at which the head is specified but is otherwise arbitrary;

$$\delta \mathrm{i} = [\delta i_x \quad \delta i_y \quad \delta i_z]^{\mathrm{T}}$$
$$= \left[\frac{\partial \delta\phi}{\partial x} \quad \frac{\partial \delta\phi}{\partial y} \quad \frac{\partial \delta\phi}{\partial z}\right]$$

is the vector of virtual hydraulic gradients: v is the vector of Darcy velocities; g is the component of the Darcy velocity in the direction of the positive outward normal to the surface S; $a = [\phi_1 \, \phi_2 \, \ldots]^{\mathrm{T}}$ is the vector of nodal heads; $\delta a = [\delta\phi_1 \, \delta\phi_2 \, \ldots]^{\mathrm{T}}$ is the vector of 'virtual' nodal heads; K is the 'flow stiffness matrix', which is known and can be calculated from the elements' configurations and properties; r is the vector of applied nodal flows, which can be calculated from the specified boundary flows on S_q and, if appropriate, S_ψ.

If the approximations, equations 5.25 and 5.26, are substituted into the equation of 'virtual work' (equation 5.24) it follows that it will be approximately satisfied if

$$\delta a^{\mathrm{T}}(Ka - r) = 0 \qquad (5.27)$$

This equation is true for any variation of 'virtual head', $\delta\phi$, that is for arbitrary values of δa, and so equation 5.27 implies that

$$Ka = r \qquad (5.28)$$

Equation 5.28 is a set of linear equations which can be solved to determine the nodal heads a.

5.6 Boundary element methods

The finite element method provides a powerful method of analysis for the determination of flow. It can deal with complex geometric configurations, a number of different material types and complex flow behavior. However, particularly if it is desired to model three-dimensional behavior, it can involve the solution of a very large set of linear equations. This disadvantage can be overcome, if the soil being considered is homogeneous, by using the boundary element method (Brebbia and Dominguez, 1989).

The boundary element method considers a slight generalization of the flow equation

$$k\nabla^2\phi = f \tag{5.29}$$

where f represents the volume of water removed from unit volume of soil in unit time.

Suppose that an arbitrary function ϕ^* satisfies certain conditions of differentiability, such that

$$\int_V \left(v_x \frac{\partial\phi^*}{\partial x} + v_y \frac{\partial\phi^*}{\partial y} + v_z \frac{\partial\phi^*}{\partial z} \right) dV = \int_V f\phi^* \, dV + \int q\phi^* \, dS \tag{5.30}$$

If, instead, the function ϕ^* satisfies the equation

$$k\Delta^2\phi^* = f^*$$

then it follows by interchanging the roles of ϕ and ϕ^* that

$$\int_V \left(v_x^* \frac{\partial\phi}{\partial x} + v_y^* \frac{\partial\phi}{\partial y} + v_z^* \frac{\phi}{\partial z} \right) dV = \int_V f^*\phi \, dV + \int_S q^*\phi \, dS \tag{5.31}$$

where

$$[v_x^* \quad v_y^* \quad v_z^*] = -k \begin{bmatrix} \frac{\partial\phi^*}{\partial x} & \frac{\partial\phi^*}{\partial y} & \frac{\partial\phi^*}{\partial z} \end{bmatrix}$$

are the Darcy velocities generated by the head ϕ^*, and q^* is the normal component of this Darcy velocity at the boundary.

Finally, if ϕ^* is chosen to be the solution corresponding to a point source at position $\boldsymbol{r}_0$ in an unbounded medium, and if f is zero so that there is no mechanism removing water from the soil, then substitution of Darcy's law and subtraction of equations 5.30 and 5.31 lead to the boundary integral equation:

$$\varepsilon\phi(\boldsymbol{r}_0) = \int_V (\boldsymbol{q}\phi^* - \boldsymbol{q}^*\phi) dS \tag{5.32}$$

where

$$\begin{aligned} \varepsilon &= 1 \quad \text{if } r_0 \text{ is within } V \\ &= \tfrac{1}{2} \text{ if } \boldsymbol{r}_0 \text{ is on a smooth portion of } S \\ &= 0 \quad \text{if } \boldsymbol{r}_0 \text{ is outside } V \end{aligned}$$

Equation 5.32 represents the head at any point, and in particular at a point on the boundary in terms of the value of head, ϕ, and the value of outward flow, q, on the boundary of the body.

The boundary element method then seeks to approximate the actual field behavior in terms of the values of ϕ and q at a certain number of nodes on the surface of the body. A systematic application of this approach (Brebbia and Dominguez, 1989; Grouch and Starfield, 1983) leads to the approximating set of equations

$$\boldsymbol{Hu} = \boldsymbol{Gq} \tag{5.33}$$

where $\boldsymbol{u} = [\phi_1 \, \phi_2 \ldots]^T$ is the vector of modal boundary heads; $\boldsymbol{q} = [q_1 \, q_2 \ldots]^T$ is the vector of nodal flows; $\boldsymbol{H}$, $\boldsymbol{G}$ are known square matrices.

At every boundary point, j, either the head ϕ_j is specified or the outward normal velocity q_j is specified, so that equation 5.33 can be used to determine all unknown heads and all unknown flows by solving a set of linear equations. Once this has been done, the head and flow on the boundary are completely defined, and so equation 5.32 can be used to determine the distribution of head throughout the body.

Chapter six

Chemical transfer by diffusion

6.1 Introduction

The basic concepts of pollutant migration by diffusion through clayey soils are presented in this chapter. Chapter 7 deals with modeling of advective–diffusive transport, Chapter 8 deals with laboratory evaluation of parameters and Chapter 9 describes observed field behavior.

In field situations where there are very low hydraulic conductivities or very small hydraulic gradients, there will be little or no fluid flow. In these cases, the mass flux through a barrier may be controlled by chemical concentration gradients. This phenomenon was discussed in Chapter 1. Since clay minerals are surrounded by adsorbed cations in amounts of 5–150 milliequivalents/100 g of solid, the replacement of exchange cations and, in some cases, anions, becomes an important factor in retardation of certain dissolved species. Therefore, retardation and the use of distribution coefficients is also discussed.

During hydraulic conductivity testing of compacted soils, diffusion into soil peds from adjacent compaction-induced macropores and fractures causes tailed breakthrough curves of effluent concentration. This requires extended testing times and increased costs for clay–leachate compatibility testing, especially if K^+ and NH_4^+ fixation occur.

6.2 Free solution diffusion (D_0)

Although the modeling techniques for prediction of pollutant migration are presented later in Chapter 7, a brief review of the simple one-dimensional transport equations at this point will serve to demonstrate the nature of diffusion and the component parts of an effective diffusion coefficient.

Maximum rates of migration by diffusion occur in bulk or free water at extreme dilution. The driving force for diffusion is a concentration gradient, or, more correctly, a chemical potential gradient, as shown schematically in Figure 6.1. The equation for the mass flux, f, is Fick's first law:

$$f = -D_0 \frac{dc}{dx} \tag{6.1}$$

where f is the mass flux produced by the concentration gradient dc/dx [$ML^{-2}T^{-1}$]; D_0 is the free solution diffusion coefficient [L^2T^{-1}]; c is the solute concentration in the fluid at any point x [ML^{-3}]; x is the distance from the high concentration interface [L].

Typical values for D_0 (Lerman, 1979) are presented in Table 6.1 for several species of cations and anions which might be encountered

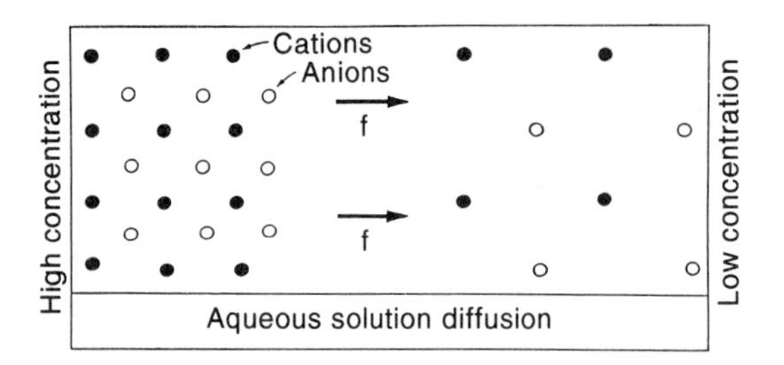

Figure 6.1 Schematic illustration of diffusion of cations and anions from high to low concentration in water. Equilibrium would be reached when a uniform concentration develops.

Table 6.1 Tracer diffusion coefficients, D_0, of selected species at infinite dilution at 25°C (Lerman, 1979)

Species	*D_0 (25°C) × 10^{10} m^2/s*	
Cations		
H^+	93.1	Note high D_0
Li^+	10.3	Monovalent
Na^+	13.3	
K^+	19.6	
NH_4^+	19.8	
Mg^{2+}	7.05	Divalent
Ca^{2+}	7.93	
Mn^{2+}	6.88	
Fe^{2+}	7.19	
Cu^{2+}	7.33	
Zn^{2+}	7.15	
Cd^{2+}	7.17	
Pb^{2+}	9.45	
Anions		
OH^-	52.7 × 10	Note high D_0
Cl^-	20.3	Monovalent
HS^-	17.3	
NO_2^-	19.1	
NO_3^-	19.0	
HCO_3^-	11.8	
SO_4^{2-}	10.7	Divalent
CO_3^{2-}	9.55	
PO_4^{3-}	6.12	
CrO_4^{2-}	11.2	

in leachate. These values, determined at infinite dilution, would normally represent the maximum values measurable in any laboratory or field situation at the nominal temperature. It is important to note that requirements for electroneutrality generally demand that a cation–anion pair diffuse together, frequently slowing one or other of the migration components.

Even D_0 is a complex coefficient, as shown by the Nernst equation (Lerman, 1979):

$$D_0 = \frac{RT}{F^2|z|}\lambda_i^0 \tag{6.2}$$

where R is the universal gas constant (= 8.314 J/K/mol); F is the Faraday constant (= 96 500 coulombs); z is the valence; T is the absolute temperature in kelvin; λ_i^0 is the limiting ionic conductance, which is maximum at infinite dilution, when the ions are most mobile, and increases with increasing temperature (cm^2/Ω equivalent).

For strong electrolytes, ionization is complete over a wide range of concentrations, and λ stays close to λ^0. For less easily dissociated species, λ decreases with increasing concentration, as many ions diffuse in molecular form, thus decreasing the diffusion coefficient. The ionic mobility, u_i, which is similar in form to D_0, is related to λ_i^0 by the equation

$$u_i = \frac{N\lambda_i^0}{F^2|z_i|} \tag{6.3}$$

where N is Avogadro's number, the number of units in a mole (= 6.022 × 10^{23}).

The effect of ionic potential, $|z|/r$, on values of D_0 calculated from equation 6.2 is shown in Figure 6.2. The plot shows a decrease in D_0 with increasing $|z|/r$ which indicates that the greater the degree of ion hydration, the lower is D_0. This is particularly well-shown by the monovalent alkalies, Li, Na, K, Rb and Cs, the latter being the largest cations with the smallest hydrated radii (Mitchell, 1993). The atomic weight also increases significantly from Li^+ to

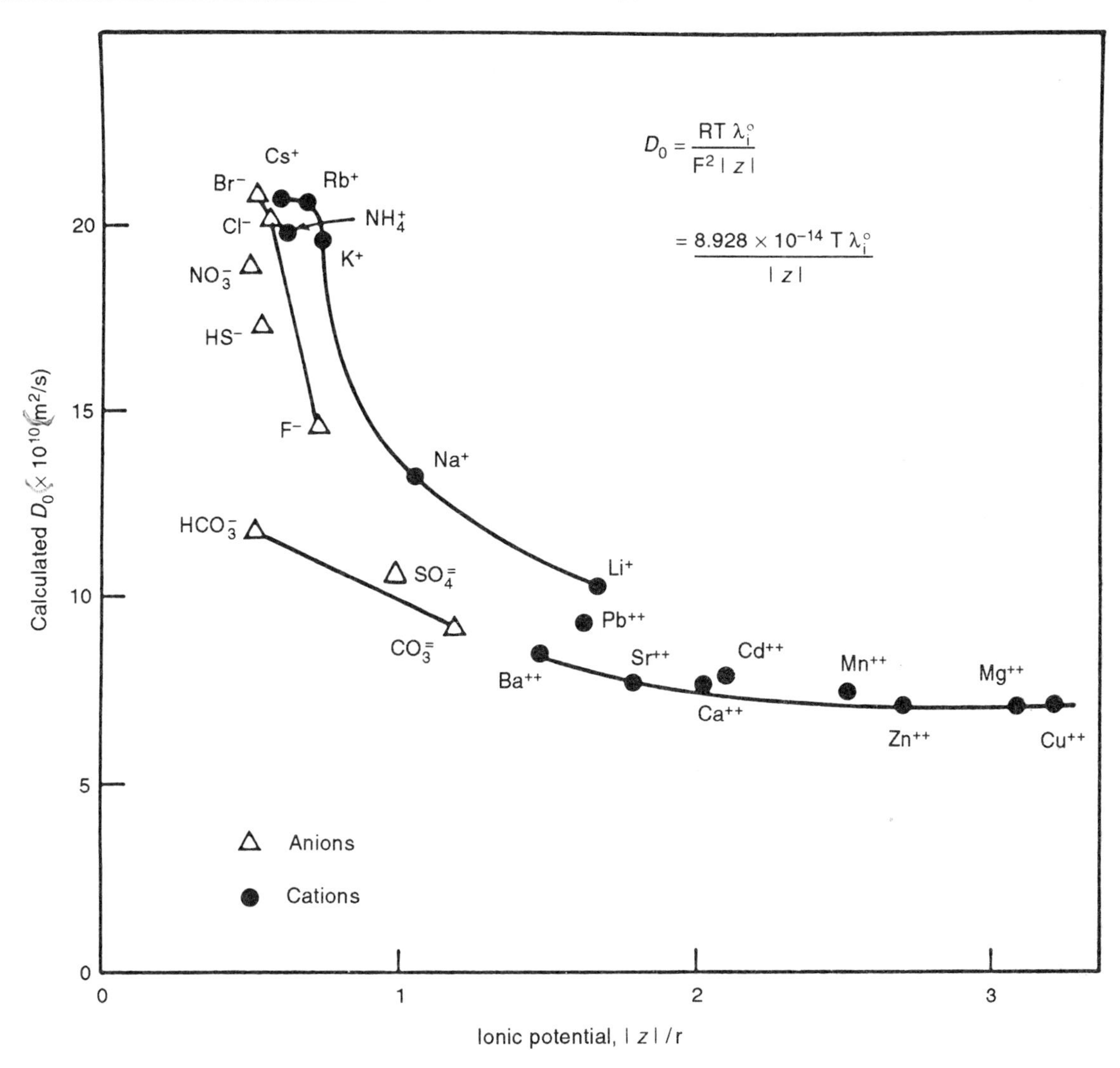

Figure 6.2 Relation between free solution diffusion coefficient, D_0, and ionic potential, $|z|/r$, at 25°C ($|z|$ is the valence, r is the ionic radius). (Quigley, Yanful and Fernandez, 1987; reproduced with permission of ASCE.)

Cs^+, another factor which could influence the rate of diffusion in a gravitational field.

The effects of temperature on D_0 are shown in Figure 6.3 for a number of species common to MSW leachate. Chloride, which is frequently considered a conservative reference species, has a relatively high diffusion coefficient. Despite the similarly high diffusion values for K^+ and NH_4^+, their rate of migration in the field may be substantially lower than implied by the diffusion coefficient above, since both are highly retarded by clays in most field situations. If landfill temperatures reach and stay at 50°C for several years, consideration should be given to the temperature in the liner, and if this exceeds 20–25°C, then the diffusion coefficient D may

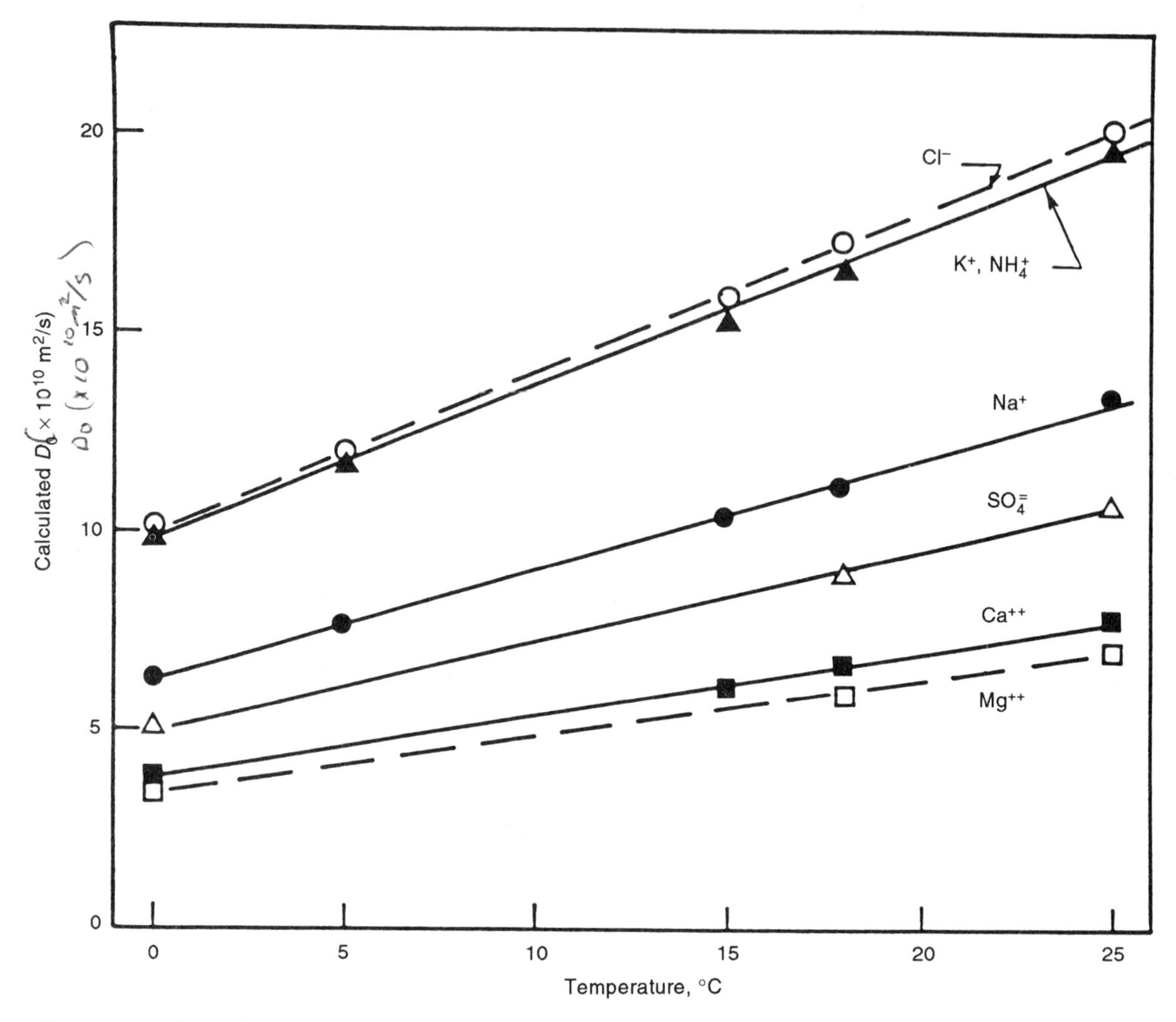

Figure 6.3 Relation between free solution diffusion coefficient, D_0, and temperature for selected species.

be higher than typical laboratory values, which are normally obtained either at 20–25°C or 10°C.

In summary, the mobility of a species in free solution, as expressed by its D_0 value, is a complex function of its mass, radius, valence and concentration–dissociation state, and the viscosity, dielectric constant and temperature of the diffusing medium.

6.3 Diffusion through soil

The presence of soil particles, particularly adsorptive clay minerals and organic matter, complicates the diffusion process. Diffusion through a network of clay particles involves the diffusive movement of the species of interest in the pore water between the clay particles, as illustrated in Figure 6.4.

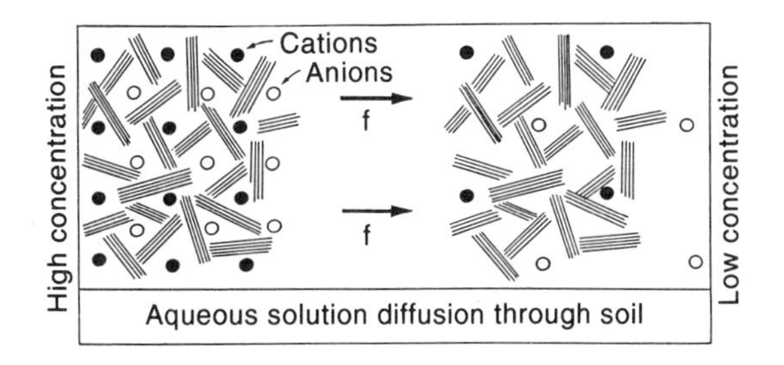

Figure 6.4 Schematic illustration of diffusing cations and anions from high to low concentration through a saturated clayey soil.

The rate of chemical movement or mass flux through a soil may be slower than by diffusion in pure water, for the following reasons:

1. tortuous flow path around particles (tortuosity, τ);
2. small fluid volume for flow (porosity, n, or volumetric water content, θ);
3. increased viscosity, especially the double-layer portion of the pore water;
4. retardation of certain species by cation and anion exchange (retardation) by both clay minerals and organics;
5. biodegradation of diffusing organics;
6. counter-osmotic flow;
7. electrical imbalance, possibly by anion exclusion.

However, certain events in a soil may also accelerate movement; for example:

1. decreased viscosity caused by any of the factors which cause double-layer contraction at constant void ratio;
2. decreased viscosity caused by K^+ fixation by vermiculite which decreases the CEC and increases the free water pore space at constant void ratio;
3. electrical imbalances such as accelerated anion migration that would electro-osmotically pull cations;
4. electrical imbalances related to cation exchange and hardness halos that might pull chloride to help balance desorbed Ca^{2+} and Mg^{2+};
5. attainment of chemical equilibrium eliminating retardation of certain species.

With these ideas in mind it is appropriate to look first at the process of steady-state diffusion, then at the process of transient diffusion.

6.4 Steady-state diffusion

Two useful examples of steady-state diffusion through a bentonitic soil were presented by Dutt and Low (1962) and Kemper and van Schaik (1966). In both papers, the concentration gradient was produced by a chloride salt across a very porous saturated montmorillonite barrier homoionized in the same cation as the diffusing salt. A sketch of the Dutt and Low experimental set-up is presented in Figure 6.5; the one-dimensional equation for a mass flux used by Kemper and van Schaik was

$$f = -D_p \frac{dc}{dx} \tag{6.4}$$

where D_p is the porous media diffusion coefficient $[L^2T^{-1}]$; D_0 is related to D_o by the relationship $D_p = D_0 W_T$, where W_T is the complex tortuosity factor, which equals unity for a pure solution and is normally less than

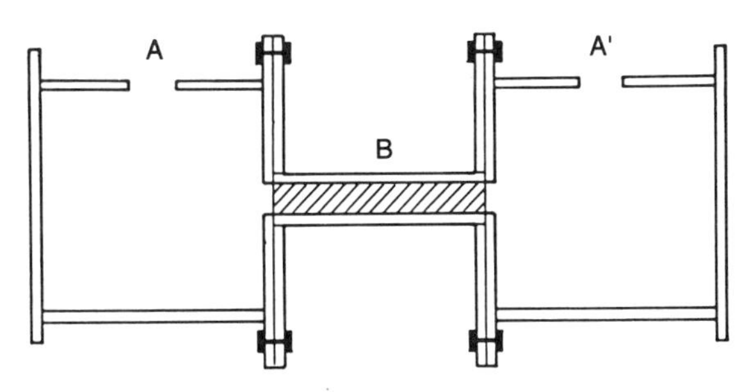

Figure 6.5 Schematic of diffusion cell used by Dutt and Low (1962).

unity in soils. In reality, many factors are lumped into W_T, as discussed later. It is most important to note that the porosity, n (which equals the volumetric water content, θ, for a saturated soil), is included in D_p in equation 6.4, possibly because the authors' slurries had high porosities of 0.85. In the case of clayey barriers and porosities close to 0.3, D and n should probably be clearly separated from one another. Thus, $D_p = \theta D_e = nD_e$ where D_e is the effective diffusion coefficient in the soil pores. Also in practical work, τ is used as the symbol for tortuosity, as subsequently defined in equation 6.7b.

The experimental procedure of the above authors, illustrated schematically in Figure 6.6, appears to be a deceptively simple way of measuring diffusion coefficients, just by measuring the mass flux, f, across the barrier. Equation 6.4 becomes

$$f = -D_p \frac{\Delta c}{\Delta x} \qquad (6.5)$$

where $\Delta c = c_2 - c_1$, and Δx is the barrier thickness (Figure 6.6).

An important observation from these two sets of experiments was a sharp drop in concentration gradient across both reservoir–soil interfaces, especially the input or high concentration

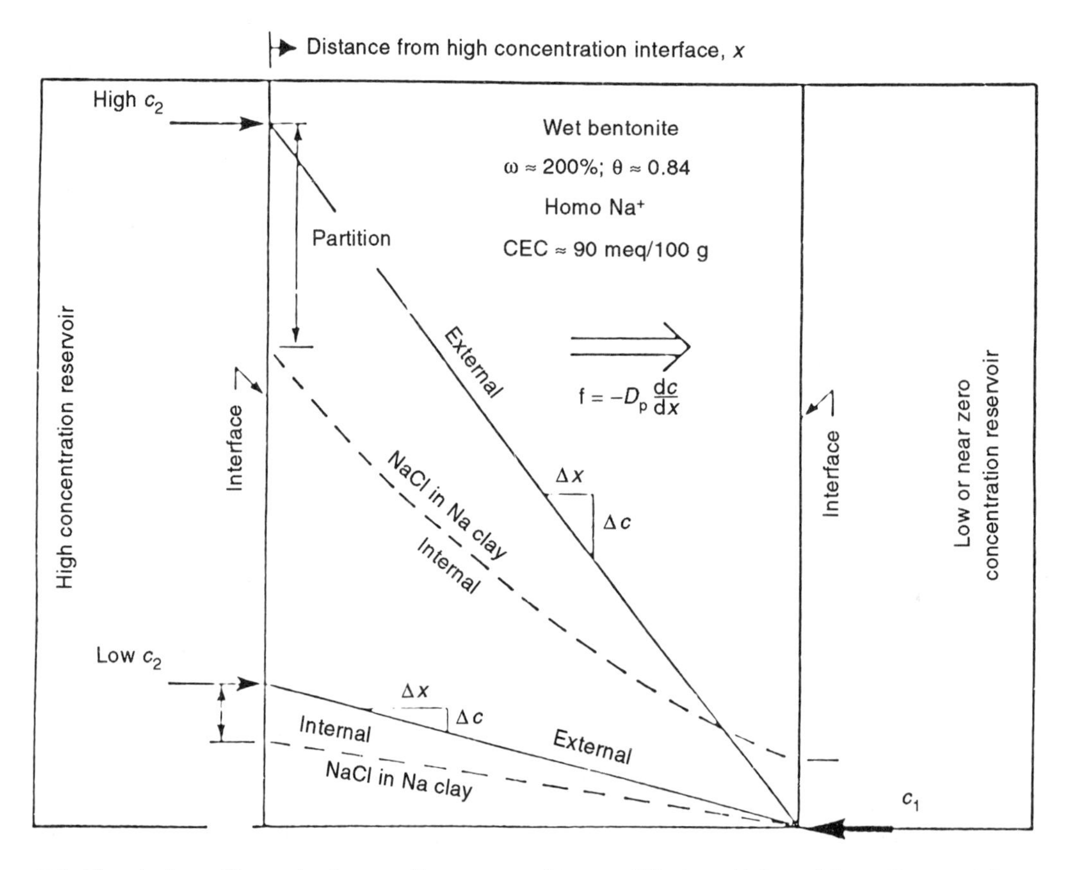

Figure 6.6 Chemical profiles and other gradients at steady-state diffusion. (Adapted from Dutt and Low, 1962; Kemper and van Schaik, 1966.)

side. Dutt and Low argued that the chloride part of the Na^+–Cl^- ion pair would tend to be rejected by the Na^+: clay$^-$ system, hence resisting diffusion (anion exclusion), and effectively producing a very low diffusion coefficient and a large concentration gradient in a thin interfacial layer.

A similar sharp concentration drop across the waste–clay interface at the Confederation Road landfill site was believed to have been observed by Crooks and Quigley (1984). This has been attributed to plugging by heavy metal precipitation, Na^+ adsorption and possibly bacterial growth. The concentration profiles were modeled by Quigley and Rowe (1986), assuming a greatly reduced diffusion coefficient in a thin interfacial zone. Interface partitioning in the field remains unresolved, however, since it has not been observed at other sites. At Confederation Road, a mat of fine black organics may have shifted the effective interface up 25 or 30 cm, thus confusing the observations.

Since the mass flux is constant at all points in a steady-state system, both external (across the barrier) and internal (within the clay) diffusion coefficients maybe calculated, and these are presented in Table 6.2 for the Dutt and Low data.

It is significant that the much larger internal D_p values were similar to those obtained by deuterium diffusion and were considered characteristic of the soil, whereas the lower external values seem more characteristic of the system.

Table 6.2 Average external and internal diffusion coefficients obtained by Dutt and Low (1962)

Salt–clay system	D_p (m^2/s), approx. values	
	External (across) barrier)	*Internal (within clay)*
LiCl/LiMont	1.3×10^{-10}	4.1×10^{-10}
NaCl/NaMont	2.1×10^{-10}	3.7×10^{-10}

Major curvature of the pore water concentration profiles at high concentrations of NaCl (Figure 6.6) was explained by Kemper and van Schaik (1966) as being caused by water movement to the low salt side (0.01 N) of their wet Na^+ bentonite (Figure 6.7) at the expense of the high salt side (1.0 N). Other experiments with more dilute NaCl systems and with $CaCl_2$ yielded little water movement and nearly linear salt concentration profiles across the barrier. Such water migration might occur in slurry walls, but seems unlikely for most natural or compacted clays, unless the soils are left completely unconfined.

Chemical concentration gradients, while in themselves easy to visualize, induce complex, coupled electrical and hydraulic gradients that affect the net flow attributable to diffusion. In Table 6.3, transfer rates calculated by Kemper and van Schaik for only two of their pure salt systems are presented. This was possible because Kemper and van Schaik employed electrode screens at the ends of the soil to assess the relative speed of Na^+ and Cl^- migration by measuring charge build-up. A positive transfer rate indicates flow from the high concentration

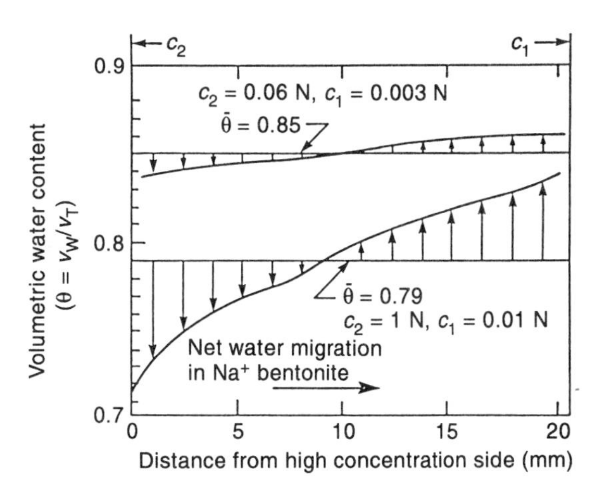

Figure 6.7 Water content profiles at steady-state diffusion for high NaCl contents and wet clays. (Adapted from Kemper and van Schaik, 1966.)

Table 6.3 Components of cation and anion movement during steady-state diffusion through soft bentonite (adapted directly from Kemper and van Schaik, 1966)

Adsorbed cation:	*Na*	*Ca*
Salt solution:	*NaCl*	$CaCl_2$
Concentrated bulk solution		
High side	3.21 g/l (0.055 N)	5.6 g/l (0.05 N)
Low side	0.18 g/l (0.003 N)	0.3 g/l (0.003 N)
Average volumetric water content, θ	0.85	0.84
Transfer rates of cations ($meq/m^2/s \times 10^{12}$)		
By diffusion	+1.8	+0.84
By electrical flow	+4.1	+0.98
By bulk flow	−3.8	−0.12
Net flow	+2.1	+1.70
Transfer rates of anions ($meq/m^2/s \times 10^{12}$)		
By diffusion	+2.7	+2.14
By electrical flow	−0.2	−0.37
By bulk flow	−0.4	−0.07
Net flow	+2.1	+1.70

to the low concentration side. The concentration gradient produces positive cation and anion migration, the rate of anion migration being greater. This collection of surplus anions (Cl^- in this case) generates a negative potential at the low concentration side of the clay. This negative potential in turn attracts hydrated cations by positive direction electro-osmotic flow, and repels anions by negative direction flow. The large, negative direction bulk flow is osmotic flow through the clay, as water from the low concentration side seeks to enter the high concentration side. The net mass flux is in the positive direction, and, while complex, all effects have resulted from the initial concentration gradient.

The porous system diffusion coefficient, D_p, calculated from Fick's law for such a system is a combination of the above factors, and might be written

$$D_p = D_0 W_T = \theta D_e \qquad (6.6a)$$

in which W_T, referred to as the complex tortuosity factor, has the following components:

$$W_T = \alpha_{ff}\gamma_e(L/L_e)^2\,\theta = \tau\theta \qquad (6.6b)$$

and hence from equations 6.6a and 6.6b,

$$D_e = \alpha_{ff}\gamma_e(L/L_e)^2\, D_0 = \tau D_0 \qquad (6.7a)$$

$$\tau = \alpha_{ff}\gamma_e(L/L_e)^2 \qquad (6.7b)$$

where α_{ff} is the decreased fluidity factor related to adsorbed double-layer water; γ_e is the electrostatic interaction factor; $(L/L_e)^2$ is the geometric tortuosity factor; θ is the volumetric water content (= n, the porosity for a saturated system), and $\tau = \alpha_{ff}\gamma_e(L/L_e)^2$ is the tortuosity factor for the clay.

In texts such as Bear (1979) the use of the term 'tortuosity' relates only to the geometric tortuosity $(L/L_e)^2$ and ignores the effects of α_{ff} and γ_e. While this may not be a bad approximation for granular materials, it is not correct to ignore α_{ff} and γ_e for clayey soils, and, as a result, the tortuosity factor for clayey soil, τ, is often less than the geometric tortuosity calculated by Bear for an ideal problem (i.e. $(L/L_e)^2 = 0.67$).

Using values of D_p calculated from their experiments (1.8 to 3.4×10^{-10} m²/s), Kemper and van Schaik calculated values for W_T which varied from about 0.07 to 0.3. Most of the variation in D_p^* was believed be caused by changes in volumetric water content, and this is shown in Figure 6.8, with θ expressed as the calculated number of water layers surrounding the bentonite particles.

An important study of nonsteady state diffusion offering further clarification of the nature of diffusion and its coupled electrical and hydraulic gradients was presented by Elrick *et al.* (1976) for Na^+ bentonite. In this study, a constant volume cell (Figure 6.9) prevented osmotic flow across the barrier (from low to high concentration), and a large pressure head developed on the high concentration side (Figure 6.10). The low concentration side of the barrier developed a positive charge due to Na^+ diffusion

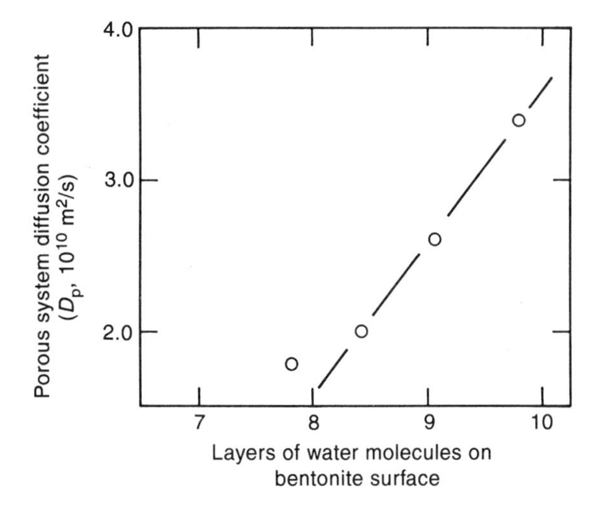

Figure 6.8 Porous system diffusion coefficient versus water content expressed as water thickness on clay particles. (Adapted from Fig. 8, Kemper and van Schaik, 1966.)

and retarded Cl^- diffusion due to exclusion (opposite effect to the open system of Kemper and van Schaik, 1966).

Silver electrodes coated with silver chloride were used to short-circuit the system at 20 hours (Figure 6.10), enabling electron flow from the high concentration side of their barrier to the low concentration side without the necessity of Cl^- migration through the barrier. The electric potential was eliminated, the rate of Na^+ diffusion, *f*, increased greatly, and the abundant water of hydration traveling with the Na^+ caused the hydraulic potential originally on the high side to reverse itself to the low concentration side (Groenevelt, 1986, personal communication). In this particular experiment, eliminating the need for chloride migration by allowing direct electron flow increased the diffusion rate. It appears, therefore, that although Cl^- is normally treated as an unretarded species, in certain experimental set-ups exclusion may occur, especially at interfaces, resulting in reduced diffusion coefficients for the migrating salt.

This nonconservative nature of chloride migration appears in other experimental measuring systems and must be understood as it affects measurement of reliable diffusion coefficients for use in field situations.

6.5 Transient diffusion

In the previous homoionic, monomineralic systems, steady-state diffusion is reached fairly rapidly compared to a natural heterogeneous soil where extensive cation exchange, precipitation–dissolution and biodegradation may occur. Use is then made of Fick's second law, which describes the time rate of change of concentration with distance, as follows:

$$\frac{dc}{dt} = D^* \frac{d^2c}{dx^2} \tag{6.8}$$

where D^* is a retarded diffusion coefficient which may include the chemical/biological

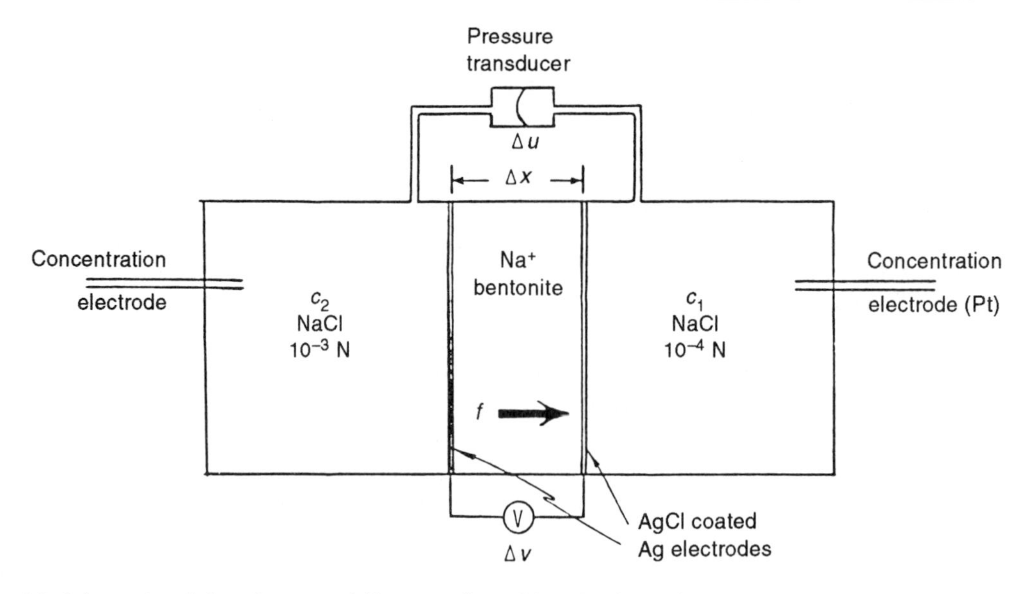

Figure 6.9 Schematic of closed system diffusion cell used by Elrick *et al.* (1976). (Osmosis not allowed.)

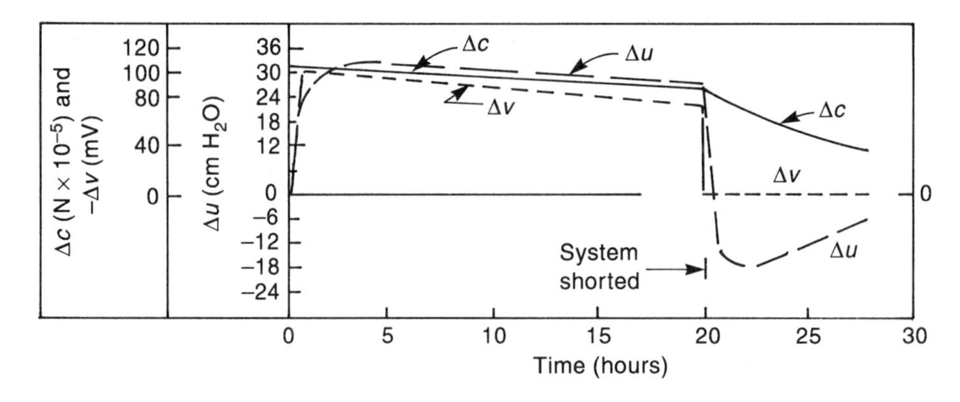

Figure 6.10 Gradients of concentration, pressure head and voltage versus time before and after electric shorting across the sample. (Adapted from Elrick *et al.*, 1976.)

retardation factors just mentioned and which itself may be described as follows:

$$D^* = \frac{D_0\tau}{R} = \frac{D_0\tau}{1 + (\varrho/\theta)\,K_d} = \frac{D_e}{1 + (\varrho/\theta)\,K_d} \tag{6.9}$$

where ϱ is the bulk dry density of the soil $[ML^{-3}]$; θ is the volumetric water content $(= V_W/V_T\ [-]$, $\theta = n$, the porosity, V_V/V_T (L^3/L^3) for 100% saturation); K_d is the distribution coefficient $[L^3M^{-1}]$ which may reflect the degree of retardation by reversible ion exchange, but may also include partitioning of organics onto organic matter in the soil and even precipitation; D_e is the effective diffusion coefficient $[L^2T^{-1}]$.

An analytical solution to equation 6.8 (Ogata,

1970) for an infinite layer with a constant surface concentration c_0 and zero advection is given by

$$\frac{c}{c_0} = \operatorname{erfc}\left(\frac{x}{2(D^*t)^{1/2}}\right) = 1 - \operatorname{erf}\left(\frac{x}{2(D^*t)^{1/2}}\right) \quad (6.10)$$

A series of relative concentration versus depth curves for chloride is plotted in Figure 6.11, along with a velocity plot of the 50% relative concentration front ($c/c_0 = 0.5$). The figure shows that the rate of diffusion is quite rapid for the first two or three years (150 → 50 mm/a), but decreases very rapidly with time, due to the decrease in dc/dx as x increases. The initial speed of diffusion makes it possible to measure values of diffusion coefficient in laboratory tests lasting only a week or so. As will be illustrated later, these short-term values are quite close to long-term values obtained from field profiles.

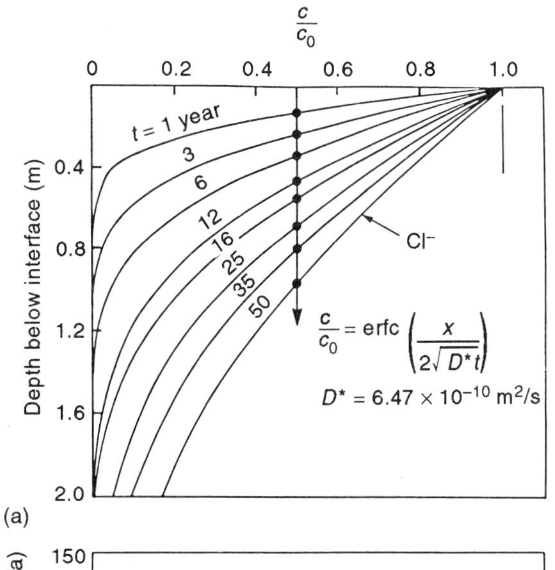

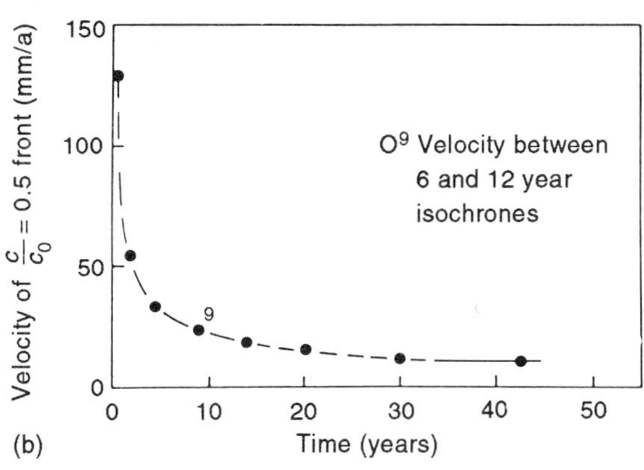

Figure 6.11 Time rate of migration by diffusion: (a) relative concentration–depth–time plots; (b) velocity of migration of $c/c_0 = 0.5$ front.

If c_0 can be kept constant, and interface partitioning is not a problem, experiments can be set up so that the concentration versus distance profile can be measured chemically or radioactively after a known period of elapsed time (e.g. Lai and Mortland, 1962; Gillham *et al.*, 1984). By assuming values for D^*, a calculated curve can be fitted to the experimental curve, and D^* selected. If a nonreactive species is also used (e.g. Cl^-, tritium deuterium), maximum values of $D^* = D_e$ are obtained and the tortuosity factor can be calculated from $D_e = D_0\tau$. Using this τ-value, retardation factors and K_d values may be estimated for the other migrating species of interest.

If c_0 varies during the period of the experiment or during field exposure, more elaborate modeling methods are required to calculate D_e and K_d (as described in Chapter 8); however, both parameters can be estimated from one test on a single soil sample.

Values of K_d may also be obtained from batch tests in which the soil is mixed with the leachate solution in question, and the loss of a species from the leachate to the soil is measured, usually by chemical analysis of the leachate. Either a linear, Langmuir or Freundlich type of plot may be used, as discussed in Chapter 1. The plot shown in Figure 6.12 is a log–log form of the Freundlich equation as follows:

$$\ln S = \ln K_f + \varepsilon \ln c_e \quad (6.11)$$

where S is the mass removed from solution per unit mass of soil solids; c_e is the concentration of dissolved solute in the leachate at equilibrium; K_f is a constant which is proportional to adsorption capacity; ε is the slope of the line which reflects the intensity of adsorption with increasing c_e.

It is noted that while K_f and ε are constant, K_d is not; it varies as the function of concentration. Thus the distribution coefficient K_d may be estimated at a given concentration level, as discussed in section 1.3.5.

The second and third lines drawn in Figure 6.12 (labeled soils 2 and 3) represent soils with greater adsorptive capacities than soil 1, and therefore higher K_d values for the species being measured. The soil 2 line is parallel to that for soil 1, indicating the same influence of liquid concentration on the amount adsorbed whereas soil 3 with a greater slope demonstrate a greater influence of liquid concentration.

Finally, adsorption isopleths of the nature shown in Figure 6.12 are only valid within certain soil and liquid concentration limits so the mixes used have to be designed in order to produce values representative of field conditions.

Adsorption isopleths are generally assumed to represent cation exchange, but frequently incorporate the effects of fixation and precipitation in the presence of carbonate at high pH values (Griffin *et al.*, 1976). An example showing the effects of pH on removal of copper from a spiked leachate is shown in Figure 6.13. In this figure the concentrations are plotted arithmetically, as distinct from the log–log plots in Figure 6.12. The amount of copper removed from solution is far greater for the carbonate-rich soil at pH 8.2 than for the carbonate-free soil at a slightly acidified leachate pH of 5.5. Since the amount removed is greater than the exchange capacity of the soil, precipitation as copper carbonate is probable. The apparent K_d values are therefore much greater for carbonate-rich clayey barriers with respect to heavy metal retardation.

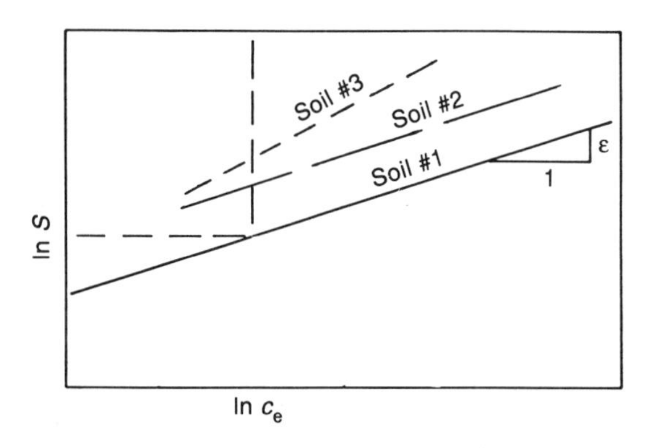

Figure 6.12 Linear log–log form of Freundlich adsorption equation.

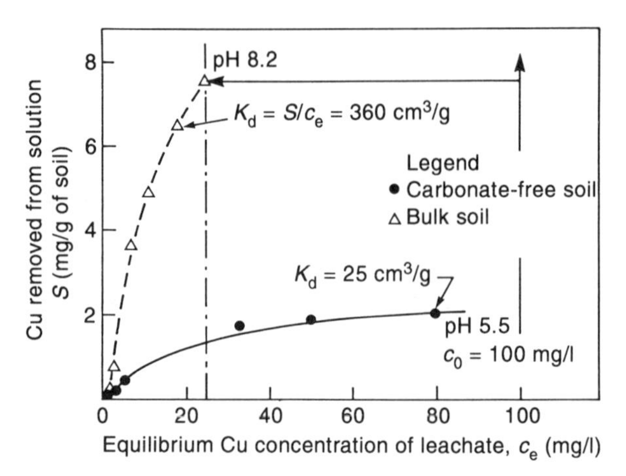

Figure 6.13 Mass of Cu removed from spiked landfill leachate at 22°C versus equilibrium concentration (soil:solution = 0.25 g:50 mL). (Adapted from Yanful, Fernandez and Quigley, 1987.)

In summary, batch tests, while useful, are usually considered as only a guide to K_d values. Nonlinearity with increasing solution concentration, pH and temperature factors, precipitation etc. all make it preferable to measure D and K_d from laboratory or field diffusion profiles.

6.6 Use of laboratory and field profiles to measure diffusion coefficient D_e

The use of laboratory and field profiles to estimate the diffusion coefficient D_e and distribution coefficient K_d is deferred to Chapter 8, after the modeling required to interpret some of these profiles has been discussed. It is noted that laboratory measurement of diffusion coefficients is not easy; however, the values so obtained appear to be quite close to values back-calculated from observed field behavior, and hence when appropriately determined, can be used with a considerable degree of confidence.

6.7 Diffusion during hydraulic conductivity testing for clay–leachate compatibility

Environmental approval often requires an assessment of clay–leachate compatibility, which is usually made by means of hydraulic conductivity testing (Chapter 4). Such tests should be run until soil–leachate equilibrium is reached, or at least until effluent concentrations equal influent values (Bowders *et al.*, 1986; Daniel *et al.*, 1984). Fracture flow and diffusion of chemicals into adjacent soil peds greatly complicates chemical control, as discussed in section 4.1, and elaborated on here.

A typical chemical control plot of relative concentration (effluent, c, divided by influent c_0) versus pore volumes is shown in Figure 6.14 for a domestic waste leachate passing through a compacted, weathered, carbonate-rich, brown, southern Ontario soil containing about 8% smectite. This figure, which is similar to the plot presented in Figure 4.5, shows that even after the passage of 12 pore volumes (PV) of leachate, potassium is still being retarded by the soil, probably by K^+ fixation on high charge smectites. This slow development of a condition of equilibrium may require testing times as long as three months on soil specimens as thin as 20 mm with high gradients (>500).

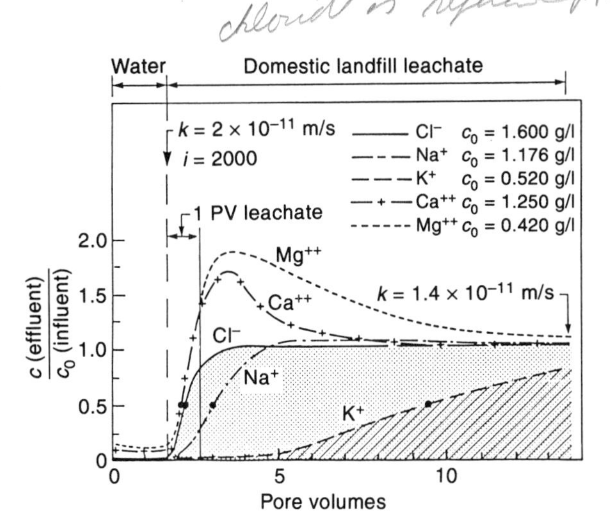

Figure 6.14 Typical curves of relative concentration of effluent obtained from high gradient hydraulic conductivity testing.

Effluent chloride concentration reaches 50% of the c_0 value at about 0.5 PV, indicating early arrival due to channel flow. Sodium is mildly retarded ($c/c_0 = 0.5$ at 1.4 PV) and K^+ is severely retarded ($c/c_0 = 0.5$ at 8 PV). Adsorption of Na^+ and K^+ from the leachate onto the clay is compensated by Ca^{2+} and Mg^{2+} desorption resulting in early arrival ($c/c_0 = 0.5$ at 0.35 PV) and effluent concentrations far above influent values.

Interped flow channels or fractures, which can occur between peds in compacted soils, may significantly affect the measured values of hydraulic conductivity (Mitchell, Hooper and Campanella, 1965). Under these circumstances, leachate flowing under a high hydraulic gradient moves primarily along the macropore channels, and attenuation of contained dissolved species is primarily by diffusion into the adjacent soil peds.

The factors affecting attenuation or retardation were described mathematically by Grisak and Pickens (1980) and demonstrated in the laboratory for fractured till by Grisak, Pickens and Cherry (1980). A summary plot of this work as it applies to clay–leachate compatibility testing in an advection dominated system is shown in Figure 6.15. At low gradients that produce low velocity flow along the channels, all attenuating mechanisms function efficiently, resulting in low c/c_0 values early in the hydraulic conductivity test (Figure 6.15(a)). At high gradients, which produce high velocity flow

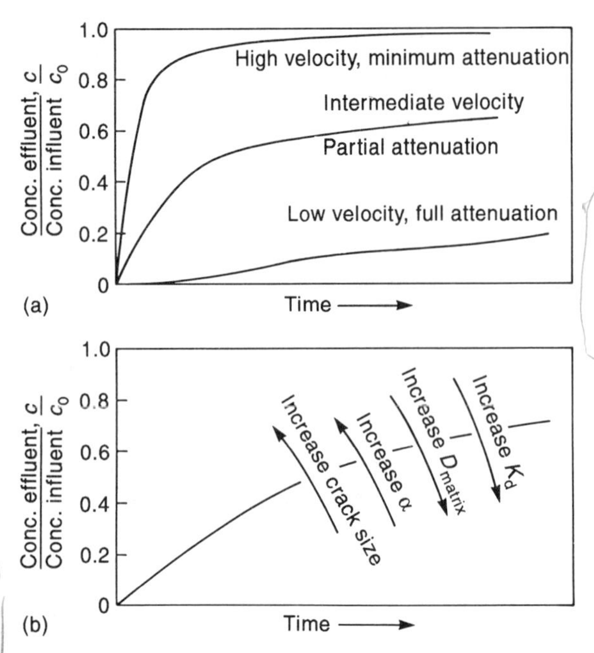

Figure 6.15 Factors affecting chemical attenuation during hydraulic conductivity testing through soils containing fractures or connected macropores: (a) fracture velocity effects; (b) controls at constant velocity (α is the dispersivity). (Adapted from Grisak and Pickens, 1980.)

along channels, minimum attenuation develops, and early breakthrough occurs, with high c/c_0 values.

The actual components of such retardation are illustrated in Figure 6.15(b). The retardation increases with an increase in the diffusion coefficient of the matrix peds, D_{matrix}, and/or in the distribution coefficient, K_d. An increase in the fracture size and/or in the dispersivity, α, of the system accelerates channel flow, causing early arrival in the effluent and inefficient clay–leachate interaction.

The important balance between the gradients employed and diffusion from the flow channels makes clay–leachate compatibility testing much more complicated than normally appreciated.

Using a method similar to that employed by Griffin *et al.* (1976) attenuation numbers have been calculated for the data shown in Figure 6.14. The chloride curve is used for the stippled reference area for input to compensate for channelling (dispersivity) factors and the attenuation number is defined as follows:

$$\text{ATN} = \frac{\text{stippled area (Cl}^-) - \text{hatched area (species)}}{\text{stippled area (Cl}^-)} \times 100$$

with all curves extrapolated to equilibrium at $c/c_0 = 1$. The hatched area is shown for K^+. Table 6.4 contains both the calculated attenuation numbers and arrival time for the 50% relative concentrations, expressed in pore volumes. The attenuation numbers reflect retardation in exactly the same sense as K_d values.

6.8 Summary and conclusions

The complexities of diffusion and its importance as both a transport mechanism and as an attenuating mechanism have been briefly discussed in this chapter. A few conclusions may be made as follows.

1. The process of steady-state diffusion expressed by Fick's first law is complex, even in homoionic, monomineralic systems, due to coupled osmotic and electro-osmotic flow, and anion exclusion, particularly at soil–liquid interfaces.
   (a) Fick's first law, $f = -nD_e\ (dc/dx)$.
   (b) For charge balance, the forward flow of cations should equal that of anions.
   (c) Concentration differences in open, steady-state diffusion systems create forward cation flow, forward cation electro-osmotic flow and reverse cation osmotic flow.
   (d) These same concentration differences create forward anion diffusion flow, reverse anion electro-osmotic flow and reverse anion osmotic flow.

Table 6.4 Attenuation numbers for selected leachate species permeated through brown Sarnia soil at $e = 0.74$

Species	*Arrival of 50% relative concentration (number of pore volumes)*	*Attenuation number relative to chloride*
Cl^-	0.52	0.0
Na^+	1.36	5.6
K^+	7.89	56.1
Ca^{2+}	0.35	–12.0
Mg^{2+}	0.35	–31.3

2. The variability in terminology leads to confusion, and one must carefully distinguish between different definitions of the diffusion coefficient and its component parts (e.g. D_p, D^*, D_e, as discussed herein). In particular, since D^* incorporates adsorption, it may increase as retardation equilibrium occurs.
3. The importance of diffusion in clay–leachate compatibility assessment by hydraulic conductivity testing was illustrated by a test which had not reached equilibrium after passage of even 12 pore volumes of municipal solid waste leachate.

(a) K^+ fixation remains a critical assessment problem because it can result in clay mineral contraction, reduced cation exchange capacity (CEC) values and double-layer contraction, all potential causes of cracking.

(b) The interrelationships of fracture flow velocity and matrix diffusion controls the effluent concentration versus time curves.

(c) K^+ fixation (by diffusion) may create testing times of up to three months, even at gradients as high as 500.

Chapter seven

Contaminant transport modeling

7.1 Introduction

A contaminant transport model consists of the governing equations together with the boundary and initial conditions. Once the model has been formulated and the appropriate parameters have been determined, it remains to find a solution to the governing equations subject to the appropriate boundary and initial conditions. The most frequently used solution techniques can be subdivided into five broad categories; namely, analytical, finite layer, boundary element, finite difference and finite element techniques.

The other techniques will be briefly discussed here, and then attention will be directed at providing some additional analytical solutions and describing the finite layer formulation.

7.1.1 Finite layer techniques

The finite layer technique is applicable to situations where the hydrostratigraphy can be idealized as being horizontally layered, with the soil properties being the same at any horizontal location within the layer. For these conditions, the governing equations can be considerably simplified by introducing a Laplace transform and a Fourier transform (the latter only being required for two- or three-dimensional problems). The transformed equations can then be readily solved. This procedure parallels that adopted in the development of many analytical solutions. The difference between finite layer solutions and analytical solutions arises from the fact that in the finite layer approach, the solution is inverted numerically rather than analytically. As a consequence, it is possible to examine more complicated and realistic situations.

Finite layer techniques have been described by Rowe and Booker (1985a,b, 1987) and are available as computer programs for 1½-D (POLLUTE: Rowe and Booker, 1994) and 2-D (MIGRATE: Rowe and Booker, 1988b) conditions.

For situations involving a clayey liner overlying a drainage layer which can be pumped, or overlying a thin natural aquifer (e.g. Figures 1.2–1.8), a reasonable initial estimate of contaminant impact can be obtained in seconds using a microcomputer and 1½-D finite layer programs such as POLLUTE (e.g. Rowe and Booker, 1994). The designation of these programs as 1½-D is intended to indicate that they consider one-dimensional transport down through aquitard layers and into the aquifer(s) or drainage layer(s), while approximately taking account of contaminant removal due to lateral flow or pumping of these permeable layers. These techniques will often provide a reasonable estimate of both the magnitude of the peak impact beneath the landfill and the time at which this occurs.

A more rigorous solution of this problem can be obtained using a full 2-D analysis (e.g. program MIGRATE). A comparison of the two approaches has been given by Rowe and Booker (1985b), and will be discussed in section 10.8. The 2-D approach allows one to consider impact at points outside the landfill. The full 2-D analysis can be readily performed on a micro-computer, but it does involve substantially more computation than the 1½-D analysis. Before performing 2-D analyses, a 1½-D analysis should always be performed to estimate the likely magnitude of impact and its time of occurrence beneath the landfill; the 2-D program can then run for appropriate times, using the 1½-D results as a reference. One of the advantages of finite layer techniques over conventional finite element methods is that it is unnecessary to determine solutions at times prior to the time period of interest.

Finite layer techniques have many of the advantages of analytical techniques. They are easy to use, and the user need not be an expert in numerical analysis. They also require minimal input, and only give results at the times and locations of interest. This technique is particularly well-suited for performing sensitivity studies to identify the potential impact of uncertainty regarding the value of key design parameters. The technique is also well-suited for performing checks on the results of numerical analyses using finite element or finite difference techniques.

Finite layer techniques are not appropriate for modeling situations where there is a complex geometry or flow pattern which cannot reasonably be idealized in terms of horizontal layers with uniform properties.

7.1.2 Boundary element techniques

The boundary element technique is suitable for solving the advection–dispersion equation (e.g. Brebbia and Skerget, 1984). Its primary advantage over finite layer methods is the ability to model more complicated geometries. To date, boundary elements have not found wide application in contaminant migration studies.

7.1.3 Finite difference and finite element techniques

There has been extensive research in the use of finite difference and finite element techniques for the analysis of contaminant migration through soils. These numerical techniques are likely to be used for:

1. calculating the steady-state flow pattern within a hydrogeologic system, thereby defining the velocity field;
2. calculating the rate of migration of contaminants by solving the advection–dispersion equation (after using velocities determined from 1).

The techniques for modeling steady-state flow are well-established (e.g. Frind, 1987) and numerous commercial software packages are available. Many of the packages will run on micro-computers; however, there is a potential danger that inappropriate finite element mesh arrangements may be used simply to fit the problem onto a microcomputer. It is essential that the suitability of any finite element mesh be checked, either against a known solution, or by comparison with results obtained by substantially refining the mesh.

Finite element techniques provide the opportunity of modeling problems with complex geometries, complicated flow patterns, heterogeneity and nonlinearity. There is a wealth of literature dealing with different algorithms which have been proposed for solving the advection–

dispersion equation. The sheer volume of literature is itself a warning that the use of these techniques is not as simple as might first appear. This is particularly so when dealing with problems where there are high advective velocities, low dispersivities and/or high contrast in dispersivity (e.g. Yeh, 1984; Allan, 1984). While good results can be obtained (e.g. Frind and Hokkanen, 1987), particular care is required to ensure the selection of a suitable computational algorithm, finite element mesh and time step.

7.1.4 General comments

All five classes of techniques discussed above have a role to play in the design of barriers. However, of these techniques, the two most useful are finite layer and finite element methods. Finite layer techniques require some idealization of the problem; however, they also provide a means of quickly assessing the implications regarding different design scenarios and different assumed parameters. The cost of performing these analyses is relatively small. The finite element method provides the most general technique for solving the governing equations and including complex geometry, velocity fields and nonlinearity. However, use of the approach requires an experienced user, as well as relatively large data preparation and computational cost.

7.2 Analytical solutions

In the preliminary investigation of contaminant migration there is often insufficient information to warrant more than a simple idealization of the physical situation. In some cases one may be able to formulate the problem in such a way that an analytical solution is possible. Analytical solutions are ideal for making quick preliminary calculations, are valuable in performing sensitivity studies and are useful in checking the results of more sophisticated analyses.

Numerous solutions have been reported in the literature (e.g. Lapidus and Amundson, 1952; Ogata and Banks, 1961; Ogata, 1970; Lindstrom *et al.*, 1967; Selim and Mansell, 1976; Rowe and Booker, 1985a; Booker and Rowe, 1987). Although these analytical solutions have sometimes been expressed in graphical form suitable for hand calculations, as discussed in section 1.7 (e.g. Ogata, 1970; Booker and Rowe, 1987), it is often more convenient to use a computer. The computation involved in such calculations is usually not particularly sophisticated, and is made easier because of the existence of efficient approximations for many mathematical functions which can often be evaluated on programmable calculators (Abramowitz and Stegun, 1964), and because of robust readily portable subroutines (e.g. Press *et al.*, 1986), which can be incorporated into relatively simple programs which can be run on a modest personal computer.

7.2.1 Solutions for deep deposits (pure diffusion)

In the absence of advection and first order decay, the equation governing contaminant migration (equation 1.15) reduces to a simple diffusion or heat flow equation. Thus the extensive literature on heat flow, Carslaw and Jaegar (1959) provides a rich source of analytical solutions. The following provides a number of additional solutions that may be found to be useful.

Figure 7.1 shows a transmissive layer ($-H/2 < z < H/2$) which is bounded by nontransmissive strata. A portion of the layer ($-a < x < +a$) has

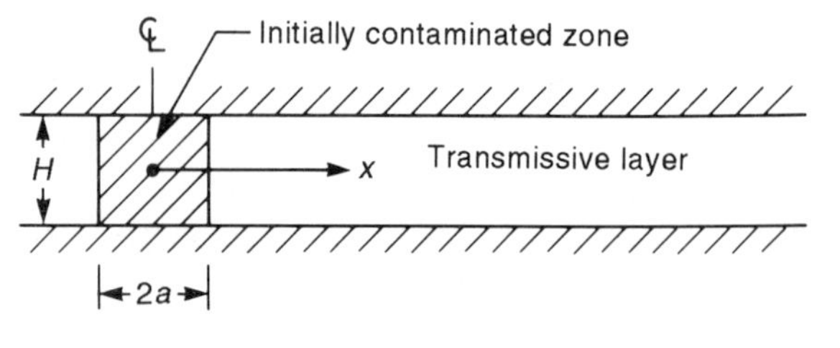

Figure 7.1 Contaminated zone in a transmissive layer.

become contaminated. If the initial concentration in the contaminated zone is c_0, then it is not difficult to show that subsequently

$$c = \frac{c_0}{2}\left[\operatorname{erfc}\left(\frac{|x| + a}{2(D^*t)^{1/2}}\right) - \operatorname{erfc}\left(\frac{|x| - a}{2(D^*t)^{1/2}}\right)\right] \tag{7.1}$$

where $D^* = nD/(n + \varrho K_d)$.

The concentration may be evaluated either by using tabulated values of the complementary error function, erfc(x), or by employing a polynomial approximation (Abramowitz and Stegun, 1964) or by using the graphical values of the function

$$\phi(X) = e^{X^2} \operatorname{erfc}(X) \tag{7.2}$$

given in Figure 1.28.

The variation of the concentration profile with time is shown in Figure 7.2, and, as would be expected, this figure shows a reduction of concentration within the initially contaminated zone as contaminant is transported by diffusion to the initially uncontaminated portion of the transmissive layer.

Figure 7.3 shows a deep deposit which contains a contaminated spherical zone ($R < a$). Initially, the contaminant concentration in the spherical zone is c_0; analysis of the diffusion equation shows that subsequently the concentration is given by

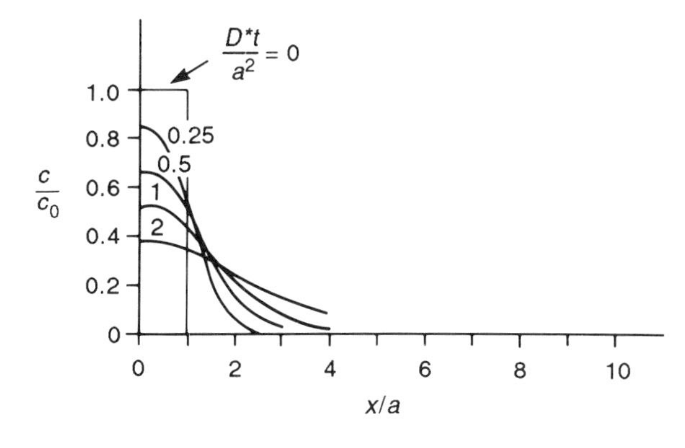

Figure 7.2 Variation of concentration with time in an initially contaminated transmissive layer.

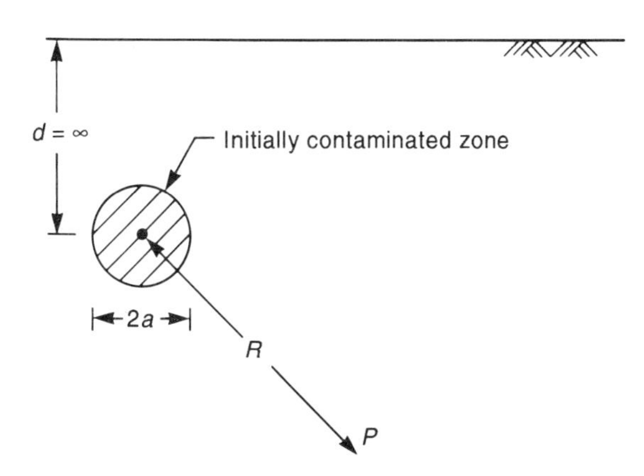

Figure 7.3 Spherical contaminated zone.

$$c = \frac{c_0}{2}\left\{\operatorname{erfc}\left(\frac{R - a}{2(D^*t)^{1/2}}\right) - \operatorname{erfc}\left(\frac{R + a}{2(D^*t)^{1/2}}\right) - 2\left(\frac{D^*t}{\pi R^2}\right)^{1/2} \exp\left(\frac{-(R - a)^2}{4D^*t}\right) - \exp\left(\frac{-(R + a)^2}{4D^*t}\right)\right\} \tag{7.3}$$

The process of diffusion from the initially contaminated zone into the initially uncontaminated zone is evident from Figure 7.4. The rate of diffusion in this case is far faster than for the one previously examined, because of three-dimensional effects.

7.2.2 Solutions for deep deposits (advection and diffusion)

The solutions derived in the previous subsection can be modified to incorporate the effect of advection. The modification depends on the following observation. Suppose that

$$c = F(x, y, z, t) \tag{7.4a}$$

is a solution of the transport equation with no

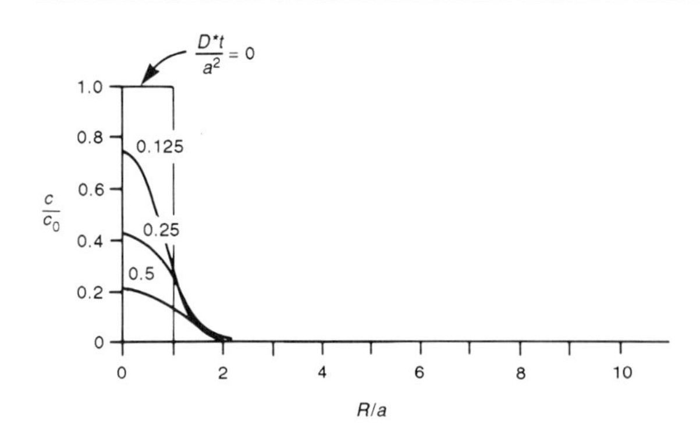

Figure 7.4 Variation of concentration with time due to an initially contaminated spherical zone.

advection ($v_x = 0$, $v_y = 0$, $v_z = 0$) and no first order (e.g. radioactive or biological) decay ($\lambda = 0$); then a solution to the equation of contaminant transport with advection and first order decay is found to be

$$c = e^{-\Lambda t} F[x - v_x^* t, y - v_y^* t, z - v_z^* t, t] \tag{7.4b}$$

where

$$\begin{aligned} v_x^* &= n v_x/(n + \varrho K_d) \\ v_y^* &= n v_y/(n + \varrho K_d) \\ v_z^* &= n v_z/(n + \varrho K_d) \\ \Lambda &= n\lambda/(n + \varrho K_d) \end{aligned}$$

The solution given by equation 7.4b obeys identical initial conditions ($t = 0$) to the solution given by equation 7.4a. There is no guarantee that the boundary conditions of the problem will be satisfied; however, in problems involving deep deposits where the boundary conditions are especially simple, the exact solution is found to be given by equation 7.4b.

To illustrate the application of this method, return to the situation shown in Figure 7.1, and suppose that there is a groundwater velocity v in the transmissive layer. The distribution of contaminant is then found to be

$$c = \tfrac{1}{2} c_0 \, e^{-\Lambda t}[\operatorname{erfc}(X_m) - \operatorname{erfc}(X_p)] \tag{7.5}$$

where

$$X_m = \frac{|x - v^* t| - a}{2(D^* t)^{1/2}}$$

$$X_p = \frac{|x - v^* t| + a}{2(D^* t)^{1/2}}$$

A comparison of the distribution of contaminant in the transmissive layer for a particular case of advection ($v^* a/D^* = 5$) and no advection ($v^* a/D^* = 0$) is shown in Figure 7.5. As would be expected from equation 7.5, the contaminant profile when advection is present is identical to the profile when there is no advection present, except that it has been translated downstream with a velocity v^*.

Similarly, it is possible to determine the solution for the initially contaminated spherical zone shown in Figure 7.3 when there is a uniform advective velocity v in the x-direction. It then follows from equation 7.3 and equation 7.4 that

$$c = \frac{c_0}{2} e^{-\Lambda t}\left[(\operatorname{erfc}(R_m) - \operatorname{erfc}(R_p)) - 2\left(\frac{D^* t}{\pi R^2}\right)^{1/2}(\exp(-R_m^2) - \exp(-R_p^2))\right] \tag{7.6}$$

$$R = [(x - v^* t)^2 + y^2 + z^2]^{1/2}$$

$$R_m = \frac{R - a}{2(D^* t)^{1/2}}$$

$$R_p = \frac{R + a}{2(D^* t)^{1/2}}$$

Again, it can be seen that the concentration profile in the advective case is identical in shape to the nonadvective case, but has been translated downgradient. The contaminant profiles, in the plane $y = 0$, along the horizontal (x) axis and vertical (z) axis are illustrated in Figure 7.6 for

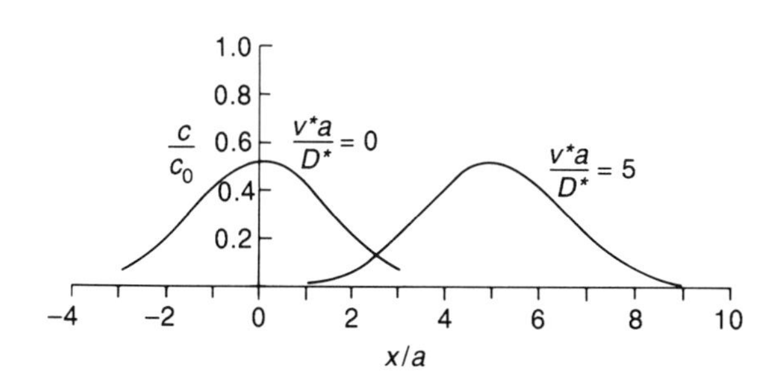

Figure 7.5 Concentration profile for different advective velocities when $D^*t/a^2 = 1$.

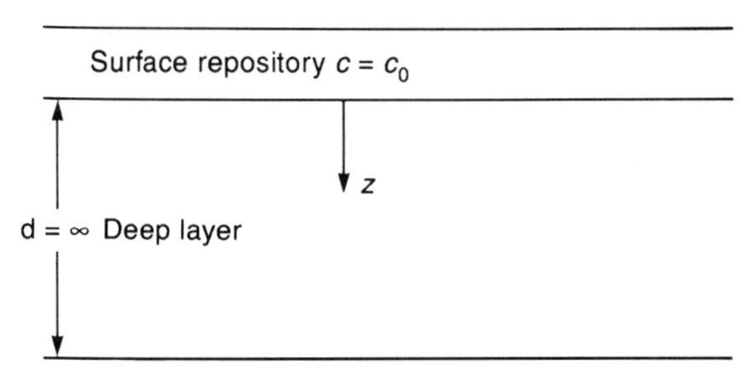

Figure 7.7 Extensive surface repository overlying a deep layer.

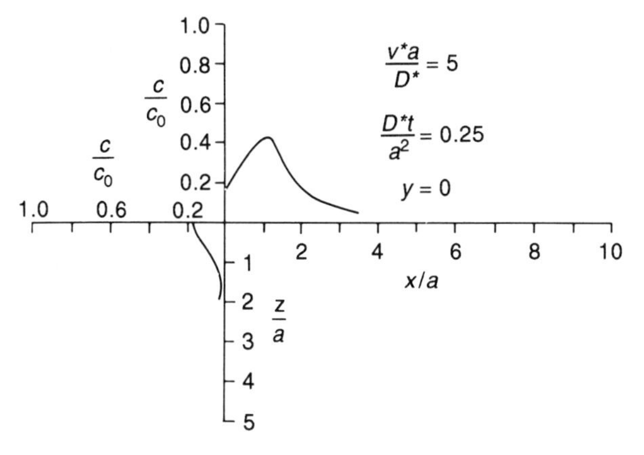

Figure 7.6 Concentration profile along horizontal and vertical axes.

the particular advective velocity $v^*a/D^* = 5$ and at the particular time $D^*t/a^2 = 0.25$.

7.2.3 Solutions for surface repositories underlain by a deep deposit

In the two previous subsections the concentration distribution near a contaminated zone within a very deep deposit was examined. In this subsection the contaminant distribution in the neighborhood of a surface repository overlying a deep deposit is examined.

The simplest situation to analyze is that of an extensive surface repository overlying a deep layer. This is shown schematically in Figure 7.7.

The simplest case to consider is that in which there is no advective transport and no first order decay, and where it can be assumed that the concentration in the repository remains constant. If it is assumed that there is no background concentration of the contaminant in the layer then the distribution of contaminant is given by the well-known expression

$$c = c_0 \operatorname{erfc}\left(\frac{z}{2(D^*t)^{1/2}}\right) \tag{7.7}$$

and is plotted in Figure 7.8.

Ogata and Banks (1961) extended this solution to take account of advection in the vertical direction and found that

$$c = \frac{c_0}{2}\left[\operatorname{erfc}\left(\frac{z - v^*t}{2(D^*t)^{1/2}}\right) + \exp\left(\frac{v^*z}{D^*}\right) \operatorname{erfc}\left(\frac{z + v^*t}{2(d^*t)^{1/2}}\right)\right] \tag{7.8}$$

This solution is shown graphically in Figure 1.25 and was discussed in section 1.7.1.

Booker and Rowe (1987) pointed out that the assumption of constant surface concentration was generally unrealistic since most landfills contain a specific mass of contaminant which is not augmented after closure. As contaminant is transported into the underlying deposit, the concentration of contaminant in the landfill

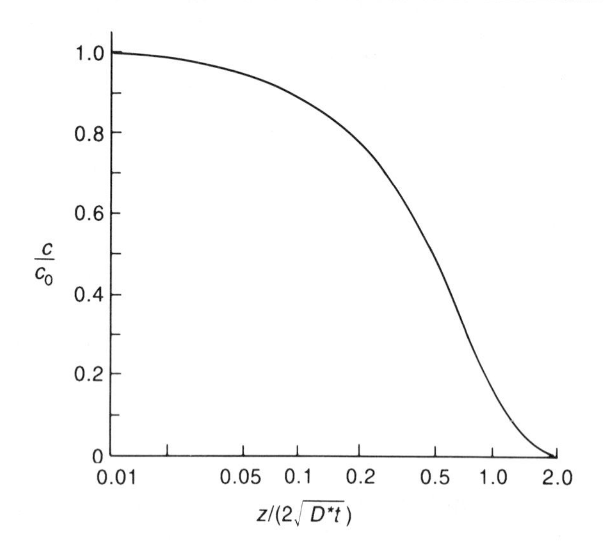

Figure 7.8 Variation of concentration for an extensive surface repository (no advection, constant surface concentration).

reduces. Booker and Rowe (1987) introduced the concept of the equivalent height of leachate H_f (see Chapter 10 for a detailed discussion), and found that in the absence of first order decay ($\lambda = 0$) for a landfill in which the initial concentration is c_0, the concentration distribution beneath the landfill is given by

$$c(z, t) = c_0 \exp(ab - b^2t)[bf(b, t) - df(d, t)]/(b - d) \tag{7.9}$$

where

$$f(b, t) = \exp(ab + b^2) \operatorname{erfc}\left(\frac{a}{2t^{1/2}} + bt^{1/2}\right)$$

$$f(d, t) = \exp(ad + d^2) \operatorname{erfc}\left(\frac{a}{2t^{1/2}} + dt^{1/2}\right)$$

and all other terms are as previously defined. The function $f(q, t)$ (for $q = b$ or d) may be evaluated directly by computer or using a hand calculator or by observing that

$$a = \frac{z}{D^{1/2}}$$

$$b = \frac{v}{2D^{1/2}}$$

$$d = \frac{(n + \varrho K_d)D^{1/2}}{H_f} - \frac{v}{2D^{1/2}}$$

$$f(q, t) = \exp(-a^2/4t)\phi(X)$$

where

$$X = \frac{a}{2t^{1/2}} + qt^{1/2}$$

and

$$\phi(X) = e^{X^2} \operatorname{erfc}(X)$$

The function $\phi(X)$ is presented graphically in Figure 7.9 and the operation of this approach was illustrated in section 1.7.2.

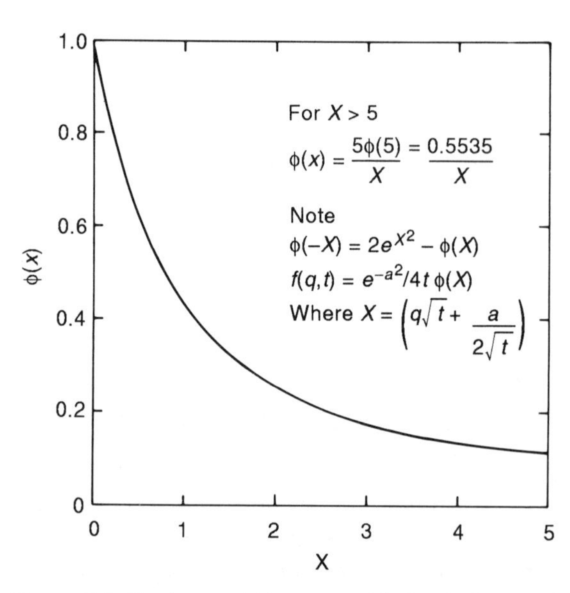

Figure 7.9 Booker and Rowe graphical solution to the advection–dispersion equation for an infinitely deep deposit with a finite mass of contaminant in the source. (Modified from Booker and Rowe, 1987; reproduced with permission of the *International Journal for Numerical and Analytical Methods in Geomechanics.*)

7.2.4 Evaluation of analytical solutions

The analytical solutions discussed in this subsection can all be expressed in terms of the complementary error function, which in turn is closely related to the error function:

$$\mathrm{erfc}(x) = 1-\mathrm{erf}(x)$$

where

$$\mathrm{erf}(x) = \frac{2}{\pi^{1/2}} \int_0^x e^{-u^2}\, du$$

The error function is extensively tabulated (Abramowitz and Stegun, 1964). The complementary error function can also be evaluated from Figure 7.9 using the relation

$$\mathrm{erfc}(X) = e^{-X^2}\phi(X)$$

Often, it is more convenient to evaluate the solutions directly, and this can be easily done by using a polynomial approximation for the error function (Abramowitz and Stegun, 1964). The evaluation can be performed on personal computers or, in many circumstances, on a programmable calculator.

7.3 Application of Laplace transforms to develop a finite layer solution for a single layer

7.3.1 Introduction

The Laplace transformation technique (Carslaw and Jaegar, 1948), provides a powerful method for the development of analytical solutions to the equations of contaminant transport. It will be shown in the following subsections that it can also be used as a most effective method of numerical solution. To illustrate this, some problems involving one-dimensional contaminant transport in the z-direction will be considered.

7.3.2 Governing equations

The equations governing advective–diffusive–dispersive transport have been discussed in Chapter 1. For ease of presentation they will be repeated here. The equation of mass balance (equation 1.14a) may be written

$$-\frac{\partial f}{\partial z} = (n + \varrho K_d)\frac{\partial c}{\partial t} + n\lambda c \qquad (7.10a)$$

If it is assumed that initially the ground is uncontaminated, then

$$c = 0 \quad \text{when } t = 0 \qquad (7.10b)$$

Finally, if it is assumed that contaminant migration is by advective diffusive transport, then

$$f = nvc - nD\frac{\partial c}{\partial z} \qquad (7.10c)$$

In equations 7.10, c is the concentration at depth z and time t [ML^{-3}]; f is the mass flux in the z-direction [$ML^{-2}T^{-1}$]; n is the effective porosity of the soil [$-$]; ϱ is the dry density of the soil [ML^{-3}]; K_d is the distribution coefficient [$M^{-1}L^3$]; v is the seepage velocity [LT^{-1}]; λ is the first order decay constant [T^{-1}].

$$\lambda = \Gamma_R + \Gamma_B + \Gamma_S$$

where Γ_R (= ln 2/(radioactive half-life)) is the radioactive decay constant [T^{-1}]; Γ_B (= ln 2/(biological decay half-life at this location)) is the biological decay constant [T^{-1}]; Γ_S is the flow removed per unit volume of soil in a permeable layer beneath a landfill [T^{-1}] (section 1.3.6)

Equations 7.10a and 7.10c may be combined to give the transport equation:

$$nD\frac{\partial^2 c}{\partial z^2} - nv\frac{\partial c}{\partial z} = (n + \varrho K_d)\frac{\partial c}{\partial t} + n\lambda c \qquad (7.10d)$$

Equation 7.10d must then be solved subject to

the initial condition (equation 7.10b) and any boundary conditions.

As mentioned in the introduction, a powerful way of solving these equations is by employing a Laplace transform:

$$\bar{\psi} = \int_0^\infty e^{-st}\, \psi(t)\mathrm{d}t \tag{7.11}$$

The variable s is called the Laplace transform parameter, and the superior bar is often used to indicate that a Laplace transform has been applied. Some typical transforms are given in Table 7.1. More detailed lists of transforms are given in Carslaw and Jaegar (1948), Abramowitz and Stegun (1964) and a very comprehensive list in Erdelyi *et al.* (1954).

If equation 7.10d is transformed using a Laplace transform, it is found that after incorporating the initial condition (equation 7.10b) that it reduces to the ordinary differential equation

$$nD\frac{\partial^2 \bar{c}}{\partial z^2} - nv\frac{\partial \bar{c}}{\partial z} = nS\bar{c} \tag{7.12a}$$

Table 7.1 Typical transforms

$\psi\ (t)$	$\bar{\psi}\ (s)$
1	$\frac{1}{s}$
e^{-bt}	$\frac{1}{s+b}$
$\frac{\partial \phi}{\partial t}$	$s\bar{\phi} - \phi(0)$
$\mathrm{erfc}\left(\frac{k}{2t^{1/2}}\right)$	$\frac{e^{-ks}}{s}\quad (k>0)$
$e^{a(k+ak)}\,\mathrm{erfc}\left(\frac{at^{1/2}+k}{2t^{1/2}}\right)$	$\frac{e^{-ks^{1/2}}}{s^{1/2}(a+s^{1/2})}\quad (k>0)$

where

$$S = \left(1 + \frac{\varrho K_d}{n}\right)s + \lambda \tag{7.12b}$$

An expression for transformed flux follows from equation 7.10c, so that

$$\bar{f} = nv\bar{c} - nD\frac{\partial \bar{c}}{\partial z} \tag{7.12c}$$

Equations 7.12 are solved, subject to any transformed boundary conditions, leading to the solution in the Laplace transform domain. The solution in the time domain is then found by inverting the Laplace transform. This procedure will be illustrated by a simple example in the following subsection.

7.3.3 Solution to a simple problem

One of the simplest situations that can be envisaged is one-dimensional diffusion in the single uniform layer of thickness H subject to the initial condition ($t = 0$) of zero concentration in the layer,

$$c(z, 0) = 0 \qquad 0 \leqslant z \leqslant H \tag{7.13a}$$

and the boundary condition of constant surface concentration at the top of the layer and zero concentration at depth H,

$$c = c_0 \quad \text{when } z = 0, t > 0 \tag{7.13b}$$

$$c = 0 \quad \text{when } z = H, t > 0 \tag{7.13c}$$

The case for which $H = \infty$ corresponds to an infinitely deep layer.

Referring to Table 7.1, the boundary conditions (equations 7.13b, c) become

$$\bar{c} = \frac{c_0}{s} \text{ when } z = 0 \tag{7.14a}$$

$$\bar{c} = 0 \quad \text{when } z = H \tag{7.14b}$$

For the sake of simplicity, attention will be restricted to the case where there is no advection and no first order decay ($v = 0$, $\lambda = 0$). The transformed transport, equation 7.12a, was developed in the previous subsection; it becomes

$$\frac{\partial^2 \bar{c}}{\partial z^2} = \frac{S}{D}\bar{c} \tag{7.15a}$$

which may be conveniently written

$$\frac{\partial^2 \bar{c}}{\partial z^2} = \alpha^2 \bar{c} \tag{7.15b}$$

where

$$\alpha^2 = \frac{S}{D} = \frac{s}{D^*}$$

$$S = \left(1 + \frac{\varrho K_d}{n}\right)s$$

$$D^* = \frac{D}{\left(1 + \dfrac{\varrho K_d}{n}\right)}$$

Equation 7.15 has the general solution

$$\bar{c} = A e^{\alpha z} + B e^{-\alpha z} \tag{7.16}$$

The constants A, B can be found from the transformed boundary conditions (equations 7.14a,b), and thus the solution in transform space is found to be

$$\bar{c} = \frac{c_0}{s}\frac{e^{\alpha z}}{(1 - e^{2\alpha H})} + \frac{c_0}{s}\frac{e^{-\alpha z}}{(1 - e^{-2\alpha H})} \tag{7.17}$$

Equation 7.17 is the solution in transform space. It is now necessary to invert the transform and find the solution in physical space. This can be done numerically using a technique developed by Talbot (1979), and this will be the approach adopted for the more complex problem considered in the following sections. In this present case it is relatively simple to obtain an analytical conversion of equation 7.17 for the case where the layer is infinitely deep ($H \approx \infty$), in which case equation 7.17 reduces to

$$\bar{c} = \frac{c_0 e^{-\alpha z}}{s} \tag{7.18a}$$

or

$$\bar{c} = c_0 \frac{e^{-ks^{1/2}}}{s} \tag{7.18b}$$

where

$$k = \frac{z}{(D^*)^{1/2}}$$

It now follows immediately from Table 7.1 that

$$c = c_0 \,\mathrm{erfc}\left(\frac{z}{2(D^* t)^{1/2}}\right) \tag{7.18c}$$

thus providing the derivation of Equation 7.7.

7.4 Contaminant transport into a single layer considering a landfill of finite mass and an underlying aquifer

Consider the situation shown schematically in Figure 7.10 of an extensive landfill overlying a clay, which in turn overlies a transmissive layer.

Since the landfill is extensive, transport will be predominantly vertical, and thus the migration of contaminant will be governed by equation 7.10d in the physical domain and equation 7.12a in the Laplace transform

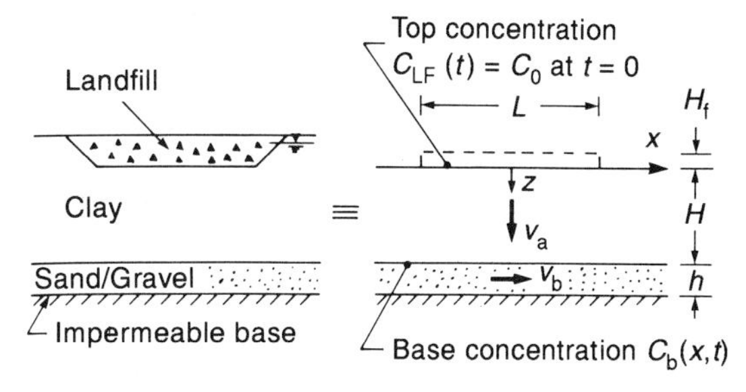

Figure 7.10 Problem description – single aquitard layer underlain by an aquifer.

domain. The general solution of equation 7.12a is

$$\bar{c} = Ae^{\alpha z} + Be^{\beta z} \tag{7.19a}$$

where

$$\alpha = \frac{v}{2D} + \left[\frac{v^2}{4D^2} + \frac{S}{D}\right]^{1/2}$$

$$\beta = \frac{v}{2D} - \left[\frac{v^2}{4D^2} + \frac{S}{D}\right]^{1/2}$$

It is not difficult to show that the transformed flux (equation 7.12c) is then given by

$$\bar{f} = nD(\beta Ae^{\alpha z} + \alpha Be^{\beta z}) \tag{7.19b}$$

It was shown in section 1.5 that it was possible to take into account the fact that the landfill contains only a finite amount of contaminant and that (equation 1.23),

$$c_T(t) = c_0 - \frac{1}{H_r}\int_0^t f_T(\tau)d\tau - \int_0^t \lambda_T c(\tau)d\tau \tag{7.20a}$$

which in Laplace transform space becomes

$$\bar{c}_T = \frac{c_0}{S_T} - \frac{\bar{f}_T}{S_T H_r} \tag{7.20b}$$

where

$$S_T = s + \lambda_T = S + \Gamma_R + \Gamma_{BT} + \frac{q_c}{H_r}$$

In the above equation, c_T is the concentration of contaminant in the landfill [ML^{-3}]; c_0 denotes the initial concentration of contaminant in the landfill [ML^{-3}]; f_T is the flux entering the clay from the landfill [$ML^{-2}T^{-1}$]; H_r is the reference height of the leachate as briefly discussed in section 1.5 (discussed in greater detail in section 10.2) [L]; Γ_R (= ln 2/(radioactive half-life)) is the radioactive decay constant [T^{-1}]; Γ_{BT} (= ln 2/(biological half-life in the landfill) is the biological decay constant [T^{-1}]; q_c is the volume of leachate collected per unit area of landfill [LT^{-1}].

It was shown in section 1.6 that it was also possible to model the underlying aquifer, and it is found that (equation 1.25b)

$$c_b(t) = \int_0^t \left[\frac{f_b(\tau)}{n_b h} - \lambda_b c_b(\tau)\right]d\tau \tag{7.21a}$$

or in Laplace transform space

$$\bar{f}_b = n_b h S_b \bar{c}_b \tag{7.21b}$$

where

$$S_b = s + \lambda_b = s + \Gamma_R + \Gamma_{Bb} + \frac{v_b}{n_b L}$$

In the above equation, n_b is the porosity of the aquifer [–]; h is the thickness of the aquifer [L]; v_b is the Darcy velocity in the aquifer [LT^{-1}]; c_b is the concentration in the aquifer [ML^{-3}]; f_b is the flux entering the aquifer from the clay [$ML^{-2}T^{-1}$]; Γ_{Rb} (= ln 2/(radioactive half-life)) is the radioactive decay constant [T^{-1}]; Γ_{Bb} (= ln 2/(biological half-life in the aquifer)) is the biological decay constant [T^{-1}]; L is the length of the landfill parallel to the direction of flow in the aquifer [L].

Substitution of equation 7.19 into equations 7.20 and 7.21 lead to the set of equations

$$\left(1 + \frac{nD\beta}{H_r S_T}\right)A + \left(1 + \frac{nD\alpha}{H_r S_T}\right)B = \frac{c_0}{S_T}$$

$$e^{\alpha H}\left(1 - \frac{nD\beta}{h n_b S_b}\right)A + e^{\beta H}\left(1 - \frac{nD\alpha}{h n_b S_b}\right)B = 0 \tag{7.22}$$

These equations can easily be solved for the constants A, B, and this together with equations 7.19 complete the solution in transform space. The solution in the physical plane is found by using the algorithm developed by Talbot (1979).

7.5 Finite layer analysis

7.5.1 One-dimensional transport of contaminant from a landfill overlying a layered deposit

The procedures developed in the previous section will now be extended to include the effects of layering of the soil beneath the landfill. The simplest case which can be considered is that of one-dimensional contaminant transport in the clayey deposit, parallel to the z-direction, as shown in Figure 7.11. It is assumed that this deposit is divided into a number of layers by node planes $z = z_0, z_1, \ldots, z_n$ and that each layer, $(z_{k-1} < z < z_k)$ may be considered homogeneous. Thus the vertical flux per unit area per unit time at a point in layer k is given by equation 7.10c

$$f_z = nv_z c - nD_{zz} \frac{\partial c}{\partial z} \tag{7.23a}$$

The equation of contaminant transport (equation 7.10d) may be written

$$nD_{zz} \frac{\partial^2 c}{\partial z^2} - nv_z \frac{\partial c}{\partial z}$$

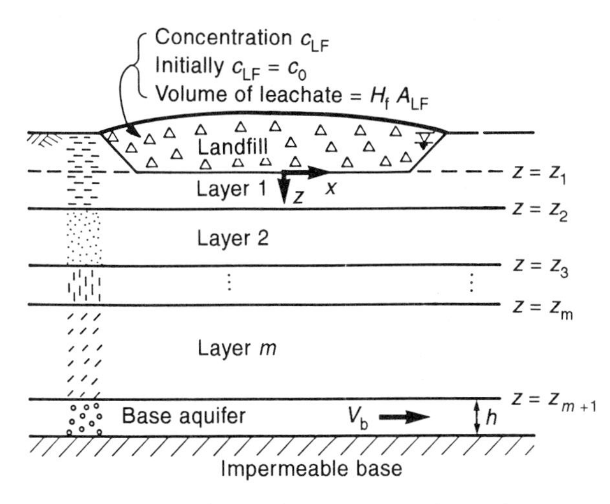

Figure 7.11 Problem description – layered system which may contain multiple aquitards and aquifers.

$$= (n + \varrho K_d) \frac{\partial c}{\partial t} - n\neq c \tag{7.23b}$$

where the subscript z denotes that this parameter applies to the z-direction, and the quantities n, v_z, D_{zz}, ϱ, K_d and λ assume constant values appropriate for the layer k under consideration. As in the previous section, it will be assumed that there is no pre-existing distribution of contaminant within the deposit, so that

$$c = 0 \quad \text{when } (z_{k-1} \leqslant z \leqslant z_k) \text{ at } t = 0$$

Equations 7.23a and 7.23b can be simplified by introducing the Laplace transform

$$(\bar{c}, \bar{f}_z) = \int_0^\infty (c, f_z)\, e^{-st}\, dt$$

yielding

$$\bar{f}_z = nv_z \bar{c} - nD_{zz} \frac{\partial \bar{c}}{\partial z} \tag{7.24a}$$

$$nD_{zz} \frac{\partial^2 \bar{c}}{\partial z^2} - nv_z \frac{\partial \bar{c}}{\partial z} = nS\bar{c} \tag{7.24b}$$

where

$$S = \left(1 + \frac{\varrho K_d}{n}\right)s + \lambda$$

Equations 7.24a,b have the solution

$$\begin{aligned} \bar{c} &= Ae^{\alpha z} + Be^{\beta z} \\ \bar{f}_z &= nD_{zz}(\beta Ae^{\alpha z} + \alpha Be^{\beta z}) \end{aligned} \tag{7.24c}$$

where A, B are constants to be determined, and

$$\begin{aligned} \alpha &= \frac{v_z}{2D_{zz}} + \left(\frac{v_z^2}{4D_{zz}^2} + \frac{S}{D_{zz}}\right)^{1/2} \\ \beta &= \frac{v_z}{2D_{zz}} - \left(\frac{v_z^2}{4D_{zz}^2} + \frac{S}{D_{zz}}\right)^{1/2} \end{aligned} \tag{7.24d}$$

If the constants A, B appearing in equation

7.24c are evaluated in terms of the concentrations at the nodal planes z_j, z_k (where $j = k - 1$), it is found that

$$\bar{c} = \bar{c}_j \left\{ \frac{e^{\alpha(z-z_k)} - e^{\beta(z-z_k)}}{e^{\alpha(z_j-z_k)} - e^{\beta(z_j-z_k)}} \right\} + \bar{c}_k \left\{ \frac{e^{\alpha(z-z_j)} - e^{\beta(z-z_j)}}{e^{\alpha(z_k-z_j)} - e^{\beta(z_k-z_j)}} \right\} \quad (7.25a)$$

and thus

$$\frac{\bar{f}_z}{nD_{zz}} = \bar{c}_j \left\{ \frac{\beta e^{\alpha(z-z_k)} - \alpha e^{\beta(z-z_k)}}{e^{\alpha(z_j-z_k)} - e^{\beta(z_j-z_k)}} \right\} + \bar{c}_k \left\{ \frac{\beta e^{\alpha(z-z_j)} - \alpha e^{\beta(z-z_j)}}{e^{\alpha(z_k-z_j)} - e^{\beta(z_k-z_j)}} \right\} \quad (7.25b)$$

It is now possible to evaluate the flux on each of the node planes, using equation 7.25b, and it is found that

$$\begin{bmatrix} \bar{f}_{zj} \\ -\bar{f}_{zk} \end{bmatrix} = \begin{bmatrix} Q_k & R_k \\ S_k & T_k \end{bmatrix} \begin{bmatrix} \bar{c}_j \\ \bar{c}_k \end{bmatrix} \quad (7.26)$$

where

$$Q_k = \frac{nD_{zz}(\beta e^{\mu\beta} - \alpha e^{\mu\alpha})}{e^{\mu\beta} - e^{\mu\alpha}}$$

$$R_k = -\frac{nD_{zz}(\beta - \alpha)}{e^{\mu\beta} - e^{\mu\alpha}}$$

$$S_k = -\frac{nD_{zz}(\beta - \alpha)e^{\mu(\beta+\alpha)}}{e^{\mu\beta} - e^{\mu\alpha}}$$

$$T_k = \frac{nD_{zz}(\beta e^{\mu\alpha} - \alpha e^{\mu\beta})}{e^{\mu\beta} - e^{\mu\alpha}}$$

where $\mu = z_k - z_{k-1}$.

Noting that both the flux f_z and concentrations c must be continuous, the layer matrices defined by equation 7.26 may be assembled for each layer in the deposit, to give equation 7.27 as shown on page 191, where $\bar{c}_T$, $\bar{f}_T$ are the values of concentration and flux at the top of the deposit; $\bar{c}_b$, $\bar{f}_b$ are the values of concentration and flux at the base of the deposit.

We now wish to solve this equation, subject to the appropriate boundary conditions. The boundary condition at the bottom of the layered system (equation 7.21) was examined in the previous section, and it follows in identical fashion that

$$\bar{f}_b = Q_{n+1}\, \bar{c}_b \quad (7.28a)$$

where $Q_{n+1} = hn_bS_b$.

The boundary condition at the top of the layered system (equation 7.20b) was also considered in the previous section, and it follows that

$$\bar{f}_T = B_0 - T_0\bar{c}_T \quad (7.28b)$$

where $B_0 = H_rc_0$ and $T_0 = H_rS_T$.

Introducing the expressions for $\bar{f}_T$, $\bar{f}_b$ into equation 7.28 then gives the complete set of equations, equation 7.29 on page 191 where c_j denotes the concentration in the layer plane $z = z_j$; Q_k, R_k, S_k and T_k are defined by equation 7.26; Q_{n+1}, B_0 and T_0 are defined by equations 7.28. Equation 7.29 can now be solved giving $\bar{c}_j$ at each node plane. The Laplace transform can then be inverted using a very efficient scheme proposed by Talbot (1979). Using this approach, an accuracy of order 10^{-6} and 10^{-10} can typically be achieved using 11 and 18 sample points respectively. For many practical problems an accuracy of 10^{-6} is more than adequate. The theory described above has been coded in program POLLUTE (Rowe and Booker, 1994) and can be run on a microcomputer.

$$\begin{bmatrix} Q_1 & R_1 & & & & \\ S_1 & T_1 + Q_2 & R_2 & & & \\ & S_2 & T_2 + Q_3 & R_3 & & \\ & & & \cdot & & \\ & & & \cdot & & \\ & & & \cdot & & \\ & & & S_{n-1} & T_{n-1} + Q_n & R_n \\ & & & & S_n & T_n \end{bmatrix} \begin{bmatrix} \bar{c}_T \\ \bar{c}_1 \\ \bar{c}_2 \\ \cdot \\ \cdot \\ \cdot \\ \bar{c}_{n-1} \\ \bar{c}_b \end{bmatrix} = \begin{bmatrix} \bar{f}_T \\ 0 \\ 0 \\ \cdot \\ \cdot \\ \cdot \\ 0 \\ -\bar{f}_b \end{bmatrix} \tag{7.27}$$

$$\begin{bmatrix} T_0 + Q_1 & R_1 & & & & \\ S_1 & T_1 + Q_2 & R_2 & 0 & & \\ 0 & S_2 & T_2 + Q_3 & R_3 & & \\ & & & \cdot & & \\ & & & \cdot & & \\ & & & \cdot & & \\ & & & S_{n-1} & T_{n-1} + Q_n & R_n \\ & & & 0 & S_n & T_n + Q_{n+1} \end{bmatrix} \begin{bmatrix} \bar{c}_T \\ \bar{c}_1 \\ \bar{c}_2 \\ \cdot \\ \cdot \\ \cdot \\ \bar{c}_{n-1} \\ \bar{c}_b \end{bmatrix} = \begin{bmatrix} B_0 \\ 0 \\ 0 \\ \cdot \\ \cdot \\ \cdot \\ 0 \\ 0 \end{bmatrix} \tag{7.29}$$

7.5.2 Finite layer analysis in two dimensions

The procedure described in the previous section can be readily generalized for contaminant transport in both the x- and z-directions. Again, it is assumed that the deposit is divided into a number of layers with node planes $z = z_0, z_1, \ldots, z_n$ and that each layer k ($z_{k-1} < z < z_k$) may be considered as homogeneous. Thus the fluxes f_x, f_z transported in the x- and z-directions within layers k are given by

$$f_x = nv_x c - nD_{xx}\frac{\partial c}{\partial x} \tag{7.30a}$$

$$f_z = nv_z c - nD_{zz}\frac{\partial c}{\partial z} \tag{7.30b}$$

For two-dimensional transport in the soil, the equation governing contaminant migration can be written as

$$(n + \varrho K_d)\frac{\partial c}{\partial t} = nD_{xx}\frac{\partial^2 c}{\partial x^2} + nD_{zz}\frac{\partial^2 c}{\partial z^2} - nv_x\frac{\partial c}{\partial x} - nv_z\frac{\partial c}{\partial z} - n\lambda c \tag{7.31a}$$

where the quantities n, v_x, v_z, D_{xx}, D_{zz}, λ, ϱ and K_d are physical constants appropriate to the layer k under consideration. For simplicity of presentation, it will be assumed that there is no initial background concentration in the deposit, so that

$$c = 0 \quad \text{when } t = 0 \tag{7.31b}$$

Equations 7.30 and 7.31 can be simplified by the introduction of a Laplace transform:

$$(\bar{c}, \bar{f}_x, \bar{f}_z) = \int_0^\infty (c, f_x, f_z)e^{-st}\,dt$$

and a Fourier transform:

$$(\bar{C}, \bar{F}_x, \bar{F}_z) = \frac{1}{2\pi} \int_{-\infty}^{\infty} (\bar{c}, \bar{f}_x, \bar{f}_z) e^{-i\xi x}\, dx$$

These transforms reduce equations 7.30 and 7.31 to

$$nD_{zz}\frac{\partial^2 \bar{C}}{\partial z^2} - nv_z\frac{\partial \bar{C}}{\partial z} = nS\bar{C} \quad (7.32a)$$

$$\bar{F}_x = nv_x\bar{C} - nD_{xx}i\xi\bar{C} \quad (7.32b)$$

$$\bar{F}_z = nv_zC - nD_{zz}\frac{\partial \bar{C}}{\partial z} \quad (7.32c)$$

where

$$S = \left(1 + \frac{\varrho K}{n}\right)s + \xi^2 D_{xx} + i\xi v_x + \lambda$$

and $\lambda = \Gamma_R + \Gamma_B + \Gamma_s$.

The solution of equations 7.32a,c is

$$\bar{C} = Ae^{\alpha z} + Be^{\beta z} \quad (7.33a)$$

$$\frac{\bar{F}_z}{nD_{zz}} = \beta Ae^{\alpha z} + \alpha Be^{\beta z} \quad (7.33b)$$

where A, B are constants to be determined, and

$$\alpha = \frac{v_z}{2D_{zz}} + \left(\frac{v_z^2}{4D_{zz}^2} + \frac{S}{D_{zz}}\right)^{1/2}$$

$$\beta = \frac{v_z}{2D_{zz}} - \left(\frac{v_z^2}{4D_{zz}^2} + \frac{S}{D_{zz}}\right)^{1/2} \quad (7.34)$$

If the constants A, B are evaluated in terms of the nodal plane concentrations, then interpolation formulae similar to equation 7.25a are found:

$$\bar{C} = \bar{C}_j\left\{\frac{e^{\alpha(z-z_k)} - e^{\beta(z-z_k)}}{e^{\alpha(z_j-z_k)} - e^{\beta(z_j-z_k)}}\right\} + \bar{C}_k\left\{\frac{e^{\alpha(z-z_j)} - e^{\beta(z-z_j)}}{e^{\alpha(z_k-z_j)} - e^{\beta(z_k-z_j)}}\right\} \quad (7.35)$$

This equation can be used to determine the node plane fluxes for the element and leads to the relationship

$$\begin{bmatrix} \bar{F}_{zj} \\ -\bar{F}_{zk} \end{bmatrix} = \begin{bmatrix} Q_k & R_k \\ S_k & T_k \end{bmatrix} \begin{bmatrix} \bar{C}_j \\ \bar{C}_k \end{bmatrix} \quad (7.36)$$

where Q_k, R_k, S_k and T_k are precisely as defined in equation 7.26, but where in this case α and β are defined by equations 7.34.

Next, consider the lower boundary. It will be assumed that the clay layer is underlain by a more permeable stratum of thickness h and porosity n_b. Assuming that the concentration is uniform across the thickness h, but may vary with position x, the general equation (equation 1.25) governing the concentration in the base aquifer was developed in section 1.6, and it becomes

$$c_b = \int_0^t \left(\frac{f_b}{hn_b} - \frac{v_b}{n_b}\frac{\partial c_b}{\partial x} + D_H\frac{\partial^2 c_b}{\partial x^2} - \lambda_b^* c_b\right) d\tau$$

where

$$\lambda_b^* = \Gamma_R + \Gamma_{Bb}$$

or on applying both Laplace and Fourier transforms,

$$\bar{F}_b = \Omega_b \bar{C}_b \quad (7.37)$$

where $\bar{C}_b$, $\bar{F}_b$ are the transformed concentration and flux at the base and

$$\Omega_b = hn_b\left(s + \lambda_b^* + \frac{i\xi v_b}{n_b} + \xi^2 D_H\right)$$

It now follows from the continuity of normal flux at the node planes that the full set of equations can be written as shown in equation 7.38 on page 193, where $\bar{C}_T$, $\bar{F}_T$ respectively denote the transformed concentration and flux at the top.

It remains to model the interaction between the layered soil and the landfill. It will be assumed that the distribution of contaminant within the landfill maintains the same spatial

$$\begin{bmatrix} Q_1 & R_1 & & & & \\ S_1 & T_1 + Q_2 & R_2 & & & \\ & S_2 & T_2 + Q_3 & R_3 & & \\ & & & S_{n-1} & T_{n-1} + Q_n & R_n \\ & & & & S_n & T_n + \Omega_b \end{bmatrix} \begin{bmatrix} \bar{C}_T \\ \bar{C}_1 \\ \bar{C}_2 \\ \bar{C}_{n-1} \\ \bar{C}_b \end{bmatrix} = \begin{bmatrix} \bar{F}_T \\ 0 \\ 0 \\ 0 \\ 0 \end{bmatrix} \quad (7.38)$$

distribution, shown in Figure 7.12. This takes into account that the concentration of contaminant will be virtually constant within the landfill, but there will be a transition zone at the edge of the landfill, where the contaminant decreases from its maximum value to a value of zero just outside the landfill. It follows that

$$c_T = p(x)\, c_{LF} \quad (7.39)$$

where $p(x)$ represents the distribution shown in Figure 7.12 and c_{LF} is the concentration away from the landfill. When equation 7.39 is transformed it becomes

$$\bar{C}_T = \frac{2}{\pi} \frac{[\cos(\xi l/2) - \cos(\xi L/2)]}{\xi^2 (L - l)} \bar{c}_{LF} \quad (7.40)$$

The equation governing mass balance in the landfill was discussed in section 1.5, and it follows in similar fashion to the derivation of equation 1.23 that

$$c_{LF} = c_0 - \frac{1}{L_{av} H_r} \int_0^t \left(\int_{-L/2}^{L/2} f_T(x, \tau) dx \right) d\tau - \lambda_T \int_0^t c_{LF}\, dt \quad (7.41a)$$

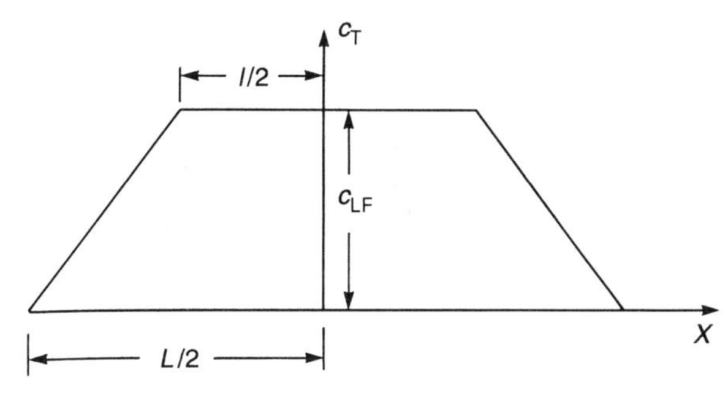

Figure 7.12 Assumed distribution of contaminant within a landfill.

where $L_{av} = (L + l)/2$, whereupon, using Fourier's inversion theorem,

$$\bar{c}_{LF} = \frac{c_0}{S_T} - \frac{2}{S_T H_r L_{av}} \int_{-\infty}^{\infty} \bar{F}_T \frac{\sin(\xi L/2)}{\xi} d\xi \quad (7.41b)$$

where $S_T = s + \lambda_T$.

At this stage, the value of c_{LF} is as yet undetermined. The key to its evaluation is to realize that although $\bar{C}_T$ and F_T cannot be individually evaluated, their ratio can be. This is most easily done by arbitrarily setting $\bar{F}_T = 1$, solving equation 7.38 for all the transformed node plane concentrations and then evaluating

$$\bar{\chi} = \frac{L_{av} \bar{F}_T}{D_{zz} \bar{C}_T} \quad (7.42)$$

Once $\bar{\chi}$ has been determined, equations 7.40, 7.41b and 7.42 can be used to show that

$$\bar{c}_{LF} = \frac{c_0}{\left(S_T + \dfrac{D_{zz}}{L_{av} H_r} \Omega_T \right)} \quad (7.43)$$

where

$$\Omega_T = \frac{8}{\pi} \int_{-\infty}^{\infty} \times \frac{\bar{\chi} \sin(\xi L/2)[\cos(\xi l/2) - \cos(\xi L/2)]}{\xi^3 (L^2 - l^2)} d\xi$$

Once $\bar{c}_{LF}$ is known, the complete solution can be obtained from equation 7.38 and the Talbot inversion.

The theory described above for two-dimensional conditions has been coded in program MIGRATE (Rowe and Booker, 1988b). The major computational effort involved is associated with the numerical inversion of the Laplace and Fourier transforms for the locations and times of interest. The Laplace transform can again be inverted using Talbot's algorithm. The Fourier transform can be efficiently inverted using 20-point Gauss quadrature. The width and number of integration subintervals which are needed to achieve a reasonable accuracy (say 0.1%) depends somewhat on the geometry and properties of the problem under consideration. These parameters can be determined from a few trial calculations for a representative point and time of interest. Similarly, it should be noted that numerical experiments are also required to determine an appropriate finite element mesh and time integration procedure if alternative finite element or finite difference codes are used.

7.5.3 One-dimensional transport with an initial contaminant distribution

In section 7.5.1 it has been assumed that:

1. there was no initial distribution of contaminant within the soil;
2. no environmental changes, such as a change in advective velocity, or a change in the general nature of conditions within the landfill or in the underlying aquifer, occur.

The analysis developed in section 7.5.1 will now be extended to incorporate these effects. Attention will be focused on case 1, since it will be shown that case 2 can always be reduced to case 1.

The configuration to be considered is that shown schematically in Figure 7.10, with the exception that it will be assumed that there is an initial distribution c_I of contaminant throughout the deposit, so that

$$c = c_I \qquad \text{when } t = 0 \tag{7.44}$$

The equation governing contaminant transport within a particular layer $z_{k-1} < z < z_k$ will again be equation 7.23b. The Laplace transform of this equation is, after incorporation of the initial condition equation 7.44, found to be

$$nD_{zz}\frac{\partial^2 \bar{c}}{\partial z^2} - nv_z\frac{\partial \bar{c}}{\partial z} = n(S\bar{c} - \phi \tag{7.45a}$$

where

$$S = \left(1 + \frac{\varrho K_d}{n}\right)s + \lambda \tag{7.45b}$$

and

$$\phi = c_I\left(1 + \frac{\varrho K_d}{n}\right) \tag{7.45c}$$

and, as before, the parameters n, v_z, D_{zz}, λ and c_I refer to the particular layer under consideration.

In order to find the solution of equation 7.45, it will be assumed that ϕ may be approximated in the form

$$\phi = Ee^{\varepsilon z} \tag{7.46}$$

The quantities E, ε can be found from the condition that $\phi = \phi_j$ when $z = z_j$ ($j = k - 1$) and $\phi = \phi_k$ when $z = z_k$ and thus

$$\varepsilon = \frac{\ln \phi_j - \ln \phi_k}{z_j - z_k}$$

$$\ln E = \frac{z_k \ln \phi_j - z_j \ln \phi_k}{z_k - z_j} \tag{7.47}$$

Equation 7.45 is now readily solved, and it is found that the transformed flux and concentration are given by

$$\bar{c} = \frac{\phi}{G} + Ae^{\alpha z} + Be^{\beta z} \tag{7.48}$$

$$\frac{\bar{f}_z}{nD_{zz}} = \frac{(\alpha + \beta - \varepsilon)}{G}\phi + \beta A e^{\alpha z} + \alpha B e^{\beta z}$$

where $G = -D_{zz}(\varepsilon - \alpha)(\varepsilon - \beta)$, and

$$\alpha = \frac{v_z}{2D_{zz}} + \left(\frac{v_z^2}{4D_{zz}^2} + \frac{S}{D_{zz}}\right)^{1/2}$$

$$\beta = \frac{v_z}{2D_{zz}} - \left(\frac{v_z^2}{4D_{zz}^2} + \frac{S}{D_{zz}}\right)^{1/2}$$

are identical to the values given in equation 7.24d. The quantities A, B can now be found in terms of the nodal values c_j, c_k, it then follows that we get equation 7.49 shown below.

There are no changes to the boundary conditions at the upper and lower surface and so if the layer matrices, equation 7.49, can be assembled, and boundary conditions added as before, to give equation 7.50 below

where

$$A_k = \frac{1}{G}[nD_{zz}(\varepsilon - \alpha - \beta)\,\phi_j + Q_k\phi_j + R_k\phi_k]$$

$$B_k = \frac{1}{G}[-nD_{zz}(\varepsilon - \alpha - \beta)\,\phi_k + S_k\phi_j + T_k\phi_k]$$

for $k = 1, 2, \ldots, n$ and $A_{n+1} = 0$.

Equation 7.50 can now be solved for the transformed node plane concentrations c_k and the solution in the physical plane found by Talbot (1979) inversion.

7.5.4 Finite layer analysis when there are environmental changes

Consider a layered system which undergoes some environmental change at time $t = t^*$. This environmental change may lead to a change in the governing differential equation, such as might arise from a change in the advective flow in the deposit (e.g. due to the failure of a primary leachate collection system with time). It might lead to a change in boundary conditions, such as that arising from the expansion of an old existing landfill by the addition of extra lifts of waste; or where the source concentration

$$\begin{bmatrix} +\bar{f}_{zj} \\ -\bar{f}_{zk} \end{bmatrix} = nD \begin{bmatrix} +(\alpha + \beta - \varepsilon)\dfrac{\phi_j}{G} \\ -(\alpha + \beta - \varepsilon)\dfrac{\phi_k}{G} \end{bmatrix} + \begin{bmatrix} Q_k & R_k \\ S_k & T_k \end{bmatrix} \begin{bmatrix} \bar{c}_j - \dfrac{\phi_j}{G} \\ \bar{c}_k - \dfrac{\phi_k}{G} \end{bmatrix} \tag{7.49}$$

$$\begin{bmatrix} T_0 + Q_1 & R_1 & & & \\ S_1 & T_1 + Q_2 & R_2 & & \\ \cdot & \cdot & \cdot\ \cdot\ \cdot & & \\ & & \cdot\ \cdot & & \\ & & \cdot\ \cdot & & \\ & & S_{n-1} & T_{n-1} + Q_n & R_n \\ & & & S_n & T_n + Q_{n+1} \end{bmatrix} \begin{bmatrix} \bar{c}_T \\ \bar{c}_1 \\ \cdot \\ \cdot \\ \cdot \\ c_n \\ \bar{c}_b \end{bmatrix} = \begin{bmatrix} B_0 + A_1 \\ B_1 + A_2 \\ \cdot \\ \cdot \\ \cdot \\ B_{n-1} + A_n \\ B_n + A_{n+1} \end{bmatrix} \tag{7.50}$$

remained relatively constant for a period of time (e.g. due to solubility limits) and then decreases with future time (e.g. Rowe and San, 1992).

In what follows, it will be shown that these types of event can be analyzed by the techniques developed above. This is facilitated by the introduction of the elapsed time

$$t' = t - t^* \tag{7.51}$$

There can be no abrupt jump in concentration in the deposit, and thus

$$c = c(t^*) \quad \text{when } t' = 0 \tag{7.52}$$

The form of the governing differential equation in any layer of the deposit does not change (it still has the form of equation 7.23b); however, the physical parameters may have changed, and thus n, D_{zz}, v_z, ϱ, Ktd and λ may also have changed values, and become n', D'_{zz}, v'_z, ϱ', K_d and λ' etc. Introduction of a Laplace transform with respect to elapsed time,

$$\bar{c}' = \int_0^\infty c e^{-s't'} \, dt' \tag{77.53}$$

leads to the transformed equations

$$n' D'_{zz} \frac{\partial^2 \bar{c}'}{\partial z^2} - n' v'_z \frac{\partial \bar{c}'}{\partial z} = n'(S'\bar{c}' - \phi') \tag{7.54a}$$

where

$$S' = \left(1 + \frac{\varrho' K'_d}{n'}\right) s' + \lambda' \tag{7.54b}$$

$$\phi = c(t^*)\left(1 + \frac{\varrho' K'_d}{n'}\right) \tag{7.54c}$$

Analysis of equation 7.54 in a specific layer will lead to a layer matrix similar in form to equation 7.49 but with any quantity Q replaced by the equivalent quantity Q'.

Consideration of conservation of mass in the landfill leads to

$$c_T = c'_0 - \frac{1}{H'_r} \int_{t^*}^{t} f_T(\tau) d\tau - \int_{t^*}^{t} \lambda'_T \, c_T(\tau) d\tau \tag{7.55}$$

where c'_0 represents the average concentration in the landfill just after $t = t^*$; λ'_T is the value of the first order decay constant; H'_r is the current reference height of leachate.

The Laplace transform of equation 7.55 is

$$\bar{f}'_T = B'_0 - T'_0 \bar{c}'_T \tag{7.56}$$

where

$$B'_0 = H'_r c'_0$$
$$T'_0 = H'_r S'_T = H'_r(s' + \lambda'_T)$$

An examination of conservation of mass in the base aquifer leads to the equation

$$c'_b - c'_{b0} = \int_{t^*}^{t} \left(\frac{f_b}{n'_b h_b} - \lambda'_b c'_b \right) dt \tag{7.57}$$

where again the prime indicates the current value of the particular parameter, and c'_{b0} denotes the concentration in the base aquifer at $t = t^*$.

The Laplace transform of equation 7.57 is

$$-\bar{f}'_b = Q'_{n+1} \bar{c}'_b - A'_{n+1} \tag{7.58}$$

where

$$Q'_{n+1} = h' n'_b S_b = h' n'_b (s + \lambda'_b)$$
$$A'_{n+1} = h' n'_b c'_{b0}$$

It is now clear that the assembly procedure for layer matrices will lead to a set of equations similar to equation 7.50 where all quantities have been replaced by the corresponding appropriate primed quantity and thus the calculation can proceed.

Thus one can model environmental changes by starting with the initial conditions (c_I at $t = 0$) and parameters and then obtain the concentration field, c_{I1}, at time $t = t^*_1$ (the time when the first change in conditions occurs) by solving equation 7.29 (if $c_I = 0$) or equation 7.50 (if $c_I \neq 0$). This concentration field c_{I1} can then be used to evaluate concentrations at time $t > t^*_1$ by solving equation 7.50, using $c_I = c_{I1}$, $t' = t - t^*_1$ and the new boundary conditions and values of D'_{zz}, v'_{zz}, n′ etc. for $t > t_1$. If there is another change in conditions at $t = t^*_2$, then the concentration field c_{12} can be evaluated at time $t = t^*_2$, as described above for $t > t^*_1$ and then for $t > t_2$ one solves equation 7.50 using $c_I = c_{I2}$, $t' = t - t^*_2$ and the next boundary conditions and values of D'_{zz}, v'_{zz}, λ' etc. for $t > t^*_2$. This process can be repeated many times.

In the finite layer formulation presented in sections 7.5.1 and 7.5.2, the number of layers was controlled by physical considerations, and any subdivision of physical layers into sublayers has no effect on the accuracy of the solution. However, the theory, as described in this section, uses an exponential approximation (equation 7.46) to interpolate between the concentration at the top and bottom of each sublayer, k, at time t^*, and hence to evaluate concentrations at $t > t^*$ to model variable environmental conditions. Since this is an interpolation function rather than the exact distribution, it follows that the number of sublayers may influence the results for $t > t^*$.

In general, the smaller the thickness of the sublayers or the more uniform the variation in concentration between the top and bottom of each sublayer, the smaller will be the error resulting from the interpolation. Thus it is prudent to do a sensitivity study of the effect of the number of sublayers when using this theory, as implemented in a computer program (e.g. POLLUTE). This is particularly important if the source concentration is changed significantly at time t^*. Under these circumstances, the most important thing is to have one thin layer near the boundary.

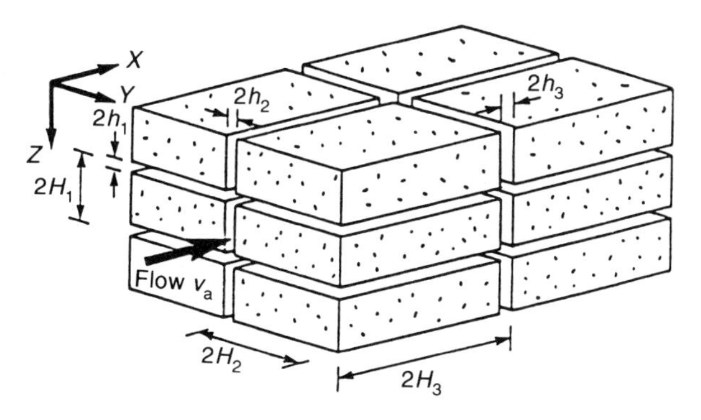

Figure 7.13 Definition of fracture geometry.

7.6 Contaminant migration in a regularly fractured medium

A major consideration in many landfill designs is the potential for contaminant migration in fractured clay or rock. For any fractured medium which can be idealized as being regularly fractured (e.g. as shown in Figure 7.13) it is possible to develop a very simple solution, as described below.

Consider an extensive landfill which is adjacent to fractured ground. It will be assumed that there are three possible sets of planar fractures. Suppose that 0_x, 0_y, 0_z are a set of cartesian reference axes. Referring to Figure 7.13, it will be assumed that the primary set of fractures are spaced $2H_1$ apart, of width $2h_1$, and parallel to the x–y-plane; the secondary set of fractures are spaced at an interval of $2H_2$, of width $2h_2$, and parallel to the x–z-plane; the tertiary set of fractures are spaced at an interval $2H_3$, of width $2h_3$, and parallel to the z–y-plane. The ground adjacent to the landfill is thus assumed to consist of a series of rectangular blocks made up of a homogeneous matrix material separated by fractures having a width far smaller than the smallest dimension of the block.

It will be assumed that the interface of the landfill and the adjacent ground is the y–z-plane ($x = 0$), and that the landfill is quite extensive, so that the predominant mechanism for transport of the leachate will be by advective–dispersive transport along fracture sets 1 and 2 accompanied by one-, two- or three-dimensional diffusion of leachate into the intact blocks.

7.6.1 Development of basic equations for a one-, two- or three-dimensional system

The net flux with components F_x, F_y, F_z may be defined by

$$F_x = \frac{h_1 H_2 F_{1x} + h_2 H_1 F_{2x}}{H_1 H_2} = \frac{h_1}{H_1} F_{1x} + \frac{h_2}{H_2} F_{2x}$$

$$F_y = \frac{h_1 H_3 F_{1y} + h_3 H_1 F_{3y}}{H_1 H_3} = \frac{h_1}{H_1} F_{1y} + \frac{h_3}{H_3} F_{3y}$$

$$F_z = \frac{h_2 H_3 F_{2z} + h_3 H_2 F_{3z}}{H_2 H_3} = \frac{h_2}{H_2} F_{2z} + \frac{h_3}{H_3} F_{3z} \tag{7.59}$$

where F_{1x}, F_{1y} are the x- and y-components of flux in the first set of fractures, with similar definitions for F_{2x}, F_{2z} and F_{3y}, F_{3z}.

Conservation of mass in the fracture system then leads to the equation

$$-\left(\frac{\partial F_x}{\partial x} + \frac{\partial F_y}{\partial y} + \frac{\partial F_z}{\partial z}\right) = n_f \left(\frac{\partial c_f}{\partial t} + \lambda_f c_f\right) + \dot{g} + \dot{q} - \dot{r} \tag{7.60}$$

where c_f is the concentration of contaminant at a point in the fracture, and

$$n_f = \frac{h_1}{H_1} + \frac{h_2}{H_2} + \frac{h_3}{H_3}$$

where λ_f is the first order decay constant of the solute ($\lambda_f = \Gamma_R + \Gamma_B + \Gamma_S$), which may result from first order radioactive decay, Γ_R, biological degradation, Γ_B or flow removed per unit volume, Γ_S; $\dot{q}$ is the rate at which the contaminant is being transported into the matrix per unit volume of matrix and fissures; $\dot{r}$ is the rate at which the contaminant is being 'injected' into the fissure system per unit volume of matrix and fissures.

Flow in the fissures is governed by

$$\begin{aligned}
F_{1x} &= v_{1x} c_f - D_{1x} \frac{\partial c_f}{\partial x} \\
F_{1y} &= v_{1y} c_f - D_{1y} \frac{\partial c_f}{\partial y} \\
F_{2x} &= v_{2x} c_f - D_{2x} \frac{\partial c_f}{\partial x} \\
F_{2z} &= v_{2z} c_f - D_{2z} \frac{\partial c_f}{\partial z} \\
F_{3y} &= v_{3y} c_f - D_{3y} \frac{\partial c_f}{\partial y} \\
F_{3z} &= v_{3z} c_f - D_{3z} \frac{\partial c_f}{\partial z}
\end{aligned} \tag{7.61}$$

where the definitions of v_{1x}, v_{1y} etc. and D_{1x}, D_{1y} etc. parallel those of F_{1x}, F_{1y} etc., and where v_{1x} etc. represent the groundwater velocity and D_{1x} etc. represent the coefficient of hydrodynamic dispersion in the referenced fracture set, and it is assumed that the fractures are open.

It follows that

$$\begin{aligned}
F_x &= v_{ax} c_f - D_{ax} \frac{\partial c_f}{\partial x} \\
F_y &= v_{ay} c_f - D_{ay} \frac{\partial c_f}{\partial y} \\
F_z &= v_{az} c_f - D_{az} \frac{\partial c_f}{\partial z}
\end{aligned} \tag{7.62}$$

where

$$v_{ax} = v_{1x}\frac{h_1}{H_1} + v_{2x}\frac{h_2}{H_2}$$

$$D_{ax} = D_{1x}\frac{h_1}{H_1} + D_{2x}\frac{h_2}{H_2}$$

with similar definitions for v_{ay}, D_{ay}, v_{az}, D_{az}.

For the special case where the groundwater velocity and coefficient of hydrodynamic dispersion is the same in both fracture sets (i.e. $v = v_{1x} = v_{2x}$, $D = D_{1x} = D_{2x}$), the above equations reduce to

$$v_a = n_f v$$
$$D_a = n_f D$$

where

$$n_f = \frac{h_1}{H_1} + \frac{h_2}{H_2}$$

From equations 7.60 and 7.62, and the assumption that there are a homogeneous layer and constant groundwater velocity in the fracture system within this layer, it follows that

$$D_{ax}\frac{\partial^2 c_f}{\partial x^2} + D_{ay}\frac{\partial^2 c_f}{\partial y^2} + D_{az}\frac{\partial^2 c_f}{\partial z^2} - v_{ax}\frac{\partial c_f}{\partial x} - v_{ay}\frac{\partial c_f}{\partial y} - v_{az}\frac{\partial c_f}{\partial z}$$
$$= (n_f + \Delta K_f)\left(\frac{\partial c_f}{\partial t} + \lambda_f c_f\right) + \dot{q} - \dot{r} \qquad (7.63)$$

where it has been assumed that the rate at which contaminant is being sorbed onto the fracture walls per unit volume is linear:

$$\dot{g} = \Delta K_f \frac{\partial c_f}{\partial t} \qquad (7.64)$$

$$\Delta = \left(\frac{1}{H_1} + \frac{1}{H_2} + \frac{1}{H_3}\right)$$

where K_f is the fracture distribution coefficient, defined by Freeze and Cherry (1979) as the mass of solute adsorbed per unit area of surface divided by the concentrations of solute in solution, and Δ represents the surface area per unit volume. Equation 7.63 can be solved by means of a Laplace transform, and it is found that

$$D_{ax}\frac{\partial^2 \bar{c}_f}{\partial x^2} + D_{ay}\frac{\partial^2 \bar{c}_f}{\partial y^2} + D_{az}\frac{\partial^2 \bar{c}_f}{\partial z^2} - v_{ax}\frac{\partial \bar{c}_f}{\partial x} - v_{ay}\frac{\partial \bar{c}_f}{\partial y} - v_{az}\frac{\partial \bar{c}_f}{\partial z}$$
$$= [(n_f + \Delta K_f) + \bar{\eta}]\,(s + \lambda_f)\bar{c}_f - \bar{\dot{r}} \qquad (7.65a)$$

where it is shown in Appendix C that the Laplace transform of $\dot{q}$ has the form

$$\bar{\dot{q}} = (s + \Delta_m)\bar{\eta}\bar{c}_f \qquad (7.65b)$$

7.6.2 An infinite fracture medium

Consider now the case of one-dimensional transport from a landfill, located at $x = 0$, into an extensively fractured zone, $x > 0$, which is initially uncontaminated. In this case, there is no injection of contaminant, so that $\dot{r} = 0$. The process is governed by the equation

$$D_{ax}\frac{\partial^2 \bar{c}_f}{\partial x^2} - v_{ax}\frac{\partial \bar{c}_f}{\partial x} = S\bar{c}_f$$

where $S = (n_f + \Delta K_f)s + n_f\lambda_f + (s + \Lambda_m)\bar{\eta}$.

The above equation has the general solution

$$\bar{c}_f = Ae^{\alpha x} + Be^{\beta x}$$

where

$$\alpha = \frac{v_{ax}}{2D_{ax}} + \left(\frac{v_{ax}^2}{4D_{ax}^2} + \frac{S}{D_{ax}}\right)^{1/2}$$

$$\beta = \frac{v_{ax}}{2D_{ax}} - \left(\frac{v_{ax}^2}{4D_{ax}^2} + \frac{S}{D_{ax}}\right)^{1/2} \qquad (7.66)$$

Recalling that the solution must remain bounded as $x \to \infty$, and the concentration in the landfill is c_{LF}, it follows that the distribution of concentration is given by

$$\bar{c} = \bar{c}_{LF} e^{\beta x} \qquad (7.67a)$$

while the flux distribution is given by

$$\bar{F}_x = \bar{c}_{LF} D_{ax} \alpha e^{\beta x} \qquad (7.67b)$$

The behavior within the landfill has been examined previously (equations 1.23 and 7.20) where it was shown that

$$\bar{c}_T = \frac{c_0}{S_T} - \frac{\bar{f}_T}{S_T H_r}$$

so that

$$\bar{c}_{LF} = \frac{c_0}{S_T + (D_{ax}/H_r)} \qquad (7.68)$$

Equations 7.67 and 7.68 define the concentration plume (in the fractures) in transformed space. The concentration at any given time and location can then be readily obtained by numerically inverting the Laplace transform using the method proposed by Talbot (1979). This has been implemented in program POLLUTE.

7.6.3 Finite layer analysis of fissured material

The solution of problems involving contaminant transport through fissured material has been illustrated by a simple example. It is not difficult to extend the finite layer methods developed in sections 7.5.1 and 7.5.2 to incorporate the behavior of fissured material. For one-dimensional transport, the finite layer equations reduce to equation 7.29, with the factor nD_{zz} appearing in equation 7.26b replaced by D_{az} and the values of α, β given by equation 7.24d replaced by those given by equation 7.66.

For two-dimensional transport, the finite layer equations reduce to equations 7.38 and 7.43 with the factor nD_{zz} appearing in equation 7.26b again replaced by D_{az} and the values of α, β given by equation 7.24a replaced by

$$\alpha = \frac{v_{ax}}{2D_{ax}} + \left(\frac{v_{ax}^2}{4D_{ax}^2} + \frac{S}{D_{ax}}\right)^{1/2}$$

$$\beta = \frac{v_{ax}}{2D_{ax}} - \left(\frac{v_{ax}^2}{4D_{ax}^2} + \frac{S}{D_{ax}}\right)^{1/2} \qquad (7.69)$$

with $S = (n_f + \Delta K_f)s + n_f\lambda_f + (s + \Lambda_m)\eta + i\xi\, v_{ax} + \xi^2 D_{ax}$. The full details are given by Rowe and Booker (1990a).

A similar approach can also be used to consider fracture media for two-dimensional contaminant transport (Rowe and Booker, 1991a). Using the finite layer approach, one can consider both fractured and unfractured layers (e.g. Rowe and Booker, 1991b).

Chapter eight

Determination of diffusion and distribution coefficients

8.1 Introduction

The key transport mechanisms and the development of the governing equations have been discussed in Chapter 1, and the basic concepts of diffusion were examined in Chapter 6. However, in order to use theoretical approaches to predict the potential impact of a waste disposal site, it is necessary to estimate appropriate design parameters. These parameters should be obtained using samples of the proposed barrier material, and using a leachate as similar as possible to that expected in the facility.

The key soil-related parameters identified in Chapter 1 were the advective velocity, diffusion coefficient, distribution/partitioning coefficient, effective porosity and dispersivity. The advective velocity will depend on the hydraulic gradient and hydraulic conductivity as already discussed. In this chapter, consideration will be focused on the other parameters with respect to clayey barriers.

8.2 Obtaining diffusion and partitioning/distribution coefficients: basic concepts

As discussed in Chapter 1, the coefficient of hydrodynamic dispersion has two components: mechanical dispersion and molecular diffusion. In this chapter attention will be focused on the movement of contaminants through clayey soils at relatively low velocities, and a technique for estimating both the coefficient of hydrodynamic dispersion, D, and the distribution (partitioning) coefficient, K_d, using a single test will be described. Under the conditions examined, the coefficient of hydrodynamic dispersion is, to all practical purposes, equal to the effective diffusion coefficient (i.e. $D \approx D_e$). In the proposed test, an undisturbed sample of soil is placed in a column and the leachate of interest is placed above the soil. Contaminant is then permitted to migrate through the specimen under the prescribed head (which may be zero). The volume of leachate above the soil will normally be selected to be sufficiently small to allow a significant drop in concentration of contaminant within the source solution. Typically the height of leachate, H_f, in the column above the clay will range from 0.05 to 0.3 m. This drop in concentration with time should be monitored.

A number of possible boundary conditions at the base of the sample may be considered. If the test is to be conducted with advective transport through the specimen, then a porous collection plate can be placed beneath the sample and the effluent collected and monitored (Figure 8.1).

The volume in the reservoir is maintained constant by addition of background reference fluid (not leachate). If there is no advective flow, then two other base boundary conditions may be considered. Firstly, the base could be an impermeable plate (Figure 8.2(a)). The alternative is to have a closed collection chamber (reservoir) similar to that for the leachate, but initially having only a background concentration of the contaminant of interest (Figure 8.2(b)). Thus, as contaminant passes through the soil, it accumulates in this collection chamber and the concentration in this reservoir can be monitored.

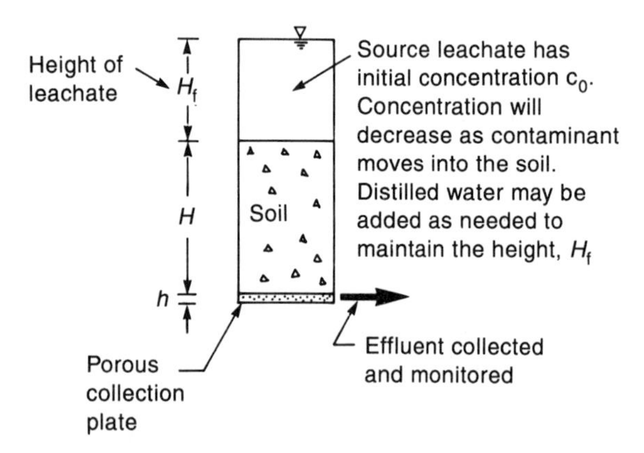

Figure 8.1 Schematic of an advection–diffusion column test with a finite mass of contaminant source boundary condition. (From Rowe, Caers and Barone, 1988; reproduced with permission of the *Canadian Geotechnical Journal.*)

Suppose that the volume of source solution (leachate) is equal to AH_f, where A is the plan area of the column and H_f is the height of the leachate in the column (e.g. Figure 8.3). At any time t, the mass of any contaminant species of interest in the source solution is equal to the concentration $c_T(t)$ in the solution, multiplied by the volume of solution (assuming here that the solution is stirred so that $c_T(t)$ is uniform throughout the solution). The principle of conservation of mass requires that at this time t, the mass of contaminant in the source solution is equal to the initial mass of the contaminant minus the mass which has been transported into the soil up to this time t. This can be written algebraically (as shown in Chapter 1) as

$$c_T(t) = c_0 - \frac{1}{H_f} \int_0^t f_T(\tau) d\tau \tag{8.1}$$

where $c_T(t)$ is the concentration in the source solution at time t [ML^{-3}]; c_0 is the initial concentration in the source solution ($t = 0$) [ML^{-3}]; H_f is the height of leachate (i.e. the volume of leachate per unit area) [L]; $f_T(\tau)$ is the mass flux of this contaminant into the soil at time τ ($ML^{-2}T^{-1}$].

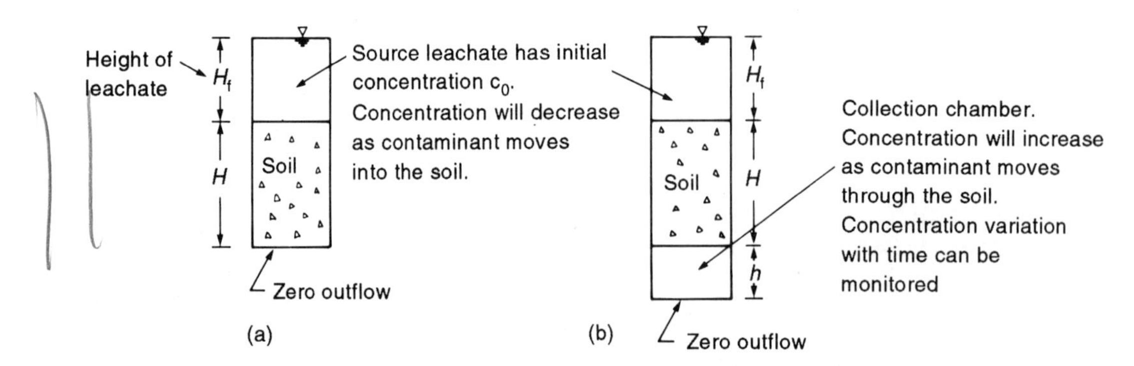

Figure 8.2 Schematic of pure diffusion test with a finite mass of contaminant source boundary condition. (Rowe, Caers and Barone, 1988; reproduced with permission of the *Canadian Geotechnical Journal.*)

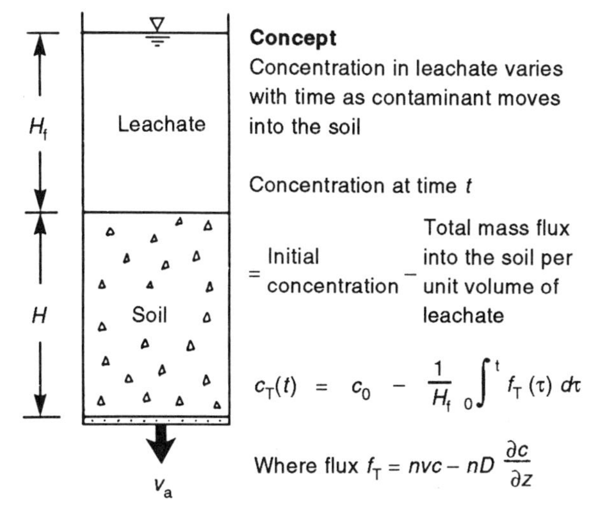

Figure 8.3 Schematic showing how the concentration of contaminant in the source varies as contaminant is transported into the soil.

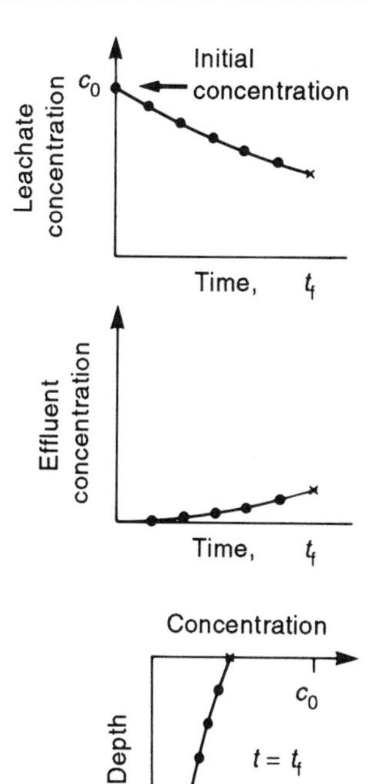

Figure 8.4 Experimental procedure used to determine the diffusion coefficient D and distribution coefficient K_d. (Modified from Rowe, 1988.)

Contaminant is allowed to migrate from the source chamber through the soil and, if present, into the collection chamber. If no additional contaminant is added to the source chamber, then the concentration of contaminant will decrease with time as mass of contaminant diffuses into the soil (Figure 8.4). The rate of decrease can be controlled by the choice of the height of leachate, H_f. Conversely, as contaminant moves into the collection chamber (Figures 8.1, 8.2(b)), the increase in mass gives rise to an increase in contaminant concentration in this reservoir (Figure 8.4). The rate of decrease in concentration in the source and increase in the collection chamber should be monitored with time. At some time t_f, the test is terminated, and the concentration profile through the soil sample may be obtained (Figure 8.4). Assuming linear sorption, theoretical models can then be used to estimate the effectivity porosity, n, effective diffusion coefficient, D_e, and product of dry density and distribution coefficient ϱK_d. This theoretical analysis has been described in detail by Rowe and Booker (1985a, 1987), as outlined in Chapter 7, and has been implemented in the computer program POLLUTE (Rowe and Booker, 1994). This approach permits accurate calculation of concentration in only a few seconds on a microcomputer, and hence it is well-suited for use in interpretation of the results of the column tests.

The sensitivity of this approach for the estimation of D and ϱK_d can be illustrated by considering the diffusive migration of a contaminant through a 10 cm thick sample given a soil porosity of 0.4, a Darcy velocity ($v_a = nv$) of 0.0 and a height of leachate ((volume of leachate)/(plan area of specimen))H_f = 0.05 m, as indicated in the insert to Figure 8.5. Taking a typical diffusion coefficient D (e.g. for Na^+) of 0.015 m^2/a, Figure 8.5(a) shows the theoretical

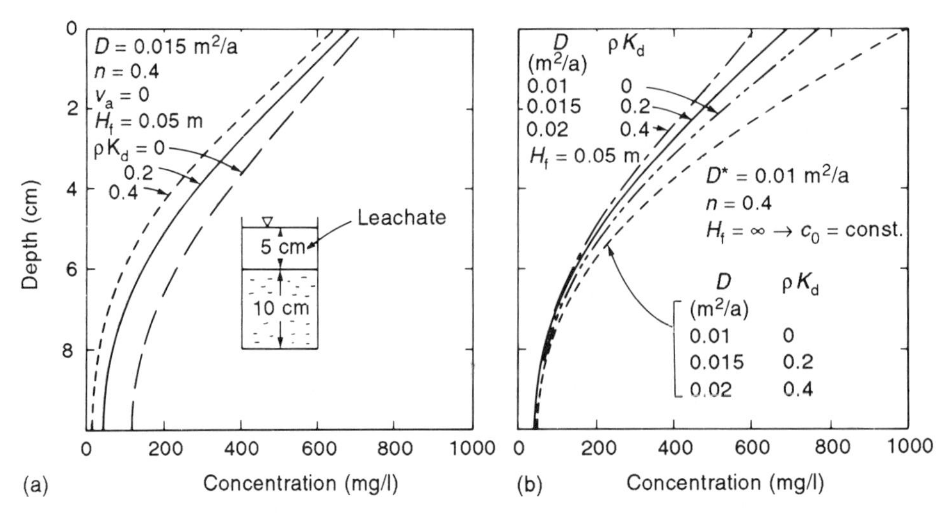

Figure 8.5 Concentration profiles with depth in a hypothetical test at 0.1 years, showing sensitivity of concentration profiles: (a) to small change in ϱK_d; (b) to different combinations of D and ϱK_d giving the same value of D^*.

concentration profile at 0.1 years for values of ϱK_d = 0, 0.2 and 0.4. It can be seen that even this relatively small difference in sorption results in a measurable difference in the concentration profiles. Figure 8.5(b) shows the theoretical concentration profile at 0.1 years, for three combinations of parameters (D, ϱK_d), viz. (0.01 m^2/a, 0), (0.015 m^2/a, 0.2) and (0.02 m^2/a, 0.4). These three combinations of the parameters correspond to the same value of D^* = 0.01 m^2/a (section 6.5 contains a discussion regarding D^*). In tests where the leachate concentration was held constant ($c_T(t) = c_0$) these three sets of parameters would give identical concentration profiles at any time. However, as is evident from Figure 8.5(b), in the proposed test where the source concentration is allowed to drop with time (H_f = 0.05 m here), these three sets of parameters give rise to different concentration profiles, illustrating the different effects of D and ϱK_d. It is for this reason that both parameters can be evaluated. The effect is even more pronounced for lower values of leachate height H_f.

8.3 Example tests for obtaining diffusion and distribution coefficients for inorganic contaminants

8.3.1 Advective–diffusive tests

A series of laboratory column tests was performed on samples of unweathered gray clay taken from beneath the Confederation Road landfill (near Sarnia, Ontario). The basic geotechnical properties and mineralogy of the soil are summarized in Table 8.1.

A schematic diagram of the apparatus used to obtain diffusion and distribution parameters in the presence of advection is shown in Figure 8.6. Details are given by Rowe, Caers and Barone (1988). The objective of this test was to simulate field conditions as realistically as possible. This involved having an applied stress and downward advective flow. As will be demonstrated, this elaborate test turned out not to be necessary, and similar results were obtained using a simple pure diffusion test.

Table 8.1 Soil description for the Confederation Road landfill (after Crooks and Quigley, 1984)

Property	*Below landfill waste*
Liquid limit (%)	~39
Plastic limit (%)	~12
Specific gravity	2.73
Moisture content (%)	~23
Mineralogy (<74 μm) (%)	
Carbonates	~34
Quartz and feldspars	12–20
Illite	23–27
Chlorite	22–26
Smectite	~1
Cation exchange capacity (<2 μm) (meq/100 g)	10.5

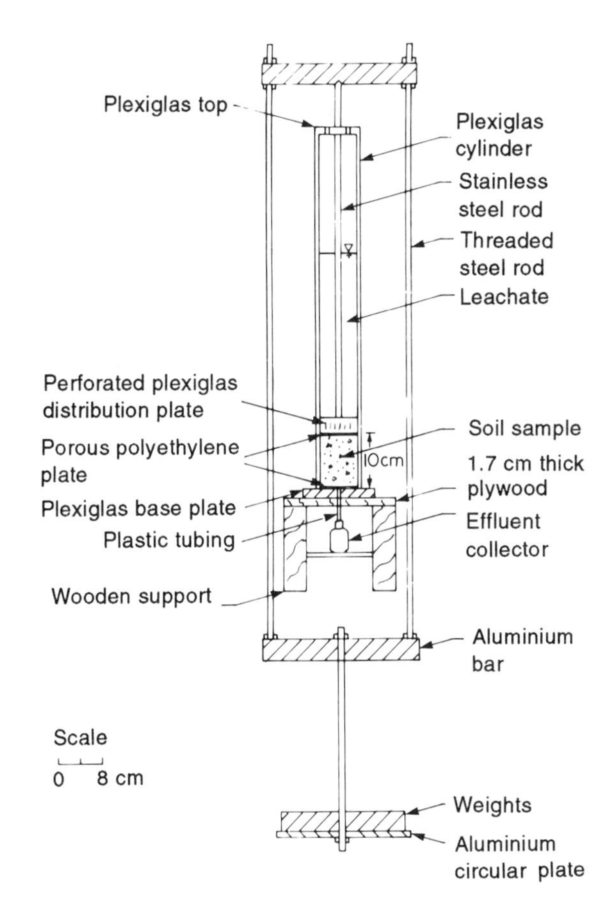

Figure 8.6 Schematic diagram of plexiglass model (advection–diffusion test).

A hanger weight system was set up and a pressure of 87 kPa was initially applied to the soil sample for two days. This pressure was considered to be large enough to provide good seating of the sample in the plexiglass column, while being well below the pre-consolidation pressure of 172 kPa (Ogunbadejo, 1973). After the first two days, the soil sample was allowed to reconsolidate for two more days at an applied pressure of 30 kPa. To prevent drying of the clay surface, a small quantity of distilled water was maintained above and below the sample (zero hydraulic gradient) during the consolidation period. After consolidation, the distilled water was replaced by salt solution above the soil, and drainage into a small polyethylene collection bottle was permitted at the bottom of the column. The source salt solution was mixed periodically to maintain a relatively uniform concentration throughout the reservoir depth. The models were maintained at a laboratory temperature of 22° ± 1°C. A similar procedure can be used with leachate as a source fluid.

The total fluid flow through the soil and into the collection bottle was monitored. To prevent a drop in height of solution in the reservoir due to seepage into the soil, a volume of distilled water equal to volume of effluent collected was added after each monitoring period. Thus, the height of leachate in the reservoir remained relatively constant. The dilution resulting from the addition of distilled water is automatically considered by equation 8.1.

Six tests (series (i), referred to as models A–F) were conducted as described above. In each test, a specified salt solution (calcium chloride, sodium chloride or potassium chloride) was placed into contact with the clay under a controlled total head for a pre-determined period, as indicated in Table 8.2. Models A–E involved a single cation source solution. Model F involved a source solution of both potassium and calcium chloride.

The effluent discharge volume was found to be linear with time over the entire test period for

Table 8.2 Characteristics of various diffusion tests

	Average background concentration (mg/l)					Source solution concentration of key species (mg/l)						
Test[a]	Cl^-	Na^+	K^+	Ca^{2+}	Mg^{2+}	Cl^-	Na^+	K^+	Ca^{2+}	Mg^{2+}	*Darcy velocity (m/a)*	*Duration of test (days)*
Series i												
A	175	180	~20	345	255	1501	975	–	–	–	0.034	86
B						1725	–	–	975	–	0.033	84
C						1725	–	–	975	–	0.030	141
D						885	–	975	–	–	0.035	105
E						885	–	975	–	–	0.025	97
F						2609	–	975	975	–	0.035	108
Series ii												
G	55	150	10	85	120	1480	955	–	–	–	0	15
H						364	–	400	–	–	0	15
I						450	–	–	250	–	0	15
J						856	–	–	–	293	0	15
Series iii												
K	55	150	10	85	120	1000	955	400	250	291	0	15

[a] A–E: single salt, 22°C; F: two salt, 22°C; G–J: single salt, 10°C; K: leachate, 10°C.

each of the six tests. Based on these discharge rates, the Darcy velocity was deduced as shown in Table 8.2. The calculated hydraulic conductivities of between 2×10^{-10} and 4×10^{-10} m/s are only marginally higher than a field value of 1.5×10^{-10} m/s obtained from falling head tests on piezometers installed in the clay below the landfill at Sarnia (Goodall and Quigley, 1977).

Figures 8.7 and 8.8 show the observed and typical best-fit theoretical matching curves for source concentration (c_T) of sodium chloride (NaCl) and potassium chloride (KCl) solutions, respectively.

The results presented in Figures 8.7 and 8.8 were both obtained for the same volume of source leachate ($H_f = 0.3$ m) and it can be seen that the decrease in chloride concentration with time is fairly similar with the minor differences reflecting a small difference in Darcy velocity. However, comparing the rate of change in concentrations with time for three ions (Na^+, K^+, Cl^-), it is apparent that there is a significant difference in diffusion coefficient (D) and distribution coefficient (K_d) for the three ions. The values of the parameters D and ρK_d deduced by fitting the theoretical curve to the observed change in concentration are summarized in Table 8.3.

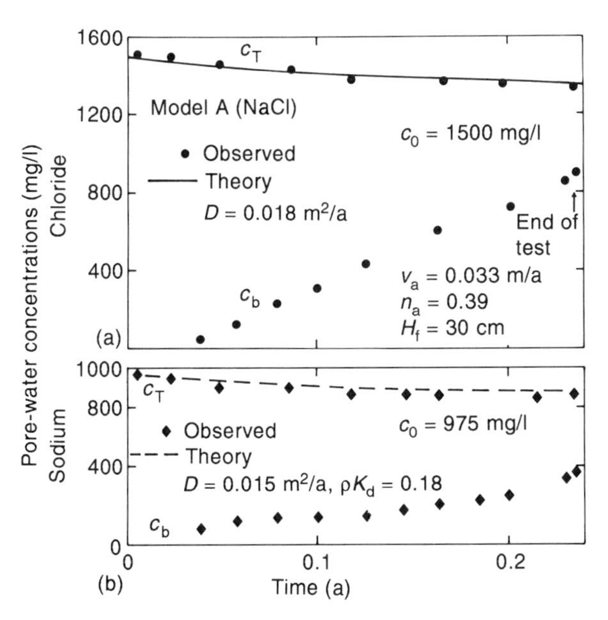

Figure 8.7 Source (c_T) and base (c_b) concentration changes over time in model A for (a) chloride, (b) sodium. (Modified from Rowe, Caers and Barone, 1988.)

The variation in source concentration (c_T) with time provides an initial means of estimating the parameters D and ρK_d; however, the variation in the concentration throughout the sample at the termination of the test provides the primary data for estimating, or checking, these parameters. Figures 8.9–8.11 show the observed anion (Cl^-) and cation concentrations with depth for models A, D and E respectively. Also shown are the theoretical curves obtained using the values of D and ρK_d deduced from the variation in source concentration with time, and given in Table 8.3.

Allowing for some small experimental scatter of data points, inspection of Figures 8.7–8.8 and 8.9–8.11 indicates that in each case the theoretical curves provide a very good fit to both the decrease in the source fluids concentration with time and the variation in concentration with depth in the soil at the end of each test. The consistency of results demonstrates the power of the analytical model (program POLLUTE), and provides some confidence in the parameters D and ρK_d for the clay and source fluids examined.

To provide an indication of parameter variation that might be expected for a given soil, a number of tests were duplicated. The diffusion coefficient, D, for chloride was deduced for each model, and ranged between 0.018 and 0.02 m^2/a, with an average value of 0.019 m^2/a. This small variation in D does not appear to be related to small differences in Darcy velocity, nor does it appear to be particularly related to the nature of the associated cation (Table 8.3). Rather, the variability from 0.018 to 0.02 m^2/a is seen as an indication of the level of repeatability that may be achieved for this type of test.

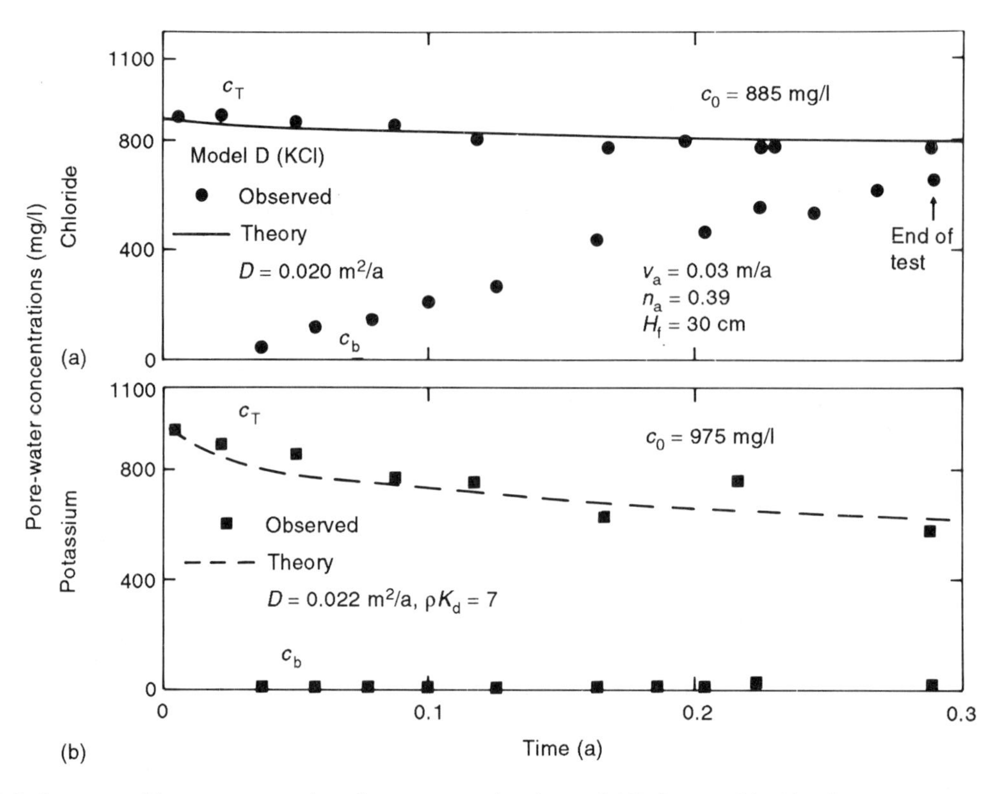

Figure 8.8 Source and base concentration changes over time in model D for (a) chloride, (b) potassium. (Modified from Rowe, Caers and Barone, 1988.)

Table 8.3 Comparisons of diffusion coefficient, D (m^2/a) and sorption parameter, ϱK_d (ϱK_d in parentheses)

Test	*Cl^-*	*Na^+*	*K^+*	*Ca^{2+}*	*Mg^{2+}*
Series i					
A	0.018 (0)	0.015 (0.18)	–	–	–
B	0.018 (0)	–	–	0.012 (2)	–
C	0.019 (0)	–	–	0.012 (2)	–
D	0.020 (0)	–	0.022 (7)	–	–
E	0.020 (0)	–	0.020 (7)	–	–
F	0.019 (0)	–	0.022 (7)	–[a]	–
Series ii					
G	0.018 (0)	0.018 (0.75)	–	–	–
H	0.019 (0)	–	0.024 (4.5)	–	–
I	0.020 (0)	–	–	0.013 (4)	–
J	0.019 (0)	–	–	–	0.012 (5)
Series iii					
K	0.024 (0)	0.014 (0.25)	0.019 (1.7)	–[a]	–[a]

[a]No good fit could be obtained.

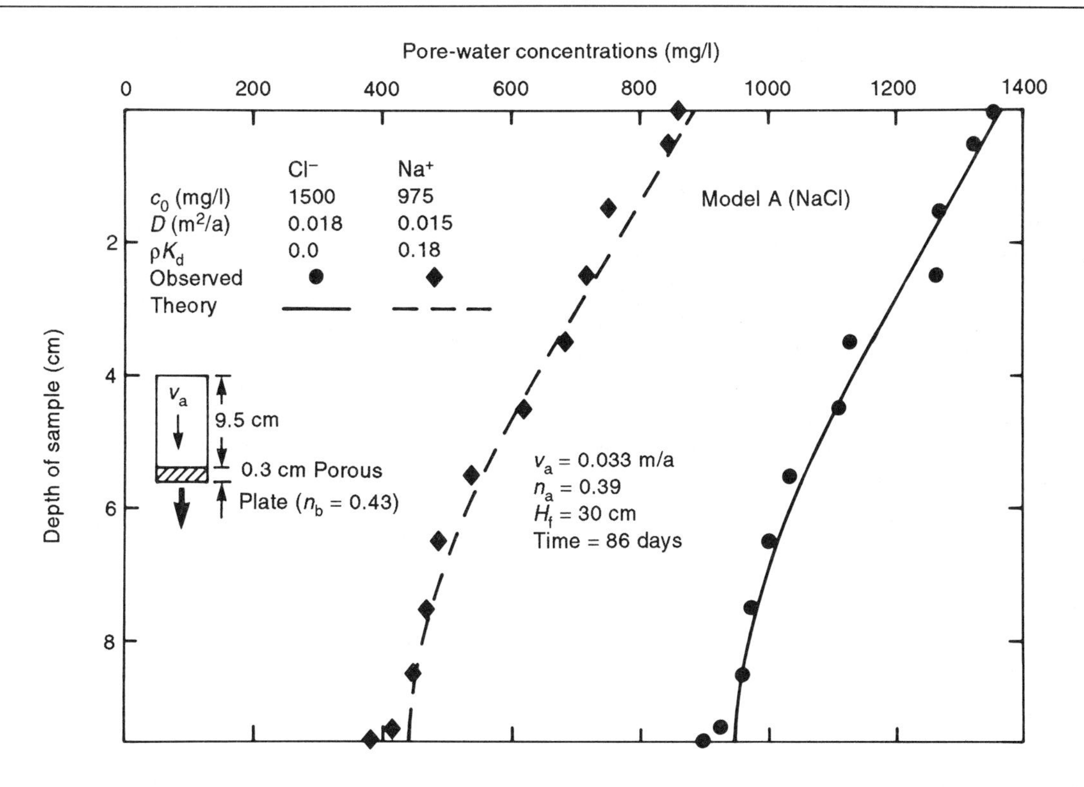

Figure 8.9 Chloride and sodium concentration versus depth of sample for model A. (Modified from Rowe, Caers and Barone, 1988.)

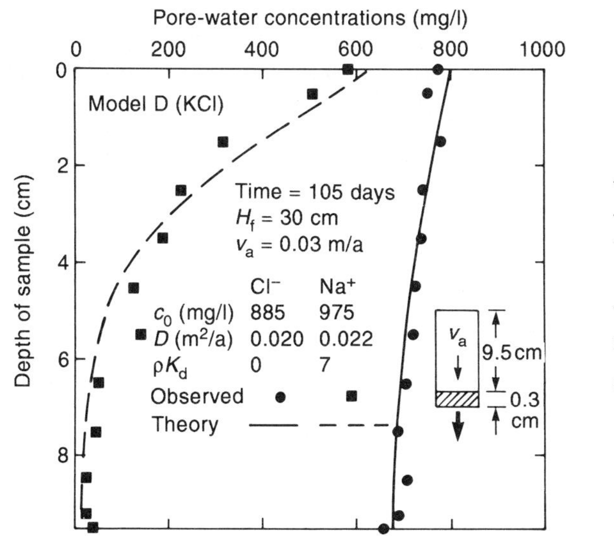

Figure 8.10 Chloride and potassium concentration versus depth of sample for model D. (Modified from Rowe, Caers and Barone, 1988.)

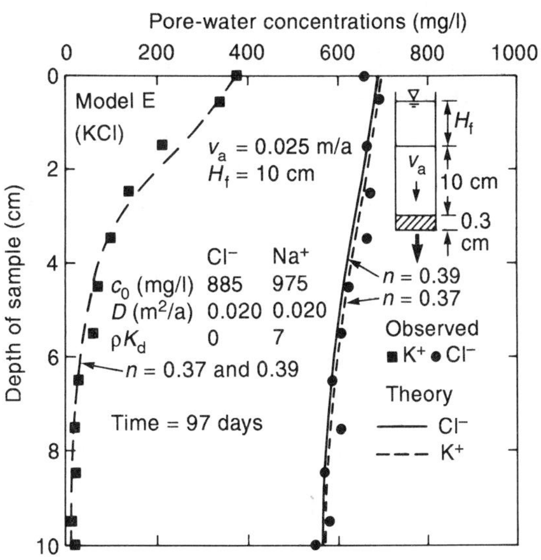

Figure 8.11 Chloride and potassium concentration versus depth of sample for model E. (Modified from Rowe, Caers and Barone, 1988.)

The application of an effective stress to the soil sample adopted in these tests is not an essential part of the proposed technique for determining the parameters D and K_d. Tests performed without the applied stress for the particular combination of clay and permeants considered herein gave similar results, both with and without the application of the effective stress. However, for some combinations of clay and permeant, shrinkage of the clay may occur in the absence of a confining stress, and this can give quite misleading results (e.g. Quigley and Fernandez, 1989). For these clays, tests should be performed at an effective stress similar to that anticipated in the field.

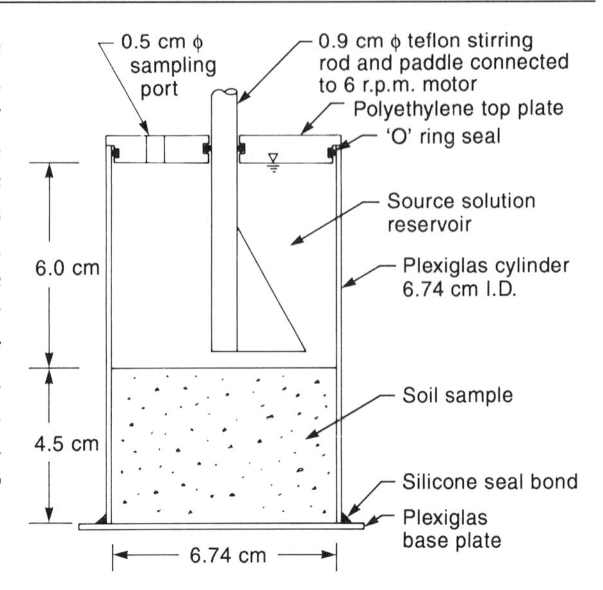

Figure 8.12 Schematic diagram of the diffusion model (simple diffusion test).

8.3.2 Pure diffusion test

In many cases, it is not necessary to perform an advection–diffusion test. Under these circumstances, a simple diffusion test can be performed for boundary conditions shown in Figure 8.2. In this test, the soil sample is placed in a plexiglass cylinder by trimming the sample to a size marginally greater than the specimen and then pressing the specimen into the cylinder, using a cutting shoe attached to the cylinder to perform the final trim. This procedure is found to work well for many clays. However, it does not work well for clays with a significant stone content because the stones tend to catch on the plexiglass and then tear the sample. In these cases the modified procedure described in section 8.3.3 may be adopted.

Once placed into the cylinder, a plexiglass base plate is placed below the sample (Figure 8.12) and the source fluid (e.g. leachate) is placed above the specimen. The container is then sealed and left for diffusion to occur. The source leachate strength is monitored periodically. After some time t, the test is disassembled, and the diffusion profile through the sample is established as shown in Figure 8.13. By adjusting parameters in a theoretical model, a fit to the data can be obtained, thereby yielding an estimate of the diffusion parameters.

8.3.3 Diffusion through soft clayey soils, or soils with significant gravel-sized particles

For some very soft soils, or some tills which have a significant gravel size, it is not practical to push the sample inside a plexiglass cylinder because of the damage (e.g. compression or tearing) that may be done to the specimen. In these cases, an alternative procedure involving the use of a latex membrane around the specimen can be used (e.g. Figure 8.14) to contain the specimen and the source fluid. Care is required to avoid leakage down the sides of the membrane. This procedure, while not as desirable as that described in section 8.3.2, has been found to work well and, under similar circumstances, gives similar results to those obtained using the set-up shown in Figure 8.12.

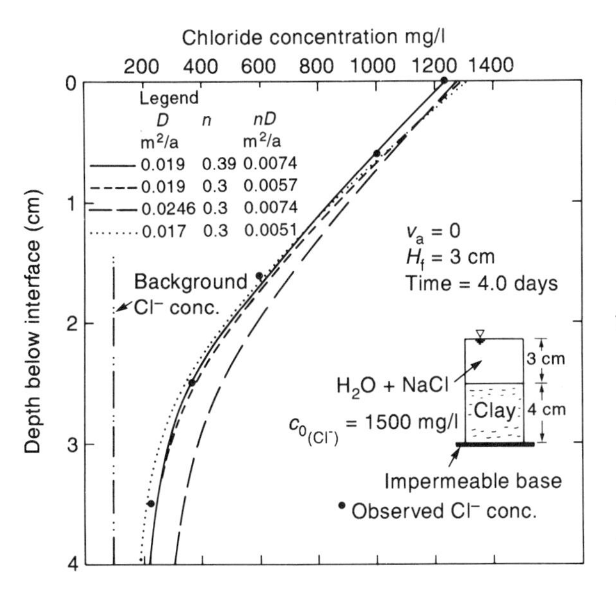

Figure 8.13 Effect of choice of effective porosity and diffusion coefficient on the comparison between observed and calculated concentration profile. (Modified from Rowe, Caers and Barone, 1988.)

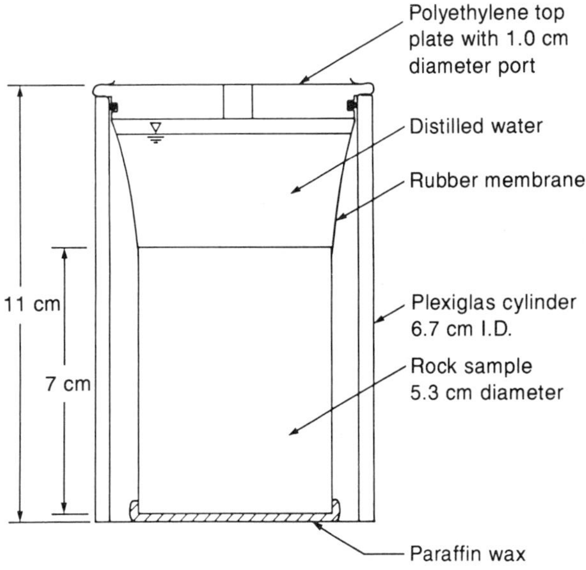

Figure 8.14 Schematic of diffusion tests with latex membrane (for rock, stiff soils or soils with significant gravel-sized particles).

In performing tests on samples that have significant gravel sizes, it is important to select samples which do not contain excessive gravel. Samples can be checked prior to testing using x-ray radiography.

8.3.4 Diffusion through sedimentary rock (shale, sandstone)

Techniques have been developed for performing diffusion tests through sedimentary rocks, as described by Barone, Rowe and Quigley (1990, 1992a). Two types of tests have been performed. If the sedimentary rock already has a relatively high concentration of the chemical species of interest (sodium, chloride etc.), then a test can be performed by encasing the rock in a latex membrane and allowing the species of interest to diffuse out of the rock into a reservoir of fluid (e.g. distilled or reference water). Figure 8.15 shows a schematic of one such test set-up, and Figure 8.16 shows the diffusion profile obtained by Barone, Rowe and Quigley (1990) after 65 days' diffusion of chloride out of a sample of Queenston shale. Based on the results of a number of these tests, the diffusion coefficient of chloride in this shale, which had a porosity of 10.6–11.3%, was between 1.4×10^{-10} m^2/s (0.0044 m^2/a) and 1.6×10^{-10} m^2/s (0.005 m^2/a) at 22°C. Similar tests were also reported by Barone, Rowe and Quigley (1992a) for Bison mudstone, which has a porosity of 21.5–25.7%, and yielded a diffusion coefficient for chloride of between 1.5×10^{-10} m^2/s (0.0047 m^2/a) and 2.0×10^{-10} m^2/s (0.0063 m^2/a) at 10°C.

If the sedimentary rock does not have a significant concentration of the chemical species of interest, then a diffusion test can be performed by placing a source 'leachate' containing the species of interest in contact with the rock, using an arrangement similar to that shown in Figure 8.15(a). The flux into and out of the sample can be monitored, and the concentration profile at the end of the test may be determined in a similar manner as that adopted for soil.

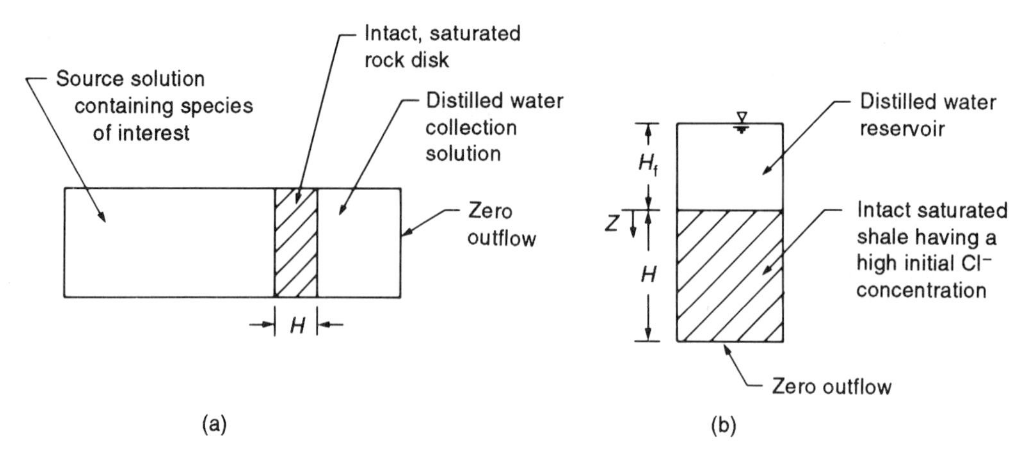

Figure 8.15 Schematic of test set-ups for obtaining diffusion coefficients in porous rock: (a) diffusion from source through the rock to a receptor; (b) back diffusion from rock with an initial high chloride concentration.

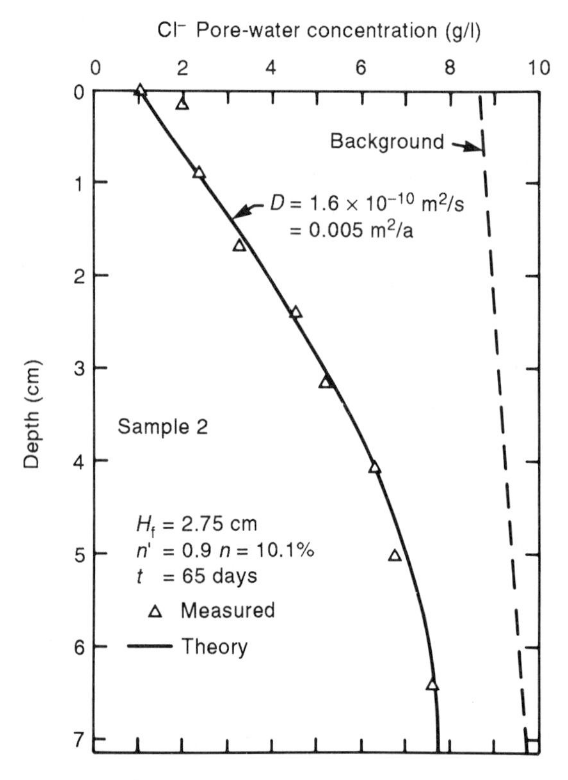

Figure 8.16 Diffusion profile for outward diffusion from a shale with an initial chloride concentration in the pore fluid. (From Barone, Rowe and Quigley, 1992a; reproduced with permission of the *Canadian Geotechnical Journal.*)

8.3.5 Comments on degree of saturation

The tests described in the previous subsection have been successfully used for both natural and compacted clayey soils, and for sedimentary rock with greater than 95% saturation. For unsaturated soils or rock, care is required in the interpretation of results in order to separate the effects of matrix suction (which may induce flow into the sample) and diffusion. One means of minimizing the effect of matrix suction is to allow the soil to equilibrate in contact with a small quantity of squeezed pore fluid from the same soil prior to performing the diffusion tests. In the case of rock, a simulated pore fluid may be used. Diffusion tests conducted on unsaturated gravels, sands and silts have been described by Rowe and Badv (1994a, b, c).

8.3.6 Comments on boundary effects

When performing diffusion tests with a set-up such as that shown in Figures 8.2(a), 8.12, 8.14 and 8.15(b), care needs to be taken not to run the test too long. For example, referring to the test reported in Figure 8.13, it can be seen that even though this test had only been run for four days, some chloride had diffused 4 cm to the bottom of the sample. When performing these tests using an impermeable base, it is desirable to terminate the test at about the time the most mobile species (typically chloride) reaches the bottom plate. If concentration is allowed to build up at the base (as seen in Figure 8.13) then accuracy may be lost; the greater the build-up the greater the potential loss in accuracy. For example, in the worst case the test could be run long enough for steady-state conditions to develop. In this case, there would be a uniform concentration profile which is independent of the diffusion coefficient; hence it is not possible to obtain a unique diffusion coefficient. On the other hand, provided that the test is terminated while there is significant concentration gradient across the sample, the diffusion and distribution coefficient can be determined uniquely to within the experimental accuracy of the concentration determination. To avoid significant test over runs, an initial estimate of the time of termination can be made by estimating the diffusion coefficient and using a model (e.g. POLLUTE) to simulate the migration. This assumes that the diffusion coefficient can be reasonably estimated prior to the test. However, even if a good estimate cannot be made initially, and a build-up in concentration does occur, the results of this test will usually give a fairly good estimate of the diffusion coefficient. This parameter can be used to estimate a better termination time for a second test, and the test can be repeated as a check. For the case shown in Figure 8.13, the diffusion coefficient was not affected by the build-up of concentration at the base.

It should be added that if one were concerned with the determination of parameters for a moderately to highly sorbed species, then it would be essential to have a length of sample such that there was not a significant build-up of either the associated anion (e.g chloride) or desorbed cations (e.g. Ca^{2+}, Mg^{2+}) at the base of the sample. The problem can be avoided by either adopting a long sample, or using a collection reservoir at the base of the sample (e.g. Figure 8.2(b)).

The results presented in this section illustrate how the proposed technique can be used to estimate relevant parameters. The values of the parameters will depend, *inter alia*, on the mineralogy and structure of the clay, the pore water chemistry of the clay, the leachate composition and the temperature at which the test is conducted.

8.4 Dispersion at low velocities in clayey soils

The Darcy velocities of between 0.025 and 0.035 m/a used in the experiments described in section 8.3.1 exceed that expected under operating conditions in most practical field applications involving clayey liners. The change in velocity from 0.025 to 0.035 did not give rise to a discernible difference in the coefficient D, and this raises the question as to whether any dispersion is evident in these tests. To provide some indication regarding the effect of the advective velocity on the coefficient D, a pure diffusion test was conducted for chloride allowing for a concentration drop in the source leachate, but zero flux at the base of the soil, as described in section 8.3.2. Tests at this scale can be performed very quickly, and the concentration profile through the sample after four days is shown in Figure 8.13. Also shown is the predicted concentration profile using the average value of $D = 0.019\ m^2/a$ obtained from the advective–dispersive tests A–E described in section 8.3.2. The prediction is in excellent agreement with the observed profile, and this suggests that the contribution of mechanical dispersion to the coefficient D at velocities of 0.035 m/a or less is negligible for this clay; hence it is appropriate here to refer to it as the diffusion coefficient.

8.5 Effective porosity

It has been shown that for some intact saturated soils with predominantly active clay minerals (e.g. montmorillonite) the effective porosity can be significantly less than the total porosity (e.g. Thomas and Swoboda, 1970; Appelt, Holtzclaw and Pratt, 1975). This is a result of anion exclusion (as discussed in section 6.4) from the highly structured double layer around negatively charged clay surfaces. For montmorillonitic soils compacted wet-of-optimum water content, the double layer may occupy a significant portion of the pore space; hence anion exclusion may result in effective porosity for a negatively charged species, such as chloride, which is significantly less than the total porosity. Based on similar arguments, one would not expect there to be a significant difference between total and effective porosity for silty clays and silty clay tills of low activity, since the double layer would only occupy a small portion of the pore space.

The authors' experience has been that the effective porosity, n, with respect to diffusion through saturated or near-saturated clayey barriers of low activity clays is often reasonably estimated on the basis of water content determined according to usual geotechnical practice (i.e. the effective porosity is essentially the same as the total porosity). However, as noted above, situations can be envisaged where the effective porosity could be less. The effective porosity can be estimated using the test procedure described in section 8.3. For example, the tests described in sections 8.3.1 and 8.3.2 were analyzed assuming a value of porosity of 0.39, although some variability in water content was observed, corresponding to a porosity ranging from 0.37 to 0.39. The calculated concentration profile using $D = 0.02$ m^2/a and both $n = 0.39$ and 0.37 are shown in Figure 8.11 for model E discussed in section 8.3.1. For potassium, the difference in the curves is not plottable. For chloride, the difference is plottable but not significant. A similar conclusion is reached if the other model tests are reinterpreted using $n = 0.37$. Thus the uncertainty as to the precise value of porosity, n, due to the small variation in water content does not significantly influence the magnitude of the parameters D and ϱK_d deduced using $n = 0.39$.

It could be argued that the effective porosity might be significantly less than the values calculated on the basis of water content, owing to the presence of immobile pore fluid or anion exclusion. To examine this possibility, an attempt was made to reinterpret the chloride tests using lower porosities (viz. $n \leqslant 0.35$); however, it was not possible to obtain a good match to the experimental data for these parameters (Figure 8.13). The discrepancies between the observed and calculated profiles increased with decreasing assumed porosity. Thus it would appear that the effective porosity of the soil tested is not significantly less than that deduced from the water content.

8.6 Distribution coefficients and nonlinearity

The approach described in the preceding sections is based on the assumption that the adsorption isopleth is linear and reversible. The species most affected by sorption in these tests is potassium. The tests involving potassium were performed at a concentration which could be expected to occupy fully the exchange sites on the clay. This represents the upper limit to which linearity of sorption could possibly occur, and greatly exceeds concentrations found in many landfill leachates in Ontario.

A series of batch tests (see Chapters 1 and 6 for more details) was performed using this Sarnia clay and distilled water spiked with KCl (i.e. the same source solution as used in tests D and E described in section 8.3.1). The isopleth shown in Figure 8.17 is linear with $\varrho K_d = 7.1$ up to a concentration of approximately 900 mg/l. This confirms that the assumption of

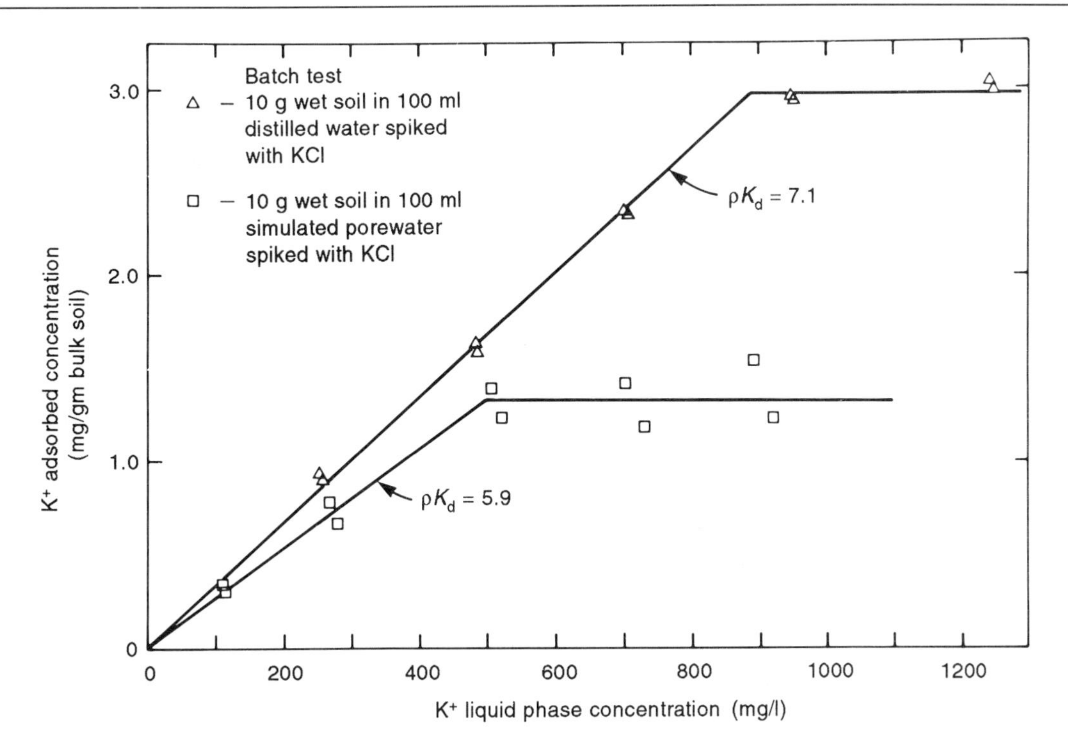

Figure 8.17 Sorption isopleth for potassium from batch tests indicating the effect of leachate chemistry on sorption of a given species. (Modified from Rowe, Caers and Barone, 1988.)

linearity adopted in the interpretation of the model tests is reasonable. The value of ϱK_d deduced from this batch test is in excellent agreement with the value back-calculated from tests D and E. The model test was also analyzed using the finite element program SFIN (Rowe and Booker, 1983), which allows direct modeling of the isopleth obtained from the batch tests. The results from the nonlinear analysis are not substantially different from those obtained using a linear isopleth for this case, since the initial concentration in the leachate is only marginally greater than the concentration at which nonlinearity occurs.

The technique described in section 8.3 provides a relatively simple means of estimating the diffusion coefficients and distribution coefficients of key contaminants as they diffuse through clayey soil. The interpretation of the test assumes that the sorption process can be reasonably approximated as being linear over the concentration range of interest. For problems where the concentration of contaminant is very high and where nonlinearity becomes important, parameters back-calculated using this procedure may not be appropriate. The validity of the linearity assumption can be checked in one of two ways.

When dealing with inorganic contaminants, batch tests can be conducted to determine the range of linearity, as discussed above and in Chapter 6. The diffusion test described in this chapter could then be performed within this range to provide values of both D and ϱK_d for the soil of interest.

The alternative to performing batch tests in

conjunction with a diffusion test is to perform at least two diffusion tests at different concentrations within the range of interest. If the linearity assumption is valid then the values of D and ϱK_d deduced from both tests will be the same. If the assumption is not reasonable, then markedly different values of ϱK_d will be back-calculated for the different concentrations and, furthermore, it may be difficult to obtain a good fit to both the variation in leachate concentration with time and the variation in concentration with depth within the sample at the end of the test.

If sorption is found to be nonlinear over practical concentration ranges, then in many cases either a Freundlich or Langmuir isopleth may be established on the basis of batch test data, as discussed in section 1.3.5. The diffusion coefficient can then be checked by performing a diffusion test and obtaining a theoretical fit to the experimental data, as described in this chapter, except that the sorption isopleth established from the batch test would be used as input for sorption parameters.

8.7 Effect of leachate composition, interaction and temperature

The migration of a particular contaminant species may be related to both the chemistry of the pore water in the soil or rock through which it will diffuse as well as the chemistry of the leachate. As discussed in Chapter 6, the diffusion coefficient can also be expected to be affected by temperature.

To illustrate the effect of chemistry, Barone *et al.* (1989) performed two series of tests (series (ii) and (iii)) using a similar Sarnia soil to that used for test series (i) which was discussed in section 8.3.1. The essential differences between these three test series are summarized in Table 8.2, and the resulting diffusion coefficients and values of the sorption parameter ϱK_d are summarized in Table 8.3.

Unless otherwise noted, a good fit could be obtained to the experimental data for all cases. For example, Figures 8.18 and 8.19 show the variation in concentration with time in the source reservoir, the variation in pore water concentration with depth in the sample and the adsorbed concentration with depth, for both a leachate test (model K) and a single salt test (models G, H), for the Na^+ and K^+ cations respectively. In these cases a good theoretical fit could be obtained. Since Na^+ and K^+ were dominant, they were adsorbed onto the clay, and other species (Mg^{2+}, Ca^{2+}) were desorbed.

The desorption of Ca^{2+} in the leachate test is evident in Figure 8.20, and as a result one cannot obtain a reasonable estimate of diffusion or sorption parameters for a simple single species model of the migration of these species. However, inspection of Figure 8.20 does show that the single species model will give a reasonable fit to the data when these species are dominant as in the case in a single salt solution.

First inspection of Table 8.3 would suggest that the diffusion coefficient of chloride is relatively insensitive to temperature and leachate concentration; however, this is not the case. Figure 8.21 shows results from a different series of tests where the soil and chemistry were held constant, and only the temperature was varied. As can be seen, the diffusion coefficient obtained for chloride at different temperatures does indeed vary, as suggested by the theoretical relationship given in Chapter 6.

Thus, the fact that the diffusion coefficient for models A–F at 22°C are similar to those for models G–J at 10°C is coincidental, and is a result of the counteracting effect of both different background chemistry and different temperatures. The effect of the difference in the clay is more evident when comparing the cation results in series (i) with the results in series (ii). For example, comparing the parameters for sodium in models A and G (both single salt tests) it can be seen that both the diffusion coefficient and sorption are higher for model G. The difference in temperature would be expected to result in a

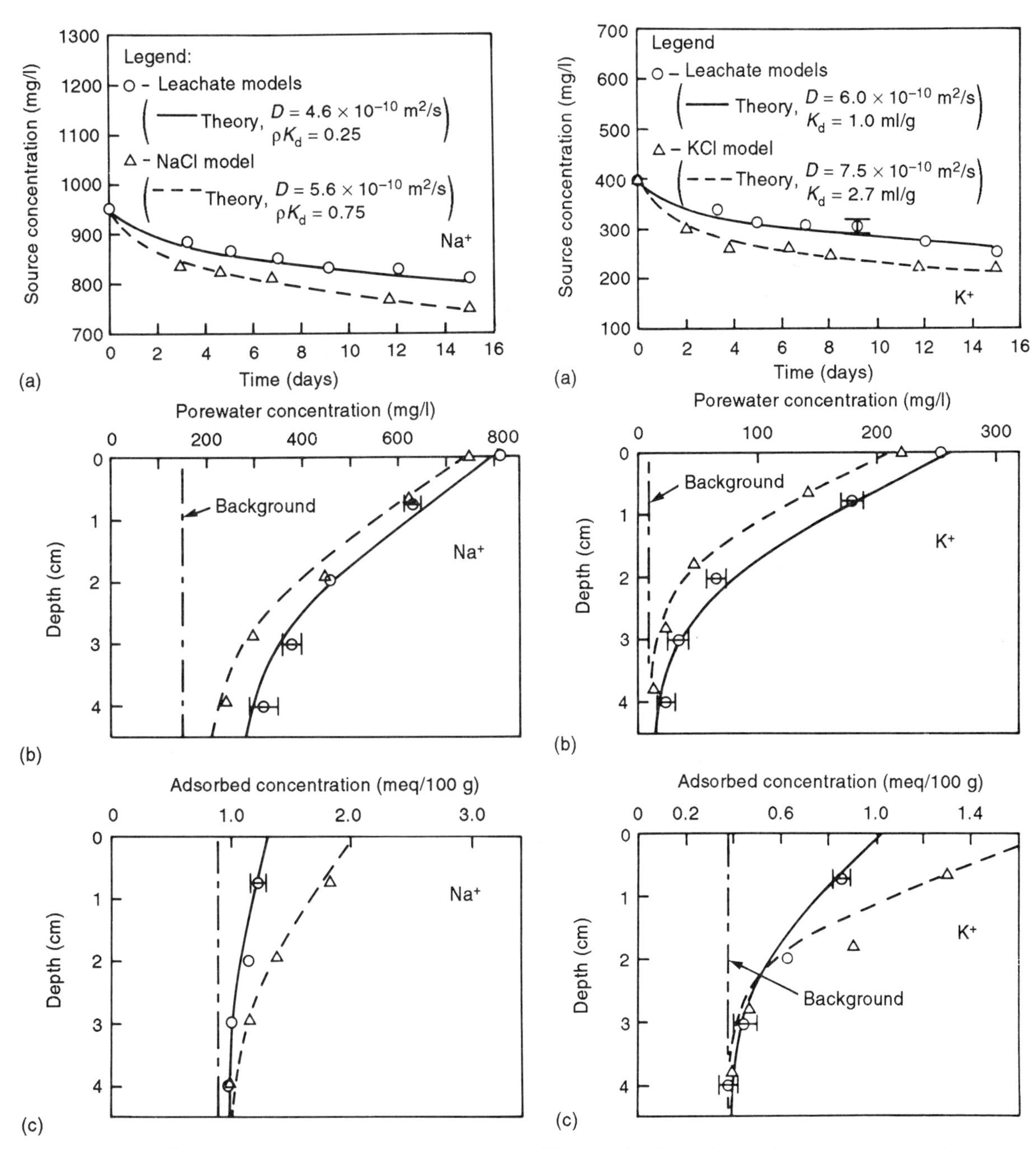

Figure 8.18 Sodium: (a) source concentration versus time; (b) pore water concentration versus depth; (c) adsorbed concentration versus depth (t = 15 days). (Modified from Barone *et al.*, 1989.)

Figure 8.19 Potassium: (a) source concentration versus time; (b) pore water concentration versus depth; (c) adsorbed concentration versus depth (t = 15 days). (Modified from Barone *et al.*, 1989.)

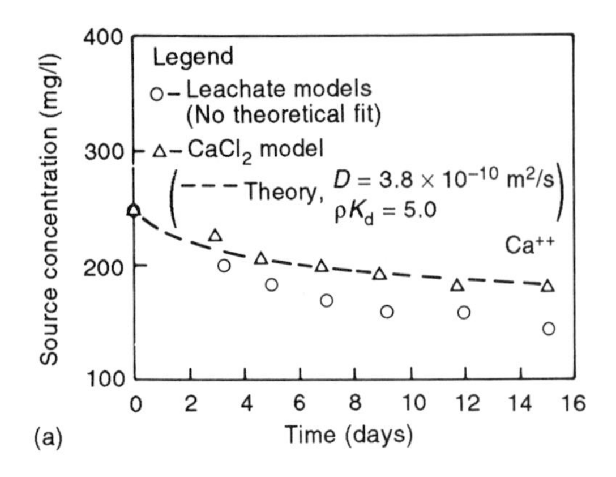

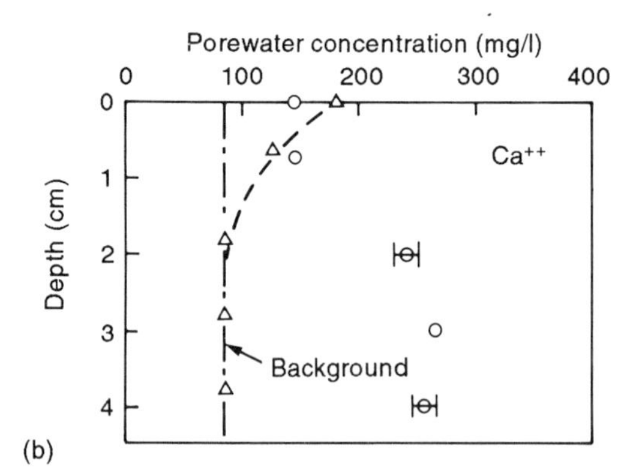

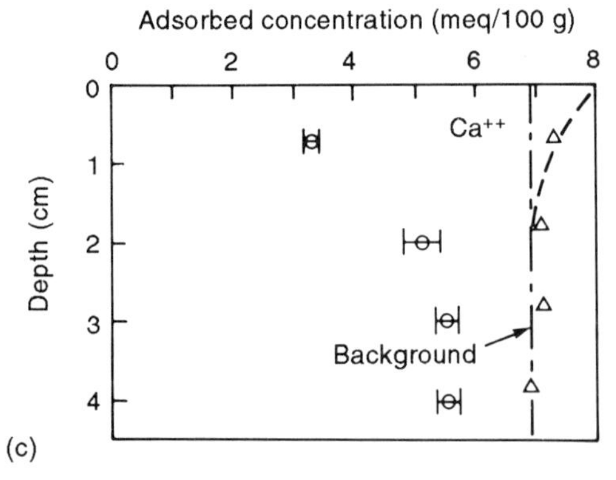

Figure 8.20 Calcium: (a) source concentration versus time; (b) pore water concentration versus depth; (c) adsorbed concentration versus depth (t = 15 days). (Modified from Barone *et al.*, 1989.)

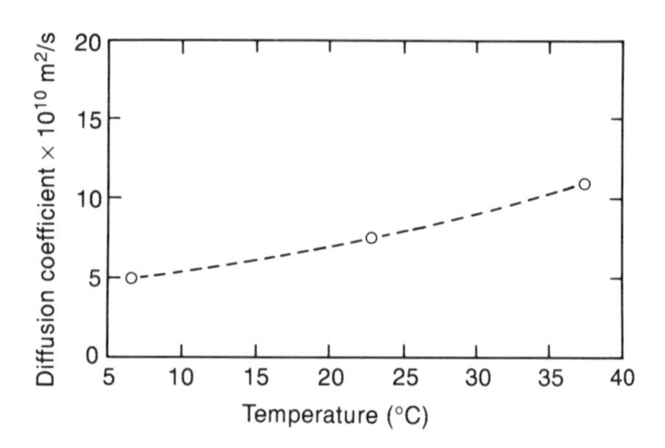

Figure 8.21 Effect of temperature on the effective diffusion coefficient, D_e, of chloride.

lower diffusion coefficient than that in model A. The fact that it is higher indicates that the physico-chemistry of the soil has an important influence on the values of D and ρK_d.

Comparing the results of tests with the single salt solution in series (ii) with those obtained with the leachate in series (iii) further emphasizes the importance of chemistry in the apparent diffusion and distribution coefficients. The effect of the additional competition for sorption sites and the requirement of electroneutrality apparent in model K resulted in a higher chloride diffusion coefficient than that obtained in the single salt tests and reduced both the diffusion coefficient and distribution coefficient of the cations. Thus even the diffusion coefficient for chloride, which is generally regarded as being conservative, does depend on the leachate and clay chemistry since, in order for there to be electroneutrality, Cl^- must diffuse in association with cations, and the migration of cations is influenced by competition for sorption sites.

It should be emphasized that tests conducted on any particular soil with a given leachate are highly repeatable, e.g. compare models B and C, or D and E, or the two tests summarized by the 'error bars' for tests with leachate shown in Figures 8.18–8.20, noting that where there are no error bars shown the results were identical to plotting accuracy. However, the results shown in Table 6.3 indicate that there are many complicating chemical–soil interactions influencing the rate of diffusion of a particular chemical species through a given clay. These factors are best captured empirically by performing tests using soil and leachate which most closely approximate that which will be used in the field application of interest.

8.8 Diffusion and sorption of organic contaminants

The techniques described in the previous sections can also be used to evaluate the diffusion and partitioning coefficients for organic chemicals; however, some modifications and considerable caution may be required. For example, with volatile organic compounds there may be difficulties in obtaining meaningful concentration profiles through the sample due to losses which occur when squeezing the pore water from the clay slices. In these cases, it is often necessary to rely on the source and collector concentrations for laboratory set-ups such as that shown in Figure 8.2(b). The source and receptor concentrations can be used to infer both the diffusion coefficient D and the partitioning coefficient K_d, but care is also required to check that the results are meaningful. This has been discussed by Barone, Rowe and Quigley (1992b), who examined the diffusion of acetone, 1,4-dioxane, aniline, chloroform and toluene through a natural clayey soil.

Barone, Rowe and Quigley (1992b) showed that good results could be readily obtained for acetone, 1,4-dioxane and analine (e.g. Figure 8.22) by fitting a theoretical curve to the measured source and collector concentrations. However, they also showed that sorption onto the walls of the apparatus could give misleading results for chloroform and toluene unless the diffusion tests were complemented by a series of batch tests, or unless the losses can be quantified. Barone, Rowe and Quigley (1992b) performed control tests which could be used to model the time-dependent losses in the source reservoir. Using these data, together with the results of the diffusion tests, Smith, Rowe and Booker (1993)

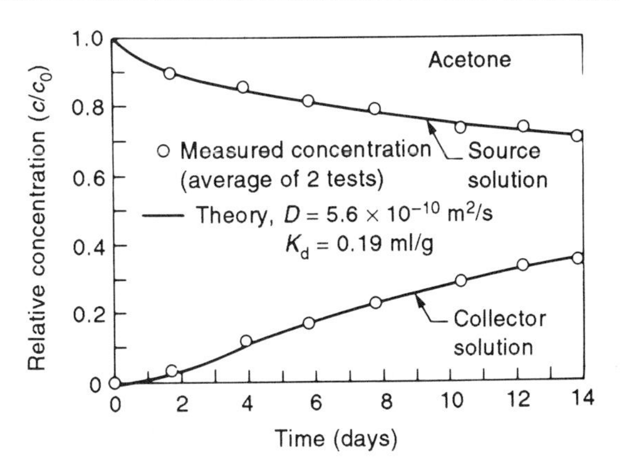

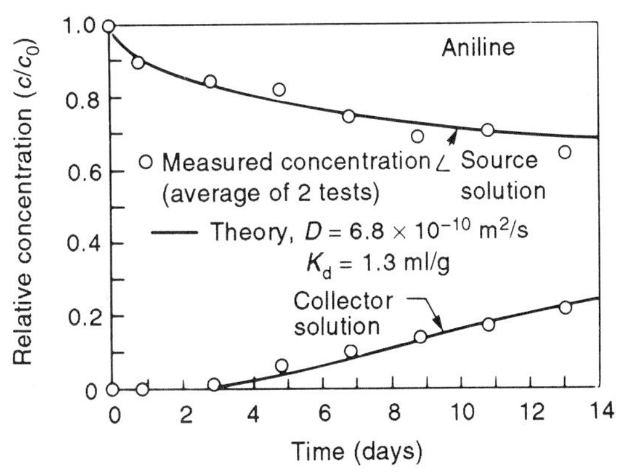

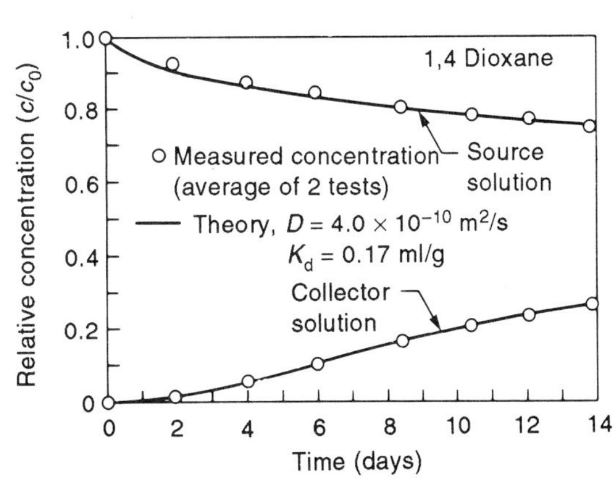

Figure 8.22 Source and collector solution concentration variation with time for three organic species. (Modified from Barone, Rowe and Quigley, 1992a.)

modified the Rowe and Booker (1985a) solution to allow for these losses, and hence to infer directly the diffusion and sorption parameters from the results of the laboratory tests.

An alternative approach is to use the results of batch tests to estimate K_d. However, batch tests may also be subject to error and the results may depend on the solids-to-water ratio (Figure 8.23). Voice, Rice and Weber (1983) attributed the sensitivity of K_d values to the solids/water ratio to microparticulate or macromolecular material being washed off the soil particles during the batch test, and then not being removed from the liquid phase during the separation procedure. These nonsettling microparticles tended to increase the capacity of the liquid phase to accommodate solute, and hence give rise to low apparent K_d values. Gschwend and Wu (1985) also studied this problem and, based on their results, it is evident that the lower the solids/water ratio, the less is the influence of the nonsettling particles, and hence the more representative the results. Combining the results of carefully performed batch tests at low solids/water ratios, with the results from diffusion tests, it is then possible to estimate the diffusion and adsorption characteristics of the organic contaminants for a given soil. Values inferred by Barone, Rowe and Quigley (1992b) for a clayey soil with a porosity of 39%, and organic carbon content, f_{oc}, of 0.58% are given in Table 8.4.

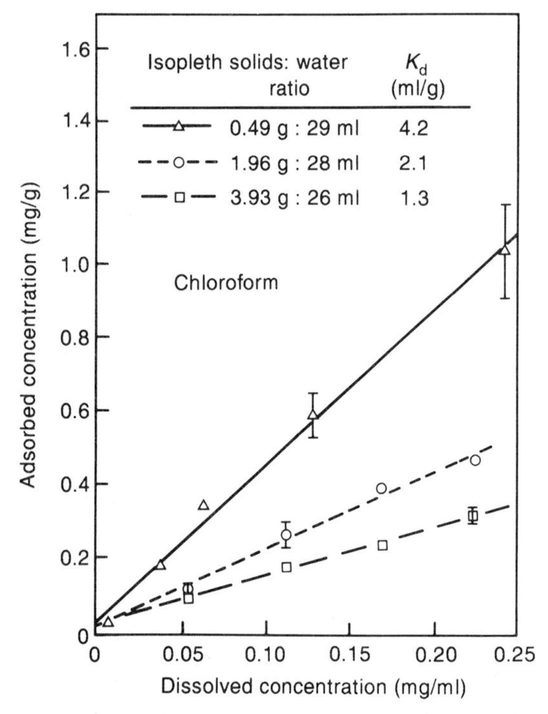

Figure 8.23 Batch test isopleths for chloroform at different solids/water ratios. (Modified from Barone, Rowe and Quigley, 1992b.)

The results reported by Barone, Rowe and Quigley (1992b) were for single organic compounds dissolved in distilled water. Tests performed by Nkedi-Kissa, Rao and Hornsby (1985) and by Quigley and Fernandez (1989, 1992) suggest that introduction of organic cosolvents may play a significant role on the interaction between an organic solute and clay minerals. This suggests that mixtures of organic liquids in water could behave differently than single species in water. Additional laboratory testing and field confirmation are needed for organic mixtures.

8.9 Use of field profiles to estimate diffusion coefficients

Field profiles of diffusive migration of various chemical species have developed by natural causes like the diffusion of Na^+ and Cl^- through clay from underlying salty bedrock, or by diffusion from landfill leachate, as will be discussed in more detail in Chapter 9. These profiles can often be used to back-calculate diffusion coefficients. For example, the Confederation Road landfill near Sarnia, Ontario has been the subject of many papers dealing with the diffusion of a variety of species into the underlying clays (Quigley and Rowe, 1986; Crooks and Quigley, 1984; Quigley, Crooks and Yanful, 1984; Goodall and Quigley, 1977).

A summary plot of the chemical profiles for a variety of species is presented in Figure 8.24 for

Table 8.4 Summary of the diffusion and adsorption coefficients at 22°C as reported by Barone, Rowe and Quigley (1992b)

Species	*D* (m^2/s)	*D* (m^2/a)	K_d *(ml/g)*	K_{oc} *(ml/g)*
Acetone	5.6×10^{-10}	0.018	0.19	33
1,4-Dioxane	4.0×10^{-10}	0.013	0.17	29
Aniline	6.8×10^{-10}	0.021	1.3	224
Chloroform	7.0×10^{-10}	0.022	4.2	724
Toluene	5.8×10^{-10}	0.018	11.3	1950

Table 8.5 Field diffusion data

	Cl^-	Na^+	K^+	Ca^{2+}	Mg^{2+}
D^* (m^2/a)	0.020	0.0072	0.0011	0.018	0.0115
D_e (m^2/a)	0.020	0.012	0.016	0.009	0.008
K_d (ml/g)	0	0.16	3.23	–0.11	–0.07
Distance to $c/c_0 = 0.5$ (m)	0.36	0.3	0.11	–	–

pore water squeezed from three closely spaced boreholes 15 years after placement of the domestic waste. The extrapolated curves are predicted from the lower 20% of the data points and extrapolated to the interface. The actual data points do not fit exactly, because the source concentration was decreasing with time within the mass of waste. The D^* values used to back-calculate these curves are presented in Table 8.5. Based on the D^* value for chloride, it is possible to infer the tortuosity, τ, and hence estimate values of the effective diffusion coefficient, D_e, for the other species as given in Table 8.5. Once the D_e value is obtained, the value of ϱK_d can then be deduced from the value of D^* for these species as discussed in Chapter 6.

The sodium, potassium and chloride ion profiles (Figure 8.24(a), actual and predicted) are in logical locations, the chloride furthest advanced, sodium slightly retarded and potassium greatly retarded. The relative positions of these three species are clearly reflected by the retarded diffusion coefficients, D^*.

Calcium and magnesium are much more difficult to deal with because, even though they are the dominant cations adsorbed on the clays, they are also intimately involved in heavy metal precipitation and dissolved CO_2 levels associated with bacterial respiration. The high D^* values used to calculate the location of the predicted curves may be more indicative of a hardness halo front than diffusion.

The distances to the position of diffusion fronts at $c/c_0 = 0.5$ are also presented in Table 8.5, yielding migration distances of 0.36, 0.30 and 0.11 m for Cl^-, Na^+ and K^+ respectively.

Finally, inferred values for K_d are also presented in Table 8.5. Conservative chloride by definition has $K_d = 0.0$ mg/l, Na^+ yields a small K_d of 0.16 ml/g, and K^+ (which tends to fix onto the clays) yields a larger value of 3.23 ml/g. The negative values for Ca^{2+} and Mg^{2+} imply desorption.

A comparison of the results back-calculated from these field profiles (Table 8.5) with values obtained from laboratory tests (Table 8.3) indicates reasonable agreement with the values back-calculated from the field falling within the range of diffusion coefficients and sorption parameters obtained from the laboratory tests.

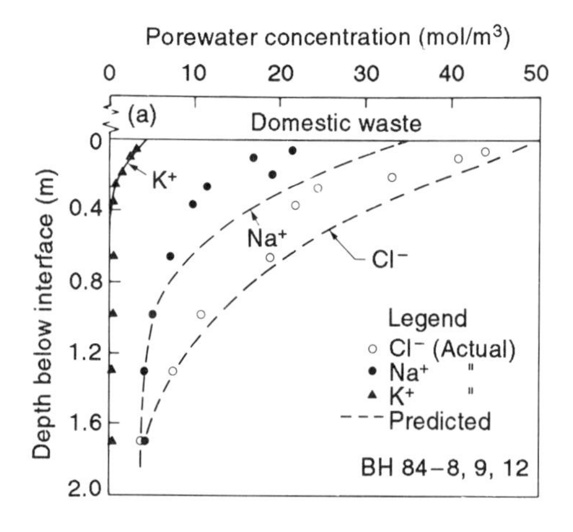

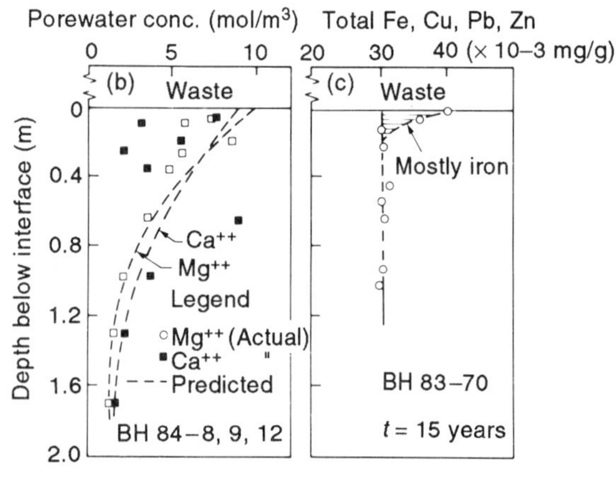

Figure 8.24 Concentration versus depth below waste–clay interface, Confederation Road landfill: (a) chloride, sodium and potassium; (b) calcium and magnesium; (c) heavy metals. (Modified from Quigley *et al.*, 1987.)

8.10 Summary and conclusions

This chapter has discussed a number of key parameters to be considered in making predictions of contaminant migration through clayey barriers. With respect to diffusion, limited data would suggest that the effective porosity of clayey soils may be close to the values deduced on the basis of water content. However, situations can be visualized where this may not be the case. Diffusion tests, such as those described in this chapter, can be used to test the hypothesis that the effective porosity of active clays is less than the bulk porosity and, if so, to estimate the effective porosity.

Techniques for estimating the diffusion and distribution/partitioning coefficients of chemical species have been described and illustrated for a number of tests. The tests are relatively easy to perform and, provided that one can obtain accurate measures of concentration, they give repeatable results for a given soil and leachate chemistry. However, both the diffusion coefficient and distribution/partitioning coefficient for a given species will depend on the chemical composition of the source leachate, and background chemistry and mineralogy of the soil. This can be true even for conservative ionic species such as chloride. Thus the diffusion and sorption coefficients should be determined for the proposed soil using a leachate as near as practicable to that expected in the field situation. Thus the complex factors influencing the diffusion of chemical species are empirically captured in the laboratory parameters.

Fortunately, despite the potential variation in diffusion coefficients due to soil and leachate composition, the range of variation is relatively small compared to that of many other parameters (like hydraulic conductivity). In engineering terms, diffusion is in fact a very predictable process. As discussed in Chapter 9, field diffusion profiles and, indeed, natural diffusion processes established over thousands of years have been shown to be consistent with diffusion coefficients which can be determined from relatively simple short-term laboratory tests, such as those discussed in this chapter.

With the foregoing caveats in mind, Table 8.6 summarizes diffusion and distribution coefficients calculated from either laboratory tests or back-calculated from observed field profiles. This table may be a useful guide when planning a testing program, but is not intended for use in the design of barriers. These values should be independently confirmed for any particular project.

Table 8.6(a) Summary of diffusion (D_e) and sorption (K_d) coefficients for various inorganic contaminant species (compiled with the assistance of Dr F. Barone)

	Soil description												
Species	*n (%)*	*Silt content (%)*	*Clay content (%)*	*PI (%)*	*CEC (meq/ 100 g)*	*Soil pH*	$D_e \times 10^{10}$ *(m^2/s)*	K_d *(ml/g)*	*Method of obtaining* K_d	*Source solution*	*Temp. (°C)*	*Source solution pH*	*Reference*
Ammonium	32	42	23	11	–	–	5.7	1.2	Diffusion test on intact soil	Leachate, c_0 (NH_4^+) = 957 mg/l	10	–	Rowe (unpublished)
Arsenic	–	25	30	–	17	6.1	–	5	One point batch test, 1:20 soil: sol'n ratio (final sol'n concn = 81 mg/l)	Distilled, deionized water spiked with KH_2ASO_4	22	–	Griffin *et al.* (1986)
	–	0	100	–	15	8.1	–	3	One point batch test, 1:20 soil: sol'n ratio (final sol'n concn = 87 mg/l)	Distilled, deionized water spiked with KH_2ASO_4	22	–	Griffin *et al.* (1986)
	–	40	15	–	5	7.5	–	3	One point batch test, 1:20 soil: sol'n ratio (final sol'n concn = 88 mg/l)	Distilled, deionized water spiked with KH_2ASO_4	22 –	–	Griffin *et al.* (1986)
Bromide	51–60		100	23	5	3.7	4.8–6.1[a]	0	–	Simulated leachate spiked with KBr, c_0(Br) = 645–1012 mg/l	23	3.7–6.7	Shackleford and Daniel (1991)
	45–47		82	42	25	6.9	18.2[b]	0	–	Simulated leachate spiked with KBr, c_0(Br) = 645–1012 mg/l	23	3.7–6.7	Shackleford and Daniel (1991)
Cadmium	–	–	86	26	8	–	–	8	Batch test, 1:12.5 soil: sol'n ratio by wt (linear range 0–5 mg/l)	MSW leachate spiked with $CdNO_3$	–	7.6	Yong and Sheremata (1991)
	51–60		100	23	5	3.7	3.2[a]–4.2	2	Batch test, 1:4 soil; sol'n ratio by wt (linear range 0–50 mg/l)	Simulated leachate spiked with CdI_2	23	3.7–6.7	Shackleford and Daniel (1991)
	45–47		82	42	25	6.9	3.0[a]–4.0	35	Batch test, 1:4 soil: sol'n ratio by wt (linear range 0–10 mg/l)	Simulated leachate spiked with CdI_2	23	3.7–6.7	Shackleford and Daniel (1991)
	–	25	30	–	17	6.1	–	70	One point batch test, 1:20 soil: sol'n ratio (final sol'n concn = 22 mg/l)	Distilled, deionized water spiked with $CdCO_3$	22	–	Griffin *et al.* (1986)
	–	0	100	–	15	8.1	–	70	One point batch test, 1:20 soil: sol'n ratio (final sol'n concn = 22 mg/l)	Distilled, deionized water spiked with $CdCO_3$	22	–	Griffin *et al.* (1986)
	–	40	15	–	5	7.5	–	29	One point batch test, 1:20 soil: sol'n ratio (final sol'n concn = 41 mg/l)	Distilled, deionized water spiked with $CdCO_3$	22	–	Griffin *et al.* (1986)
Calcium	10	43	45	27	10	8.1	3.8	1.2	Column diffusion test	$CaCl_2$ in deionized, distilled water, c_0 (Cl^-) = 975 mg/l	10	7	Rowe, Caers and Barone (1988)

Table 8.6(a) *Continued*

Species	Soil description: n (%)	Silt content (%)	Clay content (%)	PI (%)	CEC (meq/ 100 g)	Soil pH	$D_e \times 10^{10}$ (m^2/s)	K_d (ml/g)	Method of obtaining K_d	Source solution	Temp. (°C)	Source solution pH	Reference
Chloride	51–60	100		23	5	3.7	4.4–6.0[a]	0	–	Simulated leachate spiked with $ZnCl_2$, $c_0(Cl^-)$ = 231–448 mg/l	23	3.7–6.7	Shackleford and Daniel (1991)
	45–47	82		42	25	6.9	1.5–1.8[a]	0	–	Simulated leachate spiked with $ZnCl_2$, $c_0(Cl^-)$ = 231–448 mg/l	23	3.7–6.7	Shackleford and Daniel (1991)
	10	43	45	27	10	8.1	7.5	0	Column diffusion test on intact soil (no advection)	MSW leachate, $c_0(Cl^-)$ = 1000 mg/l	10	7.0	Barone *et al.* (1989)
	10	43	45	27	10	8.1	5.6	0	Column diffusion test on intact soil (no advection)	NaCl solution in distilled, deionized water	10	7.0	Barone *et al.* (1989)
	10	43	45	27	10	8.1	5.9	0	Column diffusion test on intact soil (no advection)	KCl solution in distilled, deionized water	10	7.0	Barone *et al.* (1989)
	10	43	45	27	10	8.1	6.0	0	Column diffusion test on intact soil (no advection)	$MgCl_2$ solution in distilled, deionized water	10	7.0	Barone *et al.* (1989)
	10	43	45	27	10	8.1	6.2	0	Column diffusion test on intact soil (no advection)	$CaCl_2$ solution in distilled, deionized water	10	7.0	Barone *et al.* (1989)
	10	43	45	27	10	8.1	5.7	0	Diffusion test with advection	NaCl solution	22	7.0	Rowe, Caers and Barone (1988)
	10	43	45	27	10	8.1	5.9	0	Diffusion test with advection	$CaCl_2$ solution	22	7.0	Rowe, Caers and Barone (1988)
	10	43	45	27	10	8.1	6.7	0	Diffusion test with advection	KCl_2 solution	22	7.0	Rowe *et al.* (1988)
	28	38	28	–	–	–	3.0	0	Diffusion test; no advection, intact soil	Leachate, $c_0(Cl^-)$ = 1463 mg/l	10	–	Rowe (unpublished)
	23	44	19	–	–	–	4.0	0	Diffusion test; no advection, intact soil	Leachate, $c_0(Cl^-)$ = 1463 mg/l	10	–	Rowe (unpublished)
	18	32	4	–	–	–	7.8	0	Diffusion test; no advection, intact soil	Leachate, $c_0(Cl^-)$ = 580 mg/l	10	7.0	Rowe (unpublished)
	57		85	32	–	7.0	3.3	0	Diffusion test; no advection, intact soil	Leachate, $c_0(Cl^-)$ = 1350 mg/l	10	–	Rowe (unpublished)
	47		81	21	–	–	4.0	0	Diffusion test; no advection, intact soil	Leachate, $c_0(Cl^-)$ = 1350 mg/l	10	–	Rowe (unpublished)
	33		34	8	–	–	5.3	0	Diffusion test; no advection, intact soil	Leachate, $c_0(Cl^-)$ = 14 500 mg/l	10	–	Rowe (unpublished)
	38	–	39	11	–	–	5.0	0	Diffusion test on recompacted clay	Leachate, $c_0(Cl^-)$ = 14 500 mg/l	10	–	Rowe (unpublished)
	44	–	–	17	–	–	4.0	0	Diffusion test on intact clay	Leachate, $c_0(Cl^-)$ = 1050 mg/l	10	–	Rowe (unpublished)
	37	–	–	–	–	–	5.0	0	Diffusion test on intact clay	Leachate, $c_0(Cl^-)$ = 1050 mg/l	10	–	Rowe (unpublished)
	40	–	–	10	–	–	5.0	0	Diffusion test on recompacted clay	Leachate, $c_0(Cl^-)$ = 1050 mg/l	10	–	Rowe (unpublished)
	34	55	45	8	–	–	8.0	0	Diffusion test on intact clay	Leachate, $c_0(Cl^-)$ = 1250 mg/l	10	–	Rowe (unpublished)

	35	40	24	9	–	–	5.0	0	Diffusion test on recompacted till	Leachate, $c_0(Cl^-)$ = 967 mg/l	10	–	Rowe (unpublished)
	27	48	22	6	–	–	6.0	0	Diffusion test on intact till	Leachate, $c_0(Cl^-)$ = 967 mg/l	10	–	Rowe (unpublished)
	30	48	22	6	–	–	7.0	0	Diffusion test on recompacted till	Leachate, $c_0(Cl^-)$ = 967 mg/l	10	–	Rowe (unpublished)
	20	39	15	5	–	–	5.0	0	Diffusion test on intact till	Leachate, $c_0(Cl^-)$ = 967 mg/l	10	–	Rowe (unpublished)
	30	39	15	5	–	–	6.0	0	Diffusion test on recompacted till	Leachate, $c_0(Cl^-)$ = 967 mg/l	10	7.0	Rowe (unpublished)
	32	42	23	11	–	–	4.7	0	Diffusion test on intact sand	Leachate, $c_0(Cl^-)$ = 2180 mg/l	10	–	Rowe (unpublished)
	43	46	54	18	–	–	4.5	0	Diffusion test on intact sand	Leachate, $c_0(Cl^-)$ = 1250 mg/l	10	–	Rowe (unpublished)
	41	70	30	6	–	–	7.5	0	Diffusion test on intact sand	Leachate, $c_0(Cl^-)$ = 1250 mg/l	10	–	Rowe (unpublished)
	47	53	47	19	–	–	5.0	0	Diffusion test on intact sand	Leachate, $c_0(Cl^-)$ = 1000 mg/l	10	–	Rowe (unpublished)
	36	–	40	–	–	–	2.0	0	Diffusion test on intact sand	Leachate, $c_0(Cl^-)$ = 970 mg/l	7	–	Barone (1990)
	3.4	N/A	N/A	N/A	–	–	0.6–0.8	0	Diffusion test on sandstone	NaCl solution	22	–	Rowe (unpublished)
	9.2	N/A	N/A	N/A	–	–	0.8–0.9	0	Diffusion test on mudstone	NaCl solution	22	–	Rowe (unpublished)
	10.8	N/A	N/A	N/A	–	–	1.5	0	Diffusion test on shale	Back diffusion	10	–	Barone, Rowe and Quigley (1989)
	23.8	N/A	N/A	N/A	38		1.8	0	Diffusion test on mudstone	Back diffusion	10		Barone *et al.* (1992a)
Copper	–	~43	~45	~27	~10	8.2	–	400	Batch test, 1:200 soil: sol'n ratio by wt (linear range 0–20 mg/l)	MSW leachate spiked with $Cu(NO)_3)_2$	22	7.8	Yanful *et al.* (1988b)
Iodide	51–60	100		23	5	3.7	3.5–14.7**	0	–	Simulated leachate spiked with CdI_2, $c_0(I)$ = 1089–1567 mg/l	23	3.7–6.7	Shackleford and Daniel (1991)
	45–47	82		42	25	6.9	5.3*	0	–	Simulated leachate spiked with CdI_2, $c_0(I)$ = 1089–1567 mg/l	23	3.7–6.7	Shakleford and Daniel (1991)
Lead	–	~43	~45	~27	~10	8.2	–	1900	Batch test, 1:200 soil: sol'n ratio by wt (linear range 0–5 mg/l)	MSW leachate spiked with $Pb(No_3)_2$	22	7.8	Yanful *et al.* (1988)
Potassium	51–60	100		23	5	3.7	11.7–17.7[a]	1.7	Batch test, 1:4 soil: sol'n ratio by wt	–	23	3.7–6.7	Shackleford and Daniel (1991)
	45–47	82		42	25	6.9	19.6[b]	1.1	–	–	23	3.7–6.7	
	10	43	45	27	10	8.1	6.0	1.0	Column diffusion test on intact soil	MSW leachate, $c_0(K^+) \approx$ 400 mg/l	10	7.0	Barone *et al.* (1989)
	10	43	45	27	10	8.1	7.5	2.7	Column diffusion test on intact soil	KCl sol'n in deionized, distilled water	10	7.0	Barone *et al.* (1989)
	36	–	40	–	–	–	5.0	1.0	Diffusion test on intact soil	Leachate, $c_0(K^+) \approx$ 280 mg/l	7		Barone (1990)

Table 8.6(a) *Continued*

Species	Soil description: n (%)	Silt content (%)	Clay content (%)	PI (%)	CEC (meq/100 g)	Soil pH	$D_e \times 10^{10}$ (m^2/s)	K_d (ml/g)	Method of obtaining K_d	Source solution	Temp. (°C)	Source solution pH	Reference
Sodium	32	42	40	11	–	–	2.0	0.03	Diffusion test on intact soil	$c_0(Na^+)$ = 565 mg/l	10		Rowe (unpublished)
	39	43	45	~27	~10	~8.1	4.6	0.15	Column diffusion test on relatively undisturbed soil (no advective flow)	MSW leachate, $c_0(Na^+)$ = 955 mg/l	10	7.0	Barone *et al.* (1989)
	36	–	40	–	14.8	8.7	3.7	0.3	Column diffusion lest on relatively undisturbed soil (no advective flow)	ISW leachate, $c_0(Na^+)$ = 5100 mg/l	7	9.0	Barone (1990)
	~39	~43	~45	~27	~10	~8.1	4.8	0.11	Column test (advective velocity = 0.034 m/a)	NaCl in deionized, distilled water, $c_0(Na^+)$ = 975 mg/l	22	–	Rowe, Caers and Barone (1988)
	32	42	23	11	–	–	2.0	0.03	Diffusion test on intact soil	$c_0(Na^+)$ = 565 mg/l	10	–	Rowe (unpublished)
Sulfate	36	–	40	–	–	–	2.0	0	Diffusion test on intact soil	Leachate, $c_0(SO_4^-)$ = 3300 mg/l	7	–	Barone (1990)
	32	42	23	11	–	–	2.0	0	Diffusion test on intact soil	$c_0(SO_4^-)$ = 855 mg/l	10	–	Rowe (unpublished)
Zinc	51–60	100		23	5	3.7	3.5[a]–4.5	2	Batch test, 1:4 soil: sol'n ratio by wt (linear range = 0–50 mg/l)	Simulated leachate spiked with $ZnCl_2$, c_0(Zn) = 301–374 mg/l	23	3.7–6.7	Shackleton and Daniel (1991)
	0.4	73	27	18	12	7.0	4.1	13–60	Batch test, 1:10 soil: sol'n ratio by wt and diffusion test on compacted clay	Artificial leachate from waste having high zinc, leached with MSW leachate acidified with acetic acid, c_0(Zn) = 210 mg/l	22	5.0	Quigley (unpublished)
	45–47	82		42	25	6.9	1.5[a]–2.8	35	Batch test, 1:4 soil: sol'n ratio by wt (linear range = 0–10 mg/l)	Simulated leachate spiked with $ZnCl_2$, c_0(Zn) = 301–374 mg/l	23	7.8	Shackleford and Daniel (1988)

Notes: n, porosity; PI, plasticity index; CEC, cation exchange capacity; c_0, initial source solution concentration used in the test.

[a]There were serious problems with mass balance in these tests, and the reported diffusion coefficient should be used with considerable caution.

[b]There were serious problems with mass balance in these tests, and the reported diffusion coefficient should be used with considerable caution. Highly questionable interpretation – results are reported for completeness.

Table 8.6(b) Summary of diffusion (D_e) and adsorption (K_d) coefficients for various organic species (compiled with the assistance of Dr F. Barone)

	Soil description													
Species	*n (%)*	*Silt content (%)*	*Clay content (%)*	*PI (%)*	*Soil pH*	f_{oc} *(%)*	$D_e \times 10^{10}$ *(m^2/s)*	K_d *(ml/g)*	K_{oc} *(=* K_d/f_{oc}*)*	K_f*, 1/n (ml/gl), (–)*	*Method*	*Source solution*	*Temp. (°C)*	*Reference*
Acetone	39	43	45	~27	~8.1	0.58	5.6	0.19	33	–	Column diffusion test on relatively undisturbed soil (no advective flow)	Distilled, deionized, organic free water spiked with acetone, c_0 = 300 mg/l	22	Barone *et al.* (1992b)
Aniline	39	43	45	~27	~8.1	0.58	6.8	1.3	224	–	Column diffusion test on relatively undisturbed soil (no advective flow)	Distilled, deionized, organic free water spiked with aniline, c_0 = 300 mg/l	22	Barone *et al.* (1992b)
Benzene	–	54	42	–	–	0.60	3.6	13.8	2300	–	Batch test, 1:1.8 soil: sol'n ratio max. concn = 3.5 mg/l	Organic free water spiked with benzene	20	Myrand *et al.* (1987)
	–	100	0	–	–	2.78	–	2.3	83	–	Batch test, 1:50 soil: sol'n ratio, max. concn = 900 mg/l	Organic free water spiked with benzene	25	Karickhoff *et al.* (1979)
Chloro-benzene	–	–	–	–	–	0.15	–	0.39	260	–	Batch test, 1:5 soil: sol'n ratio, max. concn = 0.02 mg/l	$CaCO_3/CO_2$ water spiked with chlorobenzene		Schwarzenbach and Westall (1981)
Chloro-form	39	43	45	~27	~8.1	0.58	7.0	4.2	724–1034	–	Combination of column diffusion test on relatively undisturbed soil plus batch test (1:60 soil: sol'n ratio)	Distilled, deionized, organic free water spiked with chloroform, c_0 = 300 mg/l	22	Barone *et al.* (1992b)
1,4-Di-chloro-benzene (DCB)	–	–	–	–	–	0.15	–	1.10	733	–	Batch test, 1:5 soil: sol'n ratio, max. concn = 0.02 mg/l	$CaCO_3/CO_2$ water spiked with 1,4-DCB	22	Schwarzenbach and Westall (1981)
	–	8	2	–	–	2.55	–	–	–	40.5, 0.618	Batch test, 1:5 soil: sol'n ratio, max. concn = 0.2 mg/l	Organic free water spiked with 1,4-DCB	21	Uchrin and Katz (1986)
	–	24	6	–	–	1.28	–	–	–	32, 0.178	Batch test, 1:5 soil: sol'n ratio, max. concn = 0.2 mg/l	Organic free water spiked with 1,4-DCB	21	Uchrin and Katz (1986)
Dichloro-methane (DCM)	35	56	31	9.2	–	0.29	8.5	1.2	410	–	Column diffusion test on recompacted soil (no advective flow)	Distilled, deionized, organic free water spiked with DCM, c_0 = 180 mg/l	22	Rowe and Barone (1991)
	32	55	29	6.3	–	0.45	8.0	1.5	330	–	Column diffusion test on recompacted soil (no advective flow)	Distilled, deionized, organic free water spiked with DCM, c_0 = 180 mg/l	22	Rowe and Barone (1991)
	31	57	22	4.5	–	0.36	8.5	1.4	390	–	Column diffusion test on recompacted soil (no advective flow)	Distilled, deionized, organic free water spiked with DCM, c_0 = 180 mg/l	22	Rowe and Barone (1991)

Table 8.6(b) *Continued*

Species	Soil description: n (%)	Silt content (%)	Clay content (%)	PI (%)	Soil pH	f_{oc} (%)	$D_e \times 10^{10}$ (m^2/s)	K_d (ml/g)	K_{oc} (= K_d/f_{oc})	K_f, $1/n$ (ml/gl), (–)	Method	Source solution	Temp. (°C)	Reference
2,4-Di-chloro-phenol (DCP)	–	8	2	–	–	2.55	–	–	–	26.4, 0.601	Batch test, 1:5 soil: sol'n ratio, max. concn = 1 mg/l	Organic free water spiked with 2,4-DCP	21	Uchrin and Katz (1986)
	–	24	6	–	–	1.28	–	–	–	8.3, 0.247	Batch test, 1:5 soil: sol'n ratio, max. concn = 1 mg/l	Organic free water spiked with 2,4-DCP	21	Uchrin and Katz (1986)
1,4-Dioxane	39	43	45	~27	~8.1	0.58	4.0	0.17	29	–	Column diffusion test on relatively undisturbed soil (no advective flow)	Distilled, deionized, organic free water spiked with 1,4-dioxane, c_0 = 300 mg/l	22	Barone *et al.* (1992b)
Penta-chloro-phenol (PCP)	–	–	–	–	4.8	2.67	–	–	–	35, 0.79	Batch test, 1:10 soil: sol'n ratio, max. concn = 0.004 mg/l	Organic free water spiked with PCP	20	Banerji *et al.* (1986)
Phenol	–	43	45	~27	~8.1	0.58	–	–	–	2.5, 0.628	Batch test 1:1 soil: sol'n ratio, max. concn = 50 mg/l	Distilled, deionized, organic free water spiked with phenol	22	Mucklow (1990)
Tetra-chloro-ethylene	–	–	–	–	–	0.15	–	0.56	373	–	Batch test, 1:5 soil: sol'n ratio, max. concn = 0.1 mg/l	$CaCO_3/CO_2$ water spiked with tetrachloroethylene	22	Schwarzenbach and Westall (1981)
Toluene	–	0	0	–	–	0.15	–	0.37	247	–	Batch test, 1:5 soil: sol'n ratio, max. concn = 0.02 mg/l	$CaCO_3/CO_2$ water spiked with toluene	22	Schwarzenbach and Westall (1981)
	39	43	45	~27	~8.1	0.588	0.58	11.3	1948–4483	–	Combination of column diffusion test on relatively undisturbed soil and batch tests (1:90 soil: sol'n ratio)	Distilled, deionized organic free water spiked with toluene, c_o = 200	22	Barone *et al.* (1992b)
	–	54	42	–	–	0.60	3.0	53.3	8883	–	Batch test, 1:1.8 soil: sol'n ratio, max. concn = 3 mg/l	Organic free water spiked with toluene	20	Myrand *et al.* (1987)
Trichloro-ethylene (TCE)	–	54	42	–	–	1.95	–	125	6410	–	Batch test, 1:1.8 soil: sol'n ratio, max. concn = 0.1 mg/l	Organic free water spiked with TCE	20	McKay and Trudell (1989)
	–	54	42	–	–	0.68	–	58.6	8617	–	Batch test, 1:1.8 soil: sol'n ratio, max. concn = 0.1 mg/l	Organic free water spiked with TCE	20	McKay and Trudell (1989)
	–	54	42	–	–	0.60	3.5	15.5	2583	–	Batch test, 1:1.8 soil: sol'n ratio, max. concn = 2 mg/l	Organic free water spiked with TCE	20	Myrand *et al.* (1987)
	–		100	–	–	1.0	–	8.3	830	–	Batch test, 1:50 soil: sol'n ratio, max. concn = 10 mg/l	Organic free water spiked with TCE	27	Acar and Haider (1990)

Notes: *n*, porosity; PI, plasticity index; f_{oc} soil organic carbon content; K_d, linear adsorption coefficient; K_f, $1/n$, Freundlich adsorption parameters; c_0, initial source solution concentration used in the diffusion test.

Chapter nine

Field studies of diffusion and hydraulic conductivity

9.1 Introduction

Examples of field diffusion are rather scarce in the literature, partly because barrier exhumations are very expensive once several meters of waste are in place, and partly because most research on pollutant migration has been devoted to plumes in granular deposits, since these impact immediately on local groundwater resources. Fortunately, geological history has provided some long-term data regarding the process of diffusion.

The purpose of this chapter is to present a variety of concentration versus depth profiles documented for clay deposits. Both long-term (~10 000 years) and short-term (three to 22 years) profiles will be discussed.

9.2 Examples of long-term field diffusion

Three sites are discussed in this section; a 10 000-year-old salt profile in 30 m of saline Leda clay near Ottawa, Ontario, a 12 000-year-old set of profiles in 35 m of freshwater clay near Sarnia, Ontario and a 12 000-year-old set of profiles in about 35 m of freshwater clay near Niagara Falls, Ontario, Canada.

9.2.1 Hawkesbury Leda clay

A geotechnical profile showing the preconsolidation state of the Hawkesbury Leda clay is presented in Figure 9.1 along with the present day salinity profile. The site was described by Quigley *et al.* (1983), and the interrupted consolidation profile was attributed to removal of ~21 m of saturated sand about 70 years after deposition by a meander of the post-glacial Ottawa River.

The importance of the σ'_p curve is that it indicates double drainage from the center of the clay at ~15 m. From this, one would logically deduce that the salt content of the entire clay layer would have been constant at the time of sand erosion about 10 000 years BP. At the present time, the groundwater conditions are almost hydrostatic, and it is difficult to see how they could ever have been much different over the past 10 000 years.

The salinity profile itself suggests that the original salt content was about 15 g/l, and that salt has diffused towards surface and been removed by surface runoff.

In addition to the salt content profile, one other profile is presented in Figure 9.2 for 'del^{18}O' (δ^{18}O) where

$$\delta^{18}O = \frac{(^{18}O/^{16}O)_{sample} - (^{18}O/^{16}O)_{SMOW}}{(^{18}O/^{16}O)_{SMOW}} \times 1000 \qquad (9.1)$$

and SMOW stands for standard mean ocean water.

The $\delta^{18}O$ is a measure of the amount of oxygen isotope ^{16}O relative to the isotope ^{18}O. For glacial water, $\delta^{18}O$ is normally −20‰ or less, and for sea water it is close to zero. The $\delta^{18}O$ values in the clay are approximately constant at −9, and reflect several factors, as follows.

1. The depositional water at the center of the clay layer probably had a $\delta^{18}O$ of −9, due to mixing of sea water and glacial ice water.
2. At the time of erosion, $\delta^{18}O$ was constant throughout the layer, due to double drainage.
3. Surface rainwater at the site has an average annual $\delta^{18}O$ of about −9‰ so that it is not possible to develop a diffusion profile similar to that for salinity.

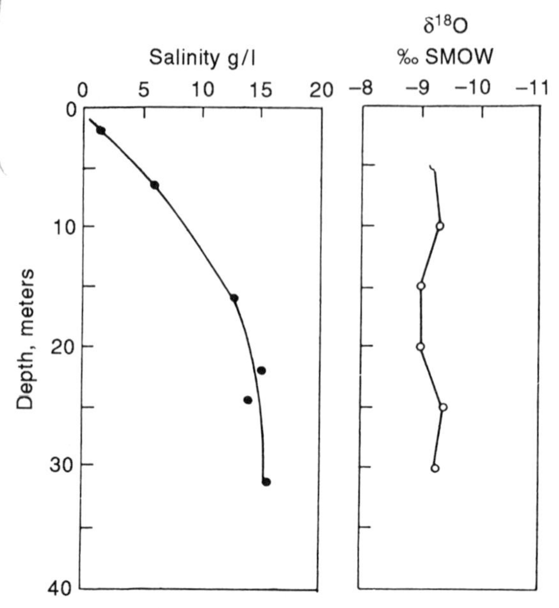

Figure 9.2 Profiles of pore water salinity and $\delta^{18}O$ in Leda clay at Hawkesbury.

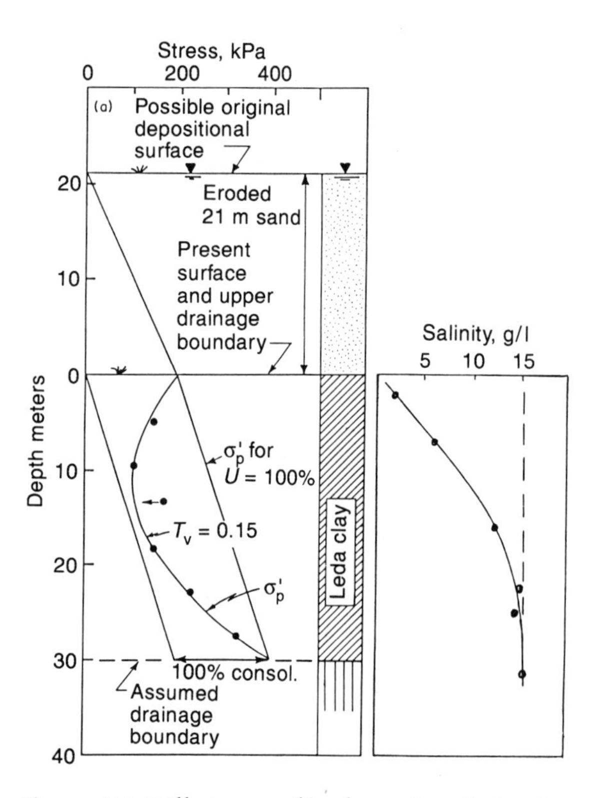

Figure 9.1 Diffusion profile for saline Leda clay, Hawkesbury, Ontario ($t \approx$ 10 000 years). (Modified from Quigley *et al.*, 1983.)

The $\delta^{18}O$ is quite important because it correlates with salinity in many ocean environments, as shown in Figure 9.3. For the Champlain Sea clays, a $\delta^{18}O$ of −9‰ would correlate with a salinity of 15 g/l, exactly that found at Hawkesbury.

Now that the reference salinity has been 'verified' as 15 g/l, one has confidence to calculate a diffusion coefficient for bulk salinity assuming no advection and t = 10 000 years, the approximate time of isostatic emergence, Champlain Sea drainage and sand erosion. The result is a diffusion coefficient of 2×10^{-10} m^2/s (0.0063 m^2/a) (Figure 9.4). This value is a little lower than most of the other available data (e.g. Table 8.6).

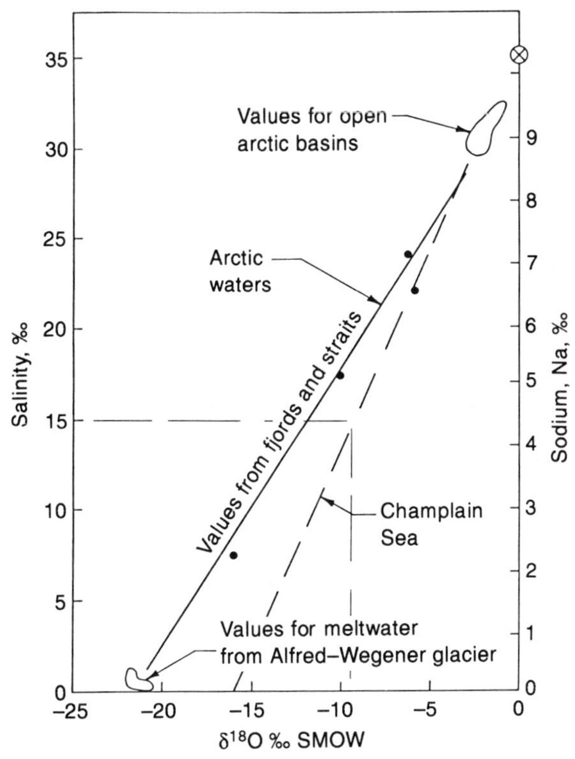

Figure 9.3 Relationship between salinity and ^{18}O of present day Arctic waters, and predicted relationship for Champlain Sea water. (Modified from Hillaire-Marcel, 1979.)

9.2.2 Sarnia water–laid clay till

Chloride and sodium diffusion profiles for the 35 m thick, freshwater, glaciolacustrine clays at Sarnia, Ontario, are presented in Figure 9.5. The important feature of these curves is an increasing concentration with depth; Cl^- increasing from zero to about 400 mg/l at depth and sodium increasing to about 220 mg/l. Since these are freshwater clays, the source of this salt must have been by diffusion from the bedrock.

The chloride profiles in Figure 9.5 were analyzed for upward diffusion of salts from the bedrock against a small, measured, downward regional flow of 0.0003 m/a (Desaulniers, Cherry and Fritz, 1981). For a diffusion coefficient of 3×10^{-10} m²/s (0.01 m²/a) and t = 10 000 years, a curve was calculated that is a reasonable fit to the actual c/c_0 for the pore water chloride (Figure 9.6).

A $\delta^{18}O$ profile was also presented by Desaulniers *et al.*, as shown in Figure 9.7,

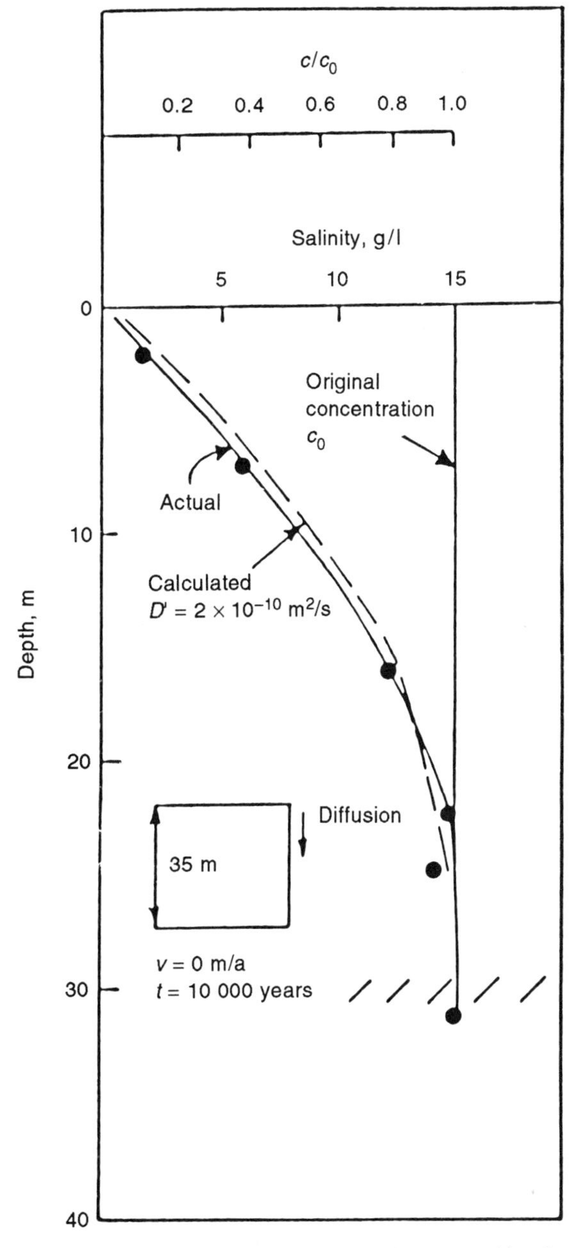

Figure 9.4 Actual and calculated diffusion profiles for bulk salinity, Hawkesbury Leda clay.

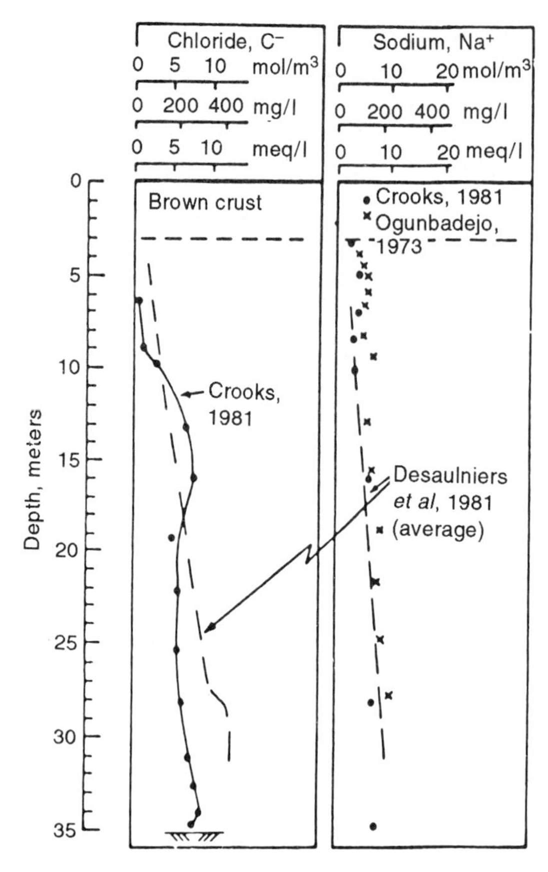

Figure 9.5 Na^+ and Cl^- profiles in thick freshwater glacial clays, Sarnia, Ontario. (From Quigley and Crooks, 1983; reproduced with permission of the *Canadian Geotechnical Journal.*)

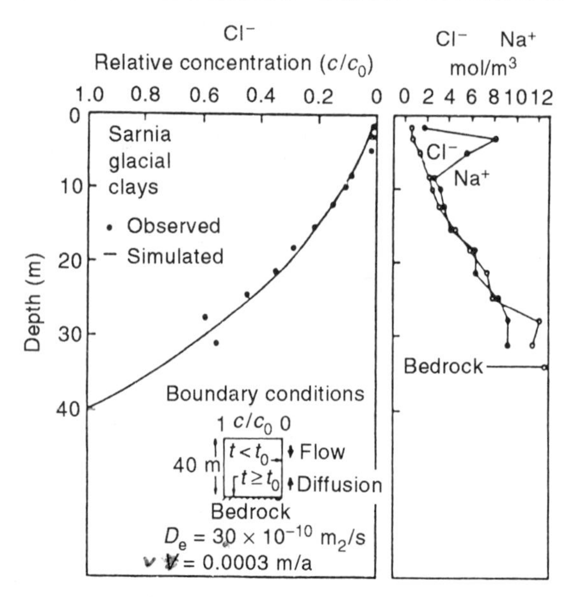

Figure 9.6 Chloride ion versus depth with concentrations calculated for upwards diffusion from bedrock. (Modified from Desaulniers, Cherry and Fritz, 1981.)

complete with calculated curves using a diffusion coefficient of 3×10^{-10} m²/s (0.01 m²/a). The $\delta^{18}O$ at surface (−10‰) represents the average annual rainfall value, and the value of −18‰ at depth reflects glacial melt water. Originally the ice-derived pore water in the glaciolacustrine sediments probably had a $\delta^{18}O$ value of −20‰ or less, and 10 000 years of diffusion have resulted in a profile which ranges from surface value of around −10‰ to −18‰ at the bottom.

Long-term diffusion profiles are difficult to find because of changing geologic and climatic conditions. Fortunately, assessment is not sensitive to a detailed knowledge of exactly when the geologic changes occur. As shown in Figure 9.7, there is little difference for the calculated c/c_0 profiles for t-values of 10 000 and 15 000 years.

9.2.3 Freshwater clay

A third and final example of a natural diffusion profile is shown for chloride in Figure 9.8 (Rowe and Sawicki, 1992). At this site, 34 m to 39 m of fine-grained glaciolacustrine deposit consisting primarily of silts and clays overlie bedrock of the Salina Formation which is composed of soft, erodible shales with evaporites (gypsum) and harder dolostone layers. Based on chemical analyses from four wells in the bedrock, the average bedrock chloride concentration is about 1400 mg/l. The observed concentration extends upward from the bedrock with decreasing concentration towards the surface. Radiocarbon dates of 12 000 to 13 000 years BP have been reported (Fullerton, 1980; Lewis, 1969) for sediments deposited in glacial

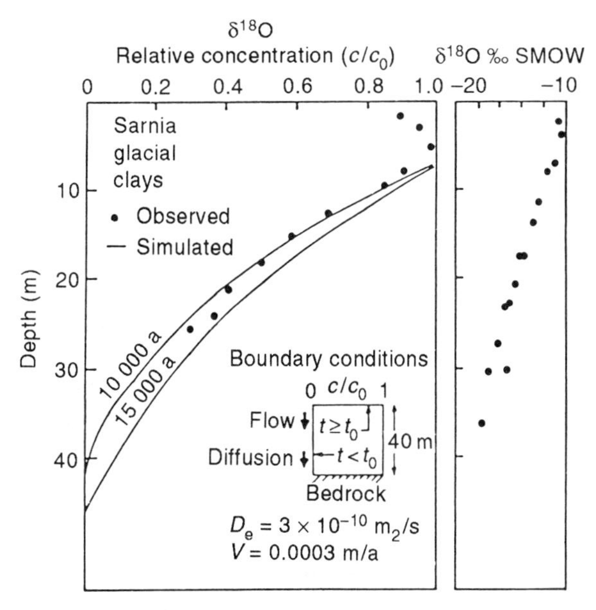

Figure 9.7 $\delta^{18}O$ versus depth and calculated dilution by downward diffusion of rainwater. (Modified from Desaulniers, Cherry and Fritz, 1981.)

Lake Warren (the source of this clay unit). Based on the observed gradients and measured hydraulic conductivity values, the downward Darcy velocity is estimated to be between 0.0004 m/a and 0.0016 m/a. Laboratory diffusion tests give a diffusion coefficient of 3.8×10^{-10} m²/s (0.012 m²/a).

Figure 9.8 shows four calculated concentration profiles for three combinations of parameters and two times. A prediction based on the measured diffusion coefficient and a Darcy velocity of 0.0004 m/a (curve 1) gives a reasonable fit to the observed concentration profile for 12 000 years of diffusion. The potential importance of downward advection is shown by comparing curve 1 (v_a = 0.0004 m/a) and curve 2 (v_a = 0.0016 m/a) for a diffusion coefficient of 3.8×10^{-10} m²/s (0.012 m²/a). Curve 2 does not provide a good fit to the data, and this implies that the downward advective velocity must have been very low for most of the past 12 000 years. To illustrate the sensitivity of results to diffusion coefficients and time, curves 3 and 4 show the predicted profile for a diffusion coefficient of 5.7×10^{-10} m²/s (0.018 m²/a) and times of 12 000 and 10 000 years. Based on these analyses, it appears that the diffusion is generally consistent with a diffusion coefficient of between 3.8×10^{-10} m²/s (0.012 m²/a) and 5.7×10^{-10} m²/s (0.018 m²/a).

9.3 Examples of short-term field diffusion

Diffusion profiles presented in this section are restricted to extensive studies carried out at the Confederation Road landfill near Sarnia, Ontario, plus one profile for the Keele Valley liner, Toronto.

9.3.1 Confederation Road landfill, Sarnia

A summary of the Confederation Road landfill site at Sarnia was presented by Quigley and

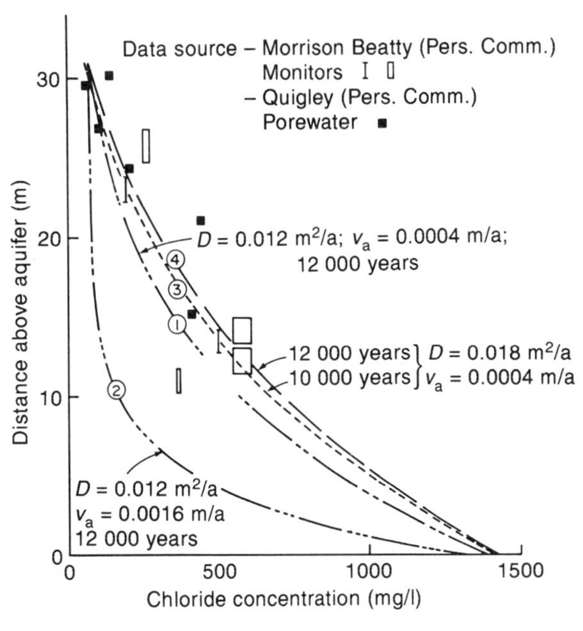

Figure 9.8 Modeling of an existing diffusion profile after 10 000–12 000 years' diffusion. (Modified from Rowe and Sawicki, 1992.)

Rowe (1986). The nature of the Confederation Road site is shown by the geotechnical data in Figures 9.9 and 9.10. The site consists of ~7.5 m of waste (including cover) placed in a 5.5 m trench excavated through a brown desiccated crust which was used for embankment borrow. The calculated effective stress profile incorporates the effect of a slight regional downward gradient which creates a downward average linearized groundwater velocity of ~0.0024 m/a (Goodall and Quigley, 1977). Figure 9.9 also shows the soil to be overconsolidated by ~90 kPa, with a slight increase near the clay–waste interface, where the moisture content decreases slightly from 23% to 21%. The nature of this crust is further illustrated in Figure 9.10. The desiccated crust is highly fissured in the upper 4 m of brown oxidized clay, and much less fissured from 4 to 6.5 m in the lower gray portion of the crust. At the waste/soil interface, which is probably within

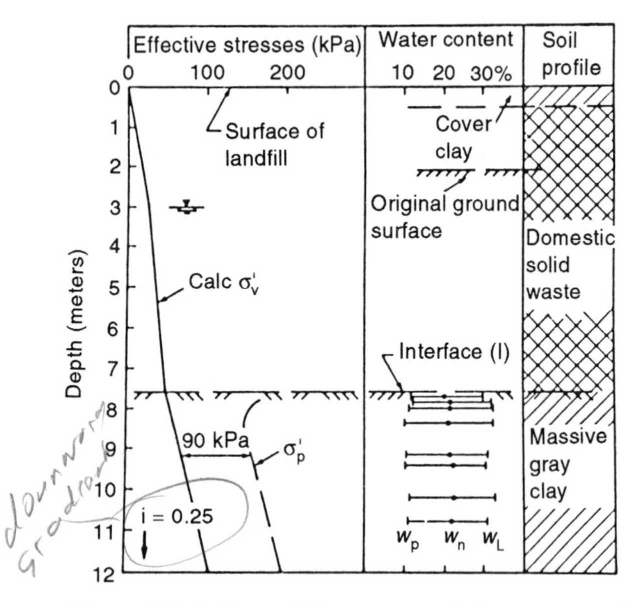

Figure 9.9 Soil conditions at Confederation Road landfill site, 1967 to present (σ'_v is the vertical effective stress; σ'_p is the pre-consolidation pressure; w_p is the plastic limit; w_n is the natural water content; w_L is the liquid limit. (From Quigley and Rowe, 1986; reproduced with permission, American Society for Testing and Materials (ASTM).)

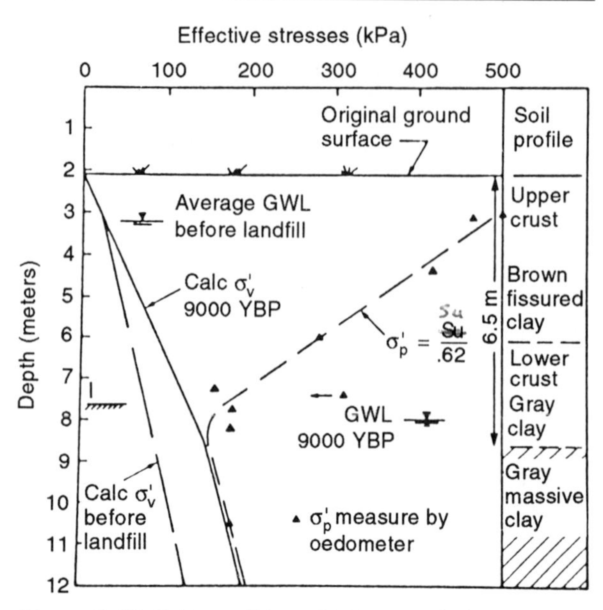

Figure 9.10 Soil conditions 9000 years before present (YBP) and before cutting landfill trench in 1966 (I is waste–clay interface; s_u is the undrained shear strength, GWL is the groundwater level; see also the legend to Figure 9.9). (From Quigley and Rowe, 1986; reproduced with permission, ASTM.)

1 m of the base of the desiccated crust, no fissures have ever been observed in the gray interfacial clay samples obtained in the 50 or more boreholes drilled at the site.

(a) Sodium chloride

Chemical profiles for Na^+ and Cl^- are presented in Figures 9.11 and 9.12 for $t = 12$ years using units of mg/l and mol/m^3. Also plotted are the estimated seepage fronts for $t = 12$ and 100 years. Both figures indicate salt migration for a distance of about 1.5 to 2.0 m in 12 years, which is far ahead of the estimated advection distance of 3 cm. This would seem to confirm the significant role that diffusion plays in the migration of chemicals. As noted in section 8.9, diffusion and distribution coefficients can be inferred from these field profiles and yield parameters as given in Table 8.5 for Cl^-, Na^+ and K^+. For chloride, the diffusion coefficient

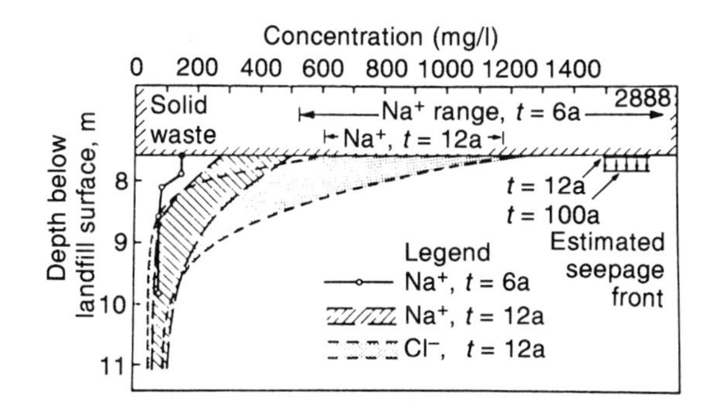

Figure 9.11 Pore fluid concentration of Na^+ and Cl^- (mg/l) in clay below waste after six and 12 years of diffusion (t = 6a and 12a). (From Quigley and Rowe, 1986; reproduced with permission, ASTM.)

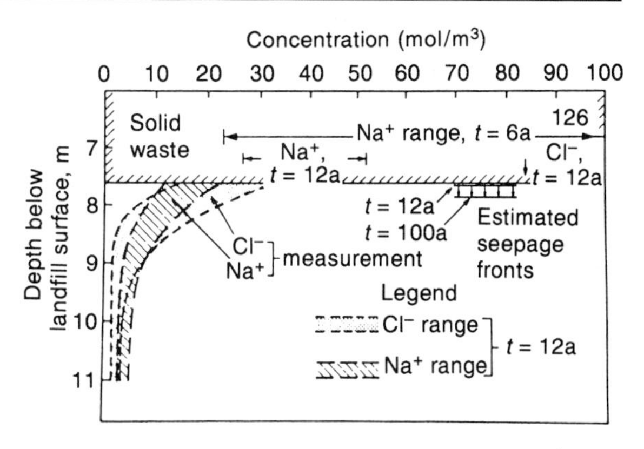

Figure 9.12 Pore fluid concentration of Na^+ and Cl^- (mol/m^3) in clay below waste at t = 12 years (12a). (From Quigley and Rowe, 1986; reproduced with permission, ASTM.)

was about 6.3×10^{-10} m^2/s (0.02 m/a), which is quite consistent with values obtained from laboratory tests on samples of the same clay (e.g. Table 8.3).

The plot in mg/l suggests that Cl^- (stippled) has migrated significantly faster than Na^+; however, this impression is an artifact of the plotting related to the higher molecular weight of chloride (35) compared to sodium (23). If the data are plotted in chemical units (mol/m^3), as shown in Figure 9.12, the two species are seen to be migrating at approximately the same rate, with Cl^- slightly in advance.

(b) Heavy metals

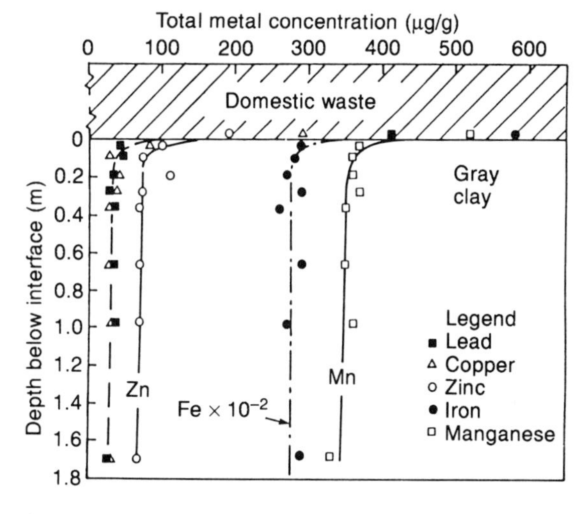

Figure 9.13 Profiles for total metals (BH 84-10, t = 16a) in clay below domestic waste. (Modified from Yanful and Quigley, 1986.)

Profiles of total metal concentration for lead, copper, zinc, iron and manganese are presented in Figure 9.13. The long vertical portions represent total background values, and the short curved sections near the interface represent metals added to the soil by diffusion from the leachate. It seems quite clear that these metals have migrated a maximum of 20 cm, and more likely only 10 cm, in 16 years. The nature of these metal compounds is shown in Figure 9.14 (Yanful, Quigley and Nesbitt, 1988). The extractions suggest that all species in the background soil exist in soil carbonate and within the lattice structure of the minerals (primarily the clay minerals). An organic phase may also be present for iron, zinc and copper. In the elevated concentration zone near the interface, carbonate dominates the mineral species, which is believed to have precipitated at the soil pH levels of ~8 and the low redox potentials of −150 mV at the interface. The interested reader

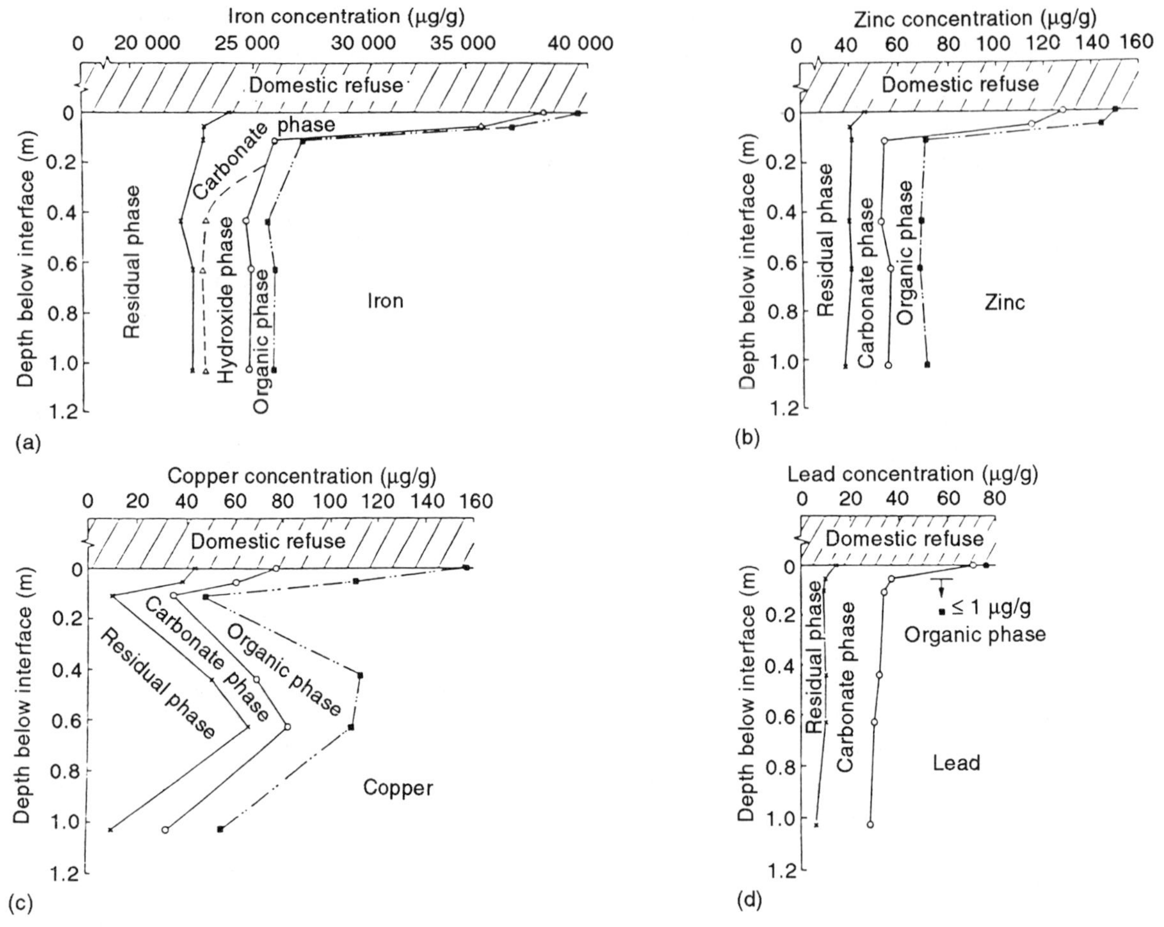

Figure 9.14 Heavy metal complexes in clayey subsoil inferred from sequential extraction, BH 83-7, $t = 15$ years: (a) iron; (b) zinc; (c) copper; (d) lead. (Modified from Yanful, Quigley and Nesbitt, 1988.)

is referred to Yanful, Quigley and Nesbitt (1988),

(c) Dissolved organic carbon (DOC)

Finally, Figure 9.15 presents a plot of dissolved organic carbon in pore fluid squeezed from the interfacial clays at $t = 16$ years. The curve indicates a migration distance of about 0.8 m compared to 1.5 m for chloride. This is encouraging, since it suggests that natural biological activities are actively degrading any organics from the leachate in a zone very close to the waste.

Another DOC curve is presented in Figure 9.16 for Confederation Road at $t = 21$ years (Quigley, Mucklow and Yanful, 1990). This curve also appears to be a diffusion profile with interface DOC values of ~100 mg/l, decreasing to ~10 mg/l at 0.9–1.0 m depth below the interface. GC-MS analysis of the pore water yielded only trace amounts of xylene and bulk phenol values of 0.4–0.6 mg/l at the interface. No other EPA priority pollutants were identified, again suggesting that all have biodegraded within the upper 1 m of clay in the 21 years since disposal.

The DOC curve itself remains a mystery in that its components are still not identified. It is

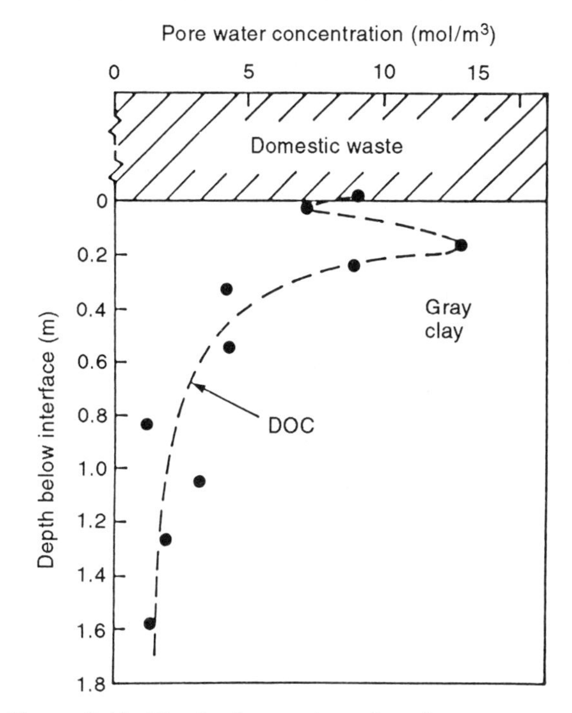

Figure 9.15 Dissolved organic carbon in pore water squeezed from samples at $t = 16$ years (16a).

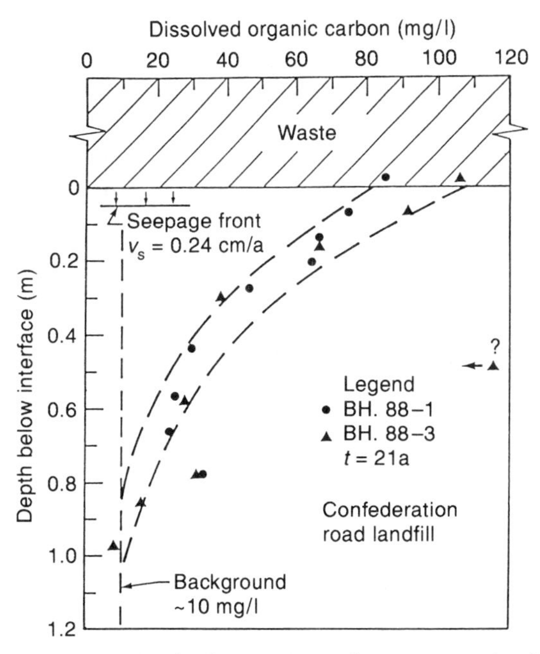

Figure 9.16 Dissolved organic carbon versus depth below the waste–clay interface at $t = 21$ years (21a). (Modified from Quigley *et al.*, 1990.)

possible that the curve does not represent migration of organic substances at all; rather it could represent biological activity that has broken down solid organics in the clay to produce lower molecular weight compounds that were removed with the water by high pressure squeezing. The background DOC available for such decay amounts to about 0.7% of the soil solids.

The phenol profile at the site is shown in Figure 9.17. Background values vary between 25 and 75 μg/l which are far above the typical drinking water objective of 2 μg/l. This indicates the importance of reference testing on clays before their use as barrier soils. Despite the amount of scatter generated by samples from three 1988 boreholes, there does appear to be a diffusion profile extending about 1 m below the interface. Since phenol levels are high in many

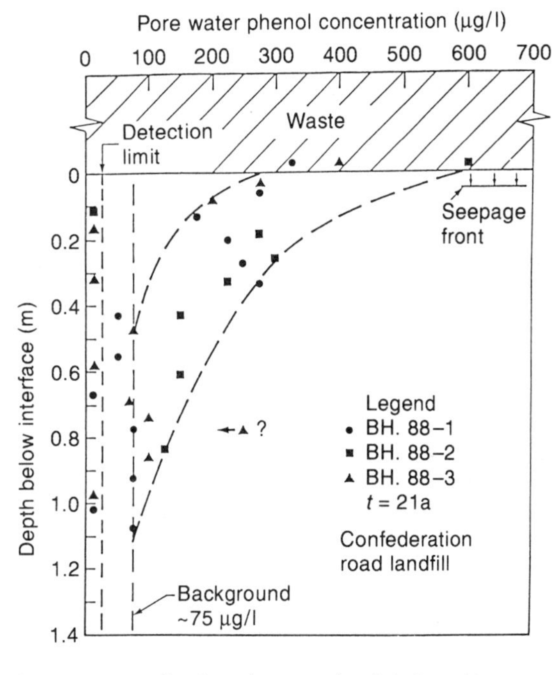

Figure 9.17 Bulk phenol versus depth below the waste–clay interface at $t = 21$ years (21a). (From Quigley, Mucklow and Yanful, 1990; reproduced with permission.)

MSW leachates, it is speculated that most of the phenol has biodegraded away in the 21 years since deposition of the waste. A half-life of about six years would be required to cause a decay from around 5000 μg/l to present values. This half-life seems reasonable, given available data on decay of phenol in soil. More research is required to define better the half-life of organics such as phenol in waste and in soil.

(d) Isotopes

In Figure 9.18, tritium analyses and fitted profiles are presented for squeeze water obtained in 1984. The central fitted curve, which also appears to be a good average curve, was calculated using a c_0 for tritium calculated for 1967 when the trench filled with rainwater. A radioactive decay was also applied to account for the short half-life of tritium (12.4 years). The curve shows migration to ~2 m, which is consistent with the distance of chloride migration.

Two curves are presented in Figure 9.19 for $\delta^{18}O$ and δ^2H (deuterium). Both curves describe diffusion of heavy water from the waste into the underlying clay which contains light water. The explanation for these profiles runs as follows.

1. The background values in the soil are in equilibrium with the local rainwater at $\delta^{18}O \approx -11‰$ SMOW.
2. Evapotranspiration from leaves results in escape of light water (i.e. light oxygen, ^{16}O, and light hydrogen, ^1H).
3. On decay in the landfill, the heavy water left in the leaves is released from the waste and migrates by diffusion downwards creating a diffusion profile.

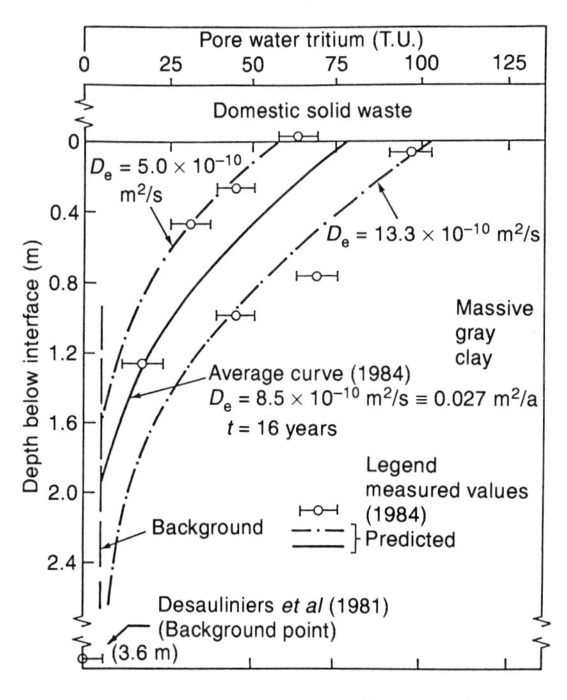

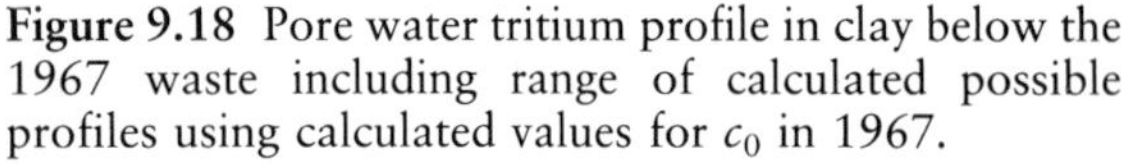

Figure 9.18 Pore water tritium profile in clay below the 1967 waste including range of calculated possible profiles using calculated values for c_0 in 1967.

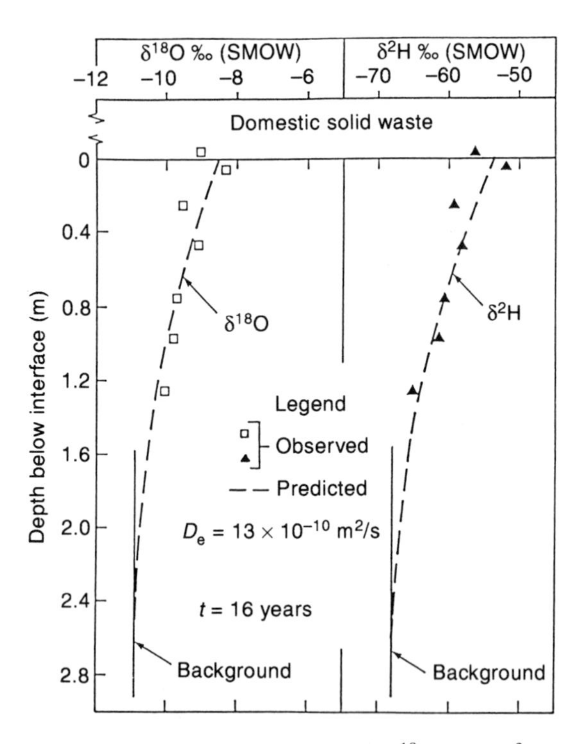

Figure 9.19 Profiles of measured $\delta^{18}O$ and δ^2H in the clay below the 1967 waste. (Modified from Yanful and Quigley, 1990.)

Another example of this phenomenon is described for a plume by Fritz, Matthess and Brown (1976).

The Confederation Road site is now 23 years old, and in certain sections appears to have been flushed with surface water whose circulation has been aided by a railway ditch along the north side of the landfill. The Na^+ and Cl^- profiles shown in Figure 9.20 were obtained on samples taken in 1984 at $t = 16$ years. They indicate an abrupt back diffusion into the source leachate from the clay soil. A simple analysis of the back diffusion profile suggests that this back diffusion event had been initiated about six to nine months prior to the time of drilling at this location. This plot demonstrates again the speed of diffusion over very short periods of time when there is a high concentration gradient. On the other hand, the diffusion profiles at depth are now migrating so slowly at $t = 20$ years ($\sim$10 mm/a) that it is no longer practical to study the rate of advance.

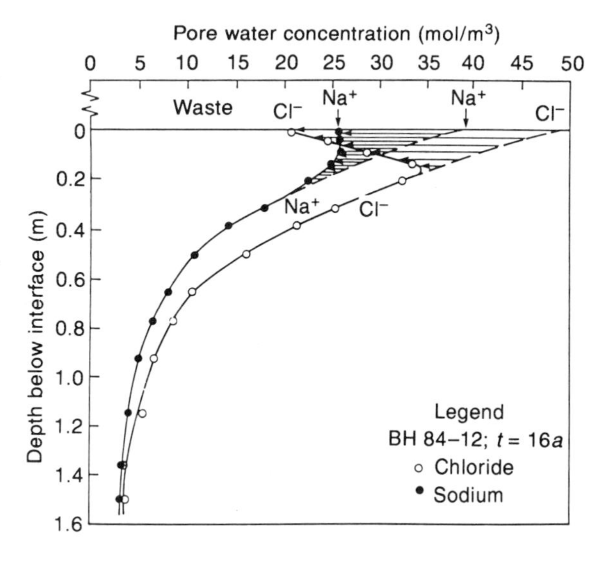

Figure 9.20 Chemical profiles for sodium and chloride showing diffusion from clay back into the waste due to site flushing.

9.3.2 Keele Valley

A chloride ion diffusion profile and a series of cation diffusion profiles have been obtained at Metropolitan Toronto's and Keele Valley landfill site (King *et al.*, 1993).

The chloride data obtained from liner samples exposed by an exhumation are presented in Figure 9.21. The water samples for analysis were centrifuged from the sand layer and squeezed from the clay liner. The results for chloride show a smooth curve starting at the top of the sand layer and extending a total distance of 70–75 cm over a period of 4.25 years. A diffusion profile fitted to the data, assuming negligible advection, yielded a field value for $D_{Cl} \approx 6.5 \times 10^{-10}$ m²/s (0.02 m²/a). This is nearly identical to values obtained in the field at the Confederation Road site (e.g. section 9.3.1) and on laboratory samples from Sarnia presented in Chapter 8 (Table 8.3).

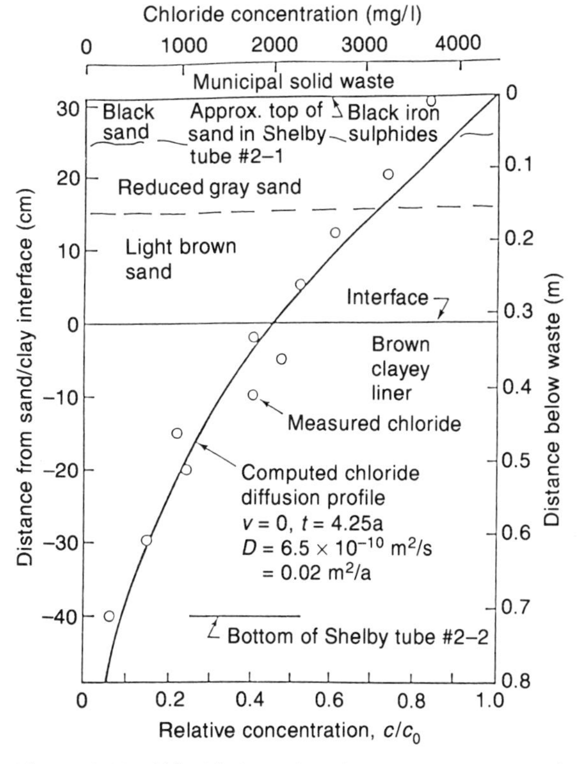

Figure 9.21 Chloride ion migration at 4.25 years, Keele Valley liner, Maple, Ontario. (Modified from Reades *et al.*, 1989.)

The most important aspect of the chloride

curve is that diffusion starts at the top of the partly clogged sand, which appears to be acting as part of the barrier, adding an extra 30 cm of thickness with respect to diffusion. This is good from the perspective of diffusive transport, but not so good from the perspective of leachate mounding. Fortunately, the hydraulic conductivity of the clay liner is so low (5×10^{-11} m/s) that advection is small, even with the leachate mounding that has occurred to date.

Cation concentration profiles (Na^+, K^+, Mg^{2+} and Ca^{2+}) presented in Figure 9.22 also demonstrate a variety of important features. Na^+ has migrated about 65 cm and is thus only slightly retarded by adsorption onto the clays. The Na^+ profile also starts at the top of the sand cushion, confirming the extra 30 cm of diffusion barrier thickness suggested by the chloride curve.

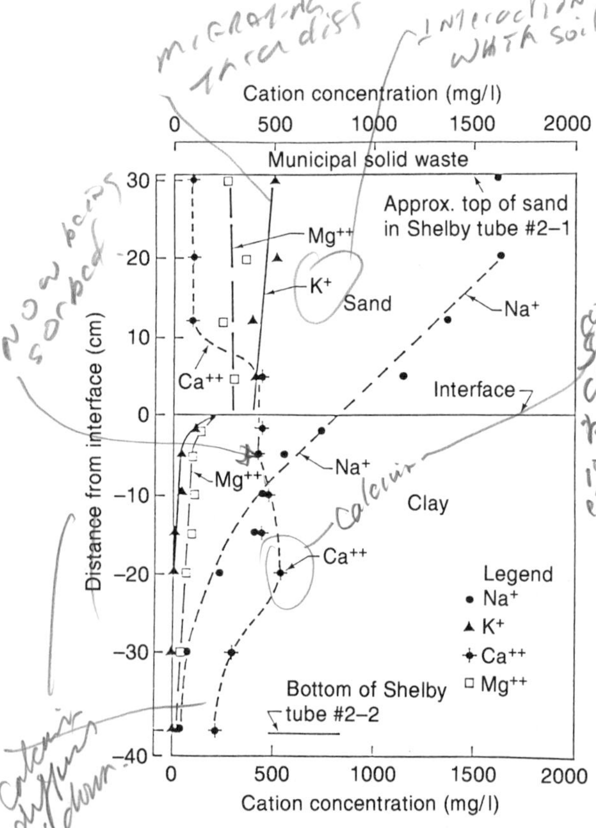

Figure 9.22 Pore water cation profiles for East Pit Shelby Tubes 2-1 and 2-2. (Reades *et al.*, 1989.)

K^+ has migrated through the sand, where it has not experienced significant retardation, but only 5 cm into the clay, demonstrating retardation by adsorption onto the liner clay. By implication, NH_4^+ has probably migrated about the same distance.

The Ca^{2+} profile forms a concentration hump throughout the depth examined. This hump is a hardness halo produced by desorption of Ca^{2+} that accompanies adsorption of Na^+, K^+ and possibly NH_4^+ and Mg^{2+}. If salinity probes are used to track the migration of salts, they should pick up the hardness halo, and thus probably suggest migration further than it has actually progressed.

Finally, a series of organic profiles reported by Barone *et al.* (1993) is presented in Figure 9.23. These profiles indicate migration of low concentrations of volatile organic liquids to depths of ~60 cm in about 4.25 years.

9.3.3 Other landfills

The authors have been involved with four field exhumations of clay liners varying in age from two to ten years. Although these data cannot be presented in this book, it is noted that diffusion is clearly the dominant transport mechanism in each case. Furthermore, reasonable predictions of migration through these liners can be made using parameters obtained from the laboratory tests described in Chapter 8.

9.4 Hydraulic conductivity of 'contaminated' clay liners

9.4.1 Confederation Road landfill

An interdisciplinary study published by Quigley *et al.* (1987) appears to be one of the very few scientific field studies performed on the hydraulic conductivity of a contaminated clay beneath a domestic waste site. A brief review is presented here since it confirms that domestic waste

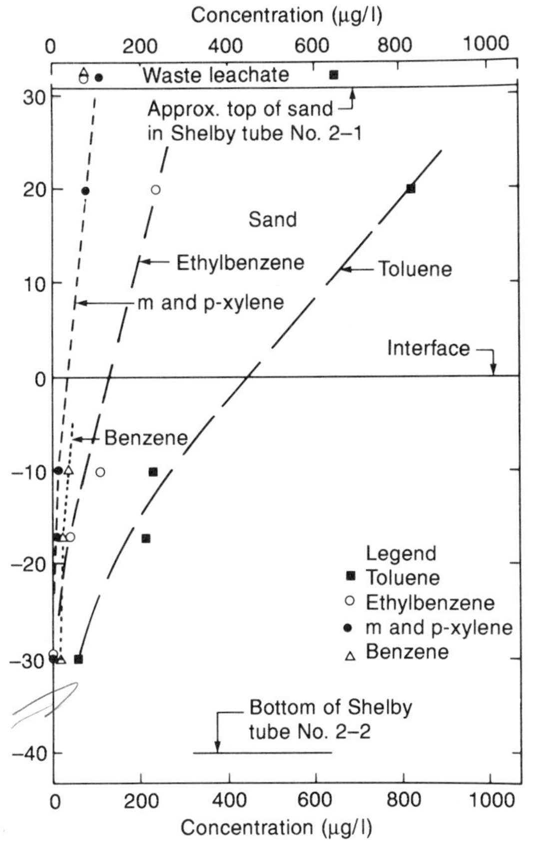

Figure 9.23 Pore water organic profiles at t = 4.25 years in the Keele Valley landfill liner. (Barone *et al.*, 1993 as adapted from Quigley, 1991.)

leachate did **not** increase the hydraulic conductivity of inactive barrier clay, confirming several laboratory studies which have employed MSW leachate as the permeant (Bowders *et al.*, 1986; Quigley *et al.*, 1988).

The site was described in section 9.3.1. The hydraulic conductivity study consisted of oedometer k-testing of three-inch (76 mm) Shelby tubes as Phase I and constant flow rate k-testing as Phase II.

A summary plot showing the oedometer k-test results, along with water content and salt profiles for BH 83-2 representing t = 15 years, is presented in Figure 9.24. The hydraulic conductivity profiles show approximately constant values, except within about 20 cm of the clay–water interface, where a slight decrease is observed. Of the two profiles presented in Figure 9.24, the hydraulic conductivity calculated for the stress range from 50 to 100 kPa (just below σ'_p) yielded values closest to the field k-values of Goodall and Quigley (1977) and the directly measured k still to be discussed. Also shown in Figure 9.24 are profiles for Cl^- and the major cations. These profiles indicate contaminant migration to ~1 m for this particular borehole (83-2). As a final comment, note that the average moisture of the clay for the two stress ranges used to calculate k are essentially constant. This is important, since it suggests that the slight drop in hydraulic conductivity at the interface is chemically controlled.

If the pore water cation concentrations are used to calculate a value of the sodium adsorption ratio (SAR), a profile like that shown in Figure 9.25 is obtained. This profile suggests up to 8% Na^+ on the interfacial clays decreasing to a background 3% Na^+ within 0.4 m of the interface. This should certainly cause at least a small reduction in hydraulic conductivity.

The results of constant flow rate k-testing on seven undisturbed tube samples from BH 84-10 (for t = 16 years) are presented in Figure 9.26. The k-test profile, obtained on samples with σ'_v = 90 kPa corresponding to the field stress, again implies a decrease in k in the zone within 20 to 30 cm of the interface. Unfortunately, the interface at this borehole location was 0.4 m higher than that at BH 83-2 used for the oedometer k-tests. This seems to include the lower part of the soil crust, since the water content w also decreased towards the interface, suggesting complete w control on k. Great care was taken in doing these tests, and pore water squeezed from adjacent soil samples was used as the permeant for the tests (Quigley, Yanful and Fernandez, 1987).

The results of mercury intrusion porosimetry and bulk bacteriology run on the interfacial

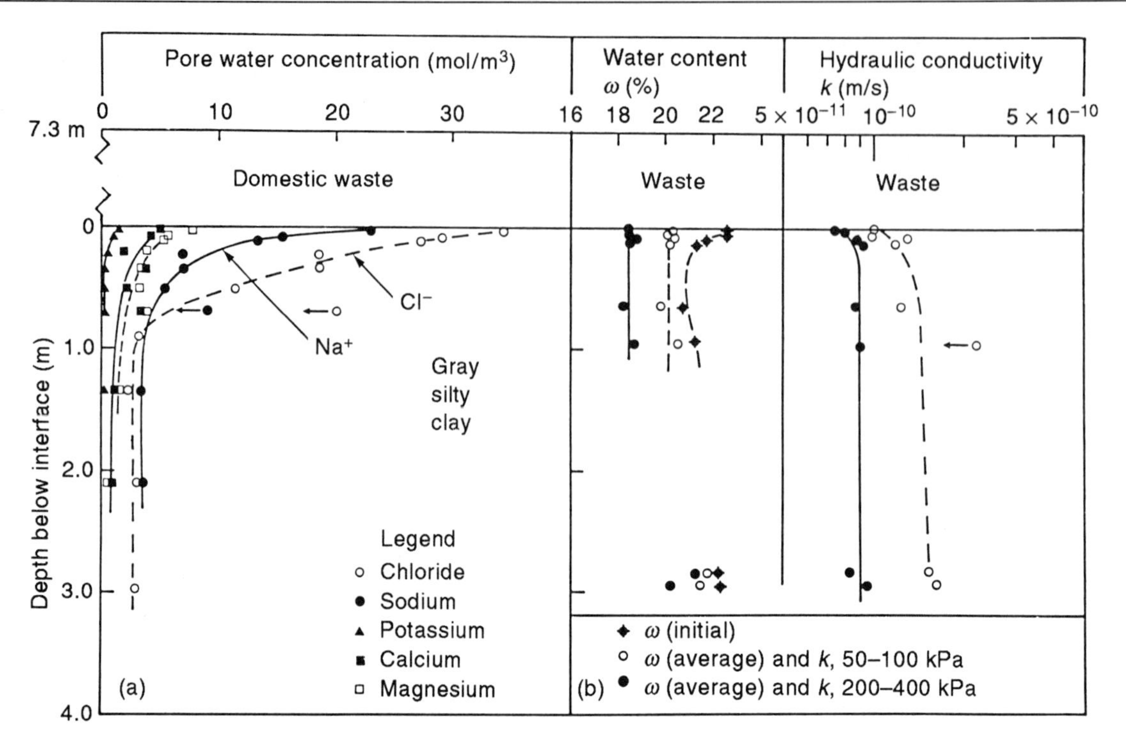

Figure 9.24 Clay–leachate interaction at interface, BH 83-2, $t = 15$ years. (a) Pore water chemistry; (b) Hydraulic conductivity and bulk water contents calculated from oedometer tests at pressures just below and above the preconsolidation pressure of 150 kPa.

clays are presented in Figures 9.27 and 9.28. Both the modal and median pore diameters plotted in Figure 9.27 for freeze-dried specimens from BH 84-3 imply a decrease in pore size near the interface. Just how much of this must be apportioned to desiccation, heavy metal precipitation or freeze-drying damage is difficult to assess. Similarly, the bulk bacteriology which was done by culturing diluted wash extracts, not natural specimen counting, also suggests the possibility of bacterial clogging at the interface.

A summary plot of hydraulic conductivity versus bulk void ratio for all of the oedometer k-tests and direct k-tests is presented in Figure 9.29. On the basis of this plot, one might conclude that domestic waste leachate has not significantly altered the hydraulic conductivity of the inactive gray clays at Sarnia. It is possible that the range in values at constant void ratio could be assigned to chemical changes. As shown in the figure this amounts to a decrease in k from about 1.4×10^{-10} to about 1.0×10^{-10} m/s at a void ratio of 0.56.

9.4.2 Three Wisconsin liners

Gordon, Huebner and Miazga (1989) have reported that clayey barriers at three Wisconsin landfill sites have significantly improved with time, based on data from field lysimeters. All three clays were residual soils derived from weathering of metamorphic rock, granite or dolomite and met the general criteria of: 50% < 200 mesh; liquid limit $w_L \geqslant 30\%$; plasticity index $I_p \geqslant 15$; and laboratory hydraulic conductivity of less than 1×10^{-9} m/s. The results of this study were summarized in Table 3.1.

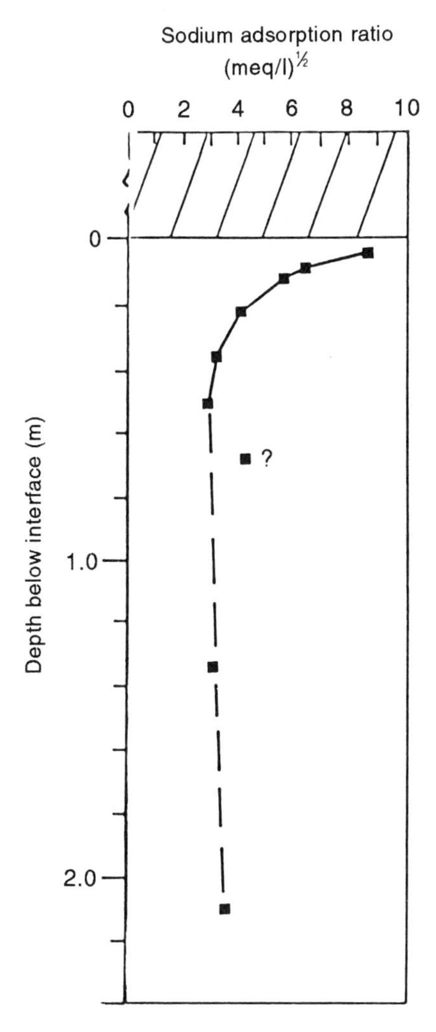

Figure 9.25 Exchangeable sodium cations on barrier clays, BH 83-2, $t = 15$ years; SAR ≈ % Na^+. (Modified from Quigley, Crooks and Yanful, 1984.)

9.4.3 Keele Valley

Reades *et al.* (1989) and later King *et al.* (1993) have reported data which are reproduced in Figure 9.30 and Table 3.1, and which clearly show that once the waste was in place, the field (lysimeter) k-values decreased to ~5 × 10^{-11} m/s. The soil used at Keele Valley is a well-graded glacial till containing 20 to 28% <2 μm sizes.

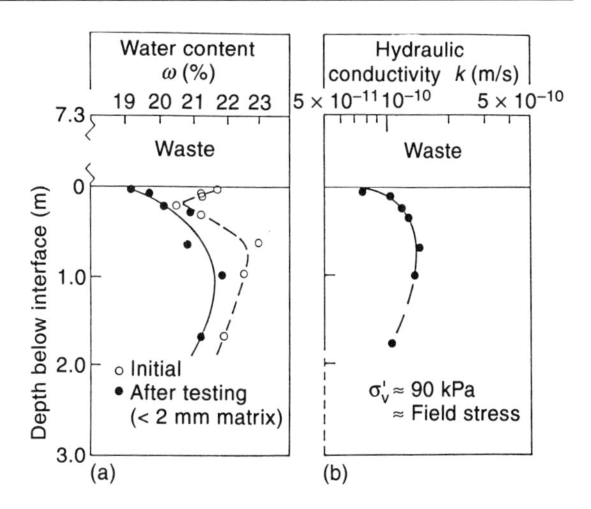

Figure 9.26 Clay–leachate interaction (BH 84-10). (a) Initial (*in situ*) and post-testing water contents of <2 mm soil matrix; (b) directly measured hydraulic conductivity using pore water as influent permeant.

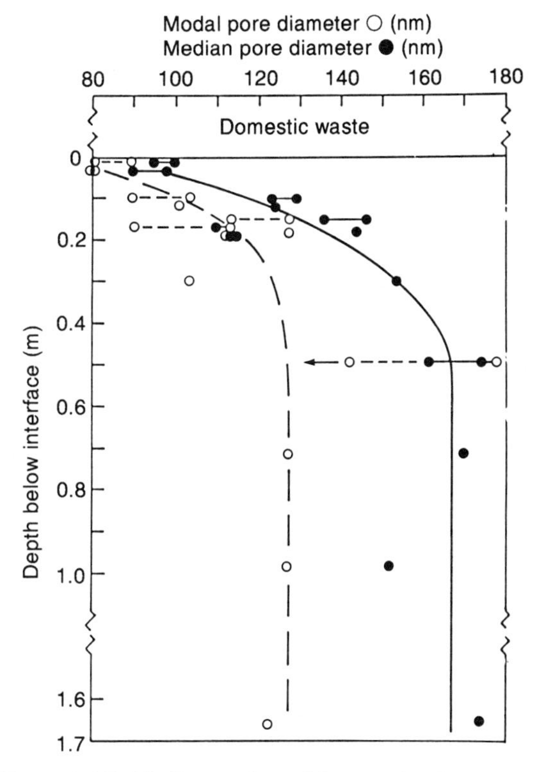

Figure 9.27 Median and modal pore size by mercury intrusion of freeze-dried samples (BH 84-3).

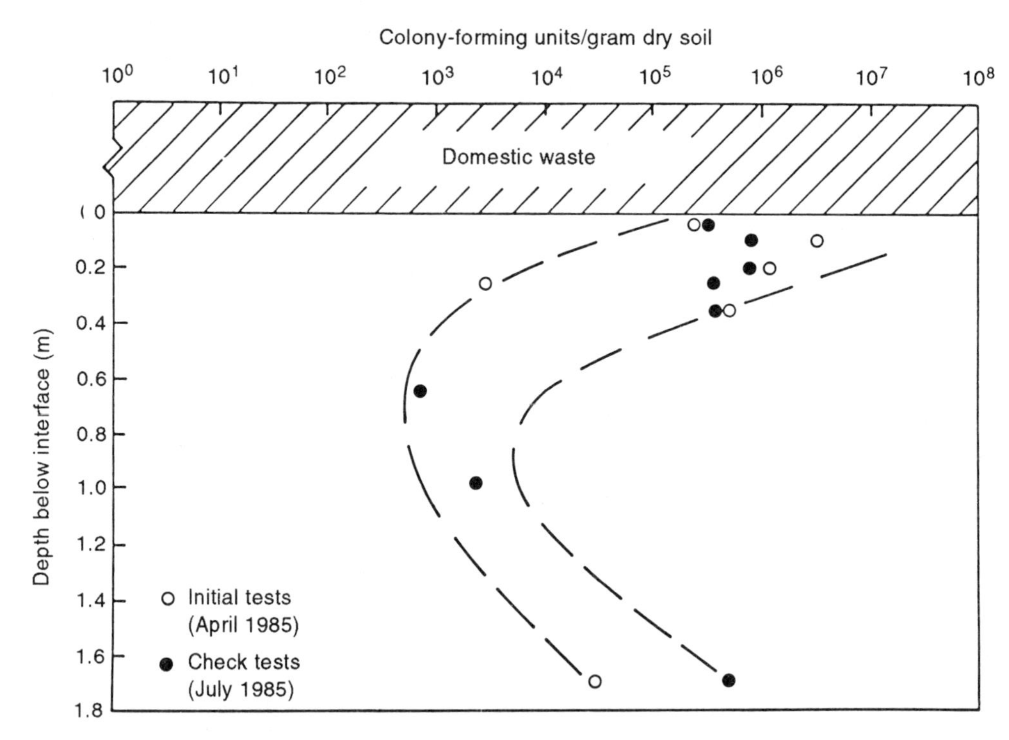

Figure 9.28 Total colony forming bacterial units/gram of dry soil from hydraulic conductivity samples (BH 84-10). (From Quigley *et al.*, 1987; reproduced with permission of the *Canadian Geotechnical Journal*.)

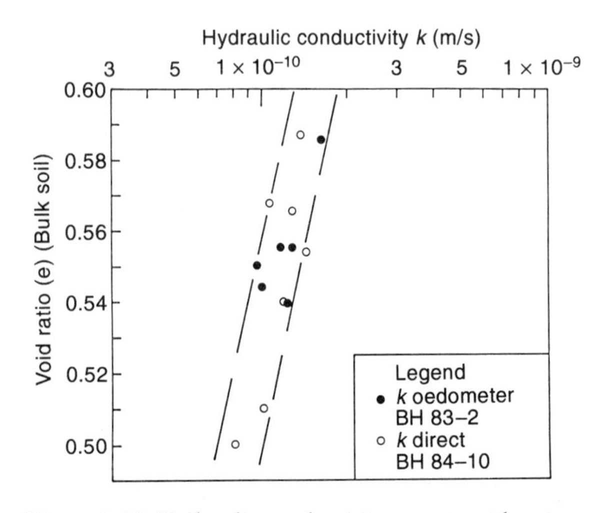

Figure 9.29 Hydraulic conductivity versus void ratio.

9.4.4 Summary

On the basis of the above field studies and several other laboratory studies (Griffin *et al.*, 1976; Daniel and Liljestrand, 1984; Schubert, Harrington and Finno, 1984; Bowders *et al.*, 1986; Quigley *et al.*, 1987), it appears that MSW leachate does not adversely affect the hydraulic conductivity of clayey barriers composed of inactive clay minerals illite and chlorite.

At least two important factors appear responsible for the improving performance of clay liners with time. The first is replacement of the Ca^{2+} and Mg^{2+} adsorbed on most natural clays by Na^{+} in the MSW leachate. About 8% of the adsorption sites appear to pick up Na^{+}. The second is consolidation caused by the increasing weight of waste applied to the wet-of-optimum liners. As noted ealier (Chapter 3), inert clayey soils properly placed at about 2–3% wet of

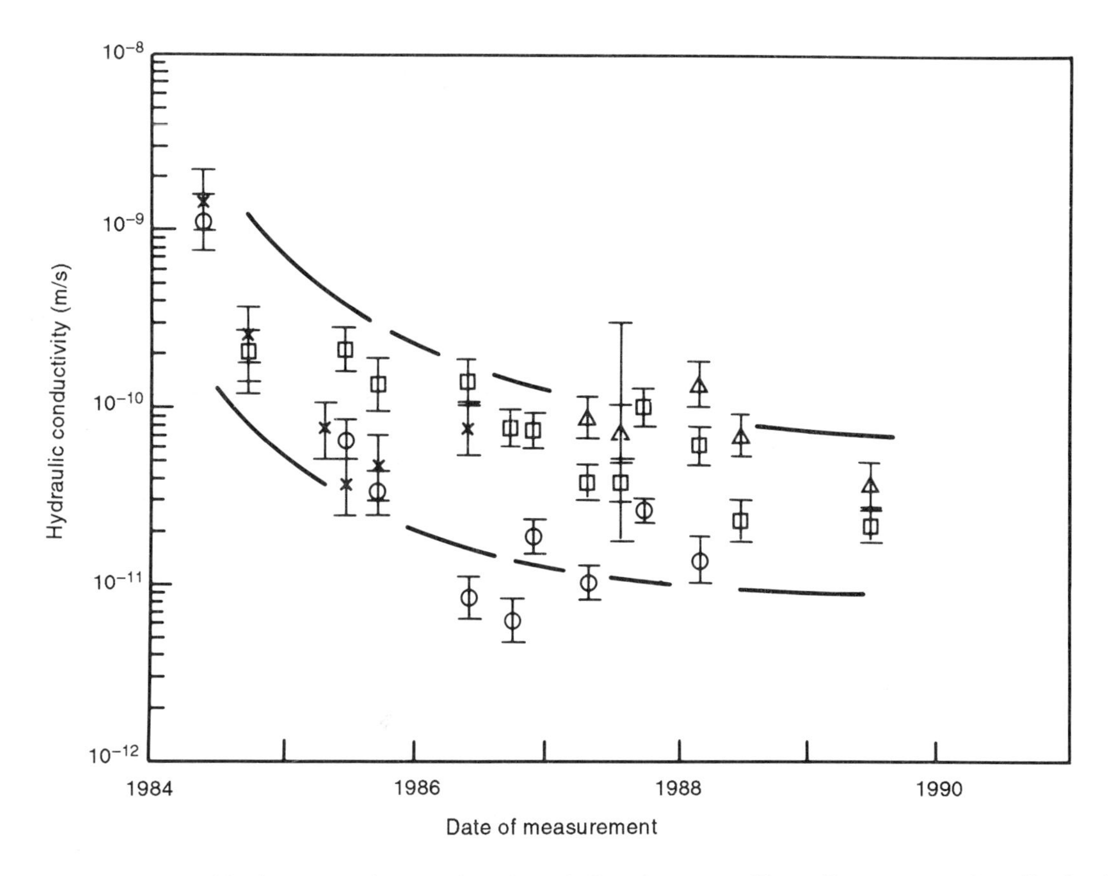

Figure 9.30 Estimated hydraulic conductivity based on shallow lysimeter effluent flow rates; Keele Valley landfill. Range bars represent possible error limits; squares, circles and triangles represent mean values within range limits. (Modified from King *et al.*, 1993.)

Standard Procter optimum have a pre-consolidation pressure of about 120 to 150 kPa. Since they are heavily remolded, the clays appear to consolidate under the combined effects of stress increase and chemical migration by diffusion, resulting in high quality liners over the long term.

It is suggested that compaction drier than 2% wet of optimum would produce a considerably stiffer clayey liner that would not be nearly as amenable to self-healing by consolidation.

Much work remains to assess the effects of various leachates on vermiculitic soils, which are prone to *c*-axis contraction by K^+ and NH_4^+ fixation with resulting decreases in CEC. Similar work has yet to be done on bentonites which are even more prone to *c*-axis contraction if highly peptized on placement. Fortunately, the bentonites are not likely to incur any decreases in CEC. Finally, much work remains to be done on bacterial activity within the mass of a barrier clay.

Chapter ten

Contaminant migration in intact porous media: analysis and design considerations

10.1 Introduction

The foregoing chapters have introduced the concepts of the various contaminant transport mechanisms, and have discussed the determination of parameters, the character and mounding of leachate and methods of analysis.

The objective of this chapter is two-fold: firstly, to discuss the mass of contaminant and its influence on the potential impact of a landfill on groundwater; secondly, to examine how some very simple finite layer models can be readily used to assess impact of contaminant migration through clay barriers for a range of situations and combinations of parameters. Many of the examples relate to clayey barriers underlain by a thin natural aquifer (e.g. Figures 1.2–1.4), since this is a common occurrence in many areas; however, similar analyses can also be performed for compacted clay liners constructed over an engineered drainage layer (e.g. Figures 1.5–1.8) or for designs involving cut-off walls or permeable surrounds (e.g. Figure 1.9).

10.2 Mass of contaminant, the reference height of leachate, H_r, and the equivalent height of leachate, H_f

For waste disposal sites such as municipal landfills, the mass of any potential contaminant within the landfill is finite. The process of collecting and treating leachate involves the removal of mass from the landfill, and hence a decrease in the amount of contaminant which is available for transport through the barrier system and into the general groundwater system. Similarly, the migration of contaminant through the barrier also results in a decrease in the mass available within the landfill. For a situation where leachate is continually being generated by percolation of water through the landfill cover, the removal of mass by either leachate collection and/or contaminant migration will result in a decrease in leachate strength with time, and there will be a decrease in concentration similar to that observed in the laboratory tests described in Chapter 8.

The impact of a waste disposal site upon groundwater quality is usually judged by monitoring the concentration of potential contaminants at a number of specific monitoring points. As will be demonstrated in this chapter, the variation in concentration with time at these points will be a function of the mass of contaminant in the system, the infiltration through the landfill cover, the proportion of leachate collected, and the proportion of leachate passing into the hydrogeologic system.

A reasonable estimate of peak concentration, c_0, of a given contaminant species can usually be estimated from past experience with similar landfills. The total mass of contaminant is more difficult to determine. Nevertheless, estimates can be made by considering the observed variation in concentration with time at landfills where leachate concentration has been monitored or by considering the results of lysimeter tests on waste.

Until fairly recently, there has been a paucity of data concerning the available mass of contaminants within landfills; however, this situation is changing now that many landfills have leachate collection systems. Given that concentration is simply mass per unit volume, the mass of a given contaminant collected in a year is equal to the average concentration multiplied by the volume of leachate collected. By monitoring how this mass varies with time, it is then possible to estimate the total mass of that species of contaminant within the landfill. In the absence of this information, studies of the composition of waste (e.g. Cheremisinoff and Morresi, 1976; Kirk and Law, 1985; Ehrig and Scheelhaase, 1993) can be used to estimate the mass of given contaminant or groups of contaminants. For example, Table 2.2 summarized an estimate of refuse composition reported by Hughes, Landon and Farvolden (1971). For contaminant species predominantly formed from breakdown or synthesis of other species (e.g. by biological action), an upper bound estimate of the mass of contaminant may be obtained from the estimates of the mass of chemicals which go to form the derived contaminant.

For the purposes of modeling the decrease in concentration in the leachate due to movement of contaminant into the collection system and through the barrier, it is convenient to represent the mass of a particular contaminant species in terms of parameters defined as the 'reference height of leachate', H_r, or the 'equivalent height of leachate', H_f, as described in the following subsections.

10.2.1 Reference height of leachate, H_r

On the simplest level, suppose that the infiltration (percolation) into the landfill is q_0, the exfiltration through the liner is q_a and the leachate collected (per unit area) q_c; then, assuming the landfill is at field capacity, continuity of flow requires that

$$q_0 = q_c + q_a \qquad (10.1)$$

If the initial mass of a contaminant species (e.g. chloride), m_{TC}, can be estimated, then the reference volume of leachate which would contain this mass at an initial concentration c_0 is

$$V_{TC} = \frac{m_{TC}}{c_0} \qquad (10.2)$$

In general, this volume will not correspond to the actual volume of leachate because it is based on the assumption that all this available mass can be quickly leached from the solid waste. It is convenient for both mathematical and physical reasons to express the volume V_{TC} in terms of a reference height of leachate, H_r, which is defined as the reference volume of leachate, V_{TC}, divided by the area, A_0, through which contaminant passes into the primary barrier:

$$H_r = \frac{V_{TC}}{A_0} \qquad (10.3a)$$

or

$$H_r = \frac{m_{TC}}{c_0 A_0} \qquad (10.3b)$$

where m_{TC} is the total mass of a contaminant species of interest [M]; c_0 is the peak concentration of that species in the landfill [ML^{-3}]; A_0 is the area through which contaminant can migrate into the underlying layer [L^2].

An equation can then be written for conservation of mass as indicated in Figure 10.1 and, as shown, this can be reduced to

$$c_T(t) = c_0 + c_r t - \frac{1}{H_r}\int_0^t f_T(c,\tau)\mathrm{d}\tau - \frac{q_c}{H_r}\int_0^t c_T(\tau)\mathrm{d}\tau \qquad (10.4a)$$

which, on substitution for the flux $f_T(c,\tau)$ (equation 1.8), becomes

$$c_T(t) = c_0 + c_r t - \frac{1}{H_r}\int_0^t \left(nvc_T(\tau) - nD\frac{\partial c}{\partial z}\right)\mathrm{d}\tau - \frac{q_c}{H_r}\int_0^t c_T(\tau)\mathrm{d}\tau \qquad (10.4b)$$

Note that as H_r approaches infinity, $c_T = c_0 + c_r t$, and for no increase in mass ($c_r = 0$), c_T becomes constant at a value of c_0.

Equation 10.4b may be used for contaminants which do not experience any first order decay (e.g. conservative contaminants such as chloride). However, some contaminant species may also experience first order biological or radioactive decay, and for these contaminants, equation 10.4b is replaced by

$$c_T(t) = c_0 + c_r t - \frac{1}{H_r}\int_0^t \Big(nvc_T(\tau)$$

$$\begin{bmatrix}\text{Mass of contaminant} \\ \text{in the landfill at time } t \\ m_t = A_0 H_r c_T(t)\end{bmatrix} = \begin{bmatrix}\text{Initial mass} \\ \text{of contaminant} \\ m_{TC} = A_0 H_r c_0\end{bmatrix} + \begin{bmatrix}\text{Increase in} \\ \text{mass deposited} \\ m_{1C} = A_0 H_r c_r t\end{bmatrix} - \begin{bmatrix}\text{Mass which has} \\ \text{passed into the} \\ \text{soil up to time } t \\ A_0\int_0^t f_T(c,\tau)d\tau\end{bmatrix} - \begin{bmatrix}\text{Mass collected} \\ \text{by the leachate} \\ \text{collection system} \\ \text{up to time } t \\ A_0\int_0^t q_c c_T(\tau)d\tau\end{bmatrix}$$

$$\therefore m_t = m_{TC} + m_{1C} - A_0\int_0^t f_T(c,\tau)d\tau - q_c\int_0^t c_T(\tau)d\tau$$

and

$$A_0 H_r c_T(t) = A_0 H_r c_0 + A_0 H_r c_r t - A_0\int_0^t f_T(c,\tau)d\tau - A_0 q_c\int_0^t c_T(\tau)d\tau$$

Where A_0 = Area of landfill through which contaminant can pass into the soil

H_r = 'Reference height of leachate'

$A_0 H_r c_r$ = Increase in mass per unit time due to deposition

$f_T(c,\tau)$ = Surface flux (mass per unit area per unit time) passing into soil

$A_0 q_c$ = Volume of leachate collected per unit time

Thus dividing by the equivalent volume of leachate ($A_0 H_r$) gives

$$c_T(t) = c_0 + c_r t - \frac{1}{H_r}\int_0^t f_T(c,\tau)d\tau - \frac{q_c}{H_r}\int_0^t c_T(\tau)d\tau$$

Figure 10.1 Conservation of mass in a landfill with contaminant inputs and leachate collection.

$$- nD\left.\frac{\partial c}{\partial z}\right|\Big)\,\mathrm{d}\tau - \left(\frac{q_c}{H_r} + \lambda_T\right)\int_0^t c_T(\tau)\mathrm{d}\tau \qquad (10.4c)$$

where λ_T is the first order decay constant, as defined in sections 1.3.6 and 1.5.

It is a relatively simple matter to formulate a mathematical model (e.g. program POLLUTE: Rowe and Booker, 1994) to incorporate equations 10.4 for the surface boundary condition, where c_0, c_r, H_r and q_c are specified boundary parameters and D, n and v are as defined in Chapter 1 (see also Appendix B) and are specified for the layers of the soil deposit.

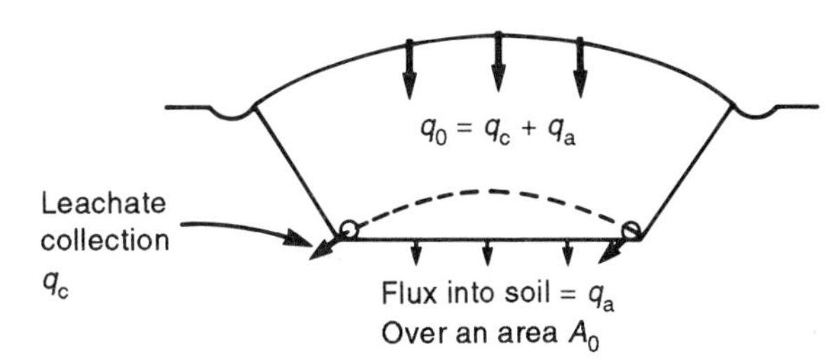

Proportion of contaminant which can pass into the soil	$= q_a/q_0$
Maximum 'initial ' concentration of contaminant species	$= c_0$
Total mass of contaminant species in the waste	$= m_{TC}$
Mass of contaminant likely to be transported into the soil:	$m_0 = m_{TC} \cdot q_a/q_0$
Equivalent volume of leachate = mass/initial/concentration:	$v_0 = m_0/c_0$
Eqivalent height of leachate = (volume)/(area ⊥ flow):	$H_f = v_0/A_0$
i.e.	$H_f = \frac{m_{TC}}{c_0 A_0} \cdot \frac{q_a}{q_0}$

Figure 10.2 Calculation of the equivalent height of leachate for a landfill with leachate collection system.

10.2.2 Equivalent height of leachate, H_f

The equation for source concentration given by equation 10.4 is the most rigorous for the case of a landfill with a leachate collection system. An alternative (approximate) formulation in place of equation 10.4b is given as

$$c_T(t) = c_0 - \frac{1}{H_f}\int_0^t \left(nvc(\tau) - nD\frac{\partial c}{\partial z}\right)\mathrm{d}\tau \qquad (10.5)$$

can also be used where the equivalent height of leachate, H_f, is given by

$$H_f = H_r\frac{q_a}{q_0} \qquad (10.6a)$$

or

$$H_f = \frac{m_{TC}}{c_0 A_0}\,\frac{q_a}{q_0} \qquad (10.6b)$$

Here, H_f can be derived as shown in Figure 10.2, and represents the mass of contaminant available for transport into the hydrogeologic system and excludes the mass that is collected by leachate collectors.

As noted above, q_0 represents the volume of leachate generated within the landfill (per unit area) and can usually be taken to be equal to the infiltration into the landfill due to percolation through the landfill cover. The quantity q_a may be defined as the average mass flux into the barrier normalized (divided) by the average concentration within the landfill, and referred to here as the normalized average flux. For situations where advection is the dominant transport mechanism, q_a is approximately equal to the Darcy velocity, v_a (i.e. $v_a = nv = ki$ where k is the hydraulic conductivity of the barrier and i is the outward hydraulic gradient in the barrier). However, for situations where diffusion is a significant transport mechanism, the determination of the normalized average flux q_a is a little more complicated, but can be readily estimated as outlined below.

The mass flux $f(\tau)$ of contaminant into the barrier beneath the landfill at time τ is given by

$$f(\tau) = nvc_T(\tau) - nD\left(\frac{\partial c(\tau)}{\partial z}\right) \qquad (10.7)$$

where c_T and $(\partial c/\partial z)$ respectively represent the concentration and concentration gradient at the boundary between the landfill and the barrier. Up to some time t, the total mass of contaminant passing into the soil is obtained by integrating equation 10.7. Thus, the average mass flux, f_a, is obtained by dividing the total mass by the period t to give

$$f_a = \int_0^t \frac{f(\tau)}{t}\,d\tau \tag{10.8}$$

The normalized average flux into the barrier over this time period is then obtained by dividing the average mass flux by the average concentration c_a:

$$c_a = \int_0^t \frac{c(\tau)}{t}\,d\tau \tag{10.9}$$

giving

$$q_a = \frac{f_a}{c_a} = \frac{\int_0^t f(\tau)d\tau}{\int_0^t c(\tau)d\tau}$$

$$= \frac{nv\int_0^t c(\tau)d\tau - nD\int_0^t \left(\frac{\partial c}{\partial z}\right)d\tau}{\int_0^t c(\tau)d\tau} \tag{10.10a}$$

and as previously discussed, for the case where D/v approaches zero (i.e. when advection governs), equation 10.10a reduces to

$$q_a = nv = v_a \tag{10.10b}$$

where v_a is the Darcy velocity through the barrier.

Using mathematical modeling (e.g. Program POLLUTE), the contaminant flux, $f(\tau)$, into the clay barrier can be determined for any combination of advection and diffusion. The average mass flux, f_a, into the liner can be automatically determined, and the normalized average flux, q_a, can then be calculated from equation 10.10a.

Although in design cases where POLLUTE is used, one would model the leachate collection explicitly (by specifying H_r and q_c, and the program would use equation 10.4 for the surface boundary condition), it is of interest to examine the impact in terms of the mass available for transport into the hydrogeologic system (using H_f), especially for generic studies such as those reported in this and the following chapters. For these situations, one can readily obtain an estimate of H_f from a simple hand calculation, as described below.

10.2.3 A simple estimate of q_a

Consider a barrier of thickness H, as shown in the insert to Figure 10.3. For the purposes of estimating H_f, it is conservative to determine the normalized average flux, q_a, assuming the following.

1. The concentration in the landfill remains constant.
2. The concentration in the aquifer is zero.

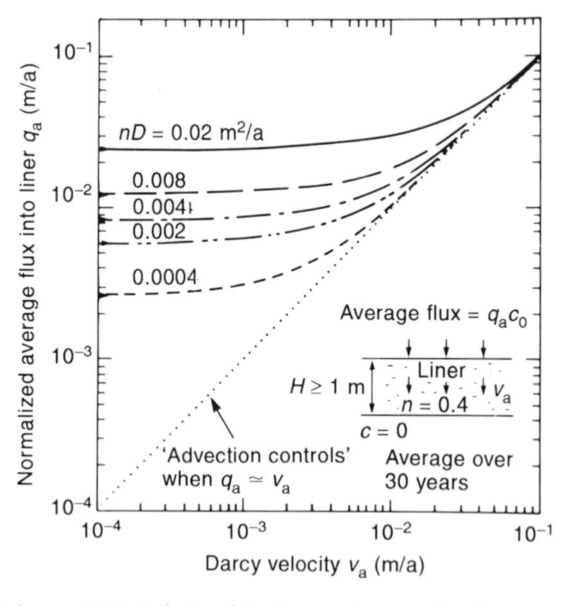

Figure 10.3 Relationship between normalized average flux into a liner and the Darcy velocity for a range of diffusion coefficients. (Modified from Rowe, 1988.)

The results obtained for this case are shown in Figure 10.3 for a barrier of thickness greater than or equal to 1 m. (For thicknesses less than 1 m, this plot may underestimate q_a.) Here, q_a is plotted against the Darcy velocity through the liner for a range of values of the product nD, where n is the porosity of the barrier and D is the effective diffusion coefficient of the contaminant being considered.

The results presented in Figure 10.3 may be conservatively used for situations where the concentration in the leachate decreases with time and/or the concentration beneath the barrier is greater than zero.

Advection controls contaminant transport when the normalized average flux, q_a, is approximately equal to the Darcy velocity, v_a. For typical situations involving clayey barriers this will be the case for Darcy velocities greater than 0.03 m/a. For velocities less than this, diffusion may noticeably affect the normalized average flux. Even if there was no flow ($v_a = 0$), contaminant would still diffuse into the barrier, and hence there is a minimum value of q_a for a given diffusion coefficient, as shown in Figure 10.3.

The results given in Figure 10.3 are based on an average flux over a 30-year period. This is considered to be a reasonable averaging period for many practical situations. The normalized average flux will in fact decrease with increasing time period t. For example, putting aside the case where there is inward flow, the minimum value of the flux will correspond to steady-state diffusion ($v = 0$) and can be calculated from equation 10.10a:

$$q_a \text{ (minimum)} = \frac{nD}{H} \qquad (10.10c)$$

Clearly, if there is flow into the landfill which opposes the outward diffusive flux, the normalized average flux would be even smaller than this (and can be calculated from equation 10.10a).

10.2.4 Example calculation of q_a/q_0

To illustrate the use of Figure 10.3, consider a clayey barrier with a porosity of 0.4 and a contaminant with an effective diffusion coefficient of 0.01 m^2/a (i.e. $nD = 0.004$ m^2/a).

1. If the Darcy velocity into the barrier were 0.03 m^3/a, then from Figure 10.3, the normalized average flux, q_a, would also be approximately 0.032 m^2/a. If the infiltration (percolation) through the landfill cover, q_0, was 0.3 m/a, then the proportion of mass entering the system, q_a/q_0, would be 0.1 (i.e. $m_0 = 0.1 m_{TC}$, or 10% of the total mass is available for transport into the barrier).
2. If the Darcy velocity was only 0.003 m/a, then from Figure 10.3, $q_a \approx 0.01$ m/a (for $nD = 0.004$ m^2/a). Thus the ten-fold decrease in Darcy velocity compared to case 1 has only reduced the normalized flux by a factor of three, and in this case diffusion has a significant influence on contaminant transport into the barrier. Again assuming $q_0 =$ 0.3 m/a, this case would correspond to $m_0 = 0.03 m_{TC}$, or 3% of the total mass being available for transport into the barrier; the remaining 97% is collected by the leachate collection system.

10.2.5 Example calculation of H_r and H_f

Suppose that an examination of the composition of typical municipal waste indicates that chloride represents less than 0.2% of the total mass of the waste in the landfill. For a proposed landfill of area A_0 of 50 ha with a total mass of waste of 2 Mt, this corresponds to a total mass of chloride in the waste $m_{TC} = 0.002 \times 2 \times 10^6$ t $= 4000$ t. Supposing that the peak concentration of chloride $c_0 = 2000$ mg/l = (2000 g/m^3), then from equation 10.3b,

$$H_r = \frac{m_{TC}}{c_0 A_0} = \frac{4000 \times 10^6}{2000 \times 50 \times 10^4} = 4 \text{ (m)}$$

and hence from equation 10.6a,

$$H_f = H_r \frac{q_a}{q_o} \text{(m)}$$

The value of q_a may be evaluated as described above. The infiltration q_0 through the landfill cover must be estimated. It should be noted that when estimating contaminant impact, it is important to be realistic in the estimation of the value of q_0. For example, it is **not** conservative to use a design value of $q_0 = 0.3$ m/a if the realistic infiltration is, say, 0.15 m/a.

Inspection of equation 10.3b or 10.6b shows that, for landfills resting on a barrier, the reference height of leachate and equivalent height of leachate are proportional to the total mass of contaminant, m_{TC}, divided by the plan area, A_0. This represents the mass per unit area, and is directly related to the height of the waste mound. Thus, for a given height of waste, H_r and H_f are the same for a landfill with an area of 50 ha and a total mass of 2 Mt as they are for a landfill with an area of 20 ha and a total mass of 0.8 Mt.

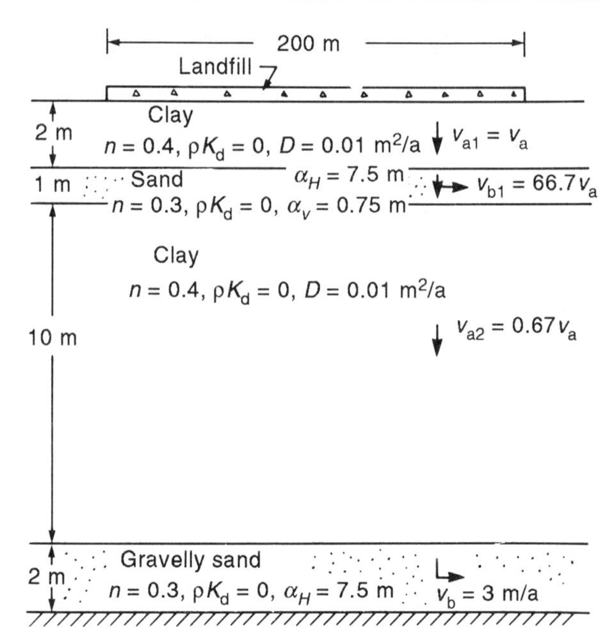

Figure 10.4 Soil profile considered in example A.

10.2.6 Effect of considering the finite mass of contaminant

To illustrate the significance of parameters such as the equivalent height of leachate H_f and the downward Darcy velocity through the barrier, consideration will be given here to the potential impact of a hypothetical landfill on groundwater quality at the site boundary, taken to be 100 m downgradient from the landfill. For this case, denoted 'example A', the assumed hydrostratigraphy is shown in Figure 10.4.

The migration of contaminant was modeled using the two-dimensional finite layer solution to the two-dimensional advection–dispersion equation for a multilayered system described in Chapter 7 (see also Rowe and Booker, 1986, 1987). The input to the model consists of the horizontal and vertical components of the Darcy velocity, the distribution coefficient and the coefficient of hydrodynamic dispersion for each layer, together with the density, porosity and thickness of each layer. In addition, it is necessary to specify the initial concentration of contaminant in the landfill and the equivalent height of leachate (which represents the mass of contaminant available for transport into the soil). It is noted that this model directly considers the variation in concentration with time within the leachate as contaminant is removed from the landfill. The model also considers the mechanism of attenuation due to diffusion of contaminant from the granular units into the clayey aquitard (above and below) as it moves from beneath the landfill towards the boundary of the site. This diffusion from the aquifer will tend to reduce concentrations at the site boundary.

It was shown above that the mass of a given contaminant available for transport into the barrier can be expressed in terms of the equivalent height of leachate, H_f. For the case shown in the inset to Figure 10.5, the concentration of a conservative contaminant in the leachate is

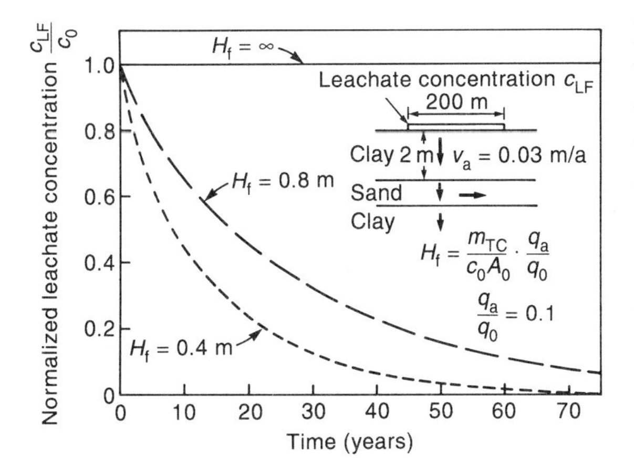

Figure 10.5 Effect of the equivalent height of leachate (H_f) on the variation in leachate concentration with time for example A. (From Rowe, 1988; reproduced with permission of the *Canadian Geotechnical Journal*.)

plotted against time for three different assumed values of H_f and for a downward Darcy velocity, v_a, of 0.03 m/a. If the mass contaminant is infinite ($H_f \rightarrow \infty$), then the concentration of contaminant within the landfill remains constant for all time. Conversely, the assumption of a constant source concentration is equivalent to assuming an infinite mass of contaminants. For the most realistic assumption of a finite mass of contaminant, the calculated concentration in the leachate decreases with time as contaminant is removed from the landfill. The rate of decline is related to the mass of contaminant, and hence H_f. For example, for $H_f = 0.8$ m, the calculated concentration in the landfill has reduced to one-third the original value after 30 years, whereas for $H_f = 0.4$ m it takes about 15 years for the same reduction to occur. It is also apparent from this, that the mass of contaminant can be back-calculated if the variation in concentration has been monitored in the leachate over a sufficiently long period.

The effect of the equivalent height of leachate upon the calculated impact of the landfill or water quality at a point x in the upper aquifer, 100 m downgradient of the landfill, is shown in Figure 10.6. If one assumes that the concentration of contaminant in the leachate remains constant (i.e. $H_f = \infty$), then the calculated concentration at point x increases until it reaches a steady-state value equal to 46% of the source concentration. When one considers a finite mass of contaminant (i.e. finite H_f) the concentration at point x increases to a peak value and then subsequently decreases. The magnitude of the peak and the time at which this occurs depends on the equivalent height of leachate, H_f. The smaller the mass of contaminant available for transport, the smaller is the impact on a downgradient monitoring point. For the case of a downward Darcy velocity of $v_a = 0.03$ m/a and $H_f = 0.4$ m, the peak concentration at point x is approximately 11% of the original source concentration in the leachate, compared with 46% of the source value obtained for $H_f = \infty$. Thus, in this case, consideration of a realistic mass of contaminant reduces the calculated potential impact on groundwater quality by a factor of four, illustrating that the assumption of a constant source

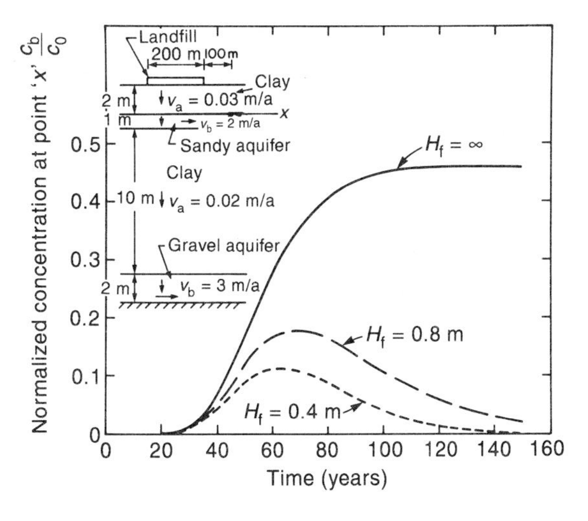

Figure 10.6 Effect of the equivalent height of leachate on the variation in concentration with time at a point within an aquifer at the site boundary for example A. (From Rowe, 1988; reproduced with permission of the *Canadian Geotechnical Journal*.)

concentration may be unrealistically conservative.

Inspection of Figure 10.3 indicates that for the Darcy velocity of 0.03 m/a, contaminant transport through the barrier is being dominated by advection, since $q_a \approx v_a$. This is a situation which might occur if there were a failure of the leachate collection system. In many landfill designs, the Darcy velocity into the barrier will be substantially smaller than 0.03 m/a, particularly while the leachate collection system is functioning. To examine the effect of this Darcy velocity, analyses were also performed for v_a = 0.003 m/a. At this velocity, diffusion dominates contaminant transport and as previously discussed with respect to Figure 10.3, the ten-fold reduction in velocity from v_a = 0.03 m/a to 0.003 m/a only reduces the normalized average flux from about 0.03 m/a to 0.01 m/a. Assuming an infiltration of 0.3 m/a, this corresponds to a three-fold reduction in the proportion of mass available for transport into the barrier (i.e from $q_a/q_0 = 0.1$ to $q_a/q_0 = 0.033$).

The calculated variation in concentration with time at the downgradient monitoring point x is shown in Figure 10.7 for an infinite mass of contaminant ($H_f = \infty \times q_a/q_0 = \infty$) and a finite mass of contaminant $H_f = 4q_a/q_0$ meters for assumed downward Darcy velocities of 0.03 m/a and 0.003 m/a. If one assumes that the concentration in the source remains constant for all time (i.e. $H_f = \infty$), then a ten-fold decrease in Darcy velocity v_a only reduces the peak concentration by about 35% from $0.46c_0$ to $0.3c_0$. However, when one considers the finite mass of contaminant (specifically $H_f = 4q_a/q_0$), then this ten-fold decrease in Darcy velocity gives rise to a more than ten-fold decrease in peak concentration from $0.11c_0$ to $0.01c_0$. The corresponding increase in the time required to reach this peak is from a little over 60 years to about 700 years. For the Darcy velocity of 0.003 m/a considered here, the assumption of a constant source concentration results in an overestimate of the peak concentration by a factor of 30 if the mass of contaminant

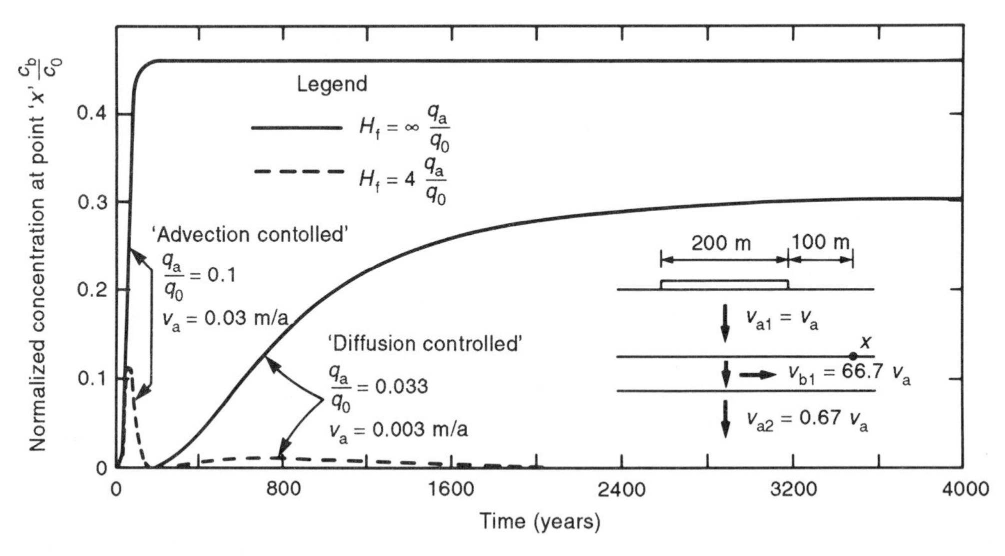

Figure 10.7 Effect of equivalent height of leachate (H_f) on the variation in concentration with time and contaminant impact at a point in the aquifer beneath the site boundary: example A. (From Rowe, 1988; reproduced with permission of the *Canadian Geotechnical Journal.*)

corresponds to $H_f = 4q_a/q_0$ meters (e.g. 4000 t of Cl^- over a site of area 50 ha at an initial source concentration of 2000 mg/l).

It may be concluded that the mass of contaminants is an important parameter to be considered when calculating attenuation of contaminants as they move into the groundwater system.

10.3 Development of a contaminant plume

Proposed landfills are often sited above aquifers. This is particularly common where the geology consists of beds of clayey or silty soil separated by relatively thin granular units which are used for water supply (e.g. many glacial deposits). In these situations it is necessary to evaluate the potential impact of the landfill on water quality in the aquifer(s) at the site boundary.

Because of factors such as the uncertainty associated with defining the groundwater system from limited data, as well as the limitations regarding the adequacy of that data, it is not reasonable to expect that one could make an accurate prediction of the exact time at which contaminant would first migrate offsite, no matter how sophisticated the theoretical model used. However, theoretical models can be particularly useful for examining the implications of different possible scenarios and different key parameters. The results from such a study can then be used in the formulation of an engineering opinion as to the potential for contamination of groundwater at the site boundaries due to a proposed landfill.

In the following sections the effect of varying different key parameters will be examined. These results were obtained for the case of a clayey barrier underlain by a natural aquifer, as shown in Figures 1.2–1.4. However, it should be emphasized again that similar calculations could be performed if there is a drainage system beneath a landfill liner (as shown in Figures 1.5–1.8).

Figures 10.8–10.10 examine various aspects of contaminant migration from a landfill (example B) with a length of 200 m parallel to the direction of flow in the underlying aquifer (insert to Figure 10.8). The Darcy velocity, v_b, in the aquifer is assumed to be 1 m/a. The landfill is separated from the 1 m thick aquifer (which has porosity $n_b = 0.3$) by 2 m of clayey soil with $n = 0.4$; diffusion coefficient $D = 0.01\ m^2/a$; downward advective velocity $v_a \approx 0$ (i.e. zero gradient or a very low hydraulic conductivity composite liner and a small gradient); no sorption for contaminant species of interest, $\rho K_d = 0$.

Figure 10.8 shows the variation in the concentration in the sand layer, c_b, with lateral position, at four times ($t = 100$, 300, 500 and 1000 years). Except at smaller times, the concentration increases approximately linearly with lateral position beneath the landfill and in all cases attains a maximum value at the 'downstream', in the sense of the direction of flow in the base strata, edge of the landfill.

Outside the landfill, the concentration decreases with increasing distance from the landfill. At smaller times this decrease is primarily due to a time lag; however, at larger times it is primarily due to diffusion of contaminant back into the adjacent clay. Thus there is a natural attenuation mechanism in the system that will ensure the maximum concentration reached at any point outside the boundaries of the landfill will never reach the maximum value calculated at the end of the landfill.

The concentration of contaminant within the sandy aquifer beneath the clay can be reduced by increasing the thickness H of the clay. If this is not possible, the concentration can also be controlled by providing a buffer zone between the landfill and the areas where the concentration of contaminant in the groundwater must not exceed a specified level.

To illustrate the effect of the attenuation in a buffering zone, Figure 10.9 shows the variation in contaminant concentration in the sand with

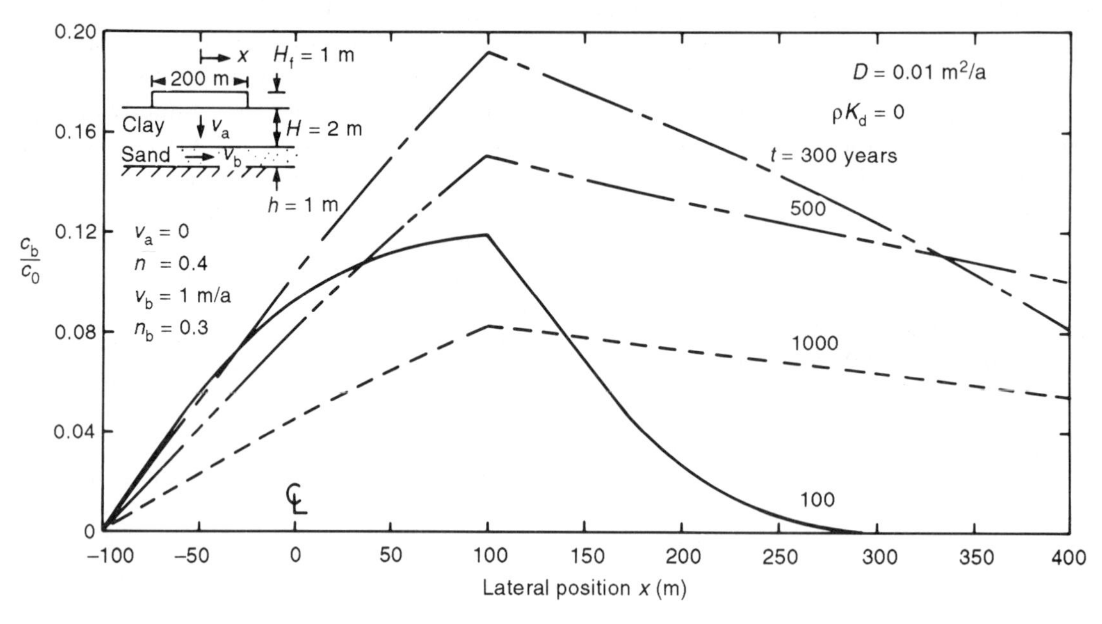

Figure 10.8 Variation in base concentration with position: example B. (Modified from Rowe and Booker, 1985b.)

time at the edge of the landfill (x = 100 m) and at a point 300 m downstream from the landfill (x = 400 m) for two values of ϱK_d. Considering firstly the curves for the case where there is no adsorption ($\varrho K_d = 0$), it is seen that the contaminant concentration at both positions increases with time until a peak value is reached, and then decreases. There is a time lag between the times at which the peak values are reached at x = 100 m and x = 400 m, due to the time required for the contaminant peak to move the 300 m between these two points. This time lag is increased above that which would be expected for purely advective contaminant movement in the sand, because of diffusion from the aquifer into the clay. Of greater interest is the fact that the magnitude of the peak concentration is substantially reduced, even for this analysis, where no horizontal dispersion in the sand layer is considered ($D_H = 0$).

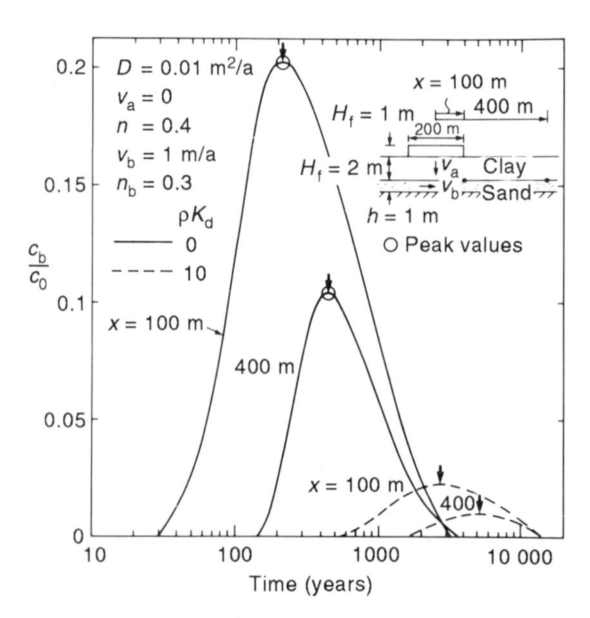

Figure 10.9 Effect of attenuation with distance from x = 100 m to x = 400 m on concentration history in an aquifer: example B. (Modified from Rowe and Booker, 1985b.)

Figures 10.8 and 10.9 have focused on the concentrations within the aquifer, since this is of primary concern in the design of a landfill separated from an aquifer by a clay layer. However, the concentration profiles within the clay can also be determined, as illustrated in Figure 10.10. Beneath the landfill (Figure 10.10(a)), the mass transport is predominantly downwards and the concentrations decrease with depth. For the case examined, the surface concentration within the landfill has reduced to about 25% of the original value after 300 years.

At locations remote from the landfill, the mass transport is also predominantly vertical, but in this case it is upwards from the aquifer into the clay, as shown in Figure 10.10(b).

10.4 Effect of base velocity

The results presented in Figure 10.9 were obtained for a specific value of the advective velocity within the aquifer (v_b = 1 m/a). This parameter is important; it is also difficult to determine in practice. What can be determined is a reasonable estimate of the range in which the velocity is expected to lie. Under these circumstances, finite layer techniques can be easily used to evaluate the effect of this uncertainty upon the expected impact. For example, Figure 10.11 shows the peak concentrations obtained at x = 100 and 400 m for analyses performed for a range of base velocities v_b.

Beneath the edge of the landfill (x = 100 m), the maximum concentration decreases monotonically with increasing base velocity due to the consequent increased dilution of the contaminant in high volumes of water. At points outside the landfill area, there is a critical velocity which

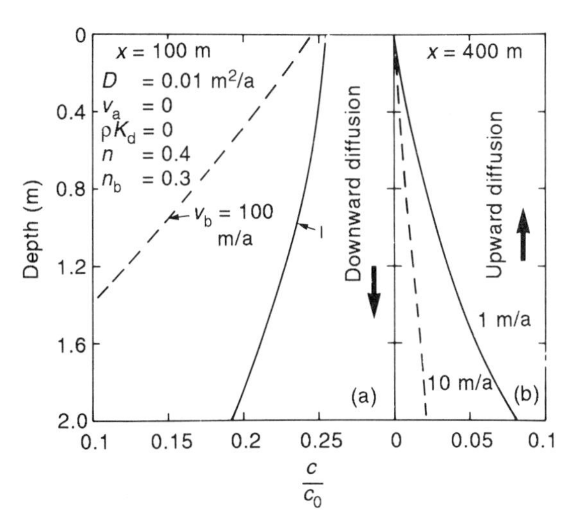

Figure 10.10 Concentration profile in the clay barrier at two positions (t = 300 years), (a) x = 100 m and (b) x = 400 m: example B. (Modified from Rowe and Booker, 1985b.)

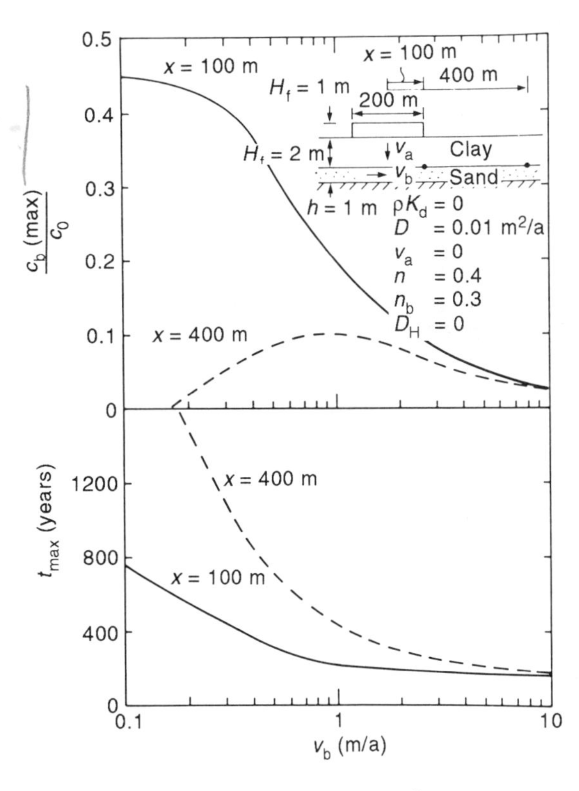

Figure 10.11 Variation in maximum base concentration with base velocity: example B. (From Rowe and Booker, 1985b; reproduced with permission of the *Canadian Geotechnical Journal*.)

gives rise to the greatest maximum concentration. As indicated by Rowe and Booker (1985b), this situation arises because of the interplay of two different attenuation mechanisms. The first of these, diffusion into the surrounding clayey soil, is dependent on the time required to reach the monitoring point. Generally, the lower the velocity v_b, the more time there is for contaminant to diffuse away, and hence the lower the maximum concentration. The second mechanism, dilution, involves decreasing contaminant concentration due to higher volumes of water (i.e. higher v_b) with which contaminant migrating from the landfill can mix.

An important practical consequence of the foregoing is that it is not necessarily conservative to design only for the maximum and minimum expected velocities in the aquifer. In performing sensitivity studies, sufficient analyses should be performed to either determine the critical velocity or, alternatively, to show that the critical velocity does not lie within the practical range of velocities for the case being considered.

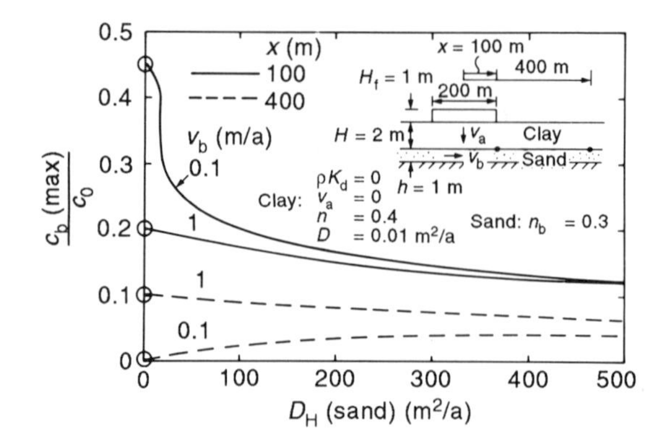

Figure 10.12 Variation in maximum base concentration with horizontal dispersion coefficient in the base layer: example B. (From Rowe and Booker, 1985b; reproduced with permission of the *Canadian Geotechnical Journal*.)

10.5 Effect of horizontal dispersivity in an underlying aquifer

In the analysis of the previous section, it was assumed that $D_H = 0$. Analyses were also performed for a range of values of velocity v_b and coefficient of dispersion D_H. In general, it was found that for horizontal Darcy velocities v_b of 1 m/a or greater, dispersion tended to reduce the maximum base concentration, although the effect was relatively small even for high values of D_H, as shown for $v_b = 1$ m/a in Figure 10.12.

For small advective velocities (i.e. less than 1 m/a), a high dispersion coefficient may have a significant effect on the peak concentration both beneath the landfill and at points outside the landfill. For example, Figure 10.12 shows the variation in the peak concentration in the aquifer with D_H at two points ($x = 100$ m, $x = 400$ m) for a base velocity $v_b = 0.1$ m/a. For these low velocities, lateral dispersion in the aquifer reduces the peak base concentration beneath the edge of the landfill, and the effect is quite significant for values of D_H between 0 and 50 m^2/a. At points remote from the landfill, lateral dispersion in the aquifer can lead to a modest increase in the predicted peak base concentration. This is because for low v_b increasing D_H increases the rate of mass transport through the aquifer and reduces the amount of diffusion that can occur into the adjacent clay layers. The magnitude of D_H encountered in the field is highly variable, although for $v_b = 0.1$ m/a the expected range would be 0.01–20 m^2/a and most probably less than 3 m^2/a. These values could be an order of magnitude higher for $v_b = 1$ m/a.

As might be expected, the peak concentration at a point in the aquifer for $v_b = 0.1$ and 1 m/a tends towards a single value as D_H becomes very large and dominates the effect of advection.

Because of nonhomogeneities present within most aquifers, the coefficient of dispersion in the sand, D_H, is a difficult parameter to determine, and there will always be considerable uncertainty regarding the precise value to use in a design. When considering concentration beneath the landfill, it is conservative to perform the analysis for $D_H = 0$. When considering points outside the landfill, analysis may be performed for $D_H = 0$ and the maximum reasonable value of D_H. For low horizontal velocities, the analysis for the high D_H may be critical; for moderate and high horizontal velocities, the analysis for $D_H = 0$ will be conservative when predicting peak impact.

10.6 Effect of thickness of an aquifer beneath the barrier

The results to be discussed in this and the following section were obtained using a 2-D analysis for the case denoted example C, as shown schematically in Figure 10.13, for parameters given in Table 10.1.

The thickness of the aquifer will have an effect on the concentration of contaminant at various locations along the aquifer. As shown in Figure 10.14, the concentration of contaminant at the edge of the landfill ($x = 100$ m) tends to decrease as the thickness of the aquifer is increased. This is primarily because of an increased dilution of contaminant which occurs for large values of h_1 (all other things being equal) due to the correspondingly higher flow in the aquifer. At points well away from the landfill (e.g. $x = 400$ m) the concentration of contaminant may increase with increasing thickness of aquifer (again, all other things being equal) because the relative diffusion into the adjacent clayey soil is reduced. The results given in Figures 10.11 and 10.14 indicate that there is a fairly complex interaction between the effects of aquifer thickness, advective velocity in the aquifer and the distance to the point of interest, on the maximum concentration expected to occur at that point.

Table 10.1 Parameters relating to example C (Figures 10.13, 10.14 and 10.15)

Layer	Symbol		Value
Landfill	L	(m)	200–1000
	H_f	(m)	1
Upper clay	D	(m^2/a)	0.01
	n		0.4
	v_a	(m/a)	0.0
	H_1	(m)	2.0
Upper sand	D_{H1}	(m^2/a)	10.0
	D_{v1}	(m^2/a)	0.2
	n		0.3
	v_{b1}	(m/a)	1.0
	h_1	(m)	0.5–1.5
Lower clay	D	(m^2/a)	0.01
	n		0.4
	v_a	(m/a)	0.0
	H_2	(m)	10.0
Lower sand	D_{H2}	(m^2/a)	10.0
	D_{v2}	(m^2/a)	0.2
	n		0.3
	v_{b2}	(m/a)	1.0
	h_2	(m)	0.3

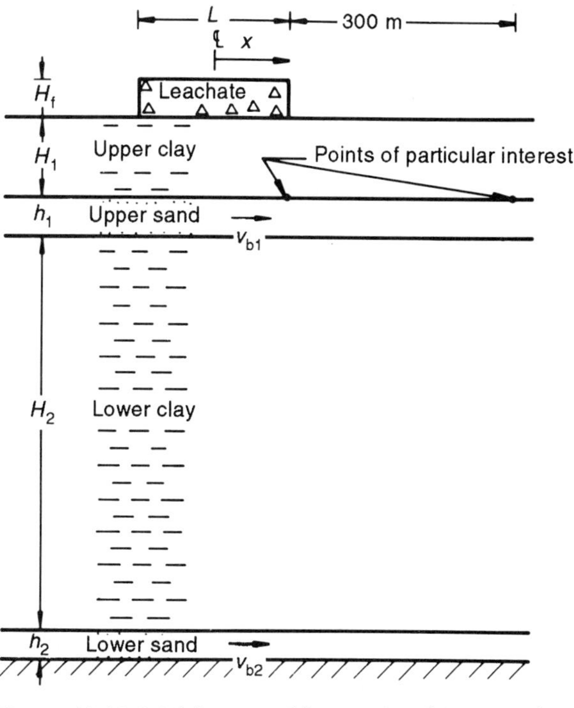

Figure 10.13 Multilayer problem analyzed in example C.

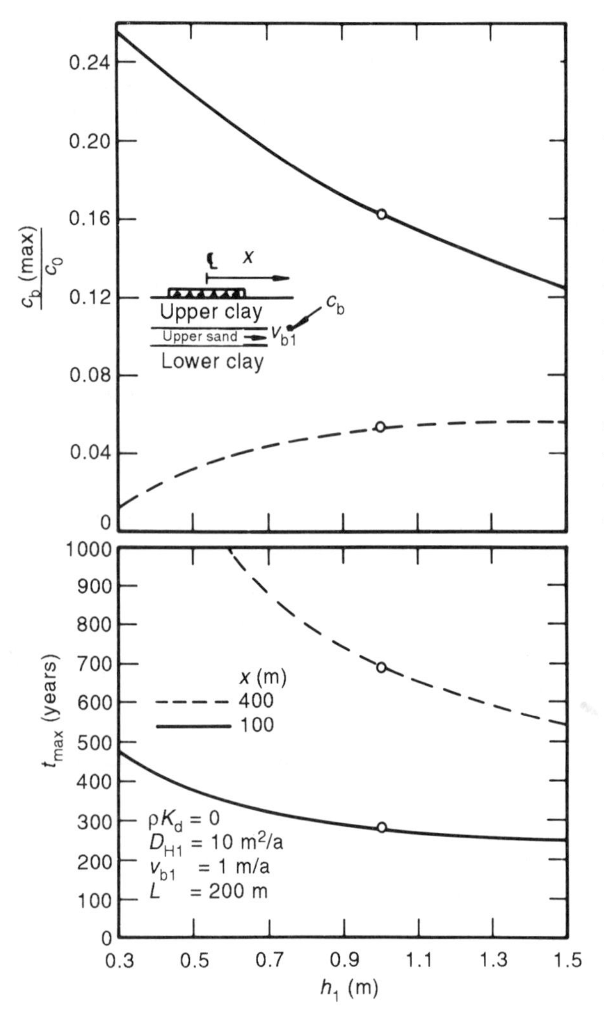

Figure 10.14 Variation in maximum concentration c_{bmax} in the upper sand layer with thickness h_1 of the sand layer: example C. (Modified from Rowe and Booker, 1986.)

10.7 Effect of landfill size on potential impact

All the foregoing results have been for a landfill 200 m long in the direction of groundwater flow (i.e. L = 200 m). For the other basic parameters given in Table 10.1, Figure 10.15 shows the variation in the maximum concentration at the edge and 300 m from the edge of a landfill of variable landfill length L. For the problem considered, increasing the length of the landfill increases the concentration at both points of interest, although the maximum concentration tends to become asymptotic to a constant value for L approaching 1000 m. The increase in concentration with L arises because of the increased mass loading of the aquifer, which arises from a large total mass of contaminant within the landfill. The tendency of the asymptote to reach a constant value for very large L arises because significant diffusion can occur into the underlying clay between the time that the contaminant enters the aquifer near the upstream edge and the time that it approaches the downstream edge, when L is large. The

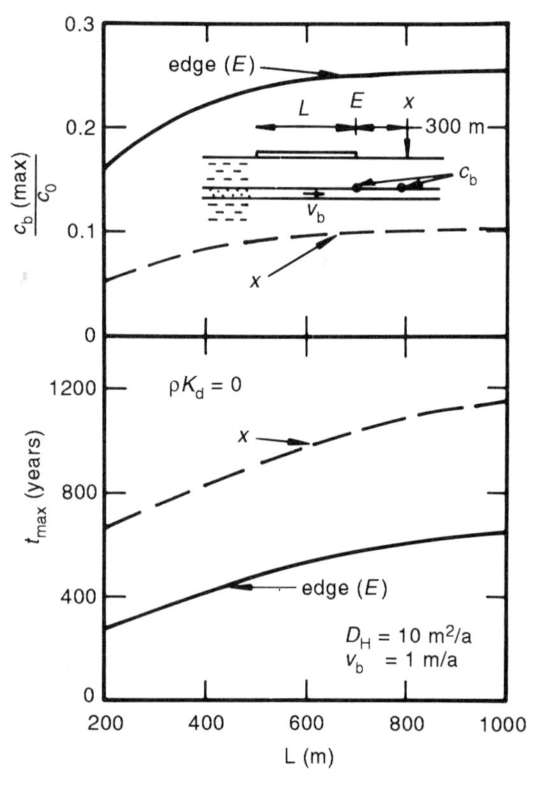

Figure 10.15 Variation in maximum concentration c_{bmax} in the upper sand layer with landfill width L: example C. (Modified from Rowe and Booker, 1986.)

value of L at which this occurs will depend on v_{b1}, h_1, H_f and the other parameters.

10.8 1½-D versus 2-D analysis and modeling of the aquifer beneath a liner

The analyses reported in the previous section were all performed using a full 2-D analysis. For 2-D conditions there are, in fact, two ways in which the aquifer can be modeled.

1. As a boundary condition (section 1.6): this approach allows for spatial variations in concentration within the horizontal plane of the aquifer, as well as advective–dispersive transport within the aquifer itself. Thus this advective–dispersive transport will depend on the horizontal velocity within the aquifer, v_b, and the coefficient of hydrodynamic dispersion in the aquifer, D_H. However, this approach does assume that the concentration in the aquifer is uniform in the vertical direction (i.e. $D_v = \infty$) and that the aquifer is underlain by an impenetrable boundary (i.e zero mass flux across this boundary). This is the approach used to get the results shown in example B (Figures 10.8–10.12).
2. As a physical layer having prescribed velocity components, v_b, and coefficients of hydrodynamic dispersion D_v, D_H: this approach allows for spatial variations in concentration both vertically and horizontally within the aquifer. This approach of treating the aquifer as a physical layer in a manner similar to the clay, but with different parameters, permits us to examine two cases where the aquifer is underlain by:
   (a) an impenetrable boundary (as assumed in method 1 above); or
   (b) an additional layer (or layers) of clay (and/or sand), as was the case in the analysis for example A (Figures 10.4–10.7) and example C (Figures 10.13–10.15).

To illustrate the effect of modeling the aquifer in different ways, a series of 2-D analyses were performed for example D using the parameters given in Table 10.2. Considering firstly case (a) where the aquifer is underlain by an impenetrable boundary, the concentration plume was calculated at t = 300 years using methods 1 and 2 above, as shown in Figure 10.16. Method 1 implicitly assumes that the concentration c_{base} is vertically uniform within the aquifer. Method 2 makes no *a priori* assumption regarding the spatial variation of concentration, and the calculated values at the top (c_{b1}) and bottom (c_{b2}) of the aquifer are both shown in Figure 10.16. For the parameters considered, there is relatively little vertical variation in concentration within the aquifer, and the results obtained by treating the aquifer as a physical layer (method 2) closely bound the results from the computationally simpler approach where the aquifer is treated as a boundary condition (method 1).

Now let us consider case (b) where the aquifer is underlain by an additional 10 m thick layer of clay resting on an impenetrable base. It is found that (Figure 10.16) the concentration plume is almost the same as that obtained for case (a) near the upstream edge of the landfill; but for case (b), diffusion out of the aquifer into the underlying clay gives rise to smaller concentrations near the downstream edge of the landfill and at points outside the landfill ($x \geqslant 100$ m). The results given at the top and bottom of the aquifer indicate that there is a small concentration gradient in the vertical direction within the aquifer.

Analyses similar to those performed to obtain Figure 10.16 were repeated for different times to give the variation in concentration with time. Figure 10.17 summarizes the results for a point beneath the downstream edge of the landfill (x = 100 m) and a point well outside the landfill (x = 400 m). For the sake of clarity, only the concentrations calculated at the top of the aquifer, c_{b1}, are shown. A number of observations

Table 10.2 Parameters relating to example D (Figures 10.16 and 10.17)

Layer	*Quantity*	*Symbol*		*Value*
Clay	Vertical Darcy velocity	v_a	(m/a)	0.0
	Porosity	n		0.4
	Sorption potential	ϱK_d		0.0
	Coefficient of hydrodynamic dispersion (horizontal and vertical)	D	(m^2/a)	0.01
Sand	Horizontal Darcy velocity	v_b	(m/a)	1.0
	Porosity	n_b		0.3
	Sorption potential	ϱK_d		0.0
	Coefficient of hydrodynamic dispersion:			
	Horizontal	D_H	(m^2/a)	1.0
	Vertical (layered case)	D_v	(m^2/a)	0.2
	Vertical (boundary condition)	D_v	(m^2/a)	∞

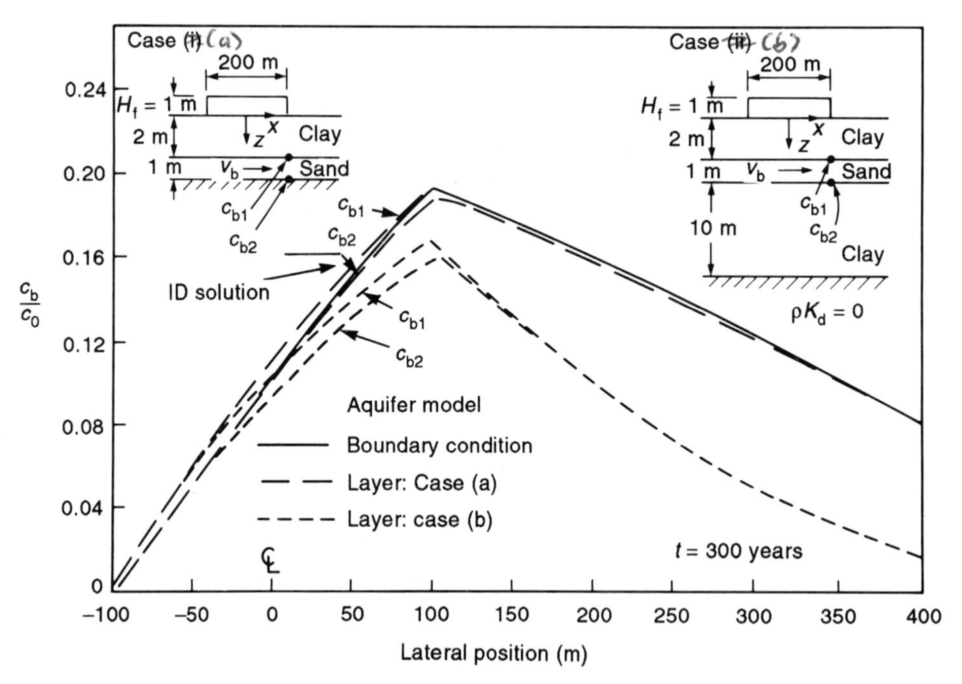

Figure 10.16 Concentration plume in aquifer at 300 years: example D. (Modified from Rowe and Booker, 1987.)

can be made regarding the results from this example.

Firstly, diffusion of contaminant into the adjacent clay gives rise to natural attenuation of contaminant plume as it advances along the aquifer (x > 100 m). Thus the maximum concentration reached at x = 400 m is less than that obtained at the edge of the landfill (x = 100 m) by more than a factor of two.

Secondly, consideration of possible diffusion

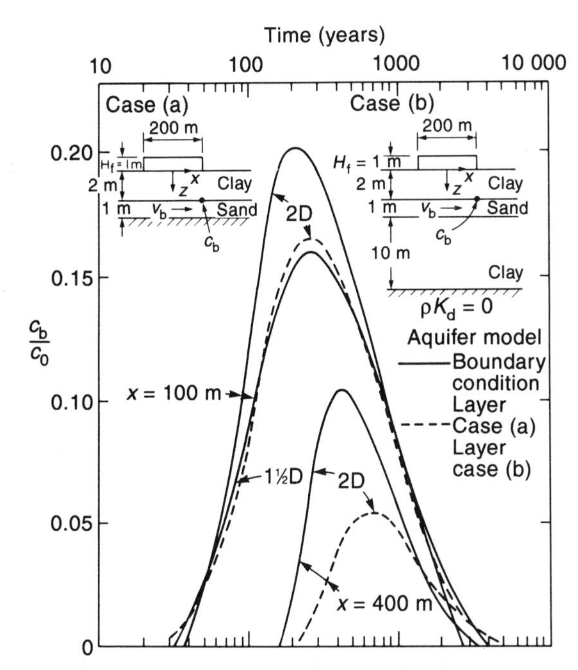

Figure 10.17 Concentration in aquifer beneath edge of landfill ($x = 100$ m) – Effect of ½-D and 2-D model: example D. (Modified from Rowe and Booker, 1987.)

into a clay layer beneath the aquifer (case (b)) gives rise to additional attenuation of contaminant concentration. At the edge of the landfill ($x = 100$ m), diffusion into the lower clay layer reduces the maximum concentration (compared to case (a)) by approximately 20%. The effect increases with distance away from the landfill and at $x = 400$ m diffusion into the lower clay reduces the maximum concentration by a factor of almost two. Thus the modeling of the aquifer as a boundary condition in a 2-D analysis provides a conservative estimate of the contaminant concentrations within the aquifer.

The 1½-D analysis allows one to analyze approximately the 2-D problems discussed above. To make the problem tractable, it is necessary to assume that the concentration within the aquifer directly beneath the landfill is spatially homogeneous at all times, and that the only mechanism for transporting mass out from directly beneath the landfill is by advection in the aquifer. Since this assumption allows one to perform a very simple 1½-D analysis, it is of some interest to compare the results from the 1½-D analysis with those of the more rigorous 2-D analysis.

The results obtained for example D are shown in Figures 10.16 and 10.17. Comparing these results with those obtained at $x = 100$ m from the 2-D analysis indicates that, for this case, the 1½-D approach slightly overestimates the time required to attain the maximum concentration and underestimates the magnitude of the maximum concentration. However, the discrepancy, which is less than 30%, may be acceptable in preliminary calculations.

Figure 10.18 shows the variation in the maximum peak concentration determined from 1½-D and 2-D analyses for a range of base velocities for case (a). It is seen that the 1½-D analysis consistently underestimates the maximum concentration, but that in general the error is quite small. For this example, the maximum error of 30% occurs for base velocities between 0.3 and 0.5 m/a. Similarly, it was found that the 1½-D analysis also gives a reasonable estimate of the time required to attain the peak base concentration beneath the landfill.

10.9 Effect of sorption

As indicated in the previous chapters, sorption can often be represented in terms of the dimensionless product ρK_d. The value of ρK_d will depend upon the soil properties, chemical reactions and their rates, and the range of concentration. Typical values lie between 0 and 100, although much higher values have been reported. As illustrated in Figure 10.9, even a modest amount of adsorption can have a significant effect on the time at which the peak impact occurs. In addition, where there is a finite mass of contaminant, sorption (e.g.

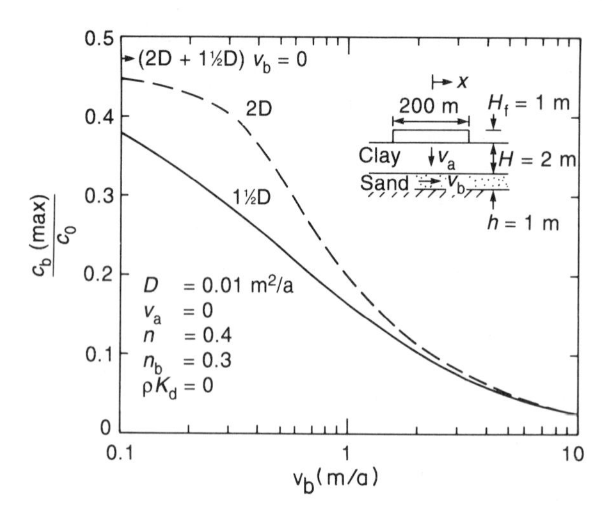

Figure 10.18 Comparison of peak concentration calculated from the 1½-D and 2-D analyses for a range of base velocities. (Modified from Rowe and Booker, 1985b.)

ϱK_d = 10 in Figure 10.9) can also significantly reduce the magnitude of peak impact. The effect of sorption on the magnitude of peak impact gets smaller as the mass of contaminant (i.e. H_f) increases (e.g. Rowe and Booker, 1985a).

10.10 Effect of liner thickness

Having established that the 1½-D approach can give a reasonable indication of contaminant impact in the aquifer, the following results were obtained using a 1½-D analysis.

In design, the thickness and known attenuation potential (ϱK_d) of the clay layer isolating the landfill from the underlying groundwater system may be used to control the maximum base concentration. Figure 10.19 shows the variation in maximum base concentration $c_{b(max)}$ with clay liner thickness H for $\varrho K_d = 0$ and $\varrho K_d = 10$. Increasing the layer thickness substantially reduces the maximum concentration ever reached at the base and increases the time required to reach this maximum. For example, Figure 10.19 shows that with $\varrho K_d =$ 0, increasing the clay liner thickness from 0.5 m to 4 m decreases the $c_{b(max)}$ by up to an order of magnitude and increases the time t_{max} by up to an order of magnitude.

The effect of layer thickness is increased by consideration of the height of leachate and the geochemical reaction. Thus, the thickness of liner required to ensure that a specified maximum base concentration is never exceeded may be significantly reduced by considering the interaction of all of these factors.

10.11 Effect of Darcy velocity in the barrier and design for negligible impact

The Darcy velocity, v_a, within the clayey barrier can be calculated, knowing the hydraulic conductivity of the barrier and the difference between the head in the landfill and in the underlying aquifer. If the head in the landfill is greater than in the aquifer, the velocity v_a is from the landfill towards the aquifer. If the head in the landfill is less than in the aquifer, there will be upward flow into the landfill.

Although seepage may be minimized, it will rarely be precisely zero, and it is of interest to examine the effect of the Darcy velocity v_a through the liner upon the maximum base concentration. Analyses were performed for a range of Darcy velocities, and the general effects of base velocity, layer thickness, equivalent heights of leachate and geochemical reaction discussed in the previous sections for $v_a = 0$ were similar for $v_a > 0$. However, the Darcy velocity did increase the magnitude of the maximum base concentration, and decrease the time required to reach this concentration, as shown in Figure 10.20 for a 2 m thick liner with $\varrho K_d = 0$. The results for other values of ϱK_d are very similar, except that the values of c_b and t_{max} differ in approximately the same ratio as they do for $v_a = 0$.

From Figure 10.20, it can be seen that for the range of parameters considered, the Darcy

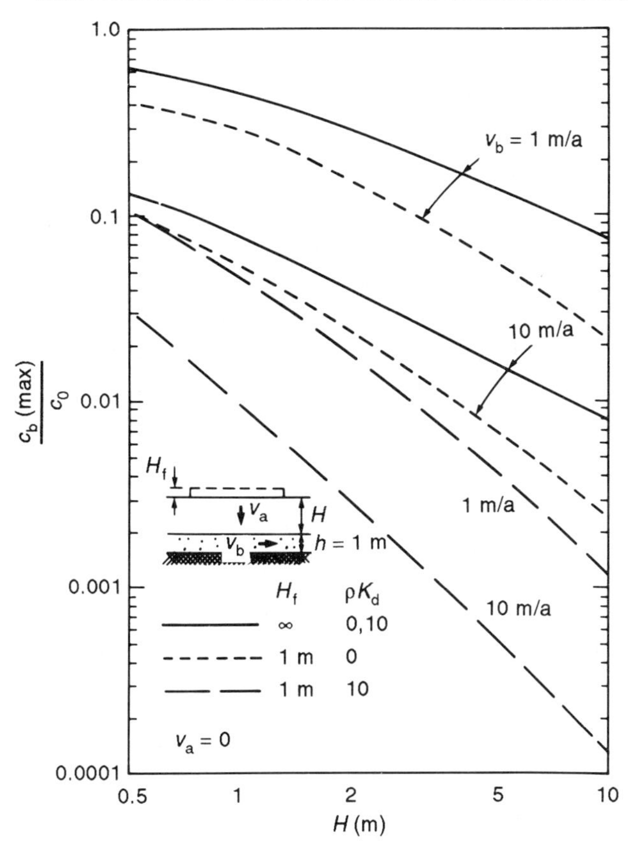

Figure 10.19 Effect of layer thickness H on maximum base concentration. (Modified from Rowe and Booker, 1985a).

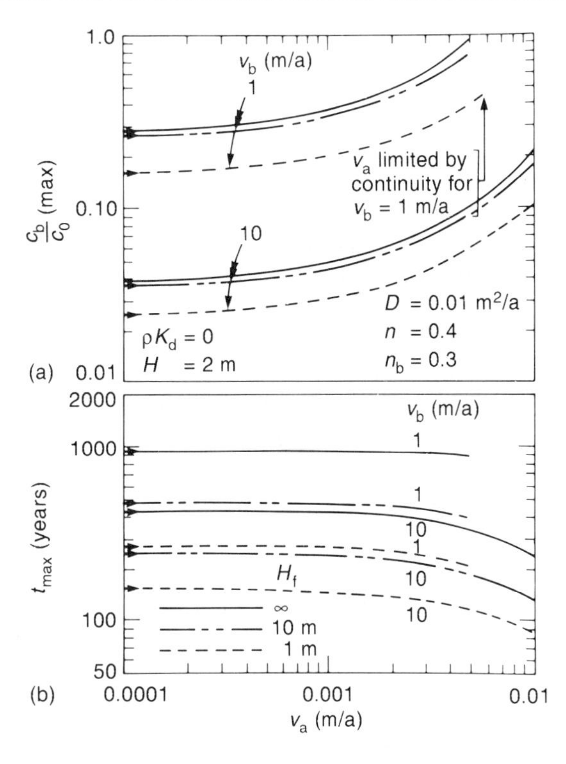

Figure 10.20 Effect of downward Darcy velocity on: (a) maximum base concentration; (b) time to reach maximum concentration. (Modified from Rowe and Booker, 1985a.)

velocity could be neglected for $v_a < 0.0004$ m/a. For v_a up to 0.001 m/a, advection increases the base concentration by up to 33%, although t_{max} is almost constant. Darcy velocities greater than 0.001 m/a may significantly affect the base concentration, with the values of $v_a = 0.01$ m/a being up to five times greater than those for $v_a = 0$. The time t_{max} is somewhat less affected by v_a, varying by less than a factor of two for the same range in Darcy velocities. This demonstrates the importance of minimizing the Darcy velocity in the design of the waste disposal site. For a clayey liner with a hydraulic conductivity of 10^{-9} m/s and a unit gradient, the Darcy velocity would be about 0.03 m/a.

The foregoing discussion has considered outward flow from the landfill. As noted earlier, contaminant migration from a landfill can be minimized if the landfill is designed so that groundwater flow is into the landfill. However, the fact that there is inward flow does not necessarily mean that there will be no contaminant migration out of the barrier. The assessment of potential impact on groundwater quality involves the following two stages.

1. Determine the Darcy velocity into the landfill.
2. Determine the diffusive movement out of the barrier.

The construction of a landfill will usually change the hydraulic characteristics of a given site. For example, a functioning leachate collection system may give rise to an average head in the leachate which is below the original groundwater level, thereby inducing inward flow at the sides of the landfill and, if the head is lowered sufficiently, at the base of the landfill. On the other hand, if significant leachate mounding were to occur due to failure of the collection system, then the head in the landfill will increase, and may exceed the head outside the barrier, thereby reversing the flow direction, and resulting in outward flow. In both cases, the mounding of the leachate must be calculated as previously discussed, and the hydraulic gradient determined for the new flow system – this will require a seepage analysis.

Having determined the advective velocities, contaminant transport modeling may then be used to assess the potential impact of contaminants moving through the barrier. If the flow is outward, then there can be no hydraulic trap, and the problem is similar to that discussed in the previous section. If the flow is inward, then analysis may be used to estimate the rate of movement and maximum extent of the contaminant movement.

If it is assumed that the concentration of contaminant within the landfill remains constant at c_0 and that the base of the barrier is flushed by flowing water, such that the concentration is zero, then the steady-state solution is given by (Al-Niami and Rushton, 1977)

$$\frac{c}{c_0} = \frac{\exp[-v(H-z)/D] - 1}{\exp(-vH/D) - 1} \tag{10.11}$$

where z is the distance from the top of the liner (positive for movement into the liner); H is the thickness of the liner; v is the groundwater velocity, and is positive for outward flow and negative for inward velocity (i.e. for the case of interest here); D is the coefficient of hydrodynamic dispersion which, for the case of inward flow, may be taken to be equal to the effective diffusion coefficient. Hence the normalized flux into the aquifer ($z = H$) is given by

$$\frac{f}{c_0} = \frac{-nv}{\exp(-vH/D) - 1} \tag{10.12}$$

where n is the porosity. Equation 10.11 may be useful for estimating the maximum extent of the contaminant plume. However, this equation **assumes** zero concentration at the base of the barrier, and hence the calculation of a small concentration near the base does not necessarily imply that the impact on an underlying aquifer would be small. For example, Figure 10.21 shows the calculated steady-state contaminant flux passing into an aquifer beneath a 1 m thick clayey barrier for a range of inward Darcy velocities and assumed diffusion coefficients.

The flux f has been divided by the initial source concentration, and the resulting normalized flux f/c_0 has units of velocity. The effective diffusion coefficient for many contaminants lies in the range $D = 0.01$–$0.02\ \mathrm{m^2/a}$ (Chapter 8). For these conditions, Figure 10.21 shows that the inward Darcy velocity would have to exceed 0.025 and 0.05 m/a for $D = 0.01$ and $0.02\ \mathrm{m^2/a}$ respectively before the outward flux was reduced to negligible levels for a 1 m thick liner.

As a rule of thumb, the impact on an underlying aquifer is likely to be negligible if the

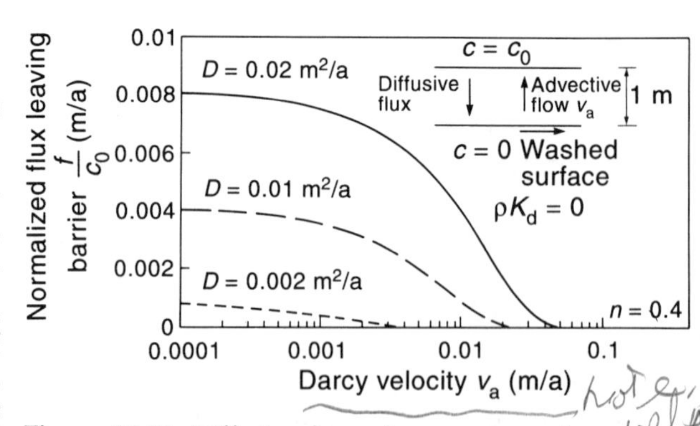

Figure 10.21 Diffusive flux of contaminant into an aquifer against an inward flow at a Darcy velocity v_a (steady-state conditions). (Modified from Rowe, 1988.)

concentration calculated from equation 10.11 is negligible for depths $z \approx 0.9H$. If the concentration is not negligible for $z \approx 0.9H$, then it is necessary to assess the potential impact of the proposed landfill upon groundwater quality using a more realistic model which explicitly considers the aquifer and the finite mass of contaminant; this can be easily done using finite layer techniques. If the calculated impact is unacceptable, then the barrier thickness must be increased and/or the inward Darcy velocity must be increased.

Chapter eleven

Migration in fractured media: analysis and design considerations

11.1 Introduction

Fractured porous media are frequently encountered adjacent to present or proposed waste disposal facilities. These fractured media may, for example, take the form of a fractured clay or till (e.g. with blocky fractures in the upper weathered zone and with predominantly vertical fractures in the unweathered clay or till), or a fractured rock (e.g. mudstone, siltstone, sandstone) which has significant horizontal and vertical fracturing. Typically, fractured media will involve a series of fractures separated by blocks of intact material which will be referred to as the matrix of the fractured soil or rock.

The primary transport mechanism for dissolved contaminants in fractured media is usually advective–dispersive transport along the fractures. In these systems, the average linearized groundwater velocity in the fractures, v_f, is related to the Darcy velocity v_a by the relationship

$$v_f = v_a/n_b$$

where n_b is the effective porosity through which flow is occurring. Thus, the fracture velocity v_f may be quite high because the fracture porosity of a fractured mass n_b is usually quite small, with a typical range of between 0.1 and 0.001% based on hydraulic conductivity–fracture opening size–fracture frequency relationships (e.g. Hoek and Bray, 1981). If conservative contaminants were to migrate along these fractures at the average linear groundwater velocity, v_f, then the rate of contaminant migration through the rock would be very fast. For example, the Burlington landfill is located in fractured Queenston shale in southern Ontario, Canada. In the environs of the landfill, the Queenston shale has a typical fracture spacing between 0.05 m and 0.35 m (Hewetson, 1985). The Burlington landfill has been generating leachate for more than 15 years. If contaminant moved at the speed of the average linear groundwater velocity, and if, for example, the velocity were 50 m/a, then 15 years after leachate entered the rock, contaminants should be readily detected at distances of up to 750 m downgradient of the landfill. However, the available field data (e.g. the Burlington Landfill Plume Delineation Study,

Gartner Lee Ltd., 1988) suggests that after 15 years this was not the case.

While it may be true that migration of conservative contaminants could occur at the rate close to that of the average linear groundwater velocity in porous media where the matrix of the media (e.g. the sand grains in a sandy aquifer) had a negligible effective porosity, it is not true that conservative contaminants (e.g. chloride) will migrate at the rate of the average linear groundwater velocity in fractured rock or soil systems where the matrix of the rock or soil between fractures has a significant effective porosity (e.g. Grisak and Pickens, 1980). This is the case at the Burlington landfill, where the Queenston shale has a matrix porosity of approximately 10%. In cases such as this, diffusion of contaminants from fractures into the adjacent matrix can control the migration of contaminants and represents an important attenuation mechanism. This phenomenon is known as matrix diffusion. The fact that diffusion can occur into or out of this type of rock was demonstrated in section 8.3.4, and for the Queenston shale the effective diffusion coefficient was found to be about 1.5×10^{-10} m^2/s for chloride.

The phenomenon of attenuation due to matrix diffusion is well-recognized (e.g. Barker and Foster, 1981; Foster, 1975; Freeze and Cherry, 1979; Grisak and Pickens, 1980; Grisak, Pickens and Cherry, 1980; Sudicky and Frind, 1982; Tang, Frind and Sudicky, 1981); Gillham and Cherry (1982) have reviewed a number of cases where matrix diffusion has been shown to decrease concentrations of species moving along fractures. For example, Foster (1975) showed that the diffusion of tritium from flowing groundwater in the fractures of porous chalk (into the pore water of the porous rock matrix) could account for a rapid decrease in tritium concentration in the fractures. Similarly, Day (1977) used this concept of matrix diffusion to account for a rapid decline in tritium concentration with depth in a fractured clay in the Winnipeg area. Finally, Grisak, Pickens and Cherry (1980) have also demonstrated that a model that considers matrix diffusion can give relatively good agreement between theoretical simulations and laboratory experiments in which a tracer solution was passed through a large column of fractured, clayey glacial till. From these studies, Gillham and Cherry (1982) concluded that

> molecular diffusion can exert a major influence on the rates and patterns of migration of contaminants in fractured argillaceous deposits. Diffusive loss of contaminants from paths of active flow in the fractures to the matrix can be a dominant mechanism of attenuation.

Matrix diffusion combined with sorption processes will result in even greater attenuation for reactive contaminants than for conservative contaminants. Some reactive contaminants may be removed from free solution by cation exchange, either at the surface of the fracture, or within the adjacent soil or rock. Organics may also be removed by preferential partitioning of contaminants on organic matter. In principle, partitioning could occur, both at the face of the fracture and within the matrix; however, relatively little is known about the effects of partitioning at the face, and it would be prudent and conservative to restrict consideration to partitioning that occurs within the soil or rock matrix.

Finite element techniques provide one means of modeling contaminant migration in fractured systems (e.g. Grisak and Pickens, 1980; Huyakorn, Lester and Mercer, 1983). These approaches potentially make it possible to analyze quite complicated two- and three-dimensional fracture networks, and may be useful when there are detailed data available concerning the distribution and characteristics of the fracture system. Frequently, however, these data are not available, and it is necessary to assess potential impact based on a knowledge of typical fracture spacings and orientations, together with some knowledge of the hydraulic gradient and hydraulic conductivity in the system. Under these circumstances, analytical or semi-analytical solutions for contaminant transport in a fractured medium may be particularly useful for quickly assessing the potential effects of uncertainty regarding key parameters like fracture spacing, fracture opening size, etc. These techniques may also be useful for benchmarking more complex numerical procedures (e.g. finite element codes).

Various investigators have developed analytical or semi-analytical solutions for contaminant transport in idealized fracture media. For example, Neretnieks (1980) and Tang, Frind and Sudicky (1981) developed a solution for one-dimensional (1-D) contaminant transport along a single fracture together with 1-D diffusion of contaminant into the matrix of the rock adjacent to the fracture. Sudicky and Frind (1982) and Barker (1982) extended this approach to consider the case of multiple parallel fractures. The aforementioned researchers all considered constant concentration at the inlet of the fractures. Moreno and Rasmuson (1986) developed an analytical equation for a constant flux boundary condition at the inlet of a single fracture.

Rowe and Booker (1988a, 1989, 1990a,b, 1991a,b) considered the situation that is encountered with landfills wherein there is a finite mass of contaminant available for transport into the groundwater systems. They produced solutions for 1-D, 2-D and 3-D fractured systems, while allowing the concentration of contaminant in the source (landfill) to decrease with time as mass is transported into the fracture network. The theory behind this solution was presented in Chapter 7. The objective of this chapter is to examine the implications of modeling both matrix diffusion and the finite mass of contaminant within the landfill.

For the purposes of illustrating the importance of various parameters, the results presented in this chapter will generally be taken as those applicable to lateral migration in a fractured shale deposit in section 11.3 and vertical migration in fractured clay or till in section 11.4. It should, however, be noted that the same issues as discussed in section 11.3 arise when considering lateral migration in fractured clay (e.g. out from the edge of a landfill), and similarly the issues raised with respect to predominantly vertical migration through fractured clay/till in section 11.4 could equally well apply to vertical migration through fractured porous rock. A more detailed discussion of the issues examined in section 11.3 can be found in Rowe and Booker (1989, 1990b). A detailed discussion of the issues raised in section 11.4 is to be found in Rowe and Booker (1990a, 1991a,b).

11.2 Numerical considerations

The analysis used in this chapter was described in Chapter 7, and the evaluation of the parameter $\bar{\eta}$ which controls matrix diffusion from the fractures is given in Appendix C as follows. For 1-D conditions,

$$\bar{\eta} = n_m R_m \left(1 - 2\sum_{j}^{\infty} \frac{s}{[s + (D_m/R_m)\alpha_j^2]} \times \frac{1}{(\alpha_j H_1)^2}\right) \tag{11.1a}$$

or

$$\bar{\eta} = n_m R_m \frac{\tanh \mu H_1}{\mu H_1} \tag{11.1b}$$

where $\mu^2 = R_m s/D_m$. For 2-D conditions,

$$\bar{\eta} = n_m R_m \left(1 - 4\sum_{j,k}^{\infty} \frac{s}{s + (D_m/R_m)(\alpha_j^2 + \beta_k^2)} \times \frac{1}{(\alpha_j H_1)^2} \frac{1}{(\beta_k H_2)^2}\right) \tag{11.2}$$

For 3-D conditions,

$$\bar{\eta} = n_m R_m \left(1 - 8\sum_{j,k,l}^{\infty} \times \frac{s}{s + (D_m/R_m)(\alpha_j^2 + \beta_k^2 + \gamma_l^2)} \frac{1}{(\alpha_j H_1)^2} \times \frac{1}{(\beta_k H_2)^2} \frac{1}{(\gamma_l H_3)^2}\right) \quad (11.3)$$

where

$$\alpha_j = (j - ½)\pi/H_1,\ j = 1, 2, \ldots, \infty$$
$$\beta_k = (k - ½)\pi/H_2,\ k = 1, 2, \ldots, \infty$$
$$\gamma_l = (l - ½)\pi/H_3,\ l = 1, 2, \ldots, \infty$$

and the definitions of H_1, H_2, H_3, h_1, h_2, h_3 are as shown in Figure 11.1, D_m, n_m and R_m are the diffusion coefficient, porosity and retardation coefficients for the matrix material between the fractures and s is the Laplace transform parameter (Chapter 7).

These equations each contain infinite sums (Σ_j^{∞} etc.) in which the spacing between fractures plays an important role. Intuitively, one would expect the 3-D and 2-D solutions to reduce to the 1-D solution as the spacing in the Y- and Z-directions (i.e. $2H_2$ and $2H_3$) is increased. Indeed, this is so, and for the case examined in Figure 11.2 the 2-D and 3-D solutions tend to approach the 1-D solution (to less than 1.5% difference in concentration) as the spacings $2H_2$ and $2H_3$ are increased up to about 1 m. Further increasing the spacing from 1 m to 10 m results in further slow convergence to the 1-D solution.

Inspection of equations 11.2 and 11.3 suggests that although the 2-D and 3-D solutions will reduce to the 1-D case (e.g. equation 11.1a), as noted above, the number of terms in the series required to obtain a given accuracy will also increase for increasing fracture spacing. Generally, good solutions can be obtained using only eight to ten terms in each series. Nevertheless, Figure 11.3 shows that the number of terms required to reach a high level of accuracy (i.e. better than 0.01%) can become quite large for widely spaced fractures. It can be seen that for spacings of $2H_2 \leqslant 1$ m (i.e. where 2-D effects are the greatest), less than 30 terms are required to attain a high accuracy. For wider spacings (i.e. as the 2-D solution reduces to the 1-D solution, as shown in Figure 11.2), the number of terms in the series increases rapidly. This generally does not pose a practical problem, for the following reasons.

1. The computation is still quite fast even on a microcomputer;
2. It is generally not necessary to evaluate the 2-D or 3-D solution for fracture spacings $2H_2$ or $2H_3$ exceeding ten times $2H_1$, since the corresponding 1-D or 2-D solutions can be used instead.

Obviously, one can program each of equations 11.1, 11.2 and 11.3 and select the equation considered most appropriate for a given problem.

In general, relatively few terms are required in the infinite series and in the case of the 1-D solution, the summation can be avoided altogether by defining $\bar{\eta}$ as given in equation 11.1b rather than 11.1a.

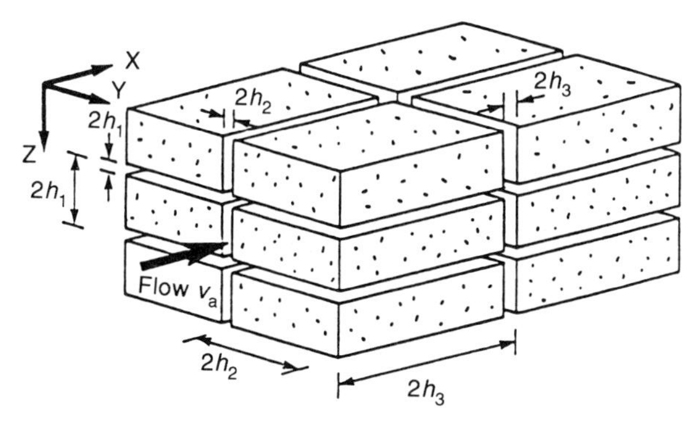

Figure 11.1 Definition of fracture geometry. (After Rowe and Booker, 1989; reproduced with permission of the *International Journal for Numerical and Analytical Methods in Geomechanics*.)

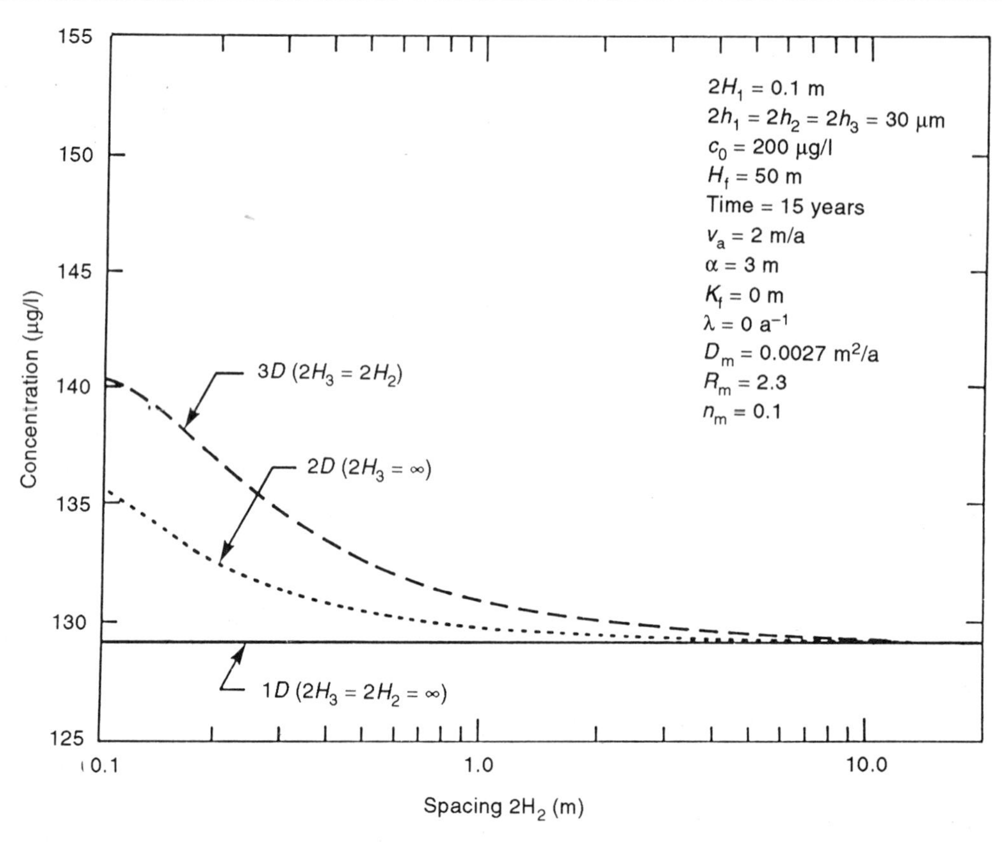

Figure 11.2 Variation in peak concentration at a point 100 m from the source as the spacing between secondary fractures ($2H_2$ and $2H_3$) is varied. (After Rowe and Booker, 1990b; reproduced with permission of the *International Journal for Numerical and Analytical Methods in Geomechanics.*)

11.3 Lateral migration through fractured media

11.3.1 Evolution of a contaminant plume: the role of matrix diffusion

To illustrate the development of a contaminant plume under 1-D and 3-D conditions, consider a slightly reactive contaminant (e.g. 1,1-dichloroethane) migrating through fractured Queenston shale. The diffusion coefficient, matrix (intact shale) porosity and retardation coefficient are estimated to be $D_m = 0.0027$ m^2/a; $n_m = 0.1$; and $R_m = (1 + \varrho K_m/n_m) = 2.3$, respectively. The dispersivity is assumed to be $\alpha = 3$ m, and the equivalent height of leachate, H_f, is taken to be a rather high value of 50 m. The initial concentration of 1,1-dichloroethane is assumed to be 200 μg/l.

The reference case involves fractured shale with an assumed hydraulic conductivity of 2×10^{-6} m/s and a lateral gradient of 0.032, giving a Darcy velocity of 2 m/a. Based on the hydraulic conductivity of 2×10^{-6} m/s, the fracture opening size was estimated for a single set of parallel fractures (based on Hoek and Bray, 1981) to be approximately 30 μm, 60 μm and 130 μm for fracture spacings of 0.01, 0.1 and 1 m respectively. These correspond to fracture porosities $n_f = 0.003$, 0.0006 and 0.00013, and

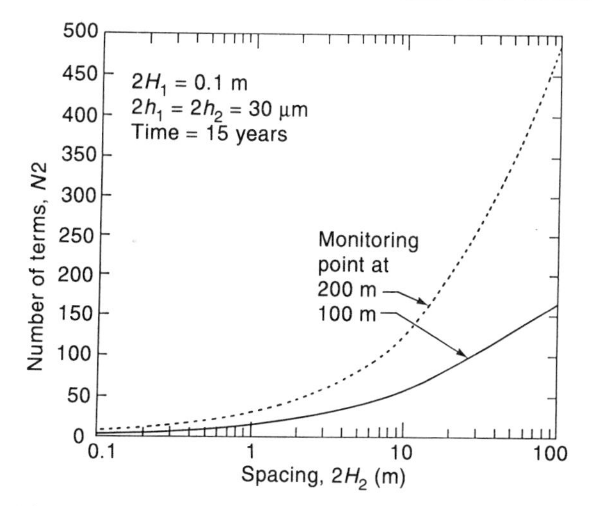

Figure 11.3 Number of terms in series required to obtain an accuracy of 0.01%. (After Rowe and Booker, 1990b; reproduced with permission of the *International Journal for Numerical and Analytical Methods in Geomechanics*.)

groundwater velocities of 667 m/a, 3333 m/a and 15 385 m/a for these three spacings.

Figures 11.4 and 11.5 show the evolution of a contaminant plume for a 1-D and 3-D fracture system respectively. In both cases, the plume is plotted at four different times. Figure 11.6 shows the variation in concentration, with time, at a monitoring point 100 m from the source for the two cases.

Firstly, inspecting Figure 11.4, it is evident that at small times (e.g. 3.9 years), the contaminant plume is restricted to a zone extending about 1000 m downgradient of the landfill. Inspecting Figure 11.6, it can be seen that the concentration at a monitoring point 100 m downgradient increases with time, until it reaches a peak value after 3.9 years. The concentration at this point then decreases for subsequent times as contaminant is flushed out of the landfill and also from the fractured rock near the landfill. This evolution of the plume is evident from Figure 11.4, where it can be seen that after 8.4 years the plume extends beyond 1000 m, and the concentration at locations between the source and the 100 m 'monitoring' point has continuously dropped below the values that were present at 3.9 years. Between points at 100 m and 160 m the concentration has also decreased below the values present at 3.9 years. Further than 160 m from the landfill there has been an increase in concentration between 3.9 and 8.4 years; the concentration at 250 m from the source reaches its peak after 8.4 years and decreases for subsequent times (e.g. plumes for 16 and 24 years). This peak is less than the peak which occurred at 100 m from the source. Thus it can be seen that as the plume spreads out, there is a decrease in the peak concentration which is attained at points away from the landfill. The peak values at points 100 m, 250 m, 500 m and 750 m are indicated in Figure 11.4 together with an envelope of peak concentration at all points up to 1000 m downgradient of the landfill. There is significant attenuation of contaminant with distance from the source and for this combination of parameters, the peak impact at 1000 m is less than 20% of the initial concentration in the leachate.

The results presented in Figures 11.4–11.6 also serve as a warning that one cannot infer the rate of movement of contaminant from isolated observations of the contaminant plume. To illustrate this, consider the rate of advance of a point where $c/c_0 = 0.5$, which is often taken as an indication of the speed of a contaminant plume. If observations were made after 3.9 years, the point where $c/c_0 = 0.5$ would be about 210 m from the source, implying a velocity of about 53 m/a. After 8.4 years, the location at which $c/c_0 = 0.5$ has moved to 250 m, implying a velocity of about 30 m/a. Thus, the velocity of the plume (defined in terms of the movement of a point where $c/c_0 = 0.5$) is not a constant. Rather, it decreases with time. Furthermore, examination of Figure 11.6 shows that when the mass of contaminant is finite, there will be up to two times at which $c/c_0 = 0.5$ at a given point. For 1-D conditions it takes less than 1 year for c/c_0 to reach 0.5 at 100 m from

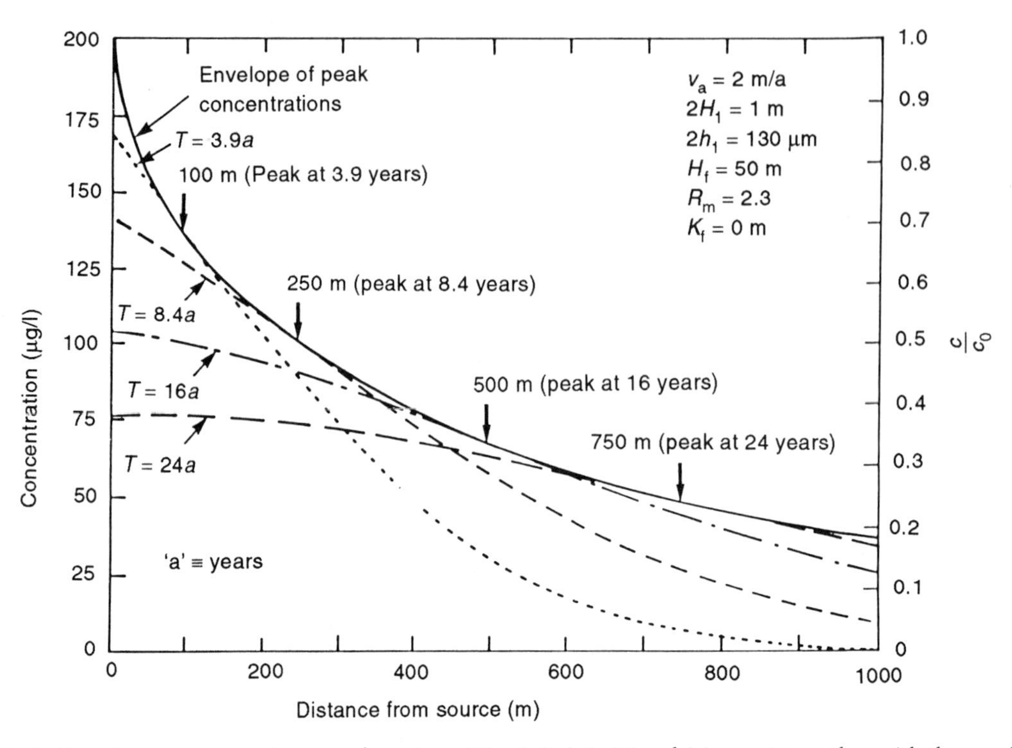

Figure 11.4 Plot of contaminant plumes at four times ($T = 3.9$, 8.4, 16 and 24 years), together with the envelope of peak concentrations. Fracture spacing 1 m, Darcy velocity 2 m/a, 1-D fracture network. (After Rowe and Booker, 1990b; reproduced with permission of the *International Journal for Numerical and Analytical Methods in Geomechanics*.)

the source. The concentration continues to rise at this point until it peaks after 3.9 years and then decreases to a value of $c/c_0 = 0.5$ again after 16 years. Thus one must be very cautious in interpreting the rate of contaminant migration from a few field observations.

The results presented in Figure 11.4 were obtained assuming a single fracture set at 1 m spacing (1-D analysis). Figure 11.5 gives the corresponding results for a 3-D fracture system with 1 m spacings in each direction. A comparison of Figures 11.4 and 11.5 shows that the additional matrix diffusion which can occur with the 3-D fracture system has a significant effect on the contaminant plume and the rate of contaminant migration. Since more contaminant is 'captured' by the matrix of the rock, with the 3-D fracture system, the contaminant plume is smaller; for example, compare the plume at 8.4 years in Figure 11.4 with that after 9.5 years in Figure 11.5. Furthermore, the plume moves slower. This is evident from a comparison of the time required to reach the peak concentration at a given point (e.g. 100 m, 250 m, 500 m or 750 m) for the 1-D and 3-D cases (Figures 11.4 and 11.5) or from Figure 11.6 which shows the variation in concentration with time at the 100 m point.

An additional consequence of the greater movement into the matrix for 3-D conditions is a much slower eventual decay of the concentration at points near the landfill, which is evident from Figures 11.4–11.6. Referring to Figure 11.6, it is seen that for the first 15 years the 1-D

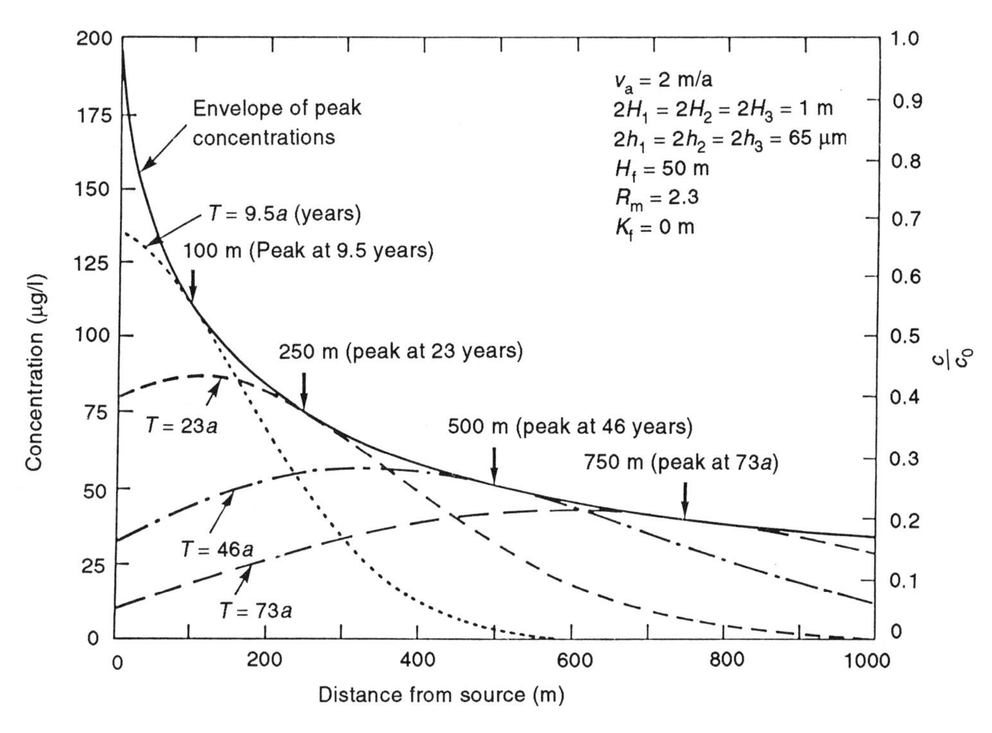

Figure 11.5 Plot of contaminant plumes at four times (9.5, 23, 46 and 73 years), together with envelope of peak concentrations, 3-D analysis. Fracture spacing 1 m, Darcy velocity 2 m/a, 3-D fracture network. (After Rowe and Booker, 1990b; reproduced with permission of the *International Journal for Numerical and Analytical Methods in Geomechanics*.)

solution gives higher concentration than the 3-D solution. After 15 years the reverse is true, because of the greater release of contaminant back into the fracture system (by diffusion from the matrix of the rock back into the fractures) which is possible in the 3-D case. Similarly, a comparison of Figures 11.4 and 11.5 shows that the 3-D plume after 23 years gives the higher concentration for the first 250 m from the source than does the 1-D solution after 24 years.

A related issue is the comparison of the envelope of peak concentration obtained from the 1-D and 3-D analyses (Figures 11.4 and 11.5). For distances less than 1000 m, the 3-D solution gives lower peak concentrations than the 1-D solution. It is, however, evident that the difference between the two solutions is quite small at the 1000 m point. The reason for this is that the effect of 1-D, 2-D and 3-D matrix diffusion is scale-dependent. The differences are greatest when the spacing between fractures and the time of interest is such that much more diffusion can occur if 3-D migration is possible than if only 1-D migration is possible. For a spacing of 1 m, this difference in the effects of diffusion will be greatest at small to moderate times (i.e. times less than the time for diffusion to penetrate the matrix fully). Thus the difference between the 1-D and 3-D cases will not be as significant for the times required for the peak contaminant to migrate 1000 m as it is, say, for the times required to migrate 100 m. For similar reasons, it is found that there is no practical difference between the 1-D and 3-D results

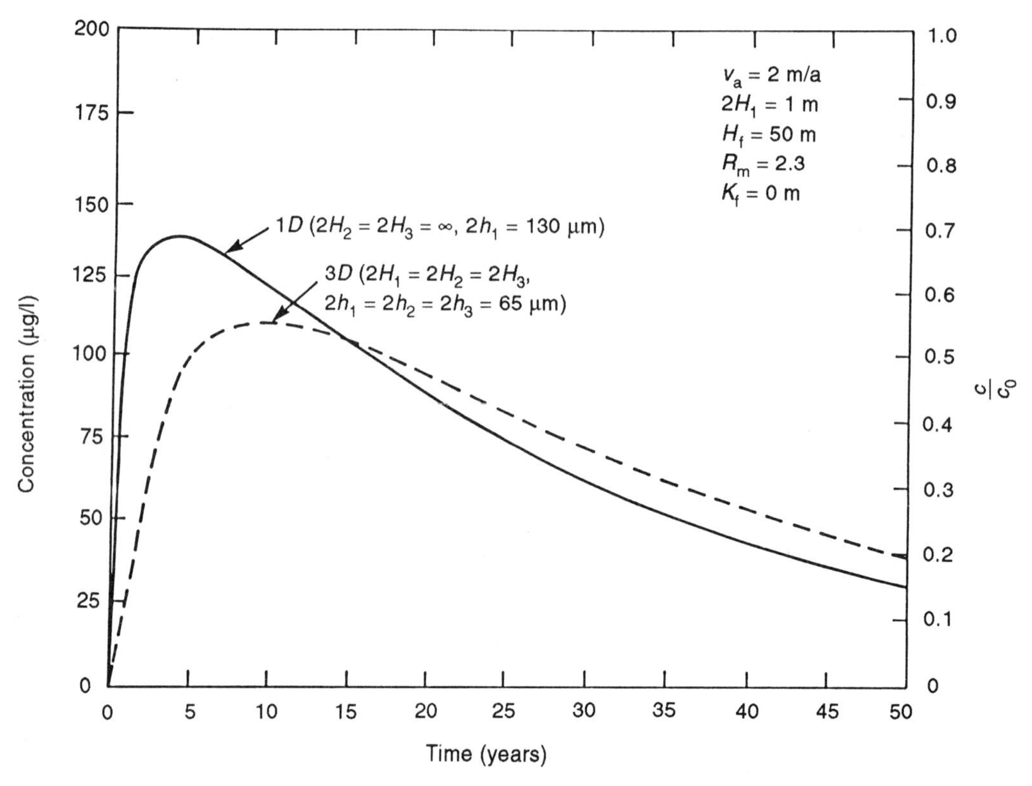

Figure 11.6 Variation in concentration with time at a point 100 m from the source for a 1-D and 3-D fracture network. (After Rowe and Booker, 1990b; reproduced with permission of the *International Journal for Numerical and Analytical Methods in Geomechanics*.)

obtained from similar calculations performed with a fracture spacing, $2H_1$, of 0.1 m.

Referring to Figures 11.4 and 11.5, it can be seen that when there is a high Darcy velocity (v_a = 2 m/a), the contaminant plume is quite extensive after a short period of time. Here, advective–dispersive transport is dominant, but matrix diffusion still does play a significant role. Referring to the curves for similar times (24 years in Figure 11.4, 23 years in Figure 11.5), it can be seen that 3-D diffusion considerably reduces the extent of the plume. In addition, the maximum concentration in the plume at 23 years (3-D analysis) exceeds the maximum at 24 years (1-D analysis) because a similar mass of contaminant is contained within a smaller volume of rock. For lower velocities (e.g. v_a = 0.2 m/a), diffusion has a more dominant role, and the time at which the peak impact reaches a given point is similar for the 1-D and 3-D cases (e.g. at 250 m it is 350 years (1-D) and 381 years (3-D)). However, since the 3-D plume is contained within a smaller area, the mass within the volume of rock where the peak impact is being monitored is greater, and consequently the impact is greater.

11.3.2 Effect of matrix diffusion coefficient and porosity

The results presented in the previous section provide some insight concerning the effect of fracture arrangement for 1-D and 3-D conditions and a reactive contaminant. Similar observations can be made for conservative contaminants which do not interact with the matrix or fracture surface (i.e. $R_m = 1$, $K_f = 0$).

Figure 11.7 shows the calculated variation in concentration with position after 30 years' migration of a conservative contaminant for a number of combinations of matrix diffusion coefficient D_m and matrix porosity, n_m. For a given fracture spacing and Darcy velocity, the velocity at which the contaminant plume moves away from the source is primarily related to the matrix porosity. For this case it can be seen that for both diffusion coefficients considered, reducing the matrix porosity by a factor of two, from 0.1 to 0.05, increased the velocity of the peak of the contaminant plume by a factor of two from about 0.7 m/a to about 1.4 m/a.

It is evident from Figure 11.7 that the peak concentration is smaller for a porosity of 0.1 than that for 0.05; however, these plots do not fully illustrate the effect that porosity has on attenuation. To illustrate this, Figure 11.8 shows the variation in concentration with time at a point 30 m from the source for the four combinations of parameters considered in Figure 11.7. This figure reinforces the earlier observation that a higher porosity reduces the speed of the contaminant plume and reduces the magnitude of the peak impact at a given point. This is because the higher matrix porosity provides

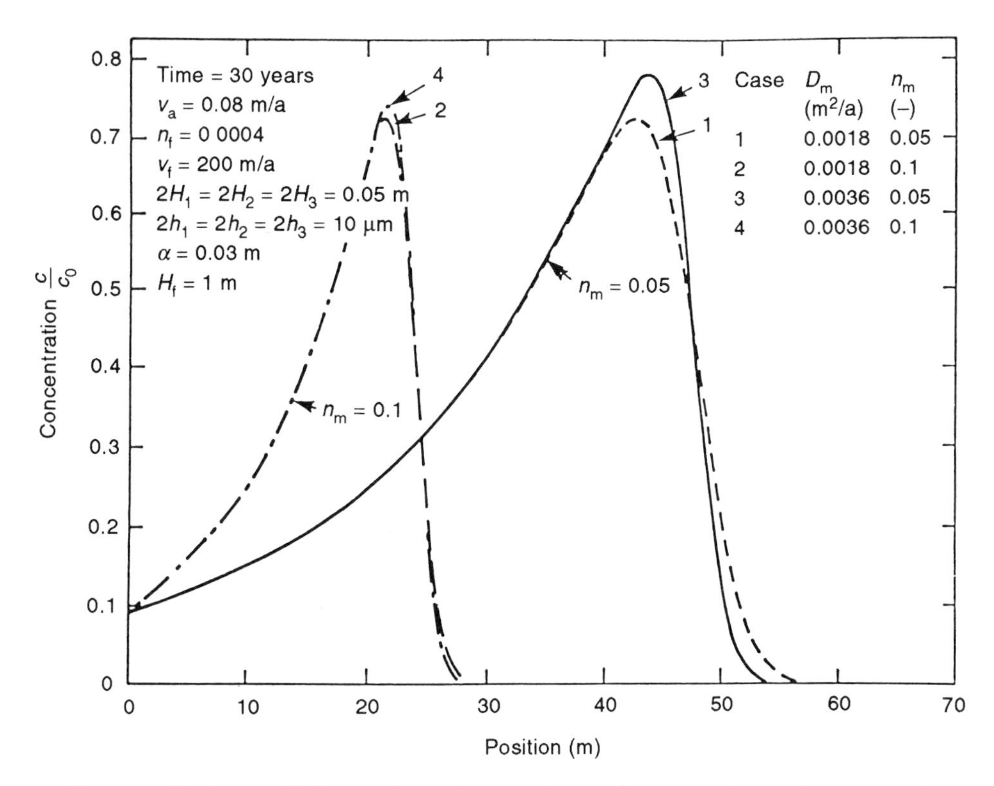

Figure 11.7 Effect of diffusion coefficient and matrix porosity on the contaminant plume: 3-D fracture system. (After Rowe and Booker, 1989; reproduced with permission of the *International Journal for Numerical and Analytical Methods in Geomechanics.*)

more fluid volume into which contaminant can diffuse from the fracture, thereby reducing the mass of contaminant within the fracture system.

The effect of diffusion coefficient on contaminant migration is related to the time required for the pore fluid between fractures to become 'chemically saturated' with contaminant. In the following discussion, this time is defined as the time required to reach a steady state for diffusion into the matrix, assuming a constant concentration in the fracture at a given point. A higher diffusion coefficient allows more rapid movement of contaminant into and out of the rock matrix. Thus as the contaminant moves along the fracture, the extent of the plume will be limited by diffusion into the rock matrix, until the concentration in the rock between fractures reaches a value equal to the concentration in the fracture. No more contaminant can then enter the rock at this point and, in fact, contaminant will diffuse out of the matrix once the concentration in the fracture begins to decrease. The time required for the matrix to become chemically saturated with contaminant at a given point is about 1 year for a single set of fractures at a spacing of 0.05 m and a diffusion coefficient of 0.0018 m^2/a, and about 0.5 years for a diffusion coefficient of 0.0036 m^2/a. Since these time scales are small compared with the 30-year period being examined here, much of the rock is chemically 'saturated' with contaminant in both cases. When chemical saturation can occur over a significant extent of the rock in the time frame of interest, the maximum

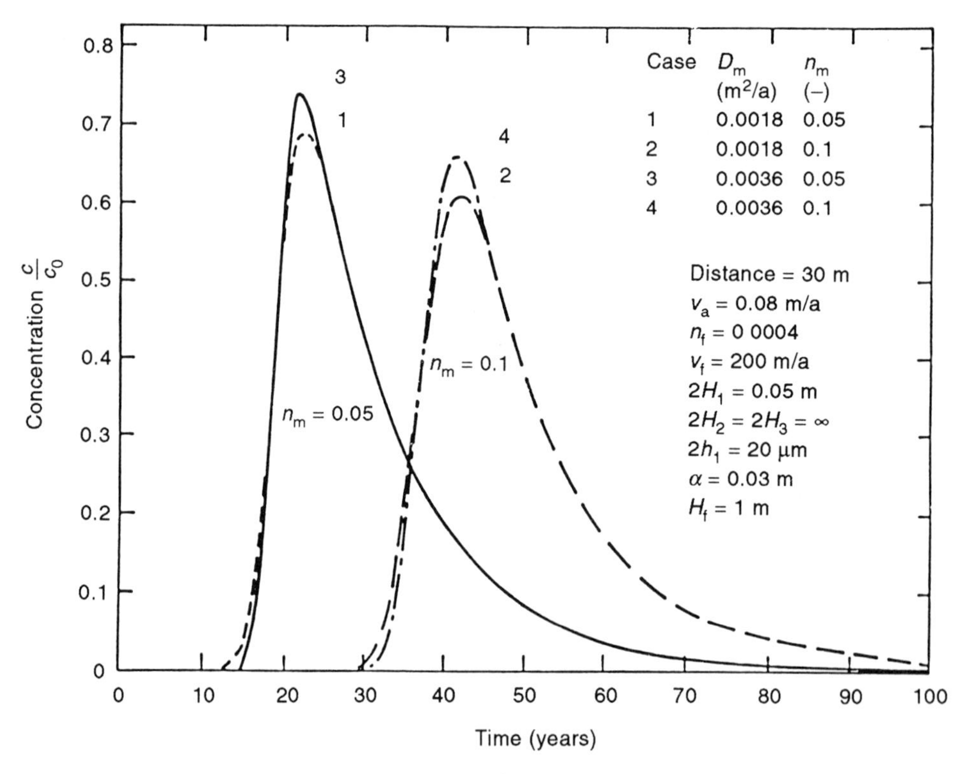

Figure 11.8 Effect of diffusion coefficient and matrix porosity on contaminant arrival times at a point 30 m from the source. (After Rowe and Booker, 1989; reproduced with permission of the *International Journal for Numerical and Analytical Methods in Geomechanics*.)

peak concentration and minimum extent of the plume will be obtained for the case where the contaminant can most quickly chemically saturate the rock between fractures. Thus, increasing the diffusion coefficient by a factor of two from 0.0018 m^2/a to 0.0036 m^2/a results in a slightly smaller plume for the higher diffusion coefficient but, since essentially the same amount of mass is involved, the peak concentration is consequently higher, as shown in Figure 11.7 or 11.8.

11.3.3 Effect of fracture spacing

As illustrated in Figures 11.7 and 11.8, the two-fold reduction in diffusion coefficient from 0.0036 to 0.0018 m^2/a does, in this case, result in a modest reduction in the magnitude of the peak concentration, but very little change in the velocity of the contaminant plume. However, this finding should not be extrapolated to other situations. For example, as the spacing between fractures increases, the time required for the rock to become chemically saturated with contaminant also increases. When the time required to saturate the rock becomes significant compared to the time frame of interest, then a higher diffusion coefficient may be expected to exert a different influence on the shape of the contaminant plume since, by definition, there will not have been time for contaminant to diffuse fully into the rock along most of the distance between the source and the point of interest. To illustrate this, Figure 11.9 shows the calculated contaminant plume for fracture spacings of 0.05 m and

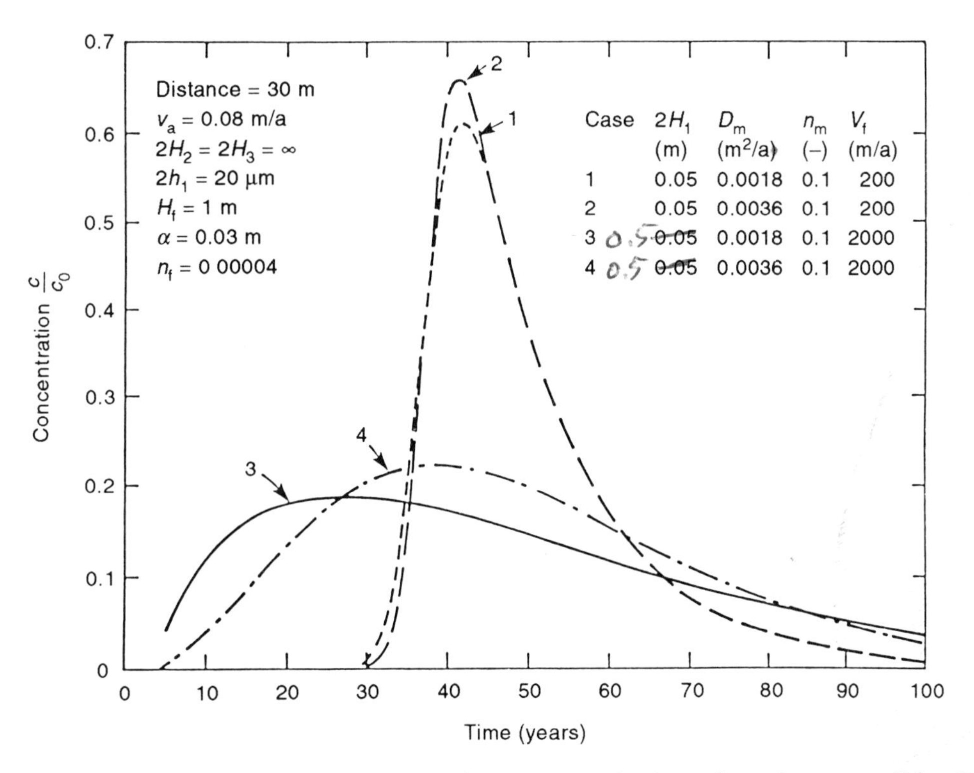

Figure 11.9 Effect of fracture spacing, $2H_1$, on arrival times at a point 30 m from the source. (After Rowe and Booker, 1989; reproduced with permission of the *International Journal for Numerical and Analytical Methods in Geomechanics*.)

0.5 m for two values of matrix diffusion coefficient (i.e. D_m = 0.0018 m^2/a and D_m = 0.0036 m^2/a).

It is evident from Figure 11.9 that the spacing between the primary set of fractures is an important parameter influencing contaminant movement. As previously noted, the time required for diffusion to cause chemical saturation of the matrix with contaminant is about 1 year and 0.5 years for D_m = 0.0018 and 0.0036 m^2/a respectively, at a spacing of $2H_1$ = 0.05 m, and about 100 years and 50 years respectively, at a spacing of $2H$ = 0.5 m.

In this case, the larger spacing between fractures ($2H_1$ = 0.5 m) also gives rise to a substantially smaller peak impact and earlier arrival of this impact than was calculated for the narrower fracture spacing ($2H_1$ = 0.05 m). This is primarily because there has not been sufficient time for chemical saturation of the rock matrix on the time scale being considered, and so the contaminant is much more spread out along the fracture for the wider spacing. Since essentially the same mass is contained in both the plumes for $2H_1$ = 0.05 m and $2H_1$ = 0.5 m, the fact that it is spread over a larger volume of rock for the wider spacing means that the peak impact is correspondingly reduced.

The effect of fracture spacing on the contaminant plume is dependent on both the mass of contaminant and the ratio of the time of interest to the time required to cause chemical saturation of the matrix between fractures. Because of the complex interaction between the effect of different parameters, care should be taken not to overgeneralize from the results of any one set of analyses. Fracture spacing is clearly an important parameter, and its effects, and interaction with other parameters such as retardation coefficient, will be discussed in more detail in section 11.3.6.

11.3.4 Effect of dispersivity

Figure 11.10 shows the calculated variation in concentration with time at a point 30 m from the contaminant source for a 3-D fracture set and three different values of dispersivity α (i.e. 0.03, 0.3 and 3 m) which correspond to 0.1%, 1% and 10% of the travel distance. All other things being equal, a small dispersivity, implying very little mechanical dispersion due to irregularities in the fracture system, gives the greatest impact. Higher dispersivities give rise to a more extensive contaminant plume and, since the mass of contaminant is spread over a greater volume of rock, a smaller peak impact at a given point.

Dispersivity is generally considered to increase with distance from the source, at least until the distance involved is large compared to the scale of the non-homogeneities of the flow system which give rise to the dispersion process (e.g. Frind, Sudicky and Schellenberg, 1987). If one considered the dispersivity to be a linear function of distance (say 1% of the travel distance: α/x = 0.01) then it is found that the peak impact at the monitoring point is reduced by more than a factor of five between the monitoring points at 30 m and 300 m.

The fracture opening size is one of the most difficult parameters to determine, and it is of interest to explore the implication of uncertainty regarding this parameter, since the fracture porosity, and hence groundwater velocity, both depend on this parameter.

Figure 11.11 shows the calculated contaminant plume after 30 years' migration for five cases involving a 3-D fracture system. All cases assume the same Darcy velocity. In cases 1, 2 and 3 the fracture opening size is varied between 10 μm and 40 μm, giving a consequent four-fold variation in groundwater velocity from 200 m/a to 50 m/a. For a given dispersivity of 0.03 m it can be seen that this four-fold variation in opening size and groundwater velocity does slightly change the contaminant plume; however, this change is of no practical significance. In each case, the peak of the contaminant plume is moving at a velocity of 0.7 m/a irrespective of

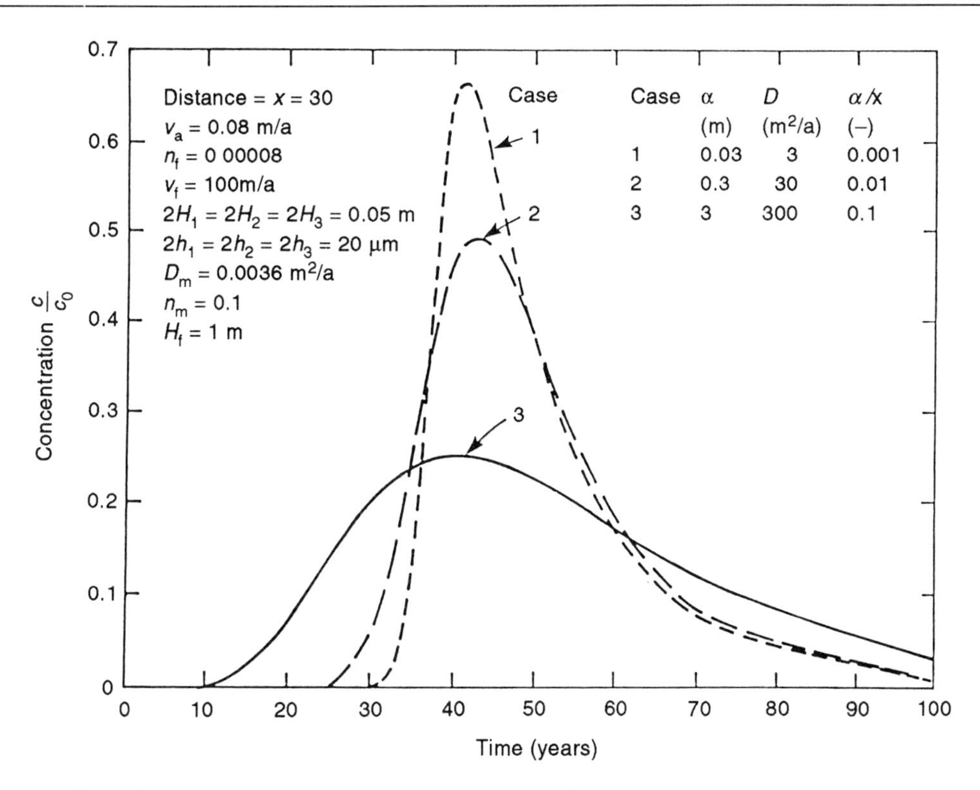

Figure 11.10 Effect of dispersivity, α, on arrival at a point 30 m from the source: 3-D fracture system. (After Rowe and Booker, 1989; reproduced with permission of the *International Journal for Numerical and Analytical Methods in Geomechanics*.)

whether the groundwater velocity is 50 or 200 m/a.

The coefficient of hydrodynamic dispersion, D, is often assumed to be linearly proportional to the groundwater velocity, v_f (i.e. $D = \alpha v_f$, where α is the dispersivity). In the comparison of cases 1, 2 and 3 above, it has been assumed that the dispersivity remained constant ($\alpha =$ 0.03 m), and hence the coefficient of hydrodynamic dispersion did vary by a factor of four due to the variation in groundwater velocity. However, it may also be reasonable to argue that really it is the coefficient of hydrodynamic dispersion that remains constant. To examine the implications of this assumption, cases 1, 4 and 5 (Figure 11.11) represent results for fracture opening sizes of 10, 20 and 40 μm and a single coefficient of hydrodynamic dispersion of 6 m²/a. In this case, it can be seen that increasing the assumed fracture opening size from 10 to 40 μm does result in a change in the contaminant plume, primarily due to the fact that for the large opening size there is (relatively speaking) a four-fold increase in the ratio of the coefficient of hydrodynamic dispersion to groundwater velocity. Nevertheless, from a practical standpoint, it can be seen that the variation in average fracture opening size, and hence groundwater velocity, considered here does not significantly influence the speed of contaminant transport or the peak impact, which varies by less than 10% for a four-fold variation in opening size. This implies that uncertainty as to the precise details concerning

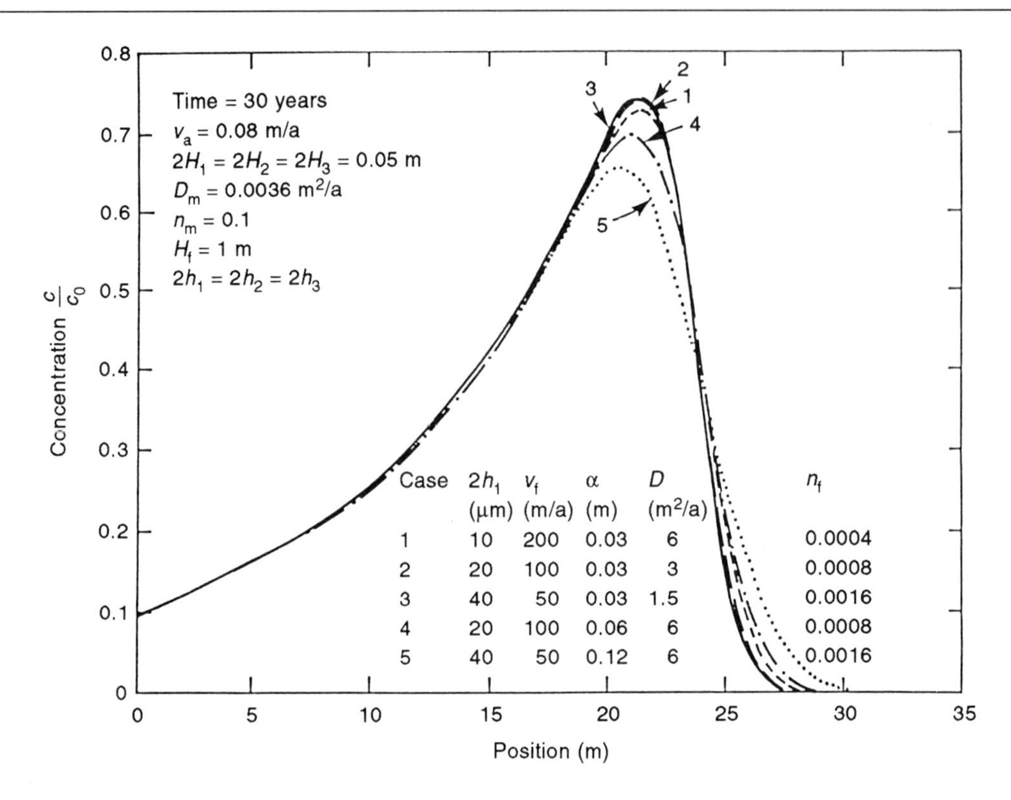

Figure 11.11 Effect of fracture opening size, $2h_1$, and dispersivity, α, on the contaminant plume. (After Rowe and Booker, 1989; reproduced with permission of the *International Journal for Numerical and Analytical Methods in Geomechanics*.)

fracture opening size is not going to have a significant effect on prediction of contaminant transport for the range of cases considered. This conclusion is valid for typical ranges of fracture opening size and spacing in a fractured media. The assumptions used in developing these results may not be applicable to situations where the fracture porosity n_f is less than 10^{-5} or greater than 0.1. The results are only applicable to the situation where almost all of the advective contaminant transport is along the fractures rather than through the matrix.

11.3.5 Effect of Darcy velocity

It is evident from the preceding discussion that the groundwater velocity is of secondary significance when predicting the rate of contaminant migration through fractured porous media. This is fortunate, since the groundwater velocity is very sensitive to changes in opening size; hence it is difficult to estimate, and may vary substantially within a given mass of rock. A more readily determined parameter is the Darcy velocity (Darcy flux) which represents the volume of water moving through the fractured rock mass per unit area per unit time. This quantity can be deduced from the measured bulk hydraulic conductivity and gradient within the fractured rock mass.

Figure 11.12 shows the effect of varying the Darcy velocity on the contaminant plume after 30 years' migration. Results are given for Darcy velocities of 0.04 and 0.08 m/a, and it can be

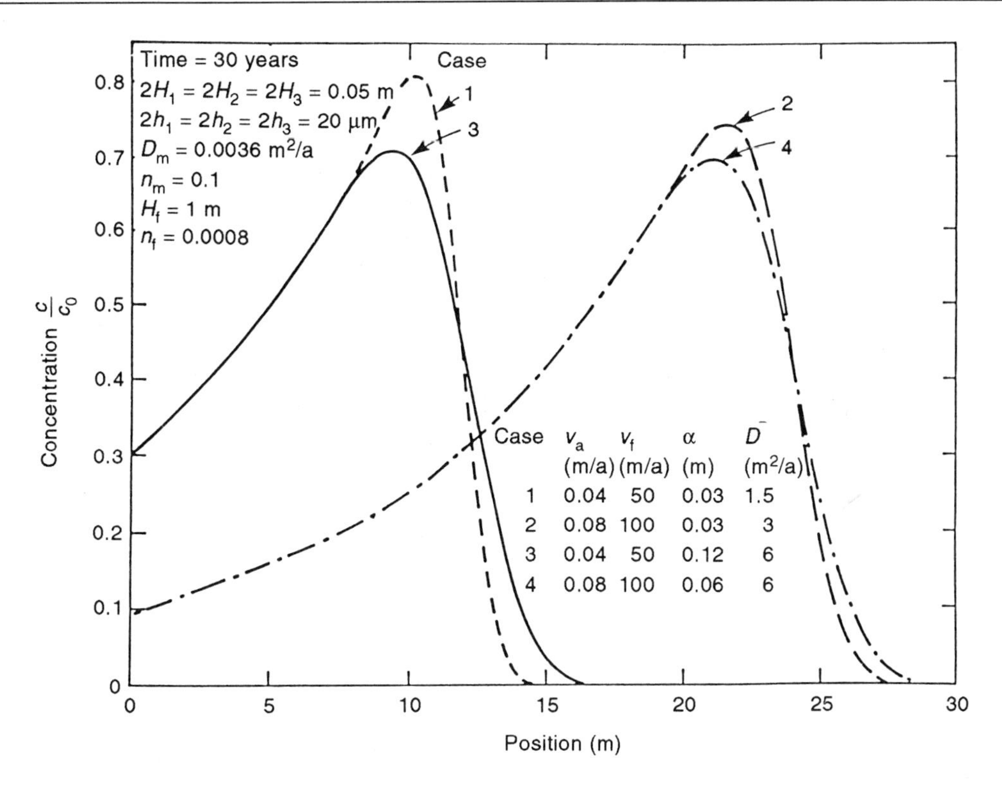

Figure 11.12 Effect of Darcy velocity on contaminant plume (constant fracture porosity). (Modified from Rowe and Booker, 1989.)

seen that doubling the Darcy velocity results in an approximate doubling of the velocity of the peak of the plume. Curves 1 and 2 show the effect of increasing the Darcy velocity while maintaining a constant dispersivity, α. Curves 3 and 4 show the effect of increasing the Darcy velocity while maintaining a constant coefficient of hydrodynamic dispersion. The assumption regarding what happens to dispersion when the Darcy velocity changes does have some impact on the contaminant plume; however, the primary variable is the Darcy velocity, and dispersion is a secondary consideration.

The results presented in Figure 11.12 were obtained by varying the Darcy velocity for a given fracture distribution and fracture porosity. Thus the groundwater velocity varies in proportion to the Darcy velocity. To emphasize the point that it is the Darcy velocity (flux) and not the groundwater velocity which is the primary control on contaminant transport, the results in Figure 11.13 were obtained by varying the Darcy velocity and fracture porosity, such that the groundwater velocity remained constant at 100 m/a. Comparing case 1 in Figures 11.12 and 11.13, it is seen that for a given dispersivity and Darcy velocity, the difference in groundwater velocity between 50 m/a and 100 m/a has no significant effect on the contaminant plume, whereas it is seen from Figure 11.13 that for a given dispersivity and groundwater velocity, the difference in Darcy velocity between 0.04 and 0.08 m/a approximately doubles the movement of the peak of the contaminant plume. Thus, it is the Darcy velocity, not the groundwater velocity, which controls the movement of the contaminant plume and the impact on groundwater quality in fractured porous media where matrix diffusion is a significant attenuation mechanism.

The effect of Darcy velocity on the peak impact at a point 100 m downgradient from a source is illustrated in Figure 11.14 for a mildly reactive contaminant (R_m = 2.3). Notice that the time at which the peak impact occurs decreases monotonically with increasing Darcy velocity; however, the magnitude peak impact decreases initially with increasing velocity and then subsequently increases for higher Darcy velocities. The velocity at which the minimum impact occurs is different for the 1-D and 3-D cases, but the trends are the same. Notice also that at low velocities the peak impact is greater for the 3-D than for the 1-D case, but for higher velocities the reverse is true. Thus there is an optimal velocity which, for a given situation, gives minimal impact. It is not normally practical to design to take advantage of this optimal situation, but it is important to recognize that an analysis for a single value of v_a could be quite misleading, in that one analysis may, fortuitously, be for an optimum or near optimal velocity. Where uncertainty exists regarding Darcy velocity, as it invariably will, it is necessary to perform analyses for a number of different velocities which define a range of values in which the actual velocity is most likely to lie.

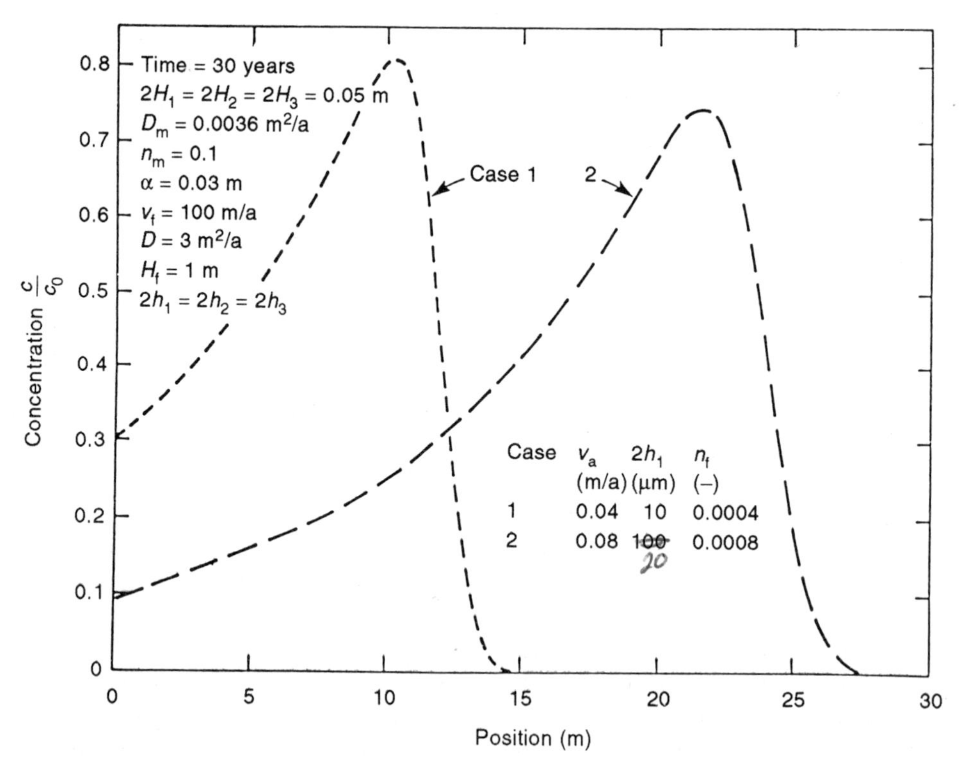

Figure 11.13 Effect of Darcy velocity on the calculated contaminant plume (constant groundwater velocity, v_f). (After Rowe and Booker, 1989; reproduced with permission of the *International Journal for Numerical and Analytical Methods in Geomechanics*.)

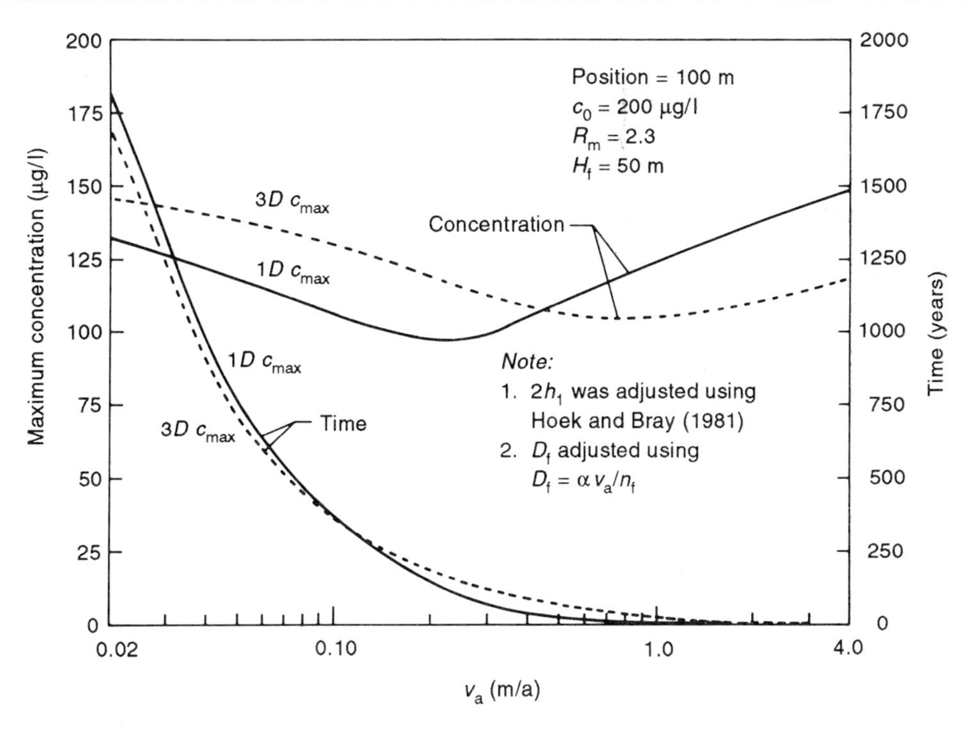

Figure 11.14 Effect of Darcy velocity on peak impact and time to peak impact for a reactive species (R_m = 2.3). (Modified from Hammoud, 1989.)

11.3.6 Effect of sorption in the rock matrix (R_m)

The results discussed in the previous sections have separately considered a contaminant either which was conservative (e.g. chloride with R_m = 1) or which was slightly retarded (1,1-dichloroethane with R_m = 2.3 for the shale being considered). Sorption can be an important attenuation mechanism, and its effects are illustrated for R_m = 1 (a conservative species), R_m = 2.3 and R_m = 10 (a reactive species) in Figure 11.15. This figure shows the calculated variation in concentration with time at a monitoring point 100 m downgradient of the contaminant source. Since R_m represents sorption which occurs within the rock matrix, the effects of R_m will be greatest when there is significant diffusion into the rock.

In Figure 11.15, the peak concentration is decreased by about a factor of two, and the time required to reach the peak is increased by about 12 years, due to a change in R_m from 1 to 10.

The effect of sorption is to remove contaminant from solution at times when the concentration is increasing. This reduces the peak impact at any given point away from the source. It is assumed here that sorption is reversible. A consequence of this assumption is that contaminant will be desorbed into solution when the contaminant concentration drops. Thus at times after the peak has passed, the concentration at a given point will be higher for R_m = 10 than for lower values of R_m, because the contaminant which had been sorbed is now being released back into the groundwater. This assumption of full reversible sorption is conservative; it may not happen in practice, since the sorption processes may not be reversible, or

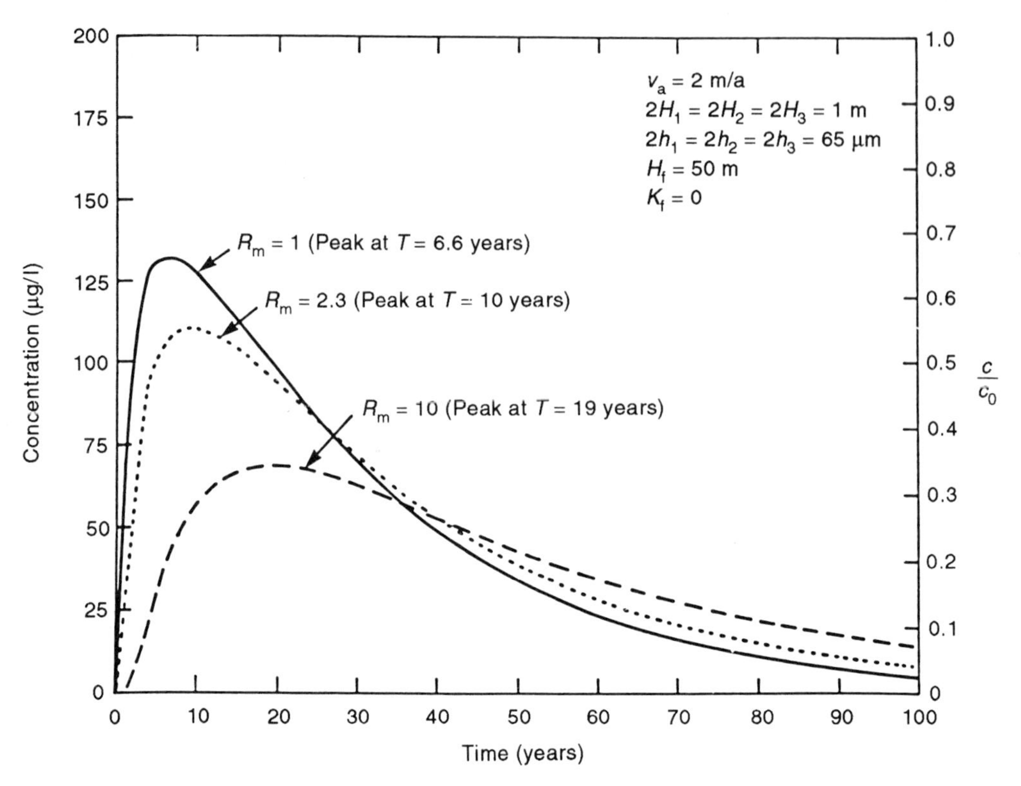

Figure 11.15 Effect of sorption in the matrix point 100 m downgradient of the contaminant source. Fracture spacing 1 m, 3-D fracture network. (After Rowe and Booker, 1990b; reproduced with permission of the *International Journal for Numerical and Analytical Methods in Geomechanics*.)

there may be a much lower desorption partitioning coefficient than sorption partitioning coefficient.

Figure 11.16 shows the effect of retardation, R_m, upon the peak concentration expected at a point 100 m from the source. Results are given for $v_a = 0.2$ m/a and two different fracture spacings ($2H_1 = 1$ m and 0.1 m). For each case, the peak impact at the 100 m monitoring point is greatest for a conservative species ($R_m = 1$) and least for a highly retarded species ($R_m = 10$). Curves are presented for two values of the equivalent height of leachate H_f (50 and 5 m), and it can be seen that sorption has a greater impact when the mass of contaminant originally in the landfill (defined in terms of H_f) is smaller.

11.3.7 Effect of the equivalent height of leachate H_f (mass of contaminant)

The results shown in Figure 11.16 indicate that the equivalent height of leachate is an important parameter and that attenuation will be considerably enhanced if the mass of contaminant available for transport into the fracture media can be minimized (e.g. by installing an effective leachate collection system). This is further demonstrated by Figure 11.17, which shows the effect of the equivalent height of leachate on the peak contaminant impact at the 100 m monitoring point. This shows that it is important to obtain a realistic, albeit conservative, estimate of the equivalent height of leachate in the

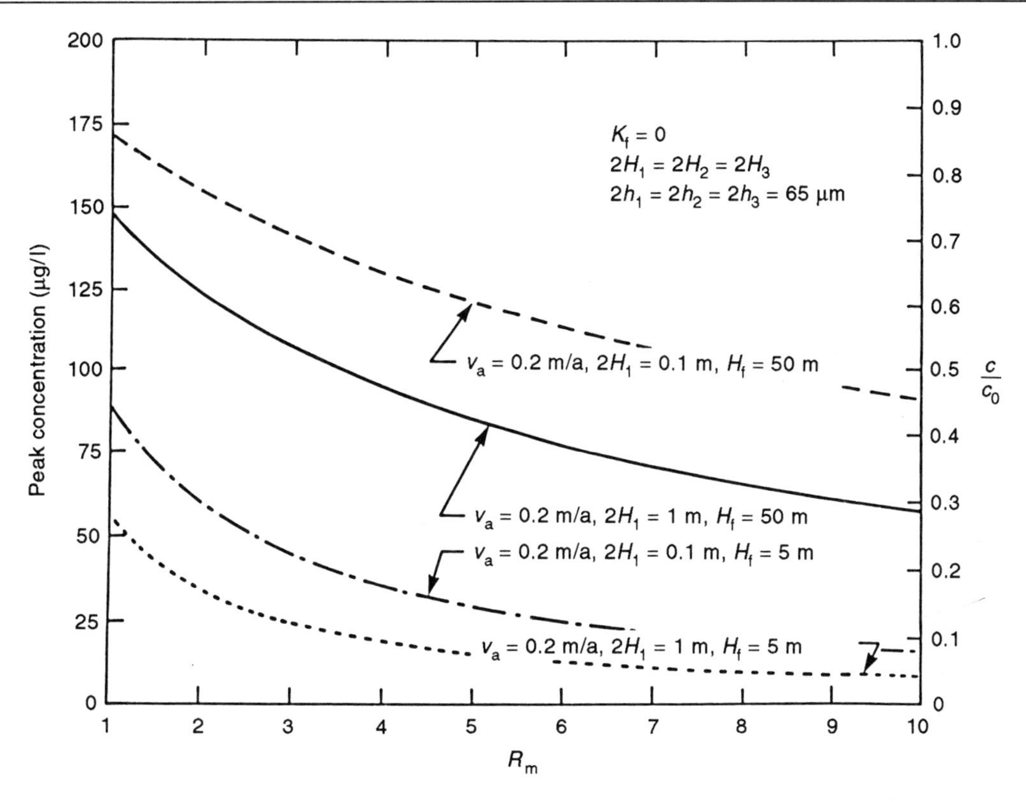

Figure 11.16 Effect of sorption in the matrix on calculated peak impact at a monitoring point 100 m downgradient of the source. (After Rowe and Booker, 1990b; reproduced with permission of the *International Journal for Numerical and Analytical Methods in Geomechanics*.)

landfill if reasonable predictions of contaminant impact are to be made.

11.3.8 Sorption onto the fracture surface

Little is known about the sorption of contaminants onto fracture surfaces. Recognizing this, Figure 11.18 shows the effect sorption onto fracture surface could have for an assumed range of values of distribution coefficient K_f. Since K_f is defined (Freeze and Cherry, 1979) as the mass of solute adsorbed per unit area [ML^{-2}] of surface divided by the concentration of solute [ML^{-3}], it has units of length [L]. It is noted that a range of values of K_f from 0 to 1 results in a decrease of up to a factor of four on the impact at a point 100 m downgradient for the landfill. More research is required to determine relevant and appropriate values of K_f which could be used in a given situation. It is, however, evident from Figure 11.18 that it is conservative to neglect sorption on fracture surfaces when predicting peak impact.

11.3.9 Case study

Over the past 35 years, two different landfills, known as the Bayview Park and Burlington landfills, have been constructed directly on fractured shale in the city of Burlington, Ontario. Neither landfill was originally constructed with a leachate collection system, although a toe drain has now been retrofitted to both landfills. A subsequent northern expansion of the Burlington landfill (Gartner Lee Ltd., 1988) is located directly on clayey till, and has a leachate underdrain system. This expansion is not expected to have led to any significant leachate egress into the shale.

Although chloride is normally used as an indicator of the extent of contaminant migration from landfills, it does not provide a good leachate diagnostic in this case, because the natural high salinity of the unpotable groundwater is such that the concentrations of chloride in the groundwater exceed values in the leachate. Diffusion tests on the shale (Barone, Rowe and Quigley, 1990; also section 8.3.4) indicate that chloride can readily diffuse out of the shale, which has a matrix porosity of between 0.1 and 0.11, with a diffusion coefficient of about 1.5 × 10^{-10} m^2/s (0.0047 m^2/a).

Since chloride cannot be used as a reliable indicator of the presence of contaminants at this site, recent investigations and monitoring have focused on organic contaminants. A number of monitors ranging in distance from 25 m to about 320 m downgradient of the Burlington landfill (Gartner Lee Ltd., 1988) have been monitored. Four organic chemicals (1,1-dichloroethane, 4 μg/l; vinyl chloride, 12 μg/l; trichloroethylene, 12 μg/l; and cis-1,2-dichloroethylene, 33 μg/l) have been detected at the monitor 25 m downgradient of the landfill (Gartner Lee Ltd., 1988). Rowe and Booker (1989) modeled the migration of the two

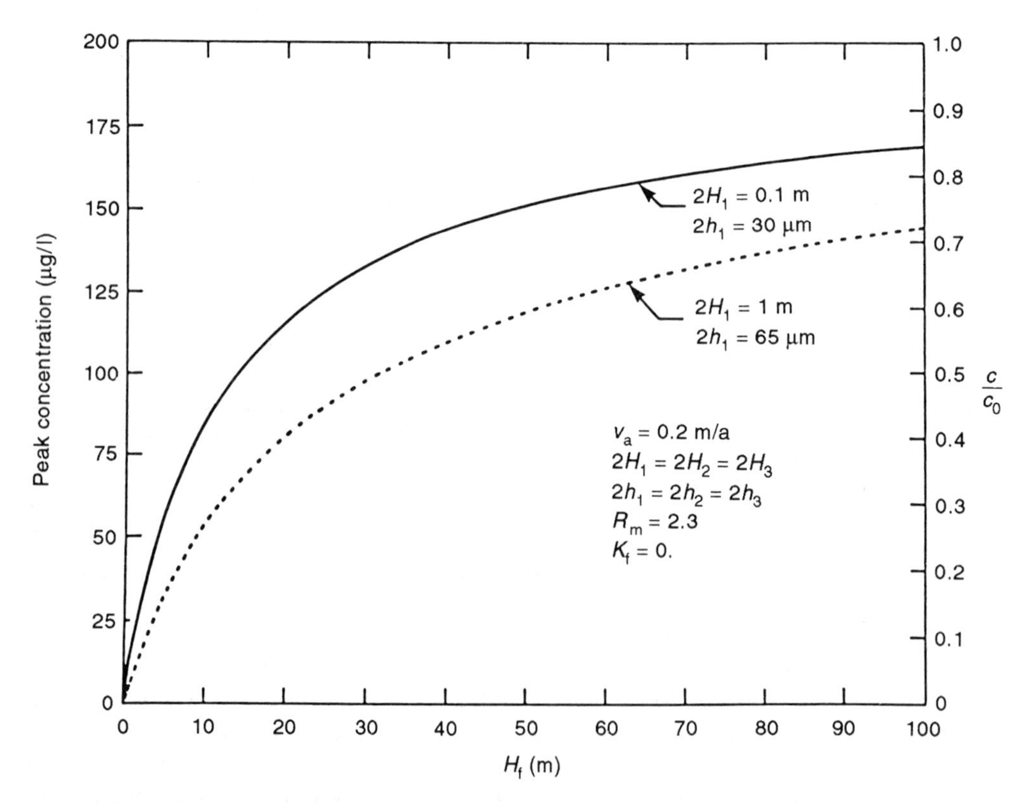

Figure 11.17 Effect of mass of contaminant (expressed in terms of the equivalent height of leachate, H_f) on contaminant impact at 100 m. (After Rowe and Booker, 1990b; reproduced with permission of the *International Journal for Numerical and Analytical Methods in Geomechanics*.)

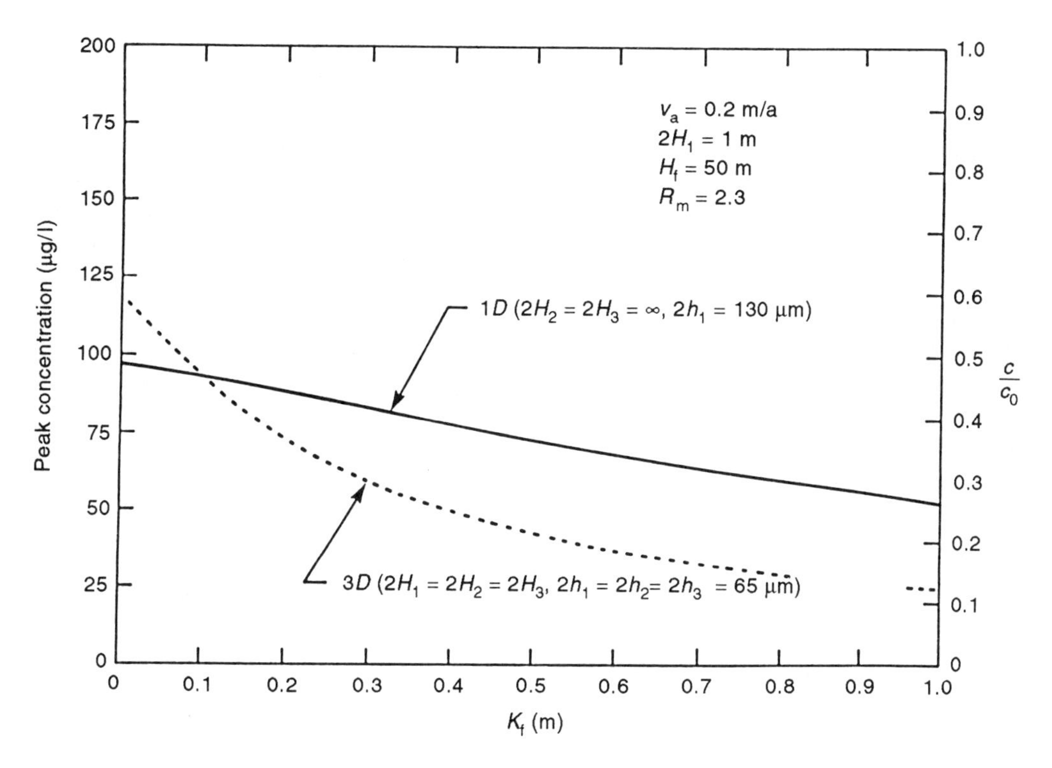

Figure 11.18 Effect of sorption on fracture surface, K_f, on contaminant impact at 100 m. (After Rowe and Booker, 1990b; reproduced with permission of the *International Journal for Numerical and Analytical Methods in Geomechanics*.)

most mobile species (vinyl chloride and 1,1-dichloroethane) and compared the calculated results with the field observations. They showed that the calculated contaminant plume which was based on a reasonable hydrogeologic evaluation of the site (a Darcy velocity of between 0.05 and 0.125 m/a and an average fracture spacing of between 0.05 and 0.3 m) was consistent with the observed, very limited, extent of the plume. These analyses indicate that contaminants would be expected to have migrated about 25 m in 15 years. On the other hand, calculations based on simple advective transport at groundwater velocities ranging from 50 m/a to 1071 m/a are not consistent with the field observations. This case demonstrates the significant role which matrix diffusion can have on the attenuation of contaminants migrating through fractured porous media.

11.3.10 Summary

For the range of parameters considered in the examination of contaminant migration through fractured shale presented in this section, the following may be concluded.

1. The rate of contaminant transport did not significantly depend on the groundwater velocity. The rate of movement and the degree of attenuation were, however, greatly dependent on the fracture spacing, matrix porosity, dispersivity, Darcy velocity (i.e.

Darcy flux) and the mass of contaminant available for transport.

2. For a given spacing between the primary fractures, consideration of the spacing between the secondary fractures was found to reduce the calculated velocity of the leading edge of the contaminant plume; however, it may either increase or decrease the calculated maximum impact of the plume depending on the specific case.
3. Any reasonable estimate of secondary fracture spacing gave results which would be sufficiently accurate for engineering purposes.
4. The larger the matrix porosity, the slower was the contaminant transport velocity and the greater the attenuation.
5. The magnitude of the matrix diffusion coefficient had a significant effect on the predicted plume when the time scale of interest was similar to, or greater than, the time required to reach a steady state, due to diffusion into the rock matrix at a point.
6. Fracture opening size was only important in that it influenced the hydraulic conductivity of the rock mass and the Darcy velocity. Any combination of fracture opening sizes which gave rise to a given hydraulic conductivity also gave rise to a contaminant impact which, for all practical purposes, was the same. The effect of uncertainty regarding the magnitude of opening size upon the groundwater velocity did not have any significant effect on the prediction of contaminant migration.
7. A theoretical examination of the evolution of a contaminant plume also demonstrated that one may not be able to infer reliably the rate of advance of a contaminant plume based on isolated observations of the contaminant plume if the source of contaminant is finite.
8. The rate of contaminant movement in a fracture system was not linearly proportional to the Darcy velocity, and may be almost independent of the groundwater velocity. The effect of changes in velocity may not be intuitively obvious, and will greatly depend on other parameters such as fracture spacing, the diffusion coefficient, retardation factor, matrix porosity and mass of contaminant. The theory presented in Chapter 7 allows the designer to examine these interactions readily and quickly.
9. The effectiveness of sorption as an attenuation mechanism was dependent on the primary, secondary and tertiary fracture spacings; however, in each case, higher retardation coefficients for sorption in the rock matrix were found to decrease the maximum contaminant impact at any specified monitoring point downgradient of the source.
10. The mass of contaminant available for transport into the fractured soil/rock was a major factor affecting the impact downgradient of the leaking landfill. By removing contaminant from the landfill, leachate collection systems reduce the mass available for contaminating the underlying soil/rock and can substantially reduce the downgradient impact of those contaminants which do escape the leachate collection system.

This section has considered lateral migration in a fractured porous media. In the following section, consideration is given to similar factors, but this time relative to vertical migration from a contaminant source (e.g. a landfill) through a fractured clay and into an underlying aquifer.

11.4 Vertical migration through fractured media and into an underlying aquifer

Rapid migration of contaminant through what was originally thought to be relatively low hydraulic conductivity clayey soils, both in the USA (e.g. at Wilsonville, IL) and Canada has awakened concern regarding the potential fracturing of stiff–very stiff clays and clayey tills. The Wilsonville problem prompted research by Herzog and colleagues (Herzog and Morse, 1986; Herzog *et al.*, 1989) which has provided evidence to suggest that the unweathered till below the obviously weathered and fractured zone at Wilsonville was also fractured to extensive depths. Similarly, recent investigations in southern Ontario (Ruland, 1988; D'Astous *et al.*, 1989; and the authors' own investigations) have indicated that clayey till which appears unweathered and unfractured in conventional borehole investigations may be fractured to depths of as much as 13 m below ground surface (e.g. to depths of up to 9 m below the weathered crust). In some cases, this may mean that there is little or no unfractured clayey till between the base of the landfill and an underlying aquifer. Careful examination of deep test pits has demonstrated that at some locations, fracturing at 1–3 m spacings may exist both above and below an upper 'permeable' zone (which could potentially act as a conduit for contaminant transport if a landfill were constructed above this unit). Similarly, hydrogeologic investigations at other sites have identified that even though there may be 4–5 m of till separating the base of a proposed landfill from an underlying permeable zone, only 1–2 m of the material may be unfractured. When fractures are identified, the question then arises as to what effect they will have on contaminant transport from a proposed landfill facility.

11.4.1 Steps to be followed in modeling fractured systems

In order to use the theory described in Chapter 7, it is necessary to follow the steps listed below.

1. Define the hydrostratigraphy of the site in terms of:
   (a) the number of layers with different properties;
   (b) the level of fracturing of each layer (i.e. the fracture spacings $2H_1$, $2H_2$, $2H_3$ and fracture opening sizes $2h_1$, $2h_2$, $2h_3$) and degree of uncertainty;
   (c) the bulk vertical and horizontal hydraulic conductivity k_z, k_x of each layer;
   (d) the hydraulic boundary conditions.
2. Estimate the horizontal and vertical Darcy velocity v_{ax}, v_{az} for each layer. Depending on the complexity of the problem, this may be done by means of a hand calculation or using a computer flow model (Chapter 5). The vertical velocities in any layer should be representative of the velocity beneath the landfill. The horizontal velocities should satisfy continuity of flow at the downgradient edge of the landfill.
3. Estimate the matrix porosity n, the dry density ρ and the distribution/partitioning coefficient K_d for each layer.
4. Estimate the coefficient of hydrodynamic dispersion D_{zz}, D_{xx} in the vertical and horizontal directions. For clay layers, this is often equal to the diffusion coefficient. For fractured layers, this is the diffusion coefficient within the matrix between fractures. For aquifer layers, this will depend on the assumed dispersivity.

   For fractured layers, also estimate fracture distribution coefficient K_f and the coefficient of hydrodynamic dispersion along the fractures in the vertical and horizontal directions, D_{xf}, D_{zf}. The former is usually zero and the latter equal to the dispersivity α, times the average groundwater velocity in the fractures (e.g. $D_{xf} \approx \alpha_x v_{xf}$; $D_{zf} \approx \alpha_z v_{zf}$).
5. Estimate the peak source concentration c_0, the equivalent height of leachate, H_f, or the reference height of leachate, H_r, and the volume of leachate collected by the collection system per unit area, q_c (e.g. section 10.2).
6. Define the boundary condition at the bottom of the deposit. Typically, a bottom aquifer is modeled as a boundary condition if it is thin and by a layer(s) if it is thick. If the lower

aquifer is modeled by layers, the bottom boundary is usually impermeable.

Once the foregoing have been estimated, the contaminant transport model can be run to calculate concentration at any specific times and locations of interest. Typically, there will be uncertainty regarding values of many of the input parameters. The implications regarding this uncertainty can be evaluated by performing a sensitivity study.

11.4.2 Definition of example case

To illustrate the potential application of the theory presented in Chapter 7, consideration will be given to the situation where the base of a hypothetical landfill is to be located in the unweathered portion of a clayey aquitard which is fractured with an average fracture spacing of 1 m. The base of the landfill is assumed to be 4 m above a 1 m thick upper permeable unit (aquifer 1) and 7 m above a 2 m thick lower permeable unit (aquifer 2) as shown in Figure 11.19.

The proposed model can readily consider situations where there is horizontal flow in aquifers beneath a landfill. Under these circumstances, flow in the aquifer provides a potential for dilution of contaminants migrating from the landfill, and the horizontal flow at the downgradient edge of the landfill is equal to the flow at the upgradient edge plus the change in flow, as required by consideration of continuity of flow, which occurs beneath the landfill. From the perspective of contaminant impact, the 'worst case' situation involves the landfill being located on a groundwater divide. This divide may have existed prior to the landfill being constructed. However, even if under natural conditions there is no groundwater divide, a landfill can change flow conditions if significant leachate mounding occurs within the landfill. The likelihood of this occurring should be evaluated for each landfill being considered.

For the purposes of this particular example, it is assumed that the landfill is 400 m wide and is to be located in a recharge zone directly over a groundwater divide. This represents worst case conditions. For the sake of definiteness, it is

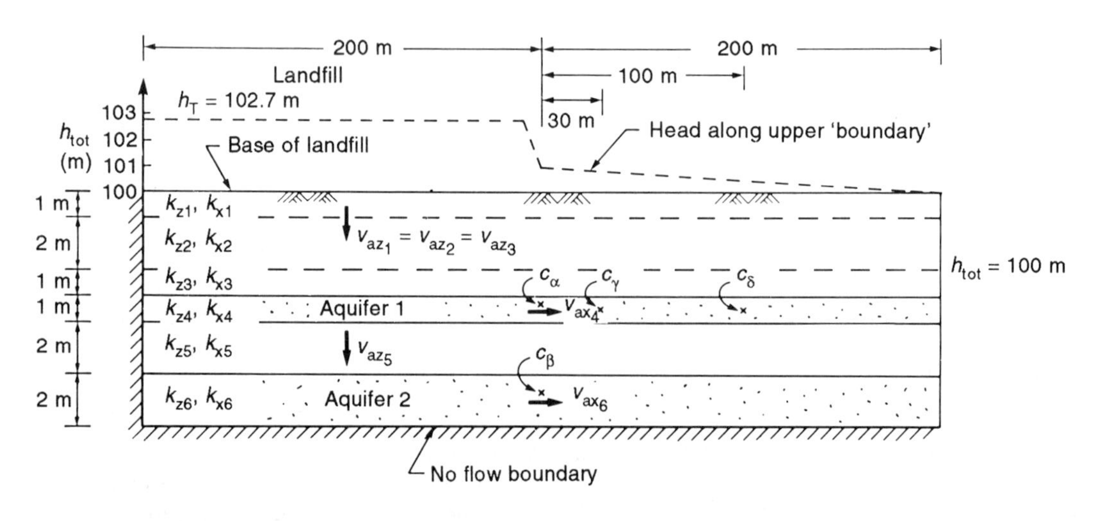

Figure 11.19 Schematic of general hydrostratigraphy considered. (After Rowe and Booker, 1991a; reproduced with permission of the *Canadian Geotechnical Journal*.)

assumed that the average (long-term) height of leachate mounding in the landfill is 2.7 m above a reference datum located 200 m downgradient of the proposed landfill, and that there are boundary heads, as shown in Figure 11.19. Other heights of mounding and boundary heads could equally have been considered. Greater levels of leachate mounding would increase the potential impact compared with the cases considered here, and lesser levels of leachate mounding would decrease the potential impact.

The symmetry associated with the assumed hydrogeologic conditions implies that the center of the landfill will be a no-flow boundary, and that only half (i.e. 200 m of the landfill) needs to be considered in flow and contaminant transport modeling. This also implies that any flow in the aquifer at the edge of the landfill arises from the landfill, and that there is no dilution due to flow entering the aquifer upgradient of the landfill.

Figure 11.19 shows the general hydrostratigraphy considered. There are potentially six different hydrostratigraphic units, with vertical and horizontal hydraulic conductivities of k_{zl}, k_{xl}. Table 11.1 summarizes the values of k_{zl} and k_{xl} assumed for the seven cases examined. Where the aquitard is assumed to be fractured it has $k_z = 10^{-9}$ m/s, $k_x = 10^{-8}$ m/s. Where the lower aquitard is not fractured, it is assumed to have $k_z = 10^{-10}$ m/s, $k_x = 10^{-9}$ m/s. The upper unfractured clay and clayey liner are assumed to be isotropic with $k_z = k_x = 2 \times 10^{-10}$ m/s.

In order to perform the contaminant transport analyses, it is necessary first to determine the Darcy velocities in each of the layers. Depending on the complexity of the problem, this can be done by hand calculation or by using computer flow models. In this case, a finite element computer flow model was used to determine the flow field for each combination of hydraulic conductivity given in Table 11.1. Based on these flow analyses, representative Darcy velocities were selected for each case, as summarized in Table 11.2. The velocities v_{az1} and v_{az5} represent average downward velocities beneath the landfill. The velocities v_{ax4}, v_{ax6} are the horizontal velocities in the upper and lower aquifer at the edge of the landfill. Figure 11.20 schematically illustrates the basic cases considered. The details will be discussed in subsequent paragraphs. Table 11.3 summarizes the contaminant transport parameters adopted for the modeling using the program MIGRATE (version 8).

The reference height of leachate, H_r, was determined assuming an average thickness, H_w, and density, ρ_w, of waste of 8 m and 625 kg/m^3 respectively. Assuming that chloride is the contaminant species to be considered, and that

Table 11.1 Assumed hydraulic conductivities k (m/s) (Figure 11.19)

Case	k_{z1}	k_{x1}	k_{z2}	k_{x2}	k_{z3}	k_{x3}	k_{z4}	k_{x4}	k_{z5}	k_{x5}	k_{z6}	k_{x6}
1	10^{-9}	10^{-8}	10^{-9}	10^{-8}	10^{-9}	10^{-8}	10^{-6}	10^{-6}	10^{-9}	10^{-8}	10^{-5}	10^{-5}
2	10^{-9}	10^{-8}	10^{-9}	10^{-8}	10^{-9}	10^{-8}	10^{-6}	10^{-6}	10^{-10}	10^{-9}	10^{-5}	10^{-5}
3	2×10^{-10}	2×10^{-10}	10^{-9}	10^{-8}	10^{-9}	10^{-8}	10^{-6}	10^{-6}	10^{-9}	10^{-8}	10^{-5}	10^{-5}
4	2×10^{-10}	2×10^{-10}	10^{-9}	10^{-8}	10^{-9}	10^{-8}	10^{-6}	10^{-6}	10^{-9}	10^{-8}	10^{-5}	10^{-5}
5	2×10^{-10}	2×10^{-10}	10^{-9}	10^{-8}	2×10^{-10}	2×10^{-10}	10^{-6}	10^{-6}	10^{-9}	10^{-8}	10^{-5}	10^{-5}
6	2×10^{-10}	2×10^{-10}	10^{-9}	10^{-8}	2×10^{-10}	2×10^{-10}	10^{-6}	10^{-6}	10^{-10}	10^{-9}	10^{-5}	10^{-5}
7	2×10^{-10}	2×10^{-10}	10^{-9}	10^{-8}	2×10^{-10}	2×10^{-10}	10^{-5}	10^{-5}	10^{-10}	10^{-9}	10^{-7}	10^{-7}

Table 11.2 Flow parameters (Darcy velocities, Figures 11.19 and 11.20)

Case	v_{az1} (m/a)	v_{az5} (m/a)	v_{ax4} (m/a)	v_{ax6} (m/a)
1	0.0085	0.0074	0.23	0.74
2	0.004	0.0024	0.32	0.24
3	0.0057	0.005	0.14	0.5
4	0.00325	0.002	0.24	0.2
5	0.0043	0.0038	0.10	0.38
6	0.00273	0.0017	0.19	0.17
7	0.0039	0.00007	0.79	0.007

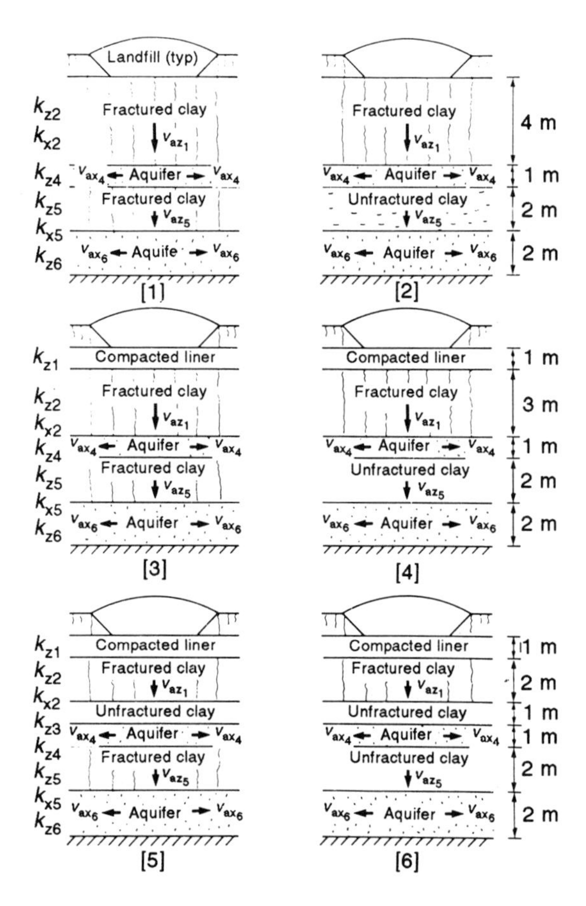

Figure 11.20 Schematic of the hydrostratigraphy for the cases considered. (After Rowe and Booker, 1991a; reproduced with permission of the *Canadian Geotechnical Journal*.)

Table 11.3 Common parameters used in transport model

Quantity	
Reference height of leachate, H_r (m)	10
Volume of leachate collected/area, q_c (m/a)	0.15
Initial concentration, c_0 (mg/l)	1000
Porosity of till/clay matrix (–)	0.4
Diffusion coefficient in matrix (m^2/a)	0.02
Distribution coefficient (m^3/kg)	0
Fracture spacing, $2H_x = 2H_y$ (m)	1
Fracture opening size, $2h_x = 2h_y$ (μm)	8.5
Coefficient of hydrodynamic dispersion along fracture (m^2/a)	0.06
Fracture distribution coefficient (m)	0
Porosity of aquifers (–)	0.3
Longitudinal dispersivity (m)	1
Transverse dispersivity (m)	0.1
Darcy velocity (m/a)	Table 11.2

it represents 0.2% ($p = 0.002$) of the waste, and that the peak concentration, c_0, is 1000 mg/l (1 kg/m^3), then H_r is given by

$$H_r = \frac{m_{TC}}{A_0 c_0} = \frac{pH_w \varrho_w}{c_0} = \frac{0.002 \times 8 \times 625}{1}$$
$$= 10 \text{ m}$$

The average volume of leachate collected (or escaping as surface seeps) per unit time per unit area is assumed to be 0.15 m^3/a/m^2. The fractured layers were considered to have orthogonal fractures at an average spacing of 1 m and average opening size of 8.5 μm.

The discussion in the following subsections relates to the particular example defined above. It should be noted that the interaction between different levels of fracturing and hydraulic conductivities in the different layers can be quite complex, as will become evident. Thus while this example serves to illustrate the application of the proposed theory and some important factors to be considered, care should be taken not to overgeneralize the results. Each landfill must be carefully evaluated as a unique situation.

11.4.3 Monitoring period and delay in impact

As illustrated in Figure 11.20 (and Table 11.1), cases 1 and 2 both consider a 4 m fractured zone with $k_z = 10^{-9}$ m/s between the landfill and the upper aquifer. Case 1 also assumes that the aquitard between the upper and lower aquifers is fractured, whereas case 2 assumes that it is not. Because of the higher transmissivity of the lower aquifer compared with the upper aquifer, and hence its ability to remove fluid from the system, a reduction in lower aquitard permeability has a significant effect on the Darcy velocity, v_{az1}, leaving the landfill (Table 11.2), which reduces from 0.0085 m/a (case 1) to 0.004 m/a (case 2). The Darcy velocity in the upper aquifer, v_{ax4}, is increased from 0.23 m/a (case 1) to 0.32 m/a (case 2) because more fluid must now be removed from this zone. The flow in the lower aquifer is only reduced by about a factor of three (i.e. from 0.74 m/a (case 1) to 0.24 m/a (case 2)), even though the hydraulic conductivity of the lower aquitard was reduced by one order of magnitude. This reduction results from the lower flux which can be passed through the unfractured aquitard.

Figure 11.21 shows the decrease in leachate strength with time determined for case 1. Also shown is a simple hand calculation of the leachate concentration based on the volume of leachate collected, q_c:

$$\frac{c_{0L}}{c_0} = \exp\left(\frac{-q_c t}{H_r}\right) \tag{11.4}$$

where c_{0L} is the concentration of the leachate at the time of interest, t; c_0 is the maximum leachate concentration; H_r is the reference height of leachate. Since the volume of leachate collected is large compared to the mass flux into the underlying soil, the simple hand calculation

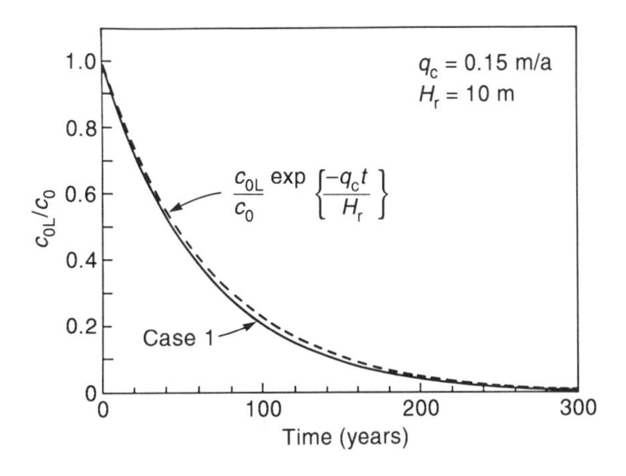

Figure 11.21 Variation in leachate with time. (After Rowe and Booker, 1991a; reproduced with permission of the *Canadian Geotechnical Journal*.)

gives a good estimate of the concentration in the landfill, compared with the more rigorous calculation performed using MIGRATE (version 8). Similar results to those for case 1 were obtained for each of the seven cases considered.

Figure 11.21 shows that for the assumed conditions, the average concentration of chloride in the leachate decreases to approximately half strength after about 45 years. After about 145 years, the leachate strength has reduced to 100 mg/l (assuming c_0 of 1000 mg/l). Assuming that the background concentration chloride in the groundwater is less than 50 mg/l, then according to policies such as Ontario's 'Reasonable Use Policy' (MoEE, 1993a), an increase in chloride concentration of up to 100 mg/l in the groundwater would be environmentally acceptable; hence one might be tempted to infer from this that there would be no need to monitor the landfill or groundwater after 145 years. That this is **not** the case is evident from Figure 11.22 which shows the calculated variation in concentration with time at two monitoring points. Assuming that for thin aquifers the monitoring wells would be screened across the entire thickness of the permeable unit, the concentrations c_α, c_β represent the average calculated concentration in the upper and lower aquifer respectively (i.e. the average of the calculated concentration at the top and bottom of each aquifer) at the edge of the landfill.

Figure 11.22 shows that if the mounding of leachate to the design elevation 102.7 m occurred quickly, then there would be no impact on either aquifer for at least 100 years. This would meet any regulatory requirements for no impact within 100 years but, eventually, there would be a significant impact. For case 1 (assuming both the upper and lower aquitards are fractured with $k_z = 10^{-9}$ m/s), the increase in chloride concentration in the upper aquifer is about 275 mg/l ($c/c_0 = 0.275$) after 140 years, even though the concentration in the leachate is only 100 mg/l at this time ($c/c_0 = 0.1$). The chloride concentration in the aquifers continues to increase until a peak impact of about 460 mg/l ($c/c_0 = 0.46$) is reached after approximately 250 years (at which time the concentration of chloride in the leachate is predicted to be less than 25 mg/l). This shows that even a fractured clayey aquitard can act as a buffer and can provide attenuation of leachate strength, since the peak concentration in the upper aquifer is only 46% of the peak value originally in the leachate. However, it is equally clear that the delay in the impact should also be considered when assessing the suitability of a site and required monitoring period; once contaminant enters the aquitard it may take a long time before it impacts on an underlying aquifer, in this case well beyond the time when the leachate strength itself has reduced to a level where it does not impose an environmental risk.

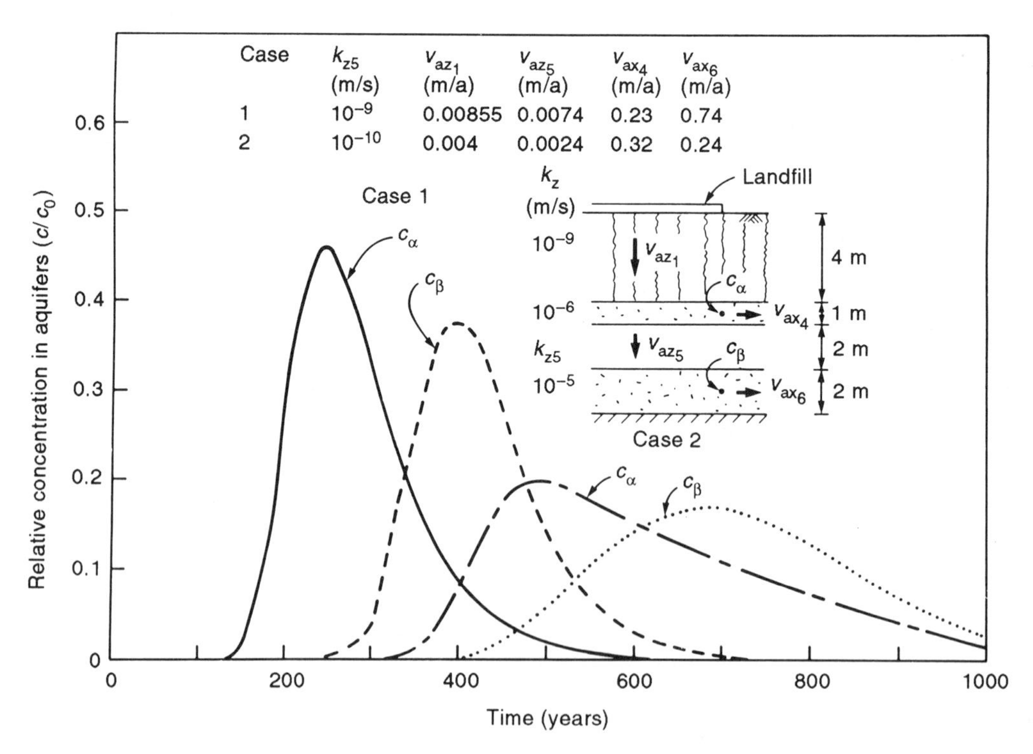

Figure 11.22 Calculated variation in concentration with time at two monitoring points, cases 1 and 2. (Modified from Rowe and Booker, 1991a; reproduced with permission of the *Canadian Geotechnical Journal*.)

The analyses for case 1 assume relatively minor fracturing of both aquitard layers (fractures with an opening size of 8.5 μm at 1 m spacing; $k_z \approx 10^{-9}$ m/s); however, even this minor fracturing can have a significant long-term effect, as shown in Figure 11.22. As discussed above, the peak impact in the upper aquifer of about 460 mg/l occurs after approximately 250 years. At this time there is negligible impact on the lower aquifer, but after another 100 years the lower aquifer is also significantly impacted, with a peak increase in chloride of 375 mg/l at about 400 years.

The only difference between case 1 and case 2 is the absence of fracturing of the lower aquitard for case 2. As discussed, this reduces the flow through both aquitards, and leads to an increase to the time of impact and a reduction in the magnitude of impact compared with case 1. Even though the average Darcy velocity through the upper aquitard is only 0.004 m/a, the peak impact of about 200 mg/l at approximately 500 years is twice that permitted in the Province of Ontario based on the 'Reasonable Use Policy', assuming a background concentration of 50 mg/l. It is also noted that there is significant impact on the lower aquifer, even though the lower aquitard is not fractured and has a hydraulic conductivity of 10^{-10} m/s.

11.4.4 The effects of fracture spacing, opening size, matrix porosity, matrix diffusion coefficient, and sorption into the matrix and onto fracture surfaces

There will always be some uncertainty regarding fracture spacing and opening size for fractured media. The bulk hydraulic conductivity will depend on both the spacing and opening size and will tend to be dominated by those fractures with a larger opening. Based on Hoek and Bray (1981), a theoretical relationship between hydraulic conductivity, k (m/s), fracture spacing, $2H$ (m) and opening size, e (m), is given by

$$k = \frac{(8.1 \times 10^5)e^3}{2H} \tag{11.5}$$

for one parallel set of fractures; and

$$k = \frac{(1.6 \times 10^4)e^3}{2H} \tag{11.6}$$

for orthogonal fractures.

The fracture spacing can be reasonably estimated (or bounded) on the basis of observations from test pits. The fracture opening size cannot usually be estimated from test pits, since the process of excavation can open the fractures, and even small changes (e.g. from 5 μm to 15 μm) can have a significant effect on hydraulic conductivity, as can be appreciated from equations 11.5 and 11.6. Thus the bulk hydraulic conductivity is usually obtained independently from field tests (e.g. section 3.2). Once the range of reasonable hydraulic conductivities is known for a given aquitard, one can then estimate fracture opening size from equation 11.5 or 11.6, based on the estimated fracture spacing.

Rowe and Booker (1990a, 1991a,b) have examined the effect of fracture spacing and matrix porosity. Based on these studies the following may be concluded.

1. Even if the bulk hydraulic conductivity is known, variations in fracture spacing will have a significant effect on the time at which contaminant impact occurs within an underlying aquifer. Often the greatest impact will occur for very closely spaced fractures; however, this is not always the case. For a given Darcy velocity, the effects of reasonable uncertainty regarding fracture spacing should be assessed by means of a sensitivity study which includes the assumption of no fractures as one limiting case.
2. For a given fracture spacing, increasing the thickness of the clayey aquitard between the landfill and the aquifer increases the time required for a given impact to reach the aquifer and decreases the maximum impact.

However, for practical ranges of fracture spacing, significant impact on the aquifer is possible, even if the fractured aquitard is up to 10 m thick.

3. The effect of uncertainty regarding Darcy velocity (e.g. as a result of uncertainty concerning hydraulic conductivity) is a critical consideration in modeling contaminant migration through the fracture system (e.g. section 11.3.5).
4. The value of effective matrix porosity used in the analysis can affect the magnitude and time of occurrence of impact on the aquifer, even if contaminant transport is through fractures.

As discussed in section 11.3.5, if the hydraulic conductivity is known from field tests, then the assumed fracture opening sizes used in the contaminant transport analysis do not have a significant effect on calculated impact for typical opening sizes in fractured aquitards.

As discussed in section 11.3.2, the matrix diffusion coefficient can also be an important parameter affecting contaminant transport through fractured systems. Often, the calculated impact will be greatest for the highest reasonable diffusion coefficient used; however, this is not always the case. When there is uncertainty concerning diffusion coefficient, the effect of that uncertainty can be assessed by a sensitivity study.

In summary, it has been found that uncertainty regarding fracture spacing, diffusion coefficient and effective porosity will influence the impact on the underlying aquifer. Because of symbiotic interaction, the effects of changing these parameters are not intuitively obvious, and sensitivity analyses should be performed for each project to assess the impact of uncertainty concerning these parameters.

As also discussed in section 11.3, sorption into the matrix and onto the surface of the fractures can significantly retard the movement of contaminants and reduce potential impact. The modeling of sorption into the matrix requires a knowledge of sorption characteristics, as discussed in section 8.6. At present, there are very few data concerning sorption onto the surface of fractures.

11.4.5 Evaluating the effects of a clay liner

Given that the impact for cases 1 and 2 is likely to be unacceptable, based on requirements for long-term protection of these aquifers, the question then arises as to what could be done to minimize impact. The magnitude of the impact could be reduced by reducing the height of leachate mounding, which is the cause of downward advective transport. For example, if the leachate collection system were to maintain a leachate level at an elevation of less than 100 m, then there would be inward flow into the landfill (a hydraulic trap), and the impact could be minimized, even though the aquitard is fractured. For case 1, the results shown in Figures 11.21 and 11.22 can be used to determine an attenuation factor $c_\alpha/c_0 = 0.46$, which is the ratio of maximum impact in the aquifer, c_α, divided by the peak source concentration c_0. Assuming that to ensure that the impact on the aquifer is not to exceed 100 mg/l, it is evident that the leachate collection system must function until the source concentration has reduced to a value c_{0L} which will not cause unacceptable impact. In this case, the concentration must reduce to $c_{0L} = 100/(c_\alpha/c_0) = 100/0.46 = 217$ mg/l. Using equation 11.4, one can then infer that the leachate collection system would have to function for about 100 years for case 1 (i.e. until the chloride concentration in the leachate c_{0L} reduced to $c_{0L}/c_0 = 217/1000 = 0.217$). For case 2, the corresponding period of operation would be about 45 years. This issue of longevity of leachate collection systems is discussed in more detail in sections 12.5, 12.6 and 12.7.

Another means of reducing impact would be to remove the upper 1 m of clayey aquitard below the proposed landfill base and rework it as a compacted clay liner. For the purposes of this example, it is assumed that the liner has a hydraulic conductivity of 2×10^{-10} m/s. Cases 3 and 4 consider a 1 m thick liner and either a fractured or unfractured lower aquitard respectively. As might be expected, installation of the liner decreases the Darcy flow from the landfill and, consequently, the impact on the aquifer (Figure 11.23). This landfill would readily meet regulatory requirements of no impact in 100 years. However, it is equally evident that this liner is not, of itself, enough, since the peak impacts shown in Figure 11.23 are still quite significant even though the time required for impact to occur is large.

11.4.6 Evaluating multiple-component aquitards

Cases 5 and 6 (Figures 11.20 and 11.24) consider the situation where the upper aquitard consists of 3 m of fractured clay or till ($k_{z2} = 1 \times 10^{-9}$ m/s) and 1 m of unfractured lacustrian clay ($k_{z3} = 2 \times 10^{-10}$ m/s). It is further assumed that the top 1 m of the fractured aquitard beneath the landfill is removed and recompacted as a liner with $k_{z1} \approx 2 \times 10^{-10}$ m/s. In case 5 the lower aquitard is assumed to be fractured ($k_{z5} \approx 1 \times 10^{-9}$ m/s); in case 6 it is not fractured ($k_{z5} = 1 \times 10^{-10}$ m/s). The presence of the unfractured clay layers in the upper aquitard serves to reduce flow and calculated impact as shown in Figure 11.24. For these cases, there is no impact within the first 100 years. The impact on the

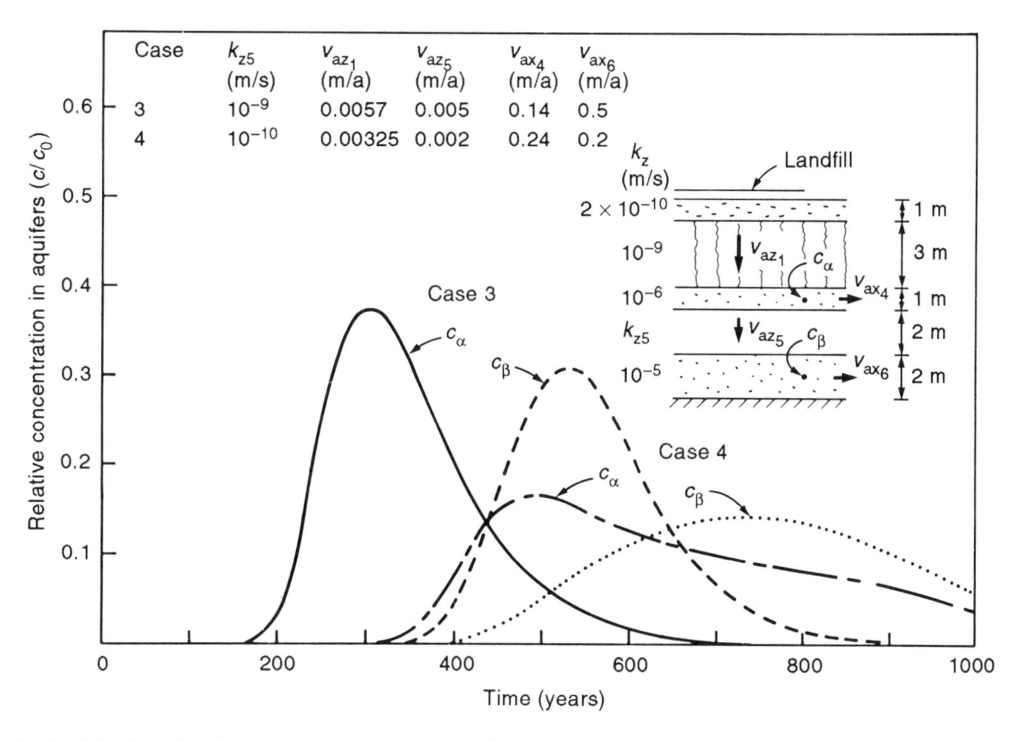

Figure 11.23 Calculated variation in concentration with time at two monitoring points, cases 3 and 4. (Modified from Rowe and Booker, 1991a; reproduced with permission of the *Canadian Geotechnical Journal*.)

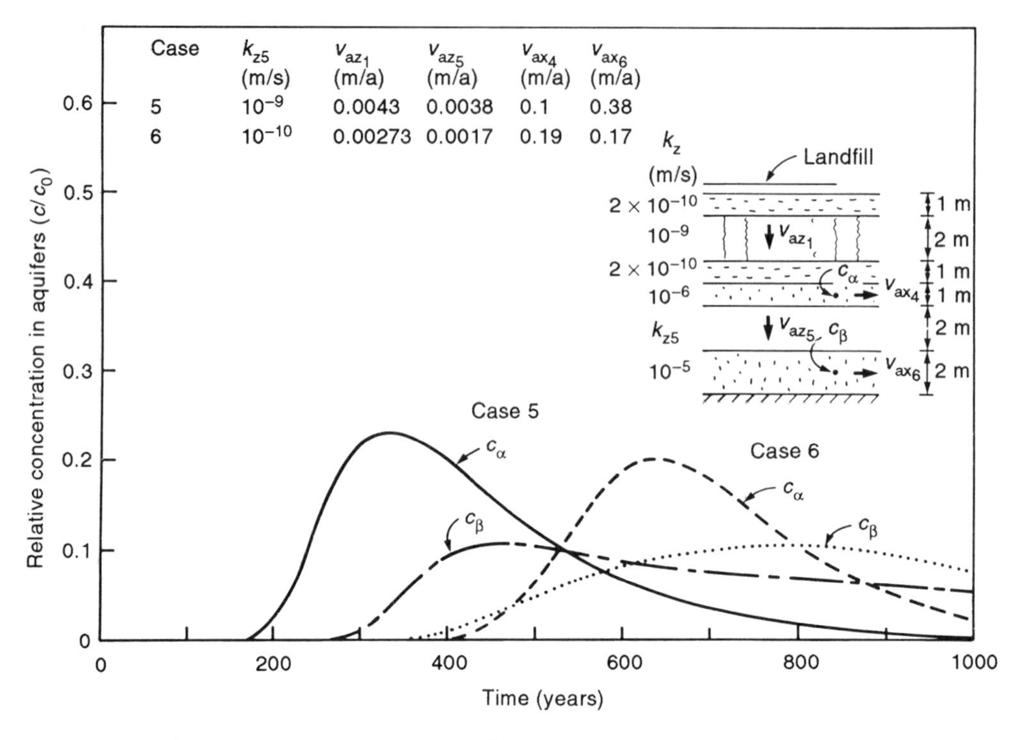

Figure 11.24 Calculated variation in concentration with time at two monitoring points, cases 5 and 6. (Modified from Rowe and Booker, 1991a; reproduced with permission of the *Canadian Geotechnical Journal.*)

lower aquifer is probably acceptable, based on 'Reasonable Use Criteria' (MoEE, 1993a); however, the impact on the upper aquifer is still unacceptable, based on these calculations.

At this point it is worth emphasizing the assumption made in modeling fractured flow; that contaminant is transported along the fractures rather than through the matrix of the fractured media. This may be a reasonable assumption when the fractures control the bulk hydraulic conductivity of the aquitard, and where there is an easy path for contaminant to pass into the fractures (e.g. cases 1 and 2). However, this assumption may not be valid when the fractured portion of the aquitard is sandwiched between two unfractured layers of clay, as for the upper aquitard in cases 5 and 6. For these cases, the modeling of migration primarily through the fractures assumes that once leachate passes through the liner, it spreads out and passes through the underlying fractures; this is equivalent to assuming that there is a thin sand layer between the liner and the fractured till which serves to distribute the 'leachate' to the fractures.

In case 6, discussed above, it was assumed that the fractures were the primary conduits for contaminant movement, and the hydraulic conductivity values used to estimate the Darcy velocities used in the contaminant migration analyses are based on this assumption. Alternatively, if there is a good contact between the fractured till and the unfractured soil above and below it, then the contaminant will not have an easy path to the fractures, and much of it is likely to pass through the matrix of the clayey soil between the fractures, rather than along the fractures. It is not a simple matter to determine what the true hydraulic conductivity would be for this case; clearly, it will lie between the limit obtained considering that the fractures are the primary conduit for fluid flow and the limit of the mass hydraulic conductivity of matrix material between the fractures. One could adjust the flows corresponding to the hydraulic conductivities associated with these two limiting cases, and then perform contaminant migration analyses for each case. This would represent the extremes of potential arrival times and impact. However, the matrix flow condition would be unconservative, since it totally neglects the presence of the fractures, and hence could not justifiably be used in estimating potential impact in an environmental assessment.

Given the uncertainty regarding what the bulk hydraulic conductivity would in fact be for this case, a conservative approach would be to assume the worst case hydraulic conductivity arising from the fractures being the primary conduit, and then use the corresponding flows to estimate potential impact, by performing contaminant transport analyses for the two cases where:

1. this flow is directed through the fractures, as assumed in Figure 11.24; or
2. this flow is through the clay matrix.

Option (2) will give earlier arrival times and impact than if the contaminant really did flow entirely through the matrix, because the flow has been estimated on the basis of fractured flow conditions. Figure 11.25 shows the effect of these two different assumptions on the

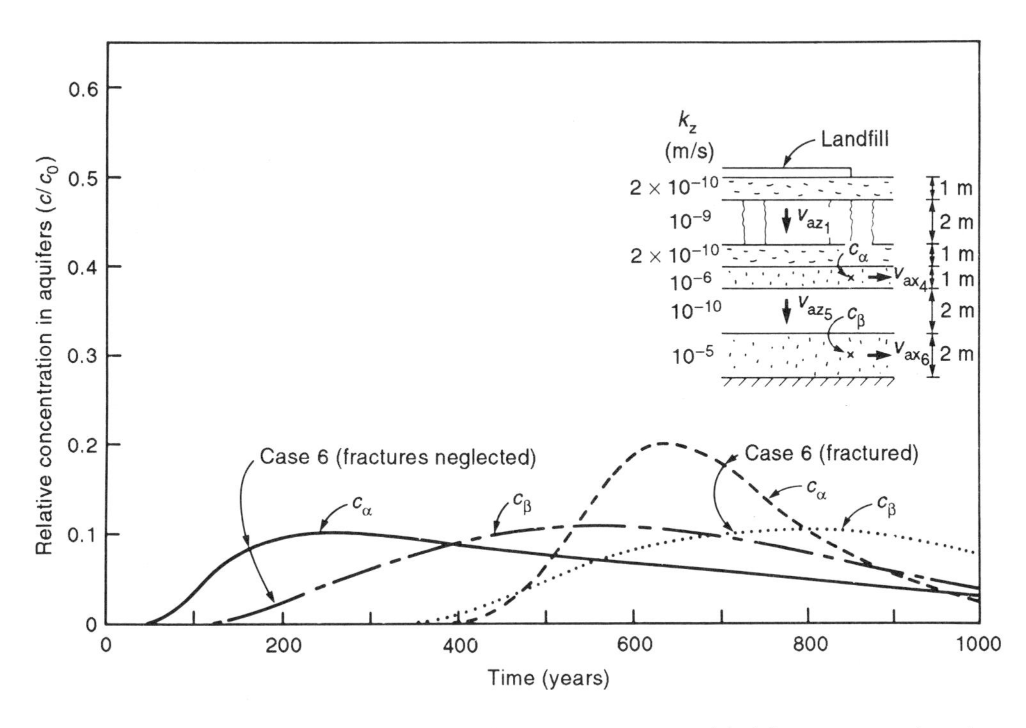

Figure 11.25 Effect of modeling strategy on calculated results, case 6. (Modified from Rowe and Booker, 1991a; reproduced with permission of the *Canadian Geotechnical Journal*.)

calculated impact for case 6. The results obtained for the 'fractured' assumption are the same as those shown in Figure 11.24. The results obtained for the 'unfractured' assumption show a faster arrival time but a lower impact on the upper aquifer than those obtained for the 'fractured' assumption. This counter-intuitive result arises primarily because matrix diffusion into the till, as contaminant moves along the fractures, provides a very effective retardation mechanism at the low Darcy velocity corresponding to case 6. If matrix diffusion were neglected, and only advective–dispersive transport along the fractures was considered, then the 'intuitive' result of very rapid arrival times would be obtained. In fact, it would only take about two days for contaminant to move through the fractured layer if matrix diffusion were neglected. However, matrix diffusion cannot be justifiably neglected, and it can provide very substantial attenuation, as is evident from Figure 11.25. As noted earlier, a contributing factor to the early arrival times for the 'no fracture' analysis is the conservative assumption that the flow through the material is equal to the flow through the fractures in the 'fractured' case. Thus, based on these two limiting cases, it is possible to make a conservative estimate of both the first arrival time, based on the 'no fractures' case, and of the peak impact, based on the 'fractured' case. Neither represents a 'prediction' of impact, since it is expected that reality lies somewhere between the results obtained from examining these limiting cases.

11.4.7 Evaluating hydraulic conductivity relationships between aquifers

The cases considered to this point have examined variations in the properties of the aquitards, and have assumed constant properties of the aquifer. Case 7 shows that the hydraulic conductivity of the aquifers is also important. Here the aquitards have the same properties as considered for case 6, but the hydraulic conductivity of the upper aquifer is increased from 10^{-6} m/s to 10^{-5} m/s and the hydraulic conductivity of the lower aquifer is decreased from 10^{-5} to 10^{-7} m/s. Thus in case 6 the lower aquifer is a significant drain for the system, taking almost 50% of the flow from the landfill, whereas in case 7 the upper aquifer is the major drain, taking almost all the flow, and this increase in the hydraulic conductivity of the upper aquifer increases the flow from the landfill, compared to case 6. Results were obtained assuming migration through fractures for the fractured till and migration through the matrix, in a similar fashion to that already discussed for case 6, and the results are shown in Figure 11.26.

Comparison of the results obtained for cases 6 and 7 shows that the higher Darcy velocity in the upper aquitard associated with case 7 gives rise to earlier impacts, compared to case 6. For the case of migration through the intact soil, the peak impact in the upper aquifer is also greater; however, this is not the case when one considers migration through the fractures. This situation arises because of the buffering effect that results from matrix diffusion from the fractures into the intact material. Thus, as has previously been discussed in section 11.3, there is a complex symbiotic relationship between the effects of Darcy velocity and matrix diffusion, and a higher Darcy velocity does not necessarily result in higher impact at a monitoring point when dealing with fractured media. When modeling contaminant migration in fractured media, it is important to examine a range of realistic possible situations in order to determine which combination of parameters provides the peak impact. This combination is not always obvious.

Assuming that regulatory requirements specify that one cannot increase chloride in the aquifer by more than about 100 mg/l, it is evident that for case 7 this requirement would be met in the lower aquifer but not in the upper aquifer at the edge of the landfill. One means of allowing additional reduction in impact is to acquire a natural attenuation zone. The impact

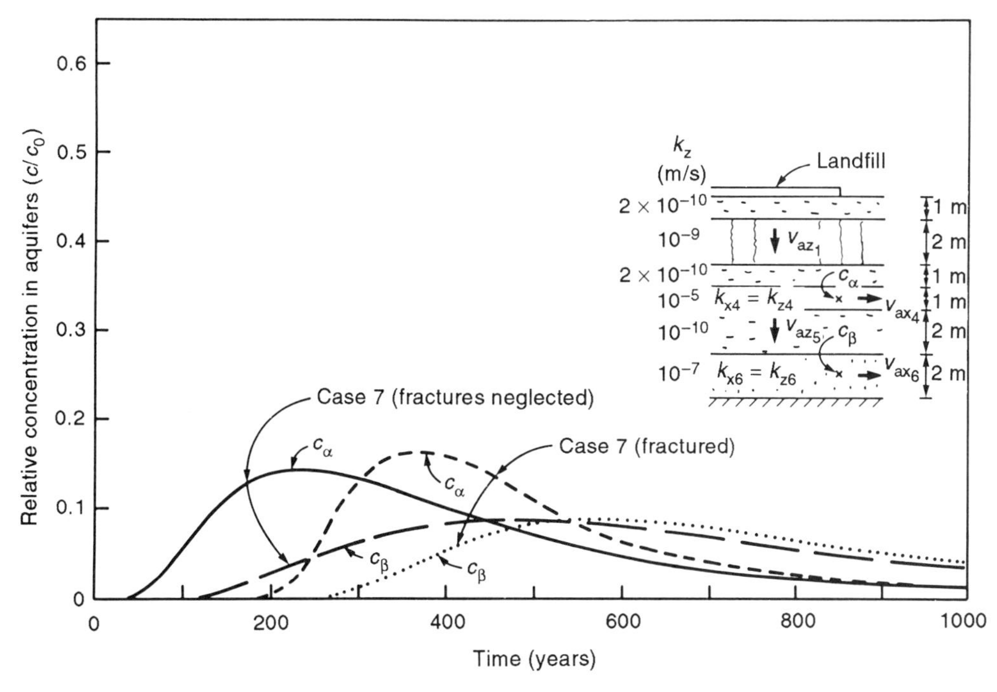

Figure 11.26 Effect of modeling strategy on calculated results, case 7. (Modified from Rowe and Booker, 1991a; reproduced with permission of the *Canadian Geotechnical Journal.*)

of the landfill on the upper aquifer decreases with distance away from the landfill, and in this particular case one could meet the 100 mg/l maximum increase in impact postulated here at a little over 100 m from the landfill.

When landfills are constructed on aquitards with a relatively low hydraulic conductivity (i.e. less than 10^{-9} m/s), the impact of that landfill upon underlying aquifers may not occur until a time long after closure of the landfill and indeed after the leachate itself has reached benign levels (e.g. Figures 11.22–11.26). The clay acts as a buffer, slowing movement into the underlying soil, but the impact, although long in coming, can be significant, and can last for decades to centuries. Regulations which only consider a limited (e.g. 100-year) period in considering potential impact fail to recognize the long-term implications of designing landfills with clay liners. For example, in case 7, if the site boundaries were 30 m from the edge of the landfill, the increased impact would exceed 100 mg/l for a period of about 225 years.

11.5 Summary

This chapter has examined contaminant migration through fractured media. Particular consideration has been given to lateral migration in fractured porous rock and vertical migration through fractured clayey aquitards, although the factors discussed are equally appropriate to lateral migration through fractured clayey aquitards (e.g. from the edge of a landfill) or vertical migration through a fractured porous rock. Attention has been focused on the migration of contaminants at concentrations typical of that encountered in domestic waste leachate. The migration of concentrated dense nonaqueous phase liquids (DNAPLs) has not been considered, and it should be noted that these may move rapidly through fractures with little or no attenuation. Concentrated DNAPLs should not be stored or disposed of in (or adjacent to) fractured media without the provision of a multiple component engineered barrier system (likely to incorporate geomembranes and compacted clay (section 4.3)).

It has been shown that diffusion of contaminants from fractures into an adjacent matrix of porous rock like shale and sandstone, or soil such as clay and till, can result in significant retardation of contaminant movement and attenuation of peak impact, compared with that which would be predicted neglecting matrix diffusion. It has also been shown that there is a symbiotic relationship between Darcy velocity, fracture spacing, matrix porosity and matrix diffusion coefficient which needs to be carefully examined on each particular project.

The development of any model invariably involves making idealizations of what, in reality, are complex situations. The model examined herein is no exception. To date, there are very few field data to allow verification of models which consider contaminant migration through fractured systems, such as those considered in this chapter. The reader should be aware of the assumptions that have been made in developing the model, and should use engineering judgement when assessing the applicability of idealized models in practical projects.

It is worth emphasizing again that when landfills are constructed on relatively 'tight' aquitards (i.e. hydraulic conductivity of 10^{-9} m/s or less), the clay acts as a buffer, slowing movement into the underlying soil, but the impact, although long in coming, can be significant, and can last for significant periods of time. Regulations which only consider a limited (e.g. 30- or 100-year) post-closure period for assessing potential impact fail to recognize the long-term implications of advective–diffusive transport. It also follows from this that one cannot terminate monitoring of a receptor aquifer simply because the leachate strength has reached benign levels. The monitoring program should be linked to predictive modeling; together, these can provide an indication of when monitoring can be terminated.

Chapter twelve

Integration of hydrogeology and engineering in the design of barriers and assessment of impact

12.1 Introduction

In order to design safe waste disposal sites (e.g. landfills), it is essential that the interaction between the natural and engineered systems be considered (Rowe, 1992). Thus the objective of this final chapter is to discuss some of these interactions, paying particular attention to factors such as the evaluation of hydrogeology in the context of different engineering designs (section 12.2); the different characteristics associated with a number of engineering designs (section 12.3); the effect of the landfill on natural groundwater levels (section 12.4); the long-term performance of leachate collection systems (sections 12.5–12.7); the effects of landfill size and the infiltration rate through the cover on the contaminating lifespan of a landfill (sections 12.6 and 12.7); and finally the modeling of the service life of primary and secondary systems (section 12.8).

12.2 Hydraulic conductivity of aquitards

It is well-recognized that in the design of systems where there will be gradients, and hence flow out of the waste disposal facility, that the bulk hydraulic conductivity of the 'barrier' or aquitard will be a key factor affecting potential impact of the landfill on an underlying aquifer. All other things being equal, the greater the bulk hydraulic conductivity, the greater will be the outward advective flow, the greater the escape of leachate and hence the greater the potential impact on the aquifer.

In these situations, a major consideration is whether the aquitard is fractured, as discussed in Chapter 11. As will be demonstrated herein, for a fractured aquitard, the presence of an unfractured layer of natural clayey soil or an artificial liner may have significant benefits. In particular, the installation of a liner over a fractured material may result in the performance of the combined system being substantially better than that of either the liner alone or fractured aquitard alone. This will be examined in section 12.2.1.

When dealing with outward gradients and situations such as shown in Figure 12.1, it is generally conservative to adopt the highest reasonable hydraulic conductivity for the purpose of impact assessment. However, when dealing with inward gradient designs (i.e. hydraulic traps) as shown in Figure 12.2 this conventional approach is no longer valid. Rather, the greatest impact is likely to occur for the lowest reasonable hydraulic conductivity, as discussed in section 12.2.2. Thus, although the hydrogeology may be exactly the same for Figures 12.1 and 12.2, the only difference being the depth of the base contours adopted in the engineering design, this difference is critical, and results in a different approach to viewing the hydrogeology and uncertainty associated with the hydraulic conductivity of both the natural and engineered components of the barrier system.

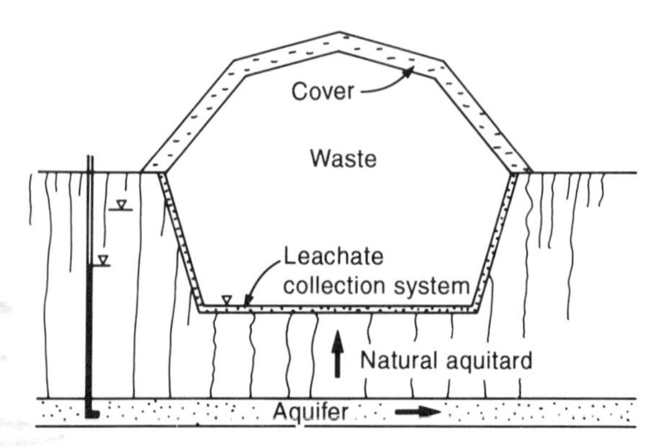

Figure 12.2 Schematic showing landfill located in a fractured aquitard – upward/inward gradients (hydraulic trap design).

12.2.1 Outward gradient design

To illustrate the potential significance of hydraulic conductivity, consideration will be given to three cases involving contaminant migration from a 200 m wide (L = 200 m) landfill separated from a 1 m thick sand aquifer (h = 1 m, n_b = 0.3) by 4 m of clayey aquitard:

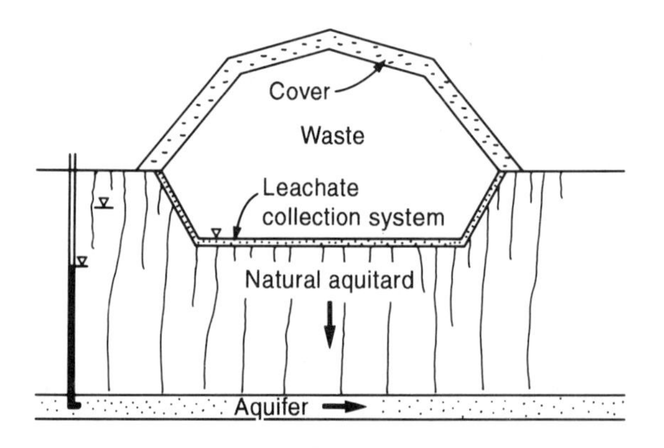

Figure 12.1 Schematic showing landfill located in a fractured aquitard – downward/outward gradients.

1. where there is 4 m of native fractured clay (H_T = 4 m, H_L = 0) between the landfill and the aquifer;
2. where the upper 1 m of fractured clay is removed and then recompacted to give a 1 m thick intact clayey liner (i.e. H_L = 1 m) underlain by 3 m (H_T = 3 m) of the original fractured clay;
3. considering only the 1 m thick recompacted clayey liner (H_L = 1 m) in isolation, any retardation of contaminant as it moves through the fractured clay being neglected, although the hydraulic conductivity of the fractured clay is considered in determining the Darcy velocity, which is, therefore, the same as for case 2.

The range of parameters considered for this example (example A) is summarized in Table 12.1. Here, it is assumed that there is a leachate

Table 12.1 Summary of parameters considered for example A

Quantity[a]	*Value*[b]
Width of landfill, L (m)	200
Reference height of leachate, H_r (m)	5
Infiltration through cover, q_0 (m/a)	0.25
Initial concentration, c_0 (mg/l)	1500
Downward Darcy velocity, v_a (m/a)	0.0001, **0.003**, 0.0058, 0.0127, 0.0166, 0.021, 0.05
Thickness of fractured clay, H_T (m)	3, 4
Porosity of clay matrix, n_m (–)	0.25, 0.3, 0.35, **0.4**
Hydraulic conductivity of fractured till, k_T (m/s)	6.3×10^{-9}
Thickness of clay liner, H_L (m)	0, **1**
Hydraulic conductivity of liner, k_L (m/s)	$\mathbf{1 \times 10^{-10}}$, 2×10^{-10}, 5×10^{-10}, 7×10^{-10}, 1×10^{-9}
Thickness of underlying aquifer, h_b (m)	1
Porosity of aquifer, n_b (–)	0.3
Horizontal Darcy velocity in aquifer, v_b (m/a)	$v_b = v_a L + 2$
Diffusion coefficient in matrix, D_m (m^2/a)	0.02
Retardation coefficient for matrix, R_m (–)	1
Coefficient of hydrodynamic dispersion along fractures, D (m^2/a)	0.06
Fracture spacing, $2H_1 = 2H_2$ (m)	1
Fracture opening size, $2h_1 = 2h_2$ (μm)	10
Diffusion coefficient through 2 mm thick geomembrane (when used) (m^2/a)	0.0001

Note: 'a' denotes 'annum' (year).
Bold values represent those used unless otherwise noted (Base Case).

collection system, that the difference in head between the leachate in the landfill and the water in the aquifer is 1 m, and that the bulk hydraulic conductivity of the fractured clay, k_T, is about 6×10^{-9} m/s. The Darcy velocity, v_a, through the liner and fractured aquitard was then calculated by first determining the harmonic mean of the liner and fractured aquitard for a range of assumed hydraulic conductivities for the liner (1×10^{-10} m/s $\leqslant k_L \leqslant 1 \times 10^{-9}$ m/s) and then multiplying by the gradient, which was held constant, since it was controlled by the height of mounding in the landfill and the head distribution in the far more permeable aquifer (this technique is discussed in section 5.2.2). As might be expected, the less permeable the liner, the smaller the Darcy flow through the system for a fixed head loss.

(a) Impact with a 1 m thick clay liner ($k_L \approx 10^{-9}$ m/s)

The calculated variation in concentration, with time, in the aquifer is shown in Figure 12.3 for three different assumed conditions. Case 1 considers 4 m of fractured clay separating the landfill from the aquifer. For the assumed conditions, the fracture porosity is 0.00002 and corresponds to a groundwater velocity in the fractures of 2500 m/a. If one were to consider simple 'plug' (i.e. advective) flow in the fractures, then full strength leachate should reach the aquifer in less than one day. However, if one considers migration along the fractures coupled with matrix diffusion into the adjacent clay, the arrival time for contaminant in the aquifer is substantially increased. For example,

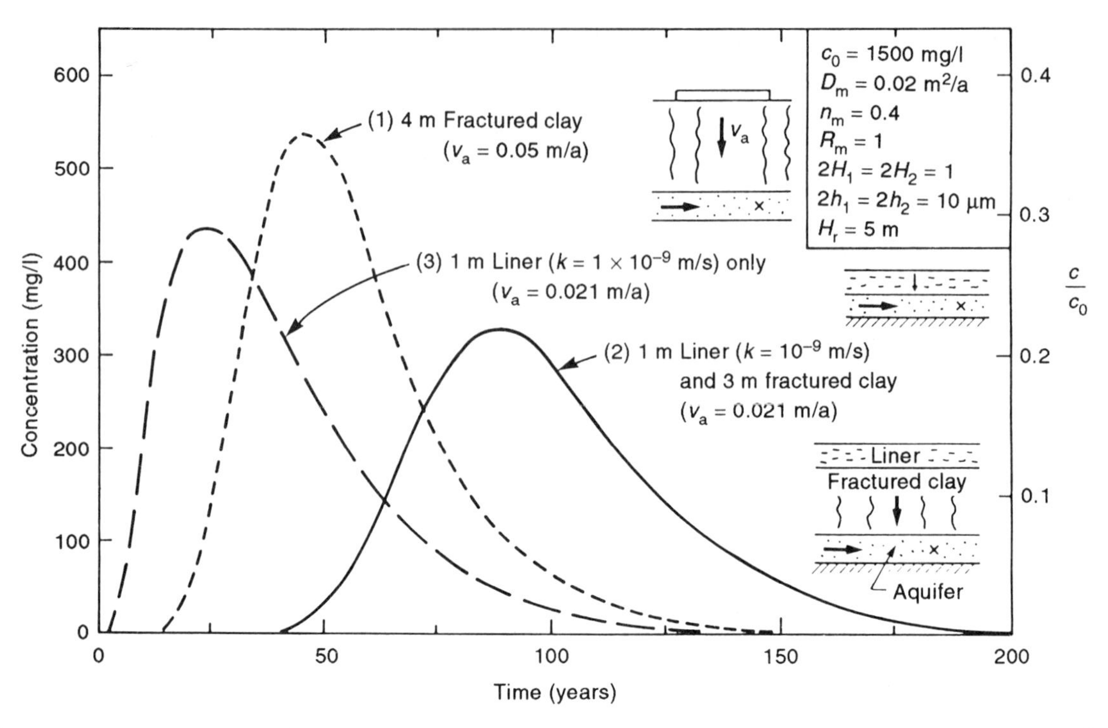

Figure 12.3 Impact on groundwater quality in the aquifer as a function of time with and without a 1×10^{-9} m/s compacted liner: example A. (Modified from Rowe and Booker, 1991b; reproduced with permission of American Society of Civil Engineers.)

it takes between 16 and 17 years for contaminant to reach a concentration of 1% of the initial source leachate (i.e. $c/c_0 = 0.01$).

The important role of matrix diffusion (i.e. the migration of contaminants between the fractures and the pores of the adjacent intact material due to diffusion from high concentration to low concentration) in retarding contaminant migration through fractured media was discussed in detail in Chapter 11. However, it is evident that despite the attenuation due to matrix diffusion, there is a substantial increase in chloride concentration in the aquifer, to a peak value of about 540 mg/l ($c/c_0 \approx 0.36$) after about 45 years. This would generally be regarded as being unacceptable.

Case 2 shows the calculated impact, assuming that the top 1 m of the fractured clay is excavated and recompacted as a liner having a hydraulic conductivity of 10^{-9} m/s. The calculations involved modeling both the migration through the 1 m thick liner, and then along the fractures for the remaining 3 m of aquitard above the aquifer. This analysis considers both the reduced flow into the subsoil due to the presence of the liner and the attenuation due to diffusion into the matrix of the fractured aquitard. As a consequence, the time for contaminants to reach the aquifer at the 1% level is increased from about 17 years to about 45 years. The peak impact is also reduced from 540 mg/l at 45 years without a liner to 328 mg/l at 88 years with **this** liner. Nevertheless, the impact is still excessive, and would generally be unacceptable. This impact may be reduced using a more effective liner, as will be discussed momentarily.

As noted above, case 2 considers attenuation

due to matrix diffusion in the fractured aquitard below the liner. Case 3 shows the results that are obtained if this attenuation is neglected. As can be seen, consideration of attenuation due to the liner alone (case 3) results in a prediction of contaminant impact which is much earlier than that obtained for case 2. (The issues of uncertainty regarding fracture spacing, opening size etc. for fractured media are discussed in Chapter 11.)

(b) Impact with a 1 m thick compacted clay liner ($k_L \approx 10^{-10}$ m/s)

The results presented in Figure 12.3 were obtained assuming a clay liner hydraulic conductivity of 10^{-9} m/s. As discussed in Chapter 9, there is substantial evidence to suggest that the bulk hydraulic conductivity of a well-designed and constructed clay liner may be substantially less than 10^{-9} m/s, and may be of the order of 10^{-10} m/s or even lower. To illustrate the benefits of a lower hydraulic conductivity of a liner, Figure 12.4 shows the calculated variation in concentration with time for three cases. The results shown for case 1 (4 m of fractured aquitard) are precisely the same as those shown in Figure 12.3 (although the time axis has been changed), and the previous discussion applies to these results. Case 2 gives the results for a 1 m thick, $k_L = 1 \times 10^{-10}$ m/s liner overlying 3 m of fractured aquitard. It can be seen that the reduced flow and mass loading on the aquifer arising from the use of the 1×10^{-10} m/s liner has a pronounced effect on the concentration with the aquifer. The time of arrival of contaminant at the 1% level in the aquifer is increased from about 17 years without the liner,

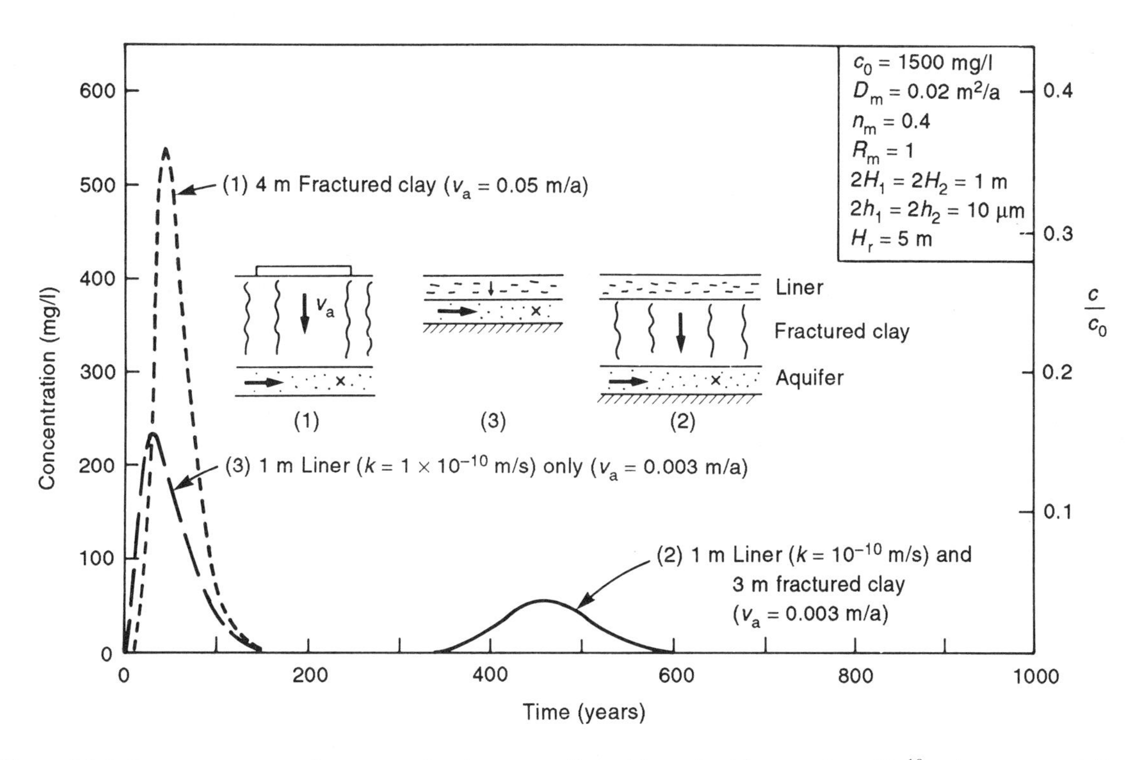

Figure 12.4 Impact on groundwater quality in an aquifer with and without a 1×10^{-10} m/s compacted liner: example A. (Modified from Rowe and Booker, 1991b; reproduced with permission of American Society of Civil Engineers.)

to about 380 years. The maximum impact is reduced by an order of magnitude from 540 mg/l at 45 years with no liner, to 54 mg/l at 458 years with a 10^{-10} m/s liner.

To illustrate the effect of considering the attenuation which occurs in the fractured aquitard, the results from case 2 may be compared with the results for case 3 which neglects any attenuation in the fractured aquitard. For the situation considered here, the analysis which neglects matrix diffusion gives arrival times much earlier and an impact much greater than that obtained when one considers matrix diffusion in the fractured aquitard.

(c) Effect of clay liner hydraulic conductivity

Comparison of the results given in Figures 12.3 and 12.4 demonstrates the importance of the hydraulic conductivity of the liner in controlling contaminant impact on the aquifer. Figure 12.5 summarizes the magnitude of peak impact for a range of assumed values of liner hydraulic conductivity, k_L. As would be expected, a lower hydraulic conductivity results in smaller impact. The ten-fold variation in k_L results in a more than six-fold decrease in impact on the aquifer and a five-fold increase, from 90 years to 450 years, in the time to reach peak impact. A further reduction in impact could be achieved by increasing the thickness of the reworked clayey liner.

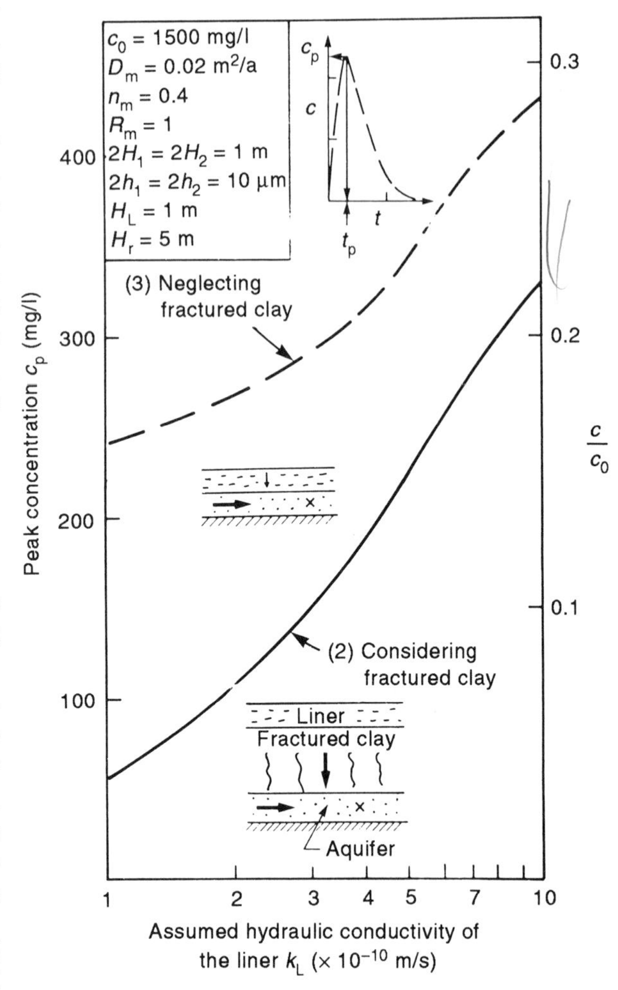

Figure 12.5 Effect of hydraulic conductivity of the liner on calculated impact in the underlying aquifer – outward gradients: example A. (Modified from Rowe and Booker, 1991b; reproduced with permission of American Society of Civil Engineers.)

(d) Effect of a composite liner

Geomembranes have gained wide acceptance as liners for waste disposal facilities in the USA and some parts of Europe. For the purposes of illustrating the potential benefit of a composite geomembrane and compacted clay liner, it is assumed here that the geomembrane is well-designed and constructed over a 1 m thick compacted clay liner which, in turn, overlies 3 m of fractured aquitard, as discussed in the previous paragraphs. The advective flow out of the landfill is controlled by the geomembrane, and is reduced to an averge 0.0001 m/a or less (i.e. less than about 3 l/ha/day) in this example. The diffusion coefficient through the 2 mm thick HDPE geomembrane is taken to be 0.0001 m²/a (see section 2.5.2 for a discussion of leakage through holes in geomembranes and diffusion through geomembranes).

Curve (i) in Figure 12.6 shows the calculated variation in concentration with time in the aquifer for this lining system. In this case, contaminant transport is controlled by diffusion through the geomembrane, the compacted clay and the natural aquitard. The fractures play no significant role, since the advective flow is so small, as a result of the composite liner. Also, the precise 'leakage' through the composite liner does not matter, provided that it is less than or equal to 0.0001 m/a, since this flow is already so low that diffusion through the geomembrane controls. The peak impact is less than 21 mg/l, and this is not reached until about 140 years. Comparing these results with those obtained using a clay liner alone (e.g. Figures 12.3–12.5), it can be seen that there is a substantial benefit to the inclusion of the geomembrane, for so long as the geomembrane continues to be a good barrier.

The service life of engineered components was discussed in section 2.8. For the purposes of this section, assume that the geomembrane has a service life of 50 years, during which the average advective velocity is controlled to 0.0001 m/a (or less), following which the membrane is assumed to degrade over 25 years, such that after 75 years the problem is reduced to that of a clay liner over the natural aquitard. The increase in chloride concentration with time, calculated using the program POLLUTE (Rowe and Booker, 1994), is plotted in Figure 12.6 for the cases of a compacted clay liner having a hydraulic conductivity for curve (ii) of 10^{-10} m/s and for curve (iii) of 10^{-9} m/s.

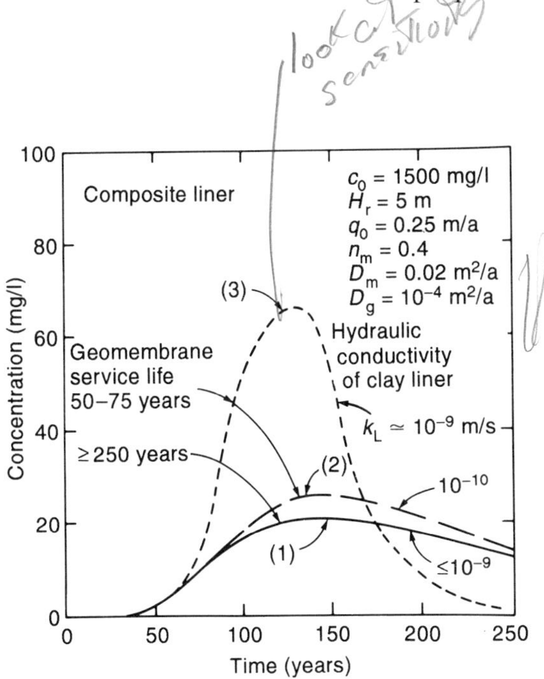

Figure 12.6 Calculated impact on an aquifer for a composite liner design: example A.

There is a clear benefit to the effective presence of the geomembrane for a period of 50–75 years; however, it is evident that with finite service life for the geomembrane, the ultimate impact is controlled by the thickness and hydraulic conductivity of the clay liner and natural aquitard system. It is also noted that a failure of the geomembrane in the 50–75-year period would not become evident from monitoring the aquitard until between about 75 and 100 years, even for a 10^{-9} m/s liner, and the peak impact would not occur until about 130 years. This raises questions as to the applicability of limited monitoring periods (e.g. 30 years post-closure) when designing systems such as this. The effect of a limited monitoring period may be even more pronounced for a larger landfill.

It has been assumed in this section that the clay liner will remain intact at its design hydraulic conductivity. As discussed in Chapter 9, this is likely to be a reasonable assumption, provided that the liner material is appropriately selected (Chapters 3 and 4) and that it is not permitted to desiccate (section 2.9.6).

12.2.2 Inward gradient

Section 12.2.1 considered the effect of varying hydraulic conductivity, assuming that the potentiometric surface in the aquifer was 1 m below the level of leachate mounding in the aquifer, and hence that there were gradients outward from the landfill to the aquifer. To illustrate the different roles played by hydraulic conductivity with inward gradients, this section considers a similar thickness of aquitard and a compacted clay liner, but assumes that the potentiometric surface in the aquifer is 1 m **above** the leachate level in the landfill (i.e. the difference in head $\Delta h = -1$ m), thereby inducing an inward gradient. For the case of inward gradients, contaminants will migrate outwards from the landfill due to the process of molecular diffusion. This will be opposed by the inward flow of water to the landfill.

(a) Hydraulic trap with permeability of natural aquitard ($k_T \approx 6 \times 10^{-9}$ m/s)

Curve (i) in Figure 12.7 shows a plot of peak concentration in the aquifer as a function of liner hydraulic conductivity for a 1 m thick liner and 3 m of natural aquitard with a hydraulic conductivity of about 6×10^{-9} m/s; this corresponds to case 2 of example A, examined in the previous section. The peak increase in concentration in the aquifer is minimal for a liner hydraulic conductivity of 1×10^{-9} m/s (10×10^{-10} m/s) because the inward flow of 0.021 m/a overcomes most of the outward diffusion.

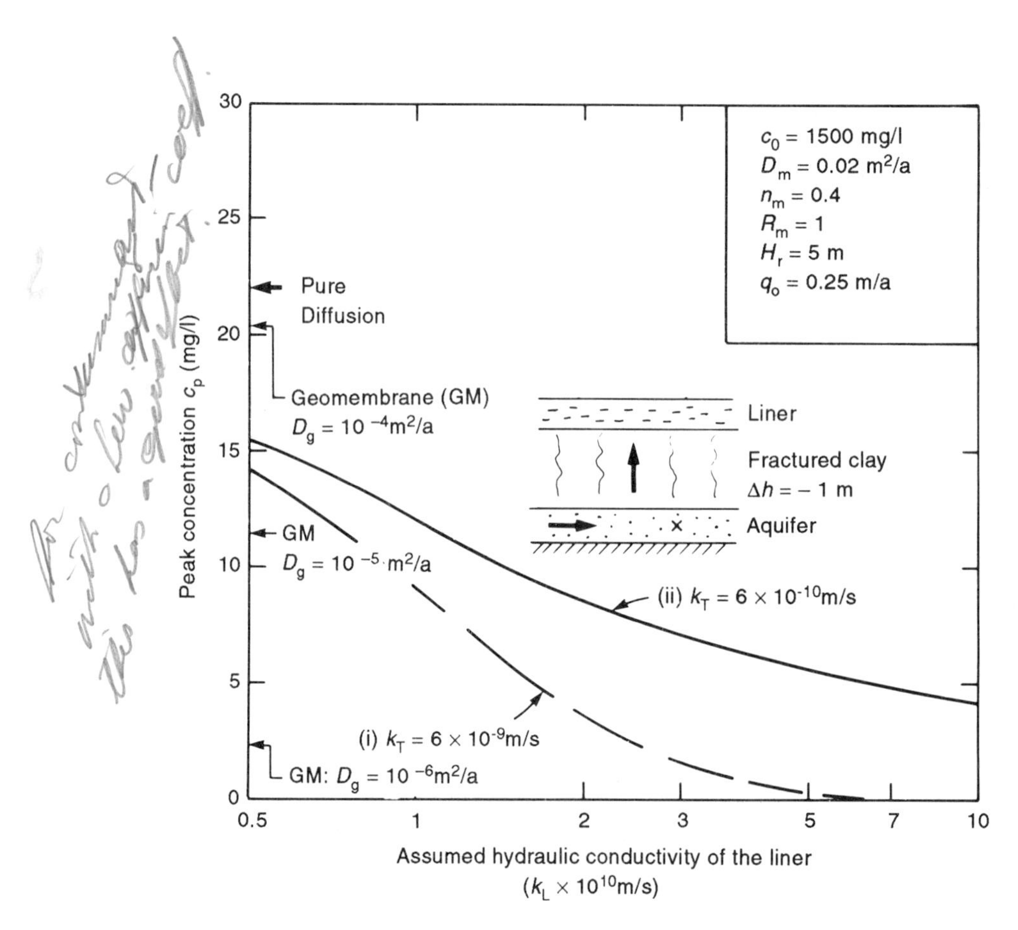

Figure 12.7 Effect of hydraulic conductivity of the liner on calculated impact in the underlying aquifer – inward gradient: example A.

As the hydraulic conductivity of the liner is reduced, so too is the inward flow, and the calculated impact increases to a maximum value for the hydraulic conductivity of 0.5×10^{-10} m/s. On the other hand, if there was no liner, and advection was controlled by flow through the fractures, then the inward flow would not counteract diffusion through the matrix and, depending on the degree of fracturing, the impact would lie between that corresponding to the hydraulic conductivity of the intact clay and that shown for pure diffusion in Figure 12.7.

(b) Hydraulic trap with permeability of natural aquitard ($k_T \approx 6 \times 10^{-10}$ m/s)

The results discussed above were obtained assuming that the fractured aquitard had a bulk hydraulic conductivity of about 6×10^{-9} m/s. While this may have been a conservative interpretation of the bulk hydraulic conductivity of the aquitard for purposes of impact assessment with outward flow, it is not conservative for assessing impact due to inward flow. If the fractures were not hydraulically significant, and a reasonable lower bound hydraulic conductivity of the natural clayey aquitard was 6×10^{-10} m/s, then greater contaminant impacts are predicted as shown by curve (ii) in Figure 12.7. This illustrates the need to consider the design being examined when assessing reasonable and conservative hydraulic conductivities to use in contaminant impact calculations.

(c) Implications of using a geomembrane/composite liner

A well-constructed geomembrane liner installed over the compacted clay liner would be likely to reduce the flow into the landfill to a negligible amount for as long as the geomembrane was functional. This would effectively destroy the hydraulic trap, except at the location of defects in the geomembrane. The maximum impact would depend on the diffusion coefficient through the geomembrane with the maximum impact of 22 mg/l being defined by the pure diffusion case. Thus by destroying what would otherwise be a hydraulic trap, the installation of the geomembrane would degrade the performance of the system, compared to the results shown in curves (i) and (ii) for the case where there was no geomembrane. Furthermore, the water pressures on the base of the geomembrane may lead to potential problems with the geomembrane, unless they are controlled while the waste is being placed.

The only difference between the 'geomembrane' case and the 'pure diffusion' case arises from the effectiveness of the geomembrane as a diffusion barrier. This is likely to be species-specific, and more research is required. However, limited available evidence would suggest that the geomembrane is likely to be a better diffusion barrier for inorganic anions such as chloride than it is for, say, manufactured solvents such as dichloromethane. Thus in the design of these systems, careful consideration should be given to the effect of diffusion coefficients if the geomembrane is to be used as a diffusion barrier. The analyses indicate that if the diffusion coefficient through the geomembrane is higher than about 10^{-4} m^2/a, then the geomembrane has negligible effect in restricting diffusive contaminant movement. Figure 12.7 shows the calculated impact, assuming pure diffusion, for three assumed values for diffusion coefficients through the geomembrane, D_g, of 10^{-4}, 10^{-5} and 10^{-6} m^2/a. At the time of writing, there is no published evidence to support a diffusion coefficient less than 5×10^{-5} m^2/a; however, the authors' current research addressing this issue indicates that for some ionic contaminants D_g may be as low as 10^{-7} m^2/a.

12.2.3 Highly engineered systems

In the previous two subsections it was shown that the critical estimate of hydraulic conductivity corresponds to a high value for outward gradients and a low value for inward gradients. For designs such as that shown in Figure 12.8

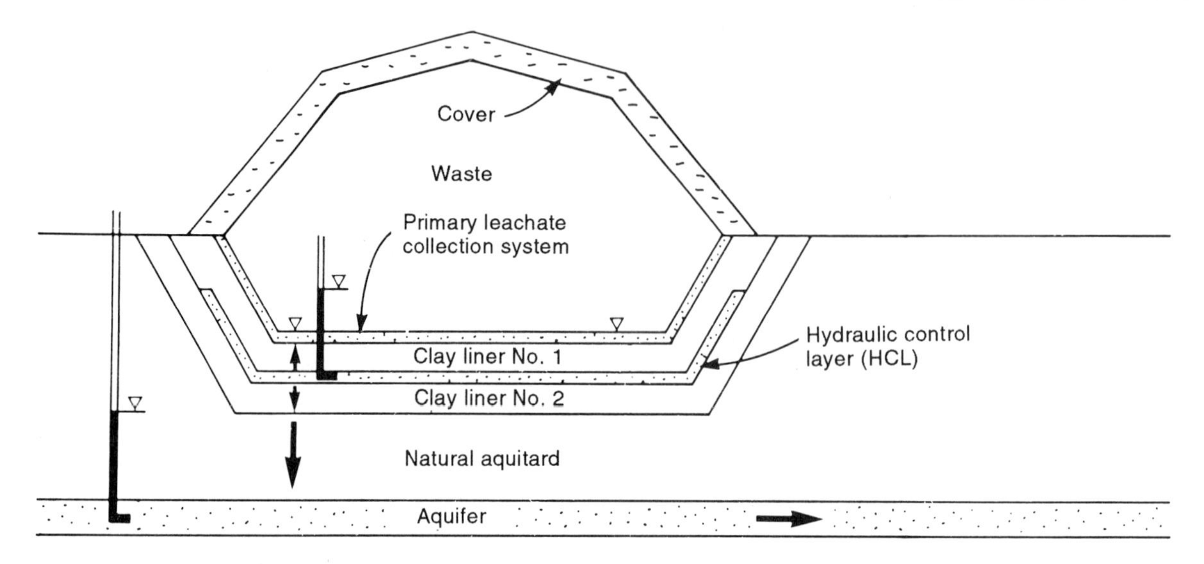

Figure 12.8 Schematic of a highly engineered system with a stone layer beneath the primary liner which can be operated as an active hydraulic control layer.

for water level (a), this is no longer the case, and it is not obvious as to what constitutes the critical hydraulic conductivity for the secondary liner (if present) and aquitard. With this design the hydraulic control layer (HCL) is pressurized to induce an inward gradient from the HCL through the primary liner, thereby inducing an engineered hydraulic trap. (See section 1.2 for a more detailed discussion of these systems.)

For the cases considered in this section, it is assumed that the potentiometric surface in the aquifer is below that in the HCL; hence there will be flow from the HCL to the aquifer. Thus contaminant, which can escape through the primary liner to the HCL by diffusion, will then be carried to the aquifer by advective–diffusive transport. If the hydraulic conductivity of the secondary liner and aquitard is low, then contaminant transport through the secondary liner–aquitard will be primarily by diffusion. This is slow, but there will be negligible dilution of contaminant in the HCL to reduce concentrations unless there is active pumping of this unit. If the hydraulic conductivity of the secondary liner and aquitard is high, then contaminants reaching the HCL by diffusion will be carried to the aquifer primarily by advection. This is fast, but there is also considerable potential for dilution of contaminant in the HCL because of the volume of water that must be injected to maintain the engineered hydraulic trap. For this type of design, the hydraulic conductivity that would give rise to the maximum impact may not be either the highest or lowest reasonable hydraulic conductivity value. Rather, it may be some intermediate value which can only be established by performing a sensitivity study, as illustrated below.

(a) Effect of secondary liner–aquitard hydraulic conductivity

Figure 12.8 shows a design involving a primary liner, a hydraulic control layer and an aquitard that may include a compacted secondary liner. This case will be referred to as 'example B'. If it is assumed that the primary liner has a hydraulic conductivity $k_1 = 10^{-10}$ m/s, Figure 12.9 shows the calculated (using the program POLLUTE) maximum increase in concentration in the aquifer for a range of values of the harmonic mean hydraulic conductivity of the aquitard, k_2, that incorporates the hydraulic conductivity of a secondary liner if present. As can be seen, the greatest impact occurs for a hydraulic conductivity which lies within the range considered, and not at either end of the range.

Assuming that the landfill has an area of 60 ha, Figure 12.9 also shows the volume of water, q_h, which must be injected to maintain the hydraulic trap as a function of the harmonic mean hydraulic conductivity of the aquitard. As might be expected, the flow increases with increasing hydraulic conductivity to a maximum of about 2.3 l/s for $k_2 = 50 \times 10^{-10} = 5 \times 10^{-9}$ m/s. For the case giving rise to the greatest contaminant impact ($k_2 \approx 4 \times 10^{-10}$ m/s), the flow rate is about 0.24 l/s.

Given that a compacted clay liner may consolidate with time, giving rise to a reduction in hydraulic conductivity, consideration should be given to the likely range in hydraulic conductivity that might be expected. Suppose, for example, the likely range in liner hydraulic conductivity was 1×10^{-10} m/s to 3×10^{-10} m/s; inspection of Figure 12.9 shows that the lowest value of k_1 for the primary liner gives the greatest impact, and thus for a likely range of between 1×10^{-10} and 3×10^{-10} m/s, one would use $k_1 = 1 \times 10^{-10}$ m/s in the contaminant impact calculations. For a 1 m thick secondary liner underlain by a 3 m aquitard having a hydraulic conductivity of 1×10^{-9} m/s, the harmonic mean hydraulic conductivity of the aquitard k_2 ranges between (a) 3×10^{-10} m/s and (b) 6.3×10^{-10} m/s for liner hydraulic conductivities of 1×10^{-10} m/s and 3×10^{-10} m/s respectively. Assuming that the hydraulic conductivities of the primary and secondary liners are the same, the corresponding maximum impacts and flow required to maintain the hydraulic traps (Figure 12.9) are 150 mg/l, 0.2 l/s for case (a) with $k_2 = 3 \times 10^{-10}$ m/s, and 67 mg/l, 0.46 l/s for case (b) with $k_2 = 6.3 \times 10^{-10}$ m/s. Thus the upper and lower limits of the range of values of hydraulic conductivity of the liner give impacts that fall on either side of the maximum impact of about 153 mg/l which could be obtained.

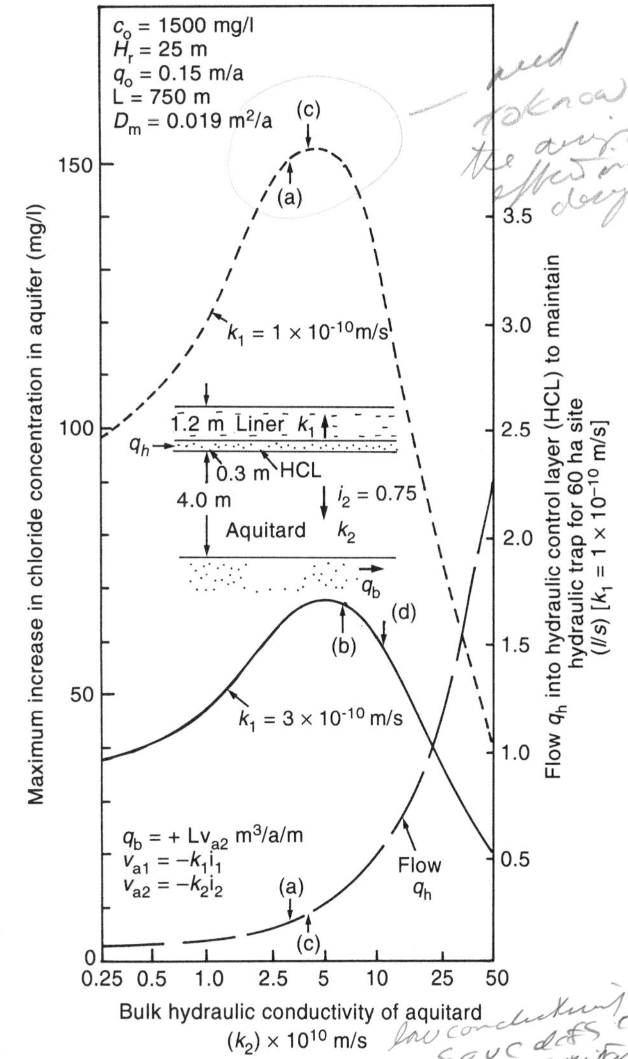

Figure 12.9 Effect of bulk hydraulic conductivity of aquitard on calculated impact in an underlying aquifer and flow required to maintain an engineered hydraulic trap: example B.

If one examines the same range of liner hydraulic conductivities, together with an assumed hydraulic conductivity of the natural aquitard of 10^{-8} m/s (rather than 10^{-9} m/s assumed above) one gets a harmonic mean hydraulic conductivity k_2 of 3.9×10^{-10} m/s (for a liner of 1×10^{-10} m/s, case (c)) which gives a maximum impact of 153 mg/l, and requires a flow $q_h = 0.24$ l/s to maintain the hydraulic trap at one end of the range of k for the liners, and a value of k_2 of 11×10^{-10} m/s (for a liner of 3×10^{-10} m/s, case (d)) which gives a maximum impact of 60 mg/l, and requires a flow $q_h = 0.68$ l/s at the other end of the range.

For the particular case examined here, a sensitivity study for reasonable uncertainty regarding the hydraulic conductivity of the liner (1×10^{-10} to 3×10^{-10} m/s) and aquitard (10^{-9} m/s to 10^{-8} m/s) gives what would be judged unacceptable impacts in the Province of Ontario. On the other hand, a single calculation based on what might conventionally be regarded as 'conservative' parameters (i.e. at the high end of the reasonable range) would correspond to $k_1 = 3 \times 10^{-10}$ m/s and $k_2 = 11 \times 10^{-10}$ m/s (for an aquitard with hydraulic conductivity of 10^{-8} m/s) giving a peak impact of only 60 mg/l (case (d) in Figure 12.9) which may well be judged acceptable for chloride. Thus what might commonly be regarded as conservative (case (d)) does in fact underestimate the potential peak impact by about 250%.

(b) Use of a composite secondary liner to minimize fluid required to maintain engineered hydraulic trap

The injection required to maintain the hydraulic trap could be reduced by including a geomembrane above the secondary liner. Assuming a well-designed and constructed geomembrane, such that the average Darcy velocity from the HCL to the aquifer is reduced to 0.0001 m/a (i.e. less than 3 l/ha/day leakage through holes in the geomembrane), the injection rate required to maintain the hydraulic trap for a primary clay liner with $k_1 = 1 \times 10^{-10}$ m/s is reduced to about 0.06 l/s. The calculated impact on the aquifer, allowing for a diffusion coefficient of 10^{-4} m^2/a in the geomembrane, is about 95 mg/l. This may be compared with the impacts calculated without a geomembrane (Figure 12.9). If desired, the potential impact on the aquifer could be reduced by adding and removing additional fluid from the hydraulic control layer. For example, the addition of 0.34 l/s would correspond to the same injection rate as for a 1.2 m thick primary clay liner (1×10^{-10} m/s) and a natural aquitard ($k \approx 6 \times 10^{-10}$ m/s) alone, but with the extraction of about 0.28 l/s from the HCL (for treatment) the impact on the aquifer is reduced to about 40 mg/l. This is about one quarter of that calculated for the same injection rate and no geomembrane, illustrating the advantage of installing the geomembrane in this case. It should be noted that the volume of fluid extracted from the HCL in this example (0.28 l/s) is small compared to the 2.9 l/s of leachate collected from the leachate collection system.

(c) Summary

The key point to be made in this section is that the design of highly engineered systems requires careful consideration of the likely range of hydraulic conductivities of barriers (including both natural aquitards and artificial liners) that can be expected in the field under the relevant stress conditions. Furthermore, it is important to recognize that when the primary design criterion is limiting long-term impact on groundwater, it is not necessarily conservative to consider either the highest or lowest reasonable hydraulic conductivity, since the maximum impact may arise from some intermediate hydraulic conductivity; thus a sensitivity analysis is required in the design of these systems. The use of a composite secondary liner to minimize the demand for water to maintain the engineered hydraulic trap should be considered. In this case, the potential for impact on the underlying aquifer may be minimized by active removal of contaminated water from the HCL when significant levels of contamination eventually break through the primary clay liner.

12.3 Some considerations in the design of highly engineered systems

In section 12.2.3, consideration was given to the effect of the hydraulic conductivity of the secondary liner–natural aquitard system for the case where a granular layer beneath the primary liner (the hydraulic control layer or HCL) was used to create an engineered hydraulic trap in which water flowed from the HCL through the primary liner and into the landfill. This section examines a number of other ways in which an engineered permeable layer beneath the primary liner can be used, and pays particular attention to the effect of the location of the potentiometric surface in an underlying aquifer, to illustrate how the hydrogeology and engineered design interrelate.

For the purposes of quantitatively illustrating a number of points, consideration will be given to the design of a hypothetical landfill with an average waste thickness of 15 m which is constructed above a natural sand aquifer (example C), as illustrated in Figures 12.10 and 12.11. For simplicity of illustration, it is assumed that the design consists of (from the waste down) a primary leachate collection system, a 1.2 m thick compacted clay liner, a 0.3 m thick granular layer (for secondary leachate collection or hydraulic control) and a 1.5 m thick secondary (natural) clay liner which is underlain by a 1 m thick granular aquifer. For comparison purposes, consideration is also given to the case where the secondary liner–natural aquitard is 8 m thick below the landfill. The landfill is assumed to be 750 m long in the direction of groundwater flow, and it is assumed that the primary piping and slope on the leachate collection system is out of the plane being considered

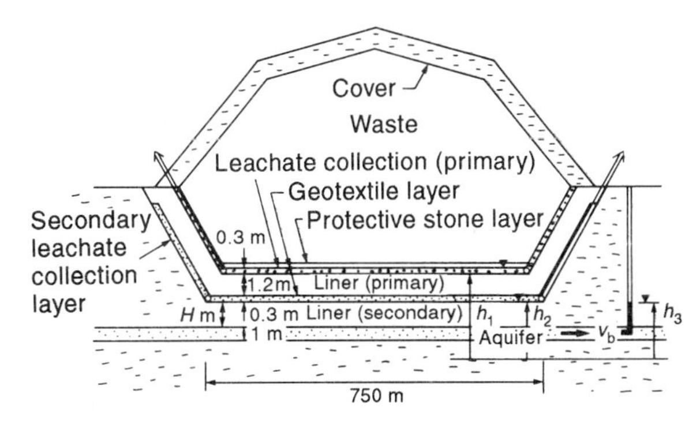

Figure 12.10 Schematic showing a primary liner underlain by a leak detection secondary leachate collection system. Advective flow is downward through the primary liner: example C.

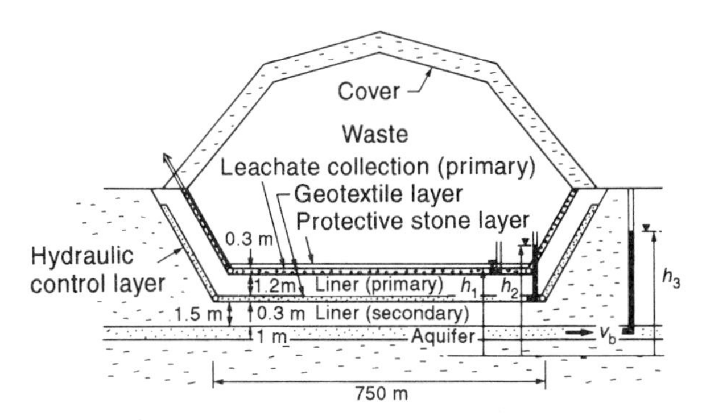

Figure 12.11 Schematic showing a primary liner underlain by a hydraulic control layer. The landfill is designed as a hydraulic trap with advective flow into the landfill: example C.

(i.e. the cross-section being examined is the critical cross-section). It is noted that some slope from the left to right of the cross-section is assumed; however, this detail is not shown on the schematics.

The basic parameters considered for example C are summarized in Table 12.2. Consideration is given to the migration of both chloride and dichloromethane. The reference height of leachate, H_r (see section 12.6 for a discussion of the significance of H_r) represents a mass equal to 0.2% and 0.0002% of the total mass of waste

Table 12.2 Summary of parameters considered for example C

Quantity	*Value*
Length of landfill, L (m)	750
Reference height of leachate, H_r (m)	12
Infiltration through cover, q_0 (m/a)	0.15
Initial chloride concentration, c_0 (mg/l)	1500
Initial dichloromethane concentration, c_0 (μg/l)	1500
Sorption parameter for dichloromethane, ϱK_d (–)	2
Primary clay liner:	
Thickness, H_1 (m)	1.2
Porosity, n (–)	0.35
Hydraulic conductivity, k_L (m/s)	3×10^{-10}
Diffusion coefficient, D_m (m^2/a)	0.019
Engineered granular collection system:	
Thickness, H_2 (m)	0.3
Porosity, n (m)	0.3
Hydraulic conductivity, k (m/s)	10^{-2}
Dispersion coefficient, D (m^2/a)	See text
Secondary liner–aquitard:	
Thickness, H_3 (m)	Variable (1.5, 8)
Porosity, n (–)	0.25
Hydraulic conductivity, k (m/s)	Variable (10^{-9}, 10^{-10})
Diffusion coefficient, D (m^2/a)	0.015
Aquifer:	
Thickness, h_b (m)	1
Porosity, n_b (–)	0.3
Hydraulic conductivity (m/s)	10^{-5}
Horizontal Darcy velocity, v_b (m/a)	Note a

[a]Fixed by continuity of flow considerations. For a hydraulic trap, $v_b = 1$ m/a is assumed; for outward flow, $v_b = 1 + v_{a2}L$; v_{a2} given in Tables 12.3–12.6.

for chloride and dichloromethane respectively. For the purposes of this analysis, it is assumed that the diffusion coefficients for chloride and dichloromethane are the same; however, dichloromethane will be retarded due to sorption ($\varrho K_d = 2$).

The leachate mound in the primary leachate collection system is assumed to be 0.3 m above the top of the compacted clay liner (i.e. $h_1 =$ 10 m, measuring head relative to an arbitrary datum and taking positive upwards). The head in the secondary leachate collection/hydraulic control layer, h_2, and in the aquifer, h_4, will vary depending on the hydrogeologic conditions being considered (Figures 12.10 and 12.11).

All the analyses reported herein were performed using a finite layer contaminant transport model described in Chapter 7, as implemented in the computer program POLLUTE (Rowe and Booker, 1994).

12.3.1 Secondary leachate collection: clay primary liner

Potentiometric surface in aquifer below secondary collection system

For situations where the water table and potentiometric surface in the underlying aquifer are well below the base of the landfill (e.g. Figure 12.10), the construction of a permeable drainage system, which is located beneath the compacted clay liner, serves two purposes. Firstly, the drainage layer functions as a secondary leachate collection system, which can remove a portion of the leachate which escapes through the liner (some escape is to be expected through any liner system where there are downward gradients). Secondly, this layer serves to reduce the hydraulic gradient through the underlying soil.

A key question in the design of these liner systems is 'What will be the impact of the contaminant on groundwater in the underlying aquifer?'. The answer to this question will depend on the properties of the soil, the drainage system and the liner, the geometry of the landfill, the design of the primary and secondary leachate collection system, and the properties of any underlying aquifer. Each case must be considered as a unique situation; however, it is important to recognize that some contaminant migration through the natural soil must be anticipated for designs such as that shown in Figure 12.10. Even in the limiting case where no leachate mounding occurred in the secondary leachate collection system and all the leachate that escaped through the liner was collected, there would still be diffusion of contaminants into the natural soil from the secondary leachate collection layer. In most cases, there will also be downward advective contaminant transport, since not all the leachate escaping through the liner is likely to be collected. Indeed, situations can readily occur where the majority of the leachate migrating through the primary liner also migrates down through the natural soil.

Ideally, the separation between the aquifer and the landfill would be as large as possible for a design involving a secondary leachate collection system such as the one shown in Figure 12.10. However, in many practical situations the actual thickness may be quite small. Under these circumstances it cannot be assumed that even a perfectly operating liner, and primary and secondary leachate collection system will necessarily prevent contamination of groundwater in an underlying aquifer. To illustrate this, consider case 1, where it is assumed that the base of the engineered landfill is chosen to correspond to the potentiometric surface of the underlying aquifer which, for this case, is assumed to correspond to the head of h_3 = 8.2 m at the downgradient toe of the landfill. This ensures an adequate factor of safety against 'blowout' of the *in-situ* secondary clay liner (section 2.9.3) and, if there is no mounding of leachate in the secondary collection layer (i.e. $h_2 = h_3 = 8.2$ m), creates a situation where there is no inward or outward flow through the secondary liner. Thus for this case there will be downward advective transport through the primary liner ($k \approx 3 \times 10^{-10}$ m/s, $i \approx 1.25$) corresponding to a Darcy velocity of 0.012 m/a. For this scenario, all leachate should be collected, and contaminant transport through the secondary liner is by the process of molecular diffusion.

The results of analyses performed for case 1 are summarized in Table 12.3. Although this case represents perfect secondary leachate collection with no advective escape through the secondary liner, the process of molecular diffusion through the secondary liner results in significant impact in the aquifer for a thin secondary liner (case 1a). With a thick secondary liner–aquitard system (case 1b) the impact is much smaller, especially for dichloromethane, which experiences attenuation due to sorption in the natural secondary liner–aquitard. For this case, where contaminant migration is only by diffusion, dilution in the aquifer is an important

Table 12.3 Calculated impacts for clay primary liner and secondary leachate collection for example C (Figure 12.10)

	Secondary liner–aquitard			Heads[a]			Darcy velocity[c]			Peak impact in aquifer[d]			
										Chloride		Dichloromethane	
Case	Hydraulic conductivity $\bar{k}$ (m/s)	Thickness, H_3 (m)	Gradient,[a] i (–)	h_1 (m)	h_2 (m)	h_3 (m)	v_{a1} (m/a)	v_{a2} (m/a)	v_b (m/a)	Conc. (mg/l)	Time (a)	Conc. (μg/l)	Time (a)
1a	Any	1.5	0	10	8.2	8.2	0.012	0	1	240	160	75	560
1b	Any	8.0	0	10	8.2	8.2	0.012	0	1	20	800	3	5320
2a	10^{-9}	1.5	0.33	10	8.2	7.7	0.012	0.0105	8.88	610	110	210	430
2b	10^{-9}	8.0	0.33	10	8.2	5.53	0.012	0.0105	8.88	470	270	20	1740
3a	10^{-10}	1.5	0.33	10	8.2	7.7	0.012	0.001	1.79	280	160	85	540
3b	10^{-10}	8.0	0.33	10	8.2	5.53	0.012	0.001	1.79	50	730	5	4880
4a	10^{-9}	1.5	–0.33	10	8.2	8.7	0.012	–0.0105	1	5	100	2	380
4b	10^{-9}	8.0	–0.33	10	8.2	10.86	0.012	–0.0105	1	<0.01	260	<0.01	1710
5a	10^{-10}	1.5	–0.33	10	8.2	8.7	0.012	–0.001	1	175	160	55	540
5b	10^{-10}	8.0	–0.33	10	8.2	10.86	0.012	–0.001	1	5	730	<1	4880

[a]Positive is downward.
[b]Relative to fixed local datum (Figure 12.10).
[c]v_{a1}: vertical Darcy velocity (flux) through primary liner; v_{a2}: vertical Darcy velocity (flux) through secondary liner; v_b: horizontal Darcy velocity (flux) in aquifer at the downgradient end of the landfill.
[d]All concentrations greater than 5 rounded to nearest 5; all times rounded to nearest decade.
See Table 12.2 for other parameters.

attenuation mechanism. If the Darcy velocity v_b were ten times higher, the calculated impacts would be about ten times less. Thus for these systems the hydrogeologic characterization of the head and flow in the aquifer can be quite important when assessing the potential impact due to a proposed design. As discussed in Chapter 10, the effects of reasonable uncertainty can be addressed by performing a sensitivity study.

The significance of the location of the potentiometric surface in the aquifer can be demonstrated by considering two different head conditions. In cases 2 and 3, it is assumed that there is an outward (downward) gradient of 0.33 from the secondary collection system to the underlying aquifer. Cases 4 and 5 examine the reverse situation where there is an inward (upward) gradient of 0.33 from the aquifer to the secondary collection system. This latter case represents the limiting inward gradient that can be achieved, while ensuring a factor of safety against blowout of the base layer of 1.5 during construction of the secondary collection system for a unit weight of 20 kN/m^3 for the secondary barrier soil (section 2.9.3).

Relative to the pure diffusion case (case 1), an outward gradient through the secondary liner (aquitard) results in a substantial increase in potential impact and, in fact, if the hydraulic conductivity of the secondary liner is of the order of 10^{-9} m/s (case 2), then the secondary leachate collection system is not very effective, with 87% of the contaminated water escaping through the primary liner also passing through the secondary liner. If the harmonic mean hydraulic conductivity of the secondary liner–aquitard is of the order of 10^{-10} m/s (case 3), then the secondary collection system is more effective, collecting about 91% of the secondary leachate (about 9% still leaking through the secondary liner).

With inward gradients, the performance of the barrier system is substantially enhanced. For a hydraulic conductivity of the secondary liner of the order of 10^{-9} m/s (case 4) there is negligible impact on groundwater quality in the aquifer. However, it should be noted that this design extracts a substantial amount of water from the aquifer and the heads and flow considered here are those expected after consideration of landfill–hydrogeologic interaction, as discussed in section 12.4. For this case, the fluid to be collected from the secondary system is almost doubled. Thus for a 60 ha site the volume of secondary leachate to be collected increases from about 0.23 l/s to 0.43 l/s; however, this is still small compared with the volume of 2.63 l/s expected to be collected from the primary collection system. Case 5 involves a small (~10%) increase in the volume of leachate collected, but because of the low hydraulic conductivity of the secondary liner–aquitard, there is still some diffusion of contaminant from the secondary collection system to the aquifer. In this case, there is a clear advantage to having a thicker aquitard, especially for attenuating organic chemicals such as dichloromethane, as can be seen by comparing cases 5a and 5b.

12.3.2 Secondary leachate collection: composite primary liner

In the previous subsection, the advective transport through the primary compacted clay liner ($k \approx 3 \times 10^{-10}$ m/s) amounted to about 8% of the total volume of leachate generated. This created a significant impact on the underlying aquifer for a number of the cases considered. One means of reducing impact would be to place a geomembrane above the compacted clay liner to create a composite primary liner. The geomembrane may be expected to reduce the advective flow through the liner. Based on consideration of the factors discussed in section 2.5.2, the estimated average advective flux through the composite liner is 0.1 mm/a (or less) and the diffusion coefficient is taken to be 0.0001 m^2/a.

Case 6 assumes that the granular layer between the primary and secondary liner is a coarse sand. For this case, even though the advective transport is very small, there is significant potential for contaminant transport through the secondary leachate collection systems by diffusion (Table 12.4). Assuming an effective value of nD of 0.001 m^2/a in the unsaturated sand of the secondary collection layer, the contaminant impact on the aquifer with the composite is reduced by a little less than 50% for chloride and a little over 50% for dichloromethane compared with the results for a simple clay primary liner (case 1a, Table 12.3) with a 1.5 m thick secondary liner. For an 8 m thick second system the effect of the composite primary liner reduces impact by between 25% and 50%.

Relatively little research has been conducted concerning diffusion through a humid, moist granular layer such as the secondary leachate collection system. It may be anticipated that if an open granular stone was used (rather than sand) then the potential diffusion through water trapped in soil pores would be reduced (Rowe and Badv, 1994a,b,c). To illustrate the potential effect this could have, case 7 assumes that the diffusion through the secondary leachate collection system is reduced by two orders of magnitude from that in case 6. The results given in Table 12.4 show a substantial reduction in impact on the aquifer compared to case 6. It is apparent that a major role for the unsaturated secondary leachate collection system in this case is to act as a barrier to diffusion, and that any significant failure of the geomembrane which is restricting advective flow through the liner would result in significant impact on the aquifer, even if all the leachate was collected by the secondary leachate collection system, if the secondary liner is thin, as considered in case 1a (Table 12.3).

From the foregoing, it is evident that diffusion is a major consideration in the design of systems such as that shown in Figure 12.10; this is particularly true if a geomembrane liner is used to minimize advective transport through the liner system. Considerable research has been conducted into the diffusion of contaminant in clayey barriers (e.g. Chapters 6–8); however, much more research is required into the diffusion of contaminants through geomembraes and through unsaturated granular (or geosynthetic) secondary leachate collection systems, since this is likely to control impact for systems such as that shown in Figure 12.10. This research is currently in progress at the University of Western Ontario, and early results have been reported by Rowe and Badv (1994a,b,c).

12.3.3 Hydraulic control layer

As discussed in section 1.2, an alternative to using the granular layer beneath the primary liner as a secondary leachate collection system is to use it as a hydraulic control layer. For example, suppose that the potentiometric surface in the aquifer shown in Figure 12.11 is such that $h_3 > h_2 > h_1$. In this case, there is both a natural hydraulic trap (i.e. water flows from the natural soil into the hydraulic control layer) and an engineered hydraulic trap (i.e. water flows from the hydraulic control layer into the landfill). Where practical, this design has the following advantages. Firstly, since there is inward flow to the hydraulic control layer and a relatively impermeable clay liner, it may be possible to design the system such that the engineered hydraulic trap is entirely passive. That is, all water required to maintain an inward gradient is provided by the natural hydrogeologic system, and no injection of water to the hydraulic control layer is required. Secondly, because of the two-level hydraulic trap, there will be substantially greater attenuation of any contaminants that do migrate through the primary liner. Thirdly, since fluid can be injected and withdrawn from the hydraulic control layer, it is possible to control the concentration of contaminant in the layer, and hence minimize the impact at the site boundary in the event of a major failure of the leachate collection system.

There are three factors that must be considered in the design of this engineered hydraulic trap. Firstly, the head in the hydraulic control layer must be controlled such that blowout of either liner does not occur during or after construction (section 2.9.3). Secondly, the volumes of water collected by the hydraulic trap must be manageable, and the hydrogeologic system must have the capacity to provide the water required to maintain the hydraulic trap; if this is not the case, then the head in the aquifer will drop, and the effectiveness of the trap may deteriorate with time (section 12.4). Thirdly, although there is a hydraulic trap, some outward diffusion of contaminants is to be expected in most cases. Contaminant migration analyses are required to assess what impact may occur under these conditions. If the impact at the site boundary is not acceptable, then it can be reduced by pumping water through the hydraulic control layer (i.e. injecting fresh water at one end and extracting contaminated water at the other end). The volume of fluid to be pumped can be assessed by appropriate modeling. Models are available (POLLUTE and MIGRATE) which readily allow the designer to estimate potential impact as a function of the flow in the hydraulic control layer.

To illustrate the effect of a hydraulic control layer, cases 8, 9 and 10 each examine the situation where the total head, h_1, on top of the landfill liner is 10 m (i.e. 0.3 m of leachate mounding on the liner – Figure 12.11) and the total head in the aquifer is 10.9 m. This induces an inward gradient across the liner system. In each case, the primary liner is assumed to have a hydraulic conductivity of 3×10^{-10} m/s. In cases 8 and 9, the secondary liner is assumed to have a hydraulic conductivity of 10^{-9} and 10^{-10} m/s respectively, and the hydraulic control layer is assumed to be operating as a natural hydraulic trap (i.e. no human introduction or removal of water from the hydraulic control layer). Under these circumstances the head, h_2, in the hydraulic control layer is controlled by the relative hydraulic conductivities and thicknesses of the primary and secondary liner, and may be calculated as described in section 5.2.

With the higher permeability secondary liner (case 8), the flows in the system (Darcy velocities v_{a1}, v_{a2} in Table 12.5) are larger than for the lower permeability secondary liner (case 9), and hence the resistance to outward flow is also greater. The greater the inward flow, the greater the resistance to outward diffusion and, consequently, the impact for case 8 is less than for case 9. The impact for case 8 with a clay primary liner system and hydraulic control is similar to that for the system with a very efficient secondary leachate collection system and a composite (geomembrane–clay) primary liner (case 7, Table 12.4).

In case 8, both a 1.5 m and 8 m thick secondary liner–aquitard are considered. For a given head in the aquifer, the thicker layer gives rise to lower gradients and hence (all other things being equal) lower inward Darcy velocities. Thus, case 8b represents a combination of less favorable inward gradients but more favorable thickness of the secondary liner–aquitard, and the net effect of these two differences can be appreciated by comparing the results of cases 8a and 8b. It should be noted that cases 8a and 8b do not correspond to different designs for the same hydrogeology; rather, they represent two different sites where the head difference between the aquifer and the top of the primary liner just happen to be the same. If there were the same inward gradients for both clay thicknesses then the impact would always be less for the case with the thicker secondary liner–aquitard.

Cases 10 and 11 examine the behavior of an engineered hydraulic trap where water is introduced to increase the head in the hydraulic control layer to 10.9 m and 11.9 m respectively. In Case 10, there is the maximum gradient

Table 12.4 Calculated impacts for composite primary liner and secondary leachate collection for example C (Figure 12.10)

	Secondary liner–aquitard			*Heads*[b]			*Darcy velocity*[c]				*Peak impact in aquifer*[e]			
											Chloride		*Dichloromethane*	
Case	*Hydraulic conductivity* $\bar{k}$ *(m/s)*	*Thickness,* H_3 *(m)*	*Gradient,*[a] i *(–)*	h_1 *(m)*	h_2 *(m)*	h_3 *(m)*	v_{a1} *(m/a)*	v_{a2} *(m/a)*	v_b *(m/a)*	*Diff. coeff.*[d] D_{ms} *(m²/a)*	*Conc. (mg/l)*	*Time (a)*	*Conc. (μg/l)*	*Time (a)*
6a	Any	1.5	0	10	8.2	8.2	0.0001	0	1	10^{-3}	125	230	30	880
6b	Any	8.0	0	10	8.2	8.2	0.0001	0	1	10^{-3}	15	960	2	6 680
7a	Any	1.5	0	10	8.2	8.2	0.0001	0	1	10^{-5}	10	680	<2	1 600
7b	Any	8.0	0	10	8.2	8.2	0.0001	0	1	10^{-5}	2	1780	<0.2	10 200

[a]Positive is downward.
[b]Relative to fixed local datum (Figure 12.10).
[c]v_{a1}: vertical Darcy velocity (flux) through primary liner; v_{a2}: vertical Darcy velocity (flux) through secondary liner; v_b: horizontal Darcy velocity (flux) in aquifer at the downgradient end of the landfill.
[d]Diffusion coefficient in unsaturated secondary leachate collection system.
[e]All concentrations greater than 5 rounded to nearest 5; all times rounded to nearest decade.

Table 12.5 Calculated impacts for a clay primary liner and hydraulic control layer for example C (Figure 12.11)

	Secondary liner–aquitard			*Heads*[a]			*Darcy velocity*[c]				*Peak impact in aquifer*[e]			
											Chloride		*Dichloromethane*	
Case	*Hydraulic conductivity* k *(m/s)*	*Thickness,* H_3 *(m)*	*Gradient,*[a] i *(–)*	h_1 *(m)*	h_2 *(m)*	h_3 *(m)*	v_{a1} *(m/a)*	v_{a2} *(m/a)*	v_b *(m/a)*	*Water added to HCL,*[d] q_H *(l/s)*	*Conc. (mg/l)*	*Time (a)*	*Conc. (μg/l)*	*Time (a)*
8a	10^{-9}	1.5	–0.167	10	10.65	10.9	–0.005	–0.005	1	0	15	160	<5	580
8b	10^{-9}	8.0	–0.098	10	10.3	10.9	–0.00236	–0.00236	1	0	<1	660	<0.05	4480
9	10^{-10}	1.5	–0.473	10	10.19	10.9	–0.0015	–0.0015	1	0	85	190	20	670
10	10^{-10}	1.5	0	10	10.9	10.9	–0.007	0	1	0.135	40	170	10	630
11	10^{-10}	1.5	0.667	10	11.9	10.9	–0.015	0.002	2.5	0.325	10	170	3	580

[a]Positive is downward.
[b]Relative to fixed local datum (Figure 12.11).
[c]v_{a1}: vertical Darcy velocity (flux) through primary liner; v_{a2}: vertical Darcy velocity (flux) through secondary liner; v_b: horizontal Darcy velocity (flux) in aquifer at the downgradient end of the landfill.
[d]Water which must be added to maintain the hydraulic trap – assumes landfill area of 60 ha.
[e]All concentrations greater than 5 rounded to nearest 5; all times rounded to nearest decade.

across the primary liner without creating an outward gradient across the secondary liner. This reduces impact compared to the corresponding passive case (case 9) with a 10^{-10} m/s secondary liner. Case 11 relies more heavily on the induced pressure in the hydraulic control layer resisting outward movement of contaminant through the secondary liner and requiring a greater addition of fluid, q_H, to maintain the hydraulic trap.

In the design of hydraulic control layers, it is important that the transmissivity of the layer is high enough to ensure a relatively uniform head on the base of the primary and top of the secondary liners over the area of the landfill. This will often imply the need for a hydraulic conductivity greater than 0.1 m/s; generally the larger the landfill, the greater the transmissivity required. To minimize impact in the event of a failure of the primary leachate collection system, the hydraulic control layer should be designed such that it can be used in an active mode, with injection and collection of fluid, without requiring a large head difference between the manholes where fluid is injected and removed.

12.3.4 Summary

Impact on a potential aquifer will depend on both the hydraulic conductivity of the various natural and engineered components of the system (as discussed in section 12.2) and the head distribution between the aquifer and the top of the primary liner system (as discussed in this section). The following general observations can be drawn from the discussions in these two sections.

1. Advective transport out from the landfill needs to be minimized in order to minimize impact on an underlying aquifer.
2. The design of engineered systems requires careful consideration of the likely range of hydraulic conductivities of both the artificial and natural components of the barrier system.
3. For some engineered systems it is not necessarily conservative to consider either the highest or lowest hydraulic conductivity values when assessing potential contaminant impact. In many cases, a sensitivity study over the expected range may be required in order to identify the peak potential impact.
4. Based on the limited available evidence, it would appear that there can be significant diffusion of at least some contaminants through even relatively thick (2 mm) HDPE geomembranes. Under these circumstances there can be significant impact on an aquifer due to diffusion through the geomembrane, even if the volume of leachate physically leaking through defects in the geomembrane is negligible. Thus the use of a geomembrane or composite geomembrane–clay liner to minimize physical leakage is not, in itself, enough to ensure negligible impact on an aquifer. However, the geomembrane may be a very useful component of a barrier system. More research concerning diffusion through geomembranes is required.
5. Even a perfect secondary leachate collection system which collects 100% of the volume of fluid escaping through the primary liner may not be sufficient to prevent impact on an underlying aquifer.
6. The effectiveness of the secondary leachate collection system as a barrier to contaminant transport is considerably enhanced when the advection through the primary liner is minimized. This is most effectively achieved by using a composite liner composed of a geomembrane and clay layer.
7. The performance of a secondary leachate collection system as a barrier to contaminant transport can be enhanced by designing the system (where practical) to encourage inward flow from an underlying aquifer to the secondary layer. This may reduce the effectiveness of the secondary layer for the

detection of leaks through the primary liner, but it can considerably enhance the resistance to outward contaminant movement from the secondary collection system to an underlying aquifer. Thus in many cases the disadvantages of this approach are outweighed by the advantages in terms of long-term environmental protection.

8. When there will be outward gradient from a secondary system to an underlying aquifer, there is an advantage to constructing a secondary composite liner below the secondary collection system. However, due to potential diffusive transport through the secondary composite liner, its presence does not guarantee negligible impact in all cases; each case must be examined individually from the perspective of potential contaminant impact.
9. The presence of an unsaturated zone (e.g. the secondary leachate collection system) has the potential to reduce the diffusion of contaminant for systems where there is a well-constructed composite liner which minimizes leakage to negligible levels. More research is required to quantify fully the effectiveness of these unsaturated systems as diffusion barriers.
10. In situations where the potentiometric surface in an aquifer is high relative to the base contours, a natural hydraulic trap can be designed to minimize impact on an underlying aquifer. In situations where the natural trap alone is not sufficient, the introduction of a hydraulic control layer can permit long-term control of potential impact on an underyling aquifer.

12.4 Effect of an engineered facility on groundwater levels (shadow effect)

A key consideration in hydrogeologic investigations is the understanding of the groundwater flow direction, and of the natural variability in the water table and in the potentiometric surface in any underlying aquifers, and flow in these aquifers. Landfills have been designed on the basis of this information; however, it is important to recognize that the construction of a landfill may change these conditions. This is particularly true when the landfill is of large areal extent.

If a landfill is located in a recharge area, the construction of an engineered landfill involving a leachate collection system and, frequently, one or more liners, will often reduce the recharge to the underlying aquifer. Thus the landfill casts a hydraulic shadow over the aquifer (and hence the term 'shadow effect') which may lead to a drop in water levels (i.e. a lowering of the potentiometric surface) within the aquifer. The extent to which this will occur will depend on the existing gradients and transmissivity of the aquifer, the hydraulic conductivity of the engineered barrier(s) and the elevation of the base contours relative to the existing potentiometric surface in the aquifer.

The greatest potential effect of the landfill on groundwater conditions occurs when a landfill of large areal extent is to be designed as a natural hydraulic trap above an aquifer of low transmissivity. In this case not only is recharge cut off over the area of the landfill, but also water is being extracted from the aquifer to provide the inward flow necessary to have a natural hydraulic trap (e.g. Figure 12.12). This can lead to a significant drop in water levels in the aquifer and, if not considered in the design, could lead to a failure of the hydraulic trap. The magnitude of the change in water levels may be estimated using numerical models (as discussed in Chapter 5), or analytical models (Rowe and Nadarajah, 1994a).

An example of the potential significance of the shadow effect is found in the design of the Halton landfill in Ontario, Canada. The Certificate of Approval for this landfill required that it be designed as a natural hydraulic trap. Based on existing water level data, an initial set

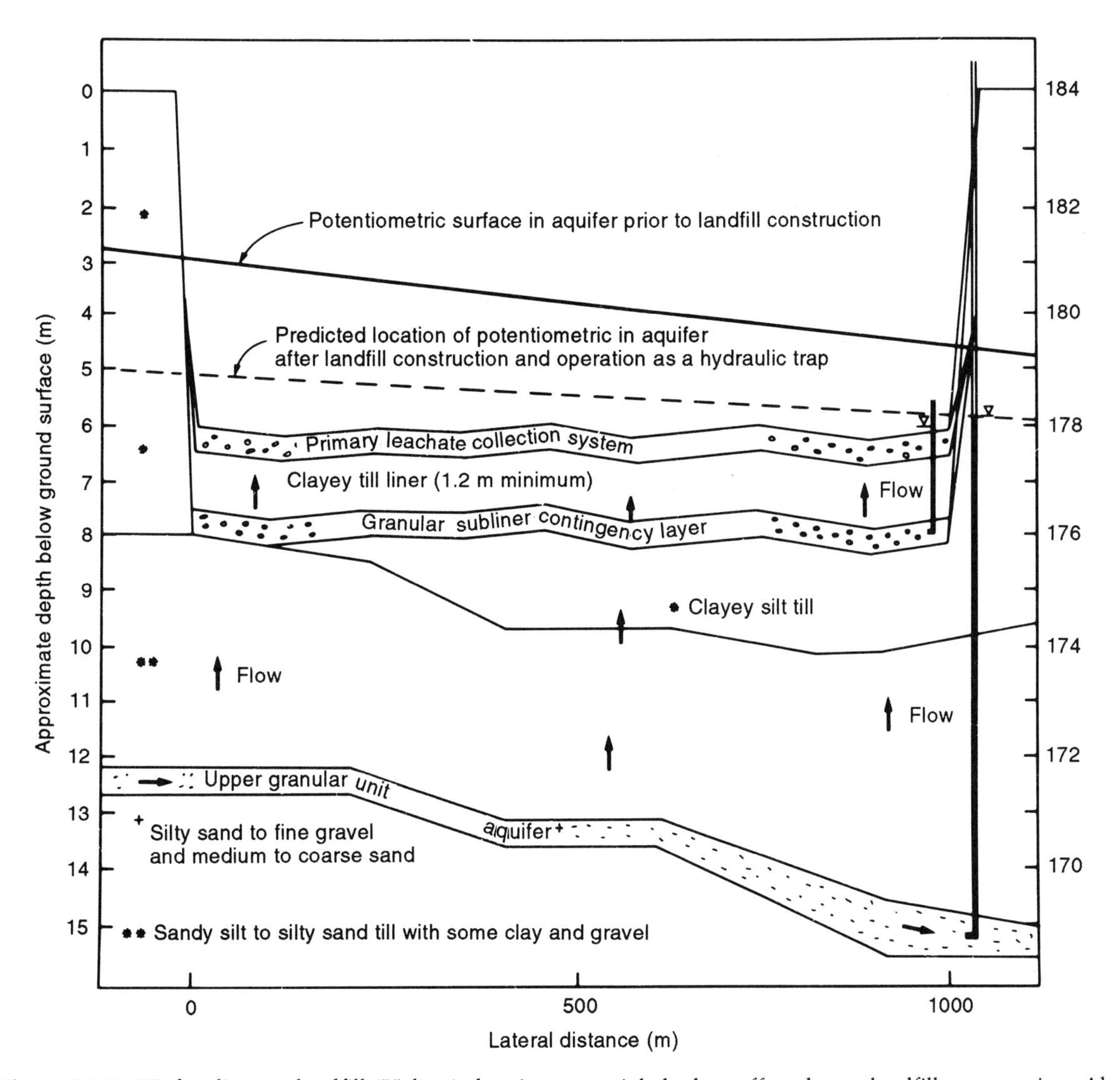

Figure 12.12 Hydraulic trap landfill (Halton) showing potential shadow effect due to landfill construction. Also shown is the subliner contingency layer.

of landfill base contours could be readily established, which provided the required hydraulic trap. However, flow modeling of this initial design, which involved a landfill length of 750 m along the flow path in the aquifer, revealed that the shadow effect could cause a lowering of the potentiometric surface in the aquifer in excess of 2 m. This would have caused a potential failure of the hydraulic trap over approximately half the landfill if the base of the landfill had been selected to provide a hydraulic trap (i.e. upward flow into the landfill) based only on the existing water levels (potentiometric surface) in the aquifer. This finding resulted in a lowering of the base contours by approximately 2 m, such that the hydraulic trap would be maintained, even after the consequent lowering of the potentiometric surface in the aquifer, due to the construction of the landfill. It is noted that the modeling showed that the flow in the aquifer was reduced by about 50% due to the operation of the landfill, compared to the pre-construction flow.

The potential significance of the shadow effect can only be assessed by an integrated examination of both the engineering design and existing hydrogeology. This will frequently involve flow modeling. In instances where the water levels in the aquifer are critical to the design, consideration should be given to the potential implications of long-term variability in water levels due to factors such as long-term changes in climate and potential developments in the recharge area for the aquifer. These effects are quite intangible, and in these cases a contingency may be required to deal with changes in the aquifer water levels. For example, in the case of the Halton landfill, a hydraulic control layer that is totally passive, referred to as the 'subliner contingency layer', was installed (Figure 12.12). Under operating conditions, no water is intended to be added or removed from this layer (i.e. the heads in the HCL will be maintained by flow from the underlying aquifer); however, in the event of some unexpected drop in the potentiometric surface in the aquifer, water may be added to the hydraulic control layer to create an engineered hydraulic trap. Alternatively, if necessary, the layer may be pumped and operated as a secondary leachate collection system. Thus, although it is not intended that this system be used, it is available for use in the event of an unexpected failure of the natural hydraulic trap due to either a change in natural conditions (e.g. lowering of the potentiometric surface due to climate change) or engineered conditions (e.g. a failure of the primary leachate collection system, as discussed in the following section).

12.5 Service life of collection systems, leachate monitoring and trigger levels

The previous sections have examined the performance of a number of combinations of engineered and hydrogeologic systems, based on the assumption that the primary leachate collection system will maintain the leachate head on the primary liner to 0.3 m or less. As discussed in section 2.4, there are numerous mechanisms which can be expected to cause a degradation in the performance of even a well-designed leachate collection system. Based on available evidence, it would seem reasonable to project a service life for a well-designed multicomponent leachate system for municipal solid waste (as discussed in section 2.4) to be 40–100 years, the service life of systems with small stone–sand blanket and/or inappropriately used geonets or geotextiles may be substantially less. Thus if the contaminating lifespan of the landfill (to be discussed in section 12.6) exceeds the service life of the primary collection system, consideration must be given to the implications of increased leachate mounding as the collection system fails. It is noted that a collection blanket may be regarded to have failed once its hydraulic conductivity drops to about 10^{-6} m/s (or lower); at this hydraulic conductivity it may still be collecting leachate; however, mounding of leachate will occur, increasing the head on the base of the landfill, and hence increasing the potential for contaminant impact on underlying groundwater.

Contingency measures for landfills, such as the installation of purge wells in an aquifer, have typically been intended to remove contaminants if they unpredictably reach the aquifer. However, if the contaminating lifespan of the landfill exceeds the service life of the primary leachate collection system, then it would be prudent to design an alternative means of controlling leachate levels once such a failure did occur, rather than rely on purge wells in an underlying aquifer to collect contaminants after they reached the aquifer. For example, by means of leachate extraction wells in the waste (Rowe and Nadarajah, 1994b), or by collecting the consequent increased exfiltration through the primary liner by means of an appropriately designed secondary leachate collection system or active operation of a hydraulic control layer.

Accepting that, at some time, failure of a leachate underdrain system is likely to occur, it is necessary to monitor leachate levels to detect this failure, and to have a trigger in terms of leachate level and concentration at which alternative leachate control would be initiated. The trigger levels of leachate mounding will vary from one landfill to another and will, *inter alia*, depend on the level of attenuation which can occur between the base of the landfill and the underlying aquifer. This in turn will depend on the engineering (e.g. the presence of a compacted clay liner, secondary leachate collection or hydraulic control layer) and the hydrogeologic characteristics of the underlying strata (e.g. the level of fracturing, the hydraulic conductivity of the different strata, etc.). Based on the available field data and engineered design, calculations can be performed to consider these factors to establish triggers in terms of leachate levels and concentrations at which leachate control measures would be initiated. These trigger levels may be based on the requirement that the consequent initiation of leachate control measures would prevent predictable, but unacceptable, impact on any underlying groundwater source. This will be illustrated in section 12.7.

12.6 Contaminating lifespan and finite mass of contaminant

As discussed in section 2.2.5, the contaminating lifespan of a landfill may be defined as the period of time during which the landfill will produce contaminants at levels that could have unacceptable impact if they were discharged into the surrounding environment. When dealing with groundwater contamination, it is necessary to consider the transport pathway (and consequent attenuation) when assessing the contaminating lifespan. This will vary from one landfill to another.

The contaminating lifespan of a landfill will depend, *inter alia*, on the mass of contaminant per unit area (i.e. the thickness and density of waste), the infiltration through the cover, leachate characteristics and the pathway for contaminant release. The thicker the waste, the greater the mass of any given contaminant and, all other things being equal, the longer the contaminating lifespan. The greater the infiltration (and hence volume of leachate collected), the shorter will be the contaminating lifespan, since there is greater opportunity for contaminant to be leached out and removed. The greater the potential for attenuation along the escape pathway, the shorter the contaminating lifespan.

The simplest case is that of a conservative contaminant species which is highly soluble and readily leached from the waste, and which does not decay due to biological activity in the landfill.

For waste disposal sites such as municipal landfills, the mass of any potential contaminant within the landfill is finite. The process of collecting and treating leachate involves the removal of mass from the landfill, and hence a decrease in the amount of contaminant that is available for transport into the general groundwater system. Similarly, the migration of contaminant through the underlying deposit also results in a decrease in the mass available within the landfill. For a situation where leachate is continually being generated (e.g. due to infiltration through the landfill cover), the removal of mass by either leachate collection and/or contaminant migration will result in a decrease in leachate strength with time.

The peak concentration, c_0, of a given contaminant species can usually be estimated from past experience with similar landfills. The total mass of contaminant is more difficult to determine. Nevertheless, estimates can be made by considering the observed variation in concentration with time at landfills where leachate concentration has been monitored or by considering the composition of the waste.

Until fairly recently, there has been a paucity of data concerning the available mass of contaminants within landfills; however, this situation is changing now that many landfills have leachate collection systems. Given that concentration is simply mass per unit volume, the mass of a given contaminant collected in a given period is equal to the concentration multiplied by the volume of leachate collected. By monitoring how this mass varies with time, it is then possible to estimate the total mass of that species of contaminant within the landfill. In the absence of this information, studies of the composition of waste (e.g. Cheremisinoff and Morresi, 1976; Kirk and Law, 1985; Hughes, Landon and Farvolden, 1971) can be used to estimate the mass of given contaminant or groups of contaminants as discussed in section 2.3. For contaminant species predominantly formed from breakdown or synthesis of other species (e.g. by biological action), an upper bound estimate of the mass of contaminant may be obtained from the estimates of the mass of chemicals which go to form the derived contaminant.

For the purposes of modeling the decrease in concentration in the leachate due to movement of contaminant into the collection system and through the barrier, it is convenient to represent the mass of a particular contaminant species in terms of the reference height of leachate, H_r, as described in section 10.2. To aid the following discussion, the definition of H_r is summarized below.

If the initial mass, m_{TC}, of a contaminant species (e.g. chloride) can be estimated, then the reference volume of leachate which would contain this mass at an initial concentration c_0 is

$$V_{TC} = \frac{m_{TC}}{c_0} \tag{12.1}$$

In general, this volume will not correspond to the actual volume of leachate because it is based on the assumption that all this available mass can be quickly leached from the solid waste. It is convenient for both mathematical and physical reasons to express the volume V_{TC} in terms of a reference height of leachate, H_r, which is defined as the reference volume of leachate divided by the area, A_0, through which contaminant passes into the primary barrier:

$$H_r = \frac{V_{TC}}{A_0} \tag{12.2}$$

or

$$H_r = \frac{m_{TC}}{c_0 A_0} \tag{12.3}$$

where m_{TC} is the total mass of a contaminant species of interest [M] ($= pm_T$, where p is the mass species of interest as proportion of total mass of waste [–] and m_T is the total mass of waste [M]); c_0 is the peak concentration of that species in the landfill [ML^{-3}]; A_0 is the area through which contaminant can migrate into the underlying layer [L^2]. H_r represents the mass of contaminant available for transport into the soil and/or collection by the leachate collection system and/or loss to surface waters through landfill seeps.

For the case where there is an infiltration q_0 through the landfill cover, and where all the leachate is collected (i.e. no migration into the underlying soils), it can be mathematically shown (Rowe, 1991a) that the process of dilution within the landfill gives rise to a decrease in the average annual leachate strength with time (recognizing that leachate concentrations may vary seasonally throughout a year) such that the concentration in the landfill at time t, $c_{0L}(t)$, is related to the peak concentration c_0 by the relationship

$$c_{0L}(t) = c_0 \exp\left(\frac{-q_0 t}{H_r}\right) \qquad (12.4a)$$

or, on rearranging terms, the time required for the leachate strength to reduce to some specified value, c_{0L}, is given by

$$t = \frac{-H_r}{q_0} \ln\left(\frac{c_{0L}}{c_0}\right) \qquad (12.4b)$$

where H_r is the reference height of leachate [L] and q_0 the infiltration [LT^{-1}] as previously defined.

For contaminant species that experience first order decay (e.g. due to biological and/or chemical processes), with a decay constant, Γ_{BT} [T^{-1}], equations 12a and 12b can be rewritten as

$$c_{0L}(t) = c_0 \exp\left[-\left(\frac{q_0}{H_r} + \Gamma_{BT}\right)t\right] \qquad (12.4c)$$

and, on rearranging,

$$t = -\left(\frac{q_0}{H_r} + \Gamma_{BT}\right)^{-1} \ln\left(\frac{c_{0L}}{c_0}\right) \qquad (12.4d)$$

Decline of contaminant concentration as implied by equations 12.4a and 12.4c has been observed for both conservative and non-conservative species. For example, Figure 12.13 shows a summary of data for chloride obtained from leaching experiments together with a curve of best fit developed by Reitzel (1990). By rearrangement of terms it can be shown that equation 12.4a is equivalent to Reitzel's empirical equation

$$c_{0L}(t) = c_0 \exp(-BX) \qquad (12.5a)$$

where B is an empirical coefficient (obtained by curve fitting); c_0 is the peak concentration (obtained by curve fitting); X is the cumulative volume of leachate removed per unit mass of waste. It can be shown that X is given by

$$X = \frac{q_0 A_0 t}{m_0} \qquad (12.5b)$$

and hence the empirical decay coefficient B in equation 12.5a is given by

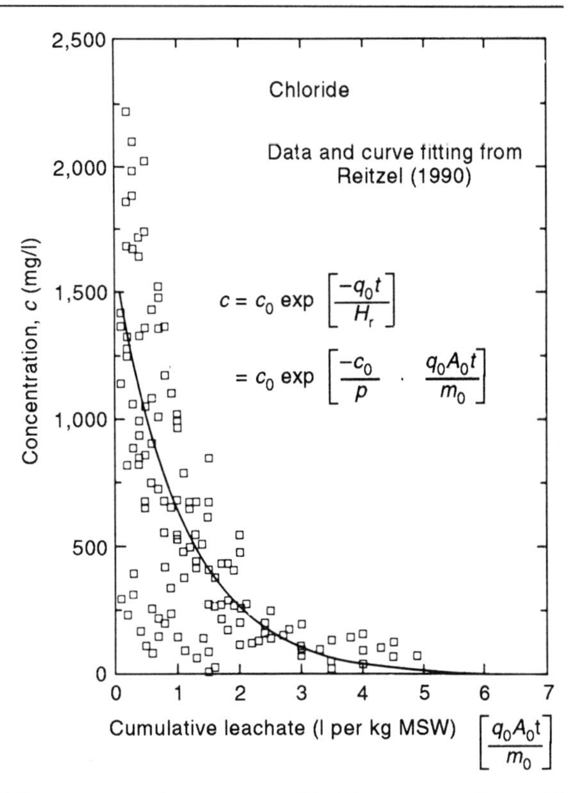

Figure 12.13 Decrease in chloride concentraion with cumulative leaching.

$$B = \frac{c_0}{p} \qquad (12.5c)$$

Based on these relationships (and adjusting Reitzel's coefficient to provide consistent units) it can be shown that Reitzel's (1990) curve of best fit to the leaching data corresponds to

$c_0 = 1614$ mg/l

$p = 0.0018$

implying that chloride represents, on average, 0.18% of the dry mass of waste examined in the lysimeter tests summarized by Reitzel (1990). This proportion of mass is approximately twice that observed by Hughes, Landon and Farvolden (1971). A similar back-analysis of data reported by Ehrig and Scheelhaase (1993) gives $p \approx 0.0019$. These values may be rounded up to

$p = 0.002$ and this value will be adopted for the following calculations.

Based on empirical work of Lu, Morrison and Stearns (1981) and equation 12.4c, the first order decay constant Γ_{BT}, and hence the first order half-life $t_{1/2}$, may be deduced for a number of species, as summarized in Table 12.6. It should be emphasized that Table 12.6 is based on limited data, and that for design purposes, these half-lives should be increased if there may be significant consequences arising from the first order decay not being as slow as implied in this table.

To illustrate the implications of equation 12.4a, consider example D, where the average thickness of waste is $H_w = 10$ m, the average density of that waste is $\varrho_{dw} = 500$ kg/m^3 and where chloride is assumed to represent 0.2% of the weight of the waste (i.e. $p = 0.2\%$), then the mass of chloride per unit area, m_{TC}/A_0, is given by $m_{TC}/A_0 = H_w \varrho_{dw} p = 10 \times 500 \times 0.002 = 10$ kg/m^2. If the peak concentration of chloride is 1000 mg/l (1 kg/m^3), then the reference height of leachate, H_r, is given by

$$H_r = \frac{m_{TC}}{A_0 c_0} = \frac{10}{1} = 10 \text{ m}$$

Table 12.6 Inferred first order decay constant and half-life (based on data by Lu *et al.*, 1981)

Parameter	c_0 *(mg/l)*	Γ_{BT} *(a^{-1})*	*Inferred half-life*[a] *(a)*
BOD_5	35 000	0.225	4.3
COD	89 000	0.192	5.5
TOC	14 000	0.260	3.6
NH_3-N	12 000	0.100	19.8
Cl	2 470	0.065	∞
SO_4	15 000	0.079	49.5
Cd	0.160	0.125	11.6
Cu	10	0.200	5.1
Cr	0.330	0.900	0.8

[a]Half-life of many of these parameters ('a' denotes 'annum') may be due to biologically induced (or accelerated) processes such as precipitation.

Assuming an average infiltration through the landfill cover of 0.15 m/a, the decrease in chloride concentration with time, simply due to dilution in the landfill, can be calculated from equation 12.4a as indicated for case (i) in Figure 12.14. For this particular example, the chloride level would reduced to 250 mg/l after approximately 90 years, and (from equation 12.4a) to 125 mg/l after about 140 years.

If the infiltration was 0.3 m/a, all other factors being equal, the concentration would decrease much faster (case (ii), Figure 12.14) and would reduce from the peak value of 1000 mg/l to 250 mg/l after about 45 years, and to 125 mg/l after about 70 years.

If chloride represents 0.1% of the waste (e.g. Hughes, Landon and Farvolden, 1971), rather than the 0.2% assumed above, then $m_{TC}/A_0 = 0.001 \times 10 \times 500 = 5$ kg/m^2 and thus $H_r = 5$ m. For an infiltration of 0.15 m/a, this gives case (iii), for which the results shown in Figure 12.14 are precisely the same as those obtained for case (ii) (since the ratio H_r/q_0 is the same). The same result would also be obtained

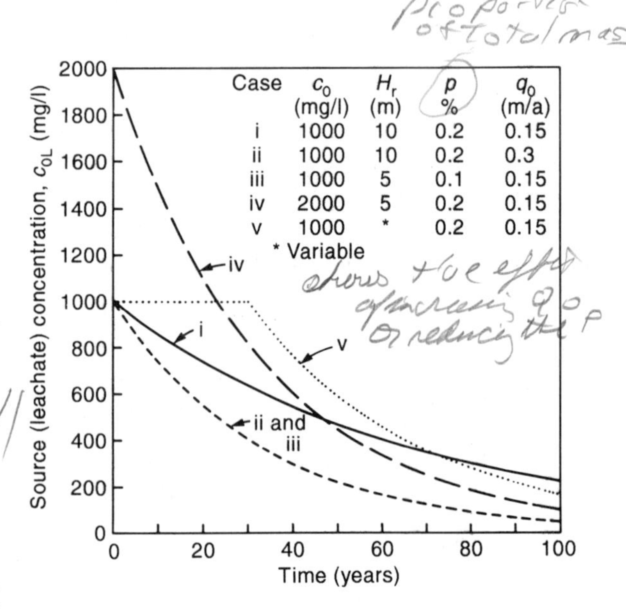

Figure 12.14 Decrease in leachate (chloride) concentration with time: example D.

for 5 m of waste if chloride represented 0.2% of the waste (since again, $m_{TC}/A_0 = 5 \times 500 \times 0.002 = 5$ kg/m^2).

If one were to assume the same total mass of chloride m_{TC} as in case (i) above, but if the peak concentration $c_0 = 2000$ mg/l (2 kg/m^3), then

$$H_r = \frac{m_{TC}}{A_0 c_0} = \frac{10}{2} = 5 \text{ m}$$

and the decrease in concentration with time given by equation 12.4a for $q_0 = 0.15$ m/a is as shown by case (iv) in Figure 12.14. In this case, the concentration decreases from the peak value of 2000 mg/l to 250 mg/l in about 70 years and to 125 mg/l in just over 90 years.

It may be argued that in some cases, not all the contaminant in the source is immediately available for transport or collection. For example, some components of a contaminant may be released due to biological breakdown of the waste over a period of years. In these cases, the source concentration may increase to a peak value and then remain relatively constant (allowing for the usual seasonal variations) for a period of time, because the removal of contaminant is being balanced by new contaminants becoming available (e.g. due to biodegradation). However, eventually all the contaminants that can be released are released, and at this point the concentration in the landfill must drop (due to consideration of conservation of mass). A similar phenomenon may be observed when the concentration in the waste is controlled by the solubility limit. This situation has been examined by Rowe and San (1992).

If a reasonable estimate can be made of the time period over which the source concentration is likely to remain relatively constant, then the behaviour can be readily modeled. To illustrate this, suppose that we consider case (v) in Figure 12.4. Here, it is assumed that the mass of contaminant ($p = 0.002$), the peak concentration and the infiltration are the same as for case (i); however, we also assume that the source remains constant for a 30-year period (due to gradual release of chloride over this period balancing the removal of chloride by the collection system etc.). After 30 years, the concentration will decrease due to mass removal no longer being balanced by contaminant release. This rate of decrease will be controlled by a new value of H_r which represents the mass of chloride still available in the landfill after 30 years. The mass removed, per unit area, over the first 30 years (for $t_c = 30$ years, $q_0 = 0.15$ m/a, $c_0 = 1000$ mg/l $= 1$ kg/m^3) is given by $m_c/A_o \simeq t_c \times q_0 \times c_0 = 30 \times 0.15 \times 1$ kg/m^2 $= 4.5$ kg/m^2. Thus the mass remaining at 30 years is

$$(m_{TC} - m_c) = H_w \varrho_{dw} p - m_c = 10 - 4.5 = 5.5 \text{ kg/m}^2$$

and hence

$$H_r = \frac{m_{TC} - m_c}{A_o c_0} = \frac{5.5}{1} = 5.5 \text{ m}$$

The corresponding rate of decrease in source concentration is shown by curve (v) in Figure 12.4. This scenario maintains a higher source concentration for a period of time, but, for the same mass as case (i), subsequently gives a more rapid decrease in concentration with time, and consequently a shorter contaminating lifespan. For example, in case (v), the concentration in the source reduces to 250 mg/l in just over 80 years (compared with 90 years for case (i)) and to 125 mg/l in about 110 years (compared with 140 years for case (i)).

In the limiting case, if the concentration in the source remained constant until all the mass was removed, this would take 66.67 years for the concentration, mass and infiltration rate considered in case (i) (i.e. $c_0 = 1000$ mg/l, $p = 0.002$, $H_w = 10$ m, $\varrho_{dw} = 500$ kg/m^3, $q_0 = 0.15$ m/a). Thus for this case the minimum contaminating lifespan is about 67 years, assuming the mass is removed at a constant rate until it is all removed.

As already discussed with respect to other uncertainties, if there is uncertainty regarding the period of time over which the source concentration may remain constant, then the effect of this uncertainty can be addressed by a sensitivity analysis. The effect of assumptions regarding the source concentrations examined in this section (Figure 12.14) on contaminant impact for an example case will be discussed in section 12.7.2.

Equation 12.4a only considers decreases in concentration due to dilution. If the contaminant experiences other decay mechanisms (e.g. biological decay), then the rate of decrease with time, as implied by equation 12.4c, will be faster than that indicated for chloride. Similarly, if contaminant is lost to the underlying stratum (e.g. due to diffusion), the rates of decrease may be faster than implied by equation 12.4a and can be calculated using contaminant transport models such as POLLUTE (Rowe and Booker, 1994).

When considering the contaminating lifespan of a landfill, it is necessary to define what is meant by unacceptable impact. In the Province of Ontario, the Ministry of the Environment and Energy has a policy (MoEE, 1993a) such that if a reasonable use for groundwater was as drinking water, then an unacceptable impact could be interpreted as an increase in contaminant which exceeds half the difference between the drinking water objective and background levels for aesthetic parameters (e.g. chloride), and a quarter of the difference between the drinking water objective and background for health-related parameters (e.g. dichloromethane). For chloride, this would mean a maximum increase at the site boundaries of 125 mg/l (or less if there are background levels of chloride in the groundwater). Based on leachate strength decay, one can estimate the time required to reach a given concentration using equation 12.4b or 12.4d. For example, the time required to reach a concentration of 125 mg/l for case (i) can be calculated from equation 12.4b:

$$t = \frac{-10}{0.15} \ln\left(\frac{125}{1000}\right)$$

$$= 138.6 \text{ years (say 140 years)}$$

If one adopts this definition of unacceptable impact, then for the examples considered in Figure 12.14 it would be necessary for the leachate collection system to operate for between a maximum of about 140 years for case (i) and a minimum of about 70 years (for cases (ii) and (iii)) before dilution of the leachate would reduce chloride to levels which are sufficiently low that they would not have an unacceptable impact if they were discharged to the environment after failure of the collection system. This calculation (i.e. equation 12.4a) assumes that dilution of leachate is the only available attenuation mechanism. The question then arises as to how much attenuation may occur as contaminants pass through this barrier and into any underlying aquifer.

Figure 12.15 shows the peak calculated

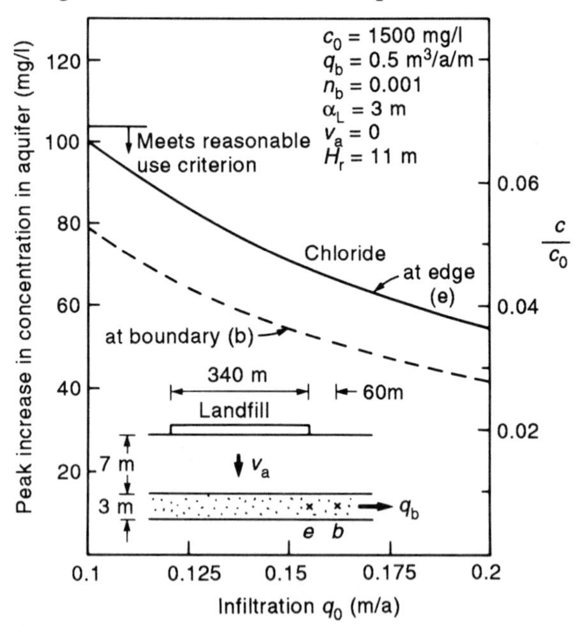

Figure 12.15 Assesment of peak chloride impact on an underlying aquifer: the effect of uncertainty regarding infiltration into the landfill: example E. (After Rowe, 1991a, reproduced with permission from *Canadian Journal of Civil Engineering*.)

increase in chloride in an aquifer at the edge of the landfill and at the site boundary for the case (example E) where contaminant transport from the landfill to the underlying aquifer is by pure diffusion. The peak impact has been calculated for a range of assumed infiltrations, q_0 (adopting $c_0 = 1500$ mg/l, $H_r = 11$ m and using the program MIGRATE (Rowe and Booker, 1988 b)). The rate of decrease in leachate strength with time which, in this case, is controlled by rate of infiltration, q_0, has a substantial effect on the maximum impact. It is also evident that the natural hydrogeologic system provides a high level of attenuation, even though the groundwater flow in the aquifer is quite small ($q_b = 0.5$ m^3/a/m).

Thus when considering contaminant impact on an underlying aquifer, the contaminating lifespan depends not only on the decrease in the leachate concentration, but also on the potential attenuation in the soils between the landfill and the aquifer. This in turn will depend on the geometry of the landfill, the base elevation of the landfill, the head difference between the leachate and underlying aquifer, the properties of the aquitard and the properties of the underlying aquifer. Of these, the most important are the hydraulic conductivity of the aquitard and the head difference between the landfill and aquifer, as will be discussed in the following section.

12.7 Failure of underdrain systems

To illustrate the concepts of contaminating lifespan of a landfill discussed in section 12.6 and the concept of developing trigger envelopes for use in monitoring leachate mounding within a landfill, consideration will be given to a number of cases involving several different hydrogeologic environments and engineered systems ranging from quite simple to highly engineered.

12.7.1 Natural barrier

Consider the two hydrogeologic systems shown in Figures 12.16(a) and 12.16(b). Figure 12.16(a) shows a landfill separated from an aquifer by a thickness H_T of fractured clay which has a bulk hydraulic conductivity k_T. In this case the fracture frequency decreases with depth, but some fractures extend through the entire thickness of the clayey till. Figure 12.16(b) shows a similar situation, except that in this case the fracture frequency decreases until at some depth, H_T, below the base of the landfill, the fractures terminate. Thus the fractured clay is underlain by an unfractured clay layer, of thickness H_B and hydraulic conductivity k_B, which in turn overlies the aquifer. In section 12.7.2, consideration will be given to the effect of installing a compacted clay liner, as shown in Figures 12.16(c) and (d).

For the purpose of the following discussion, it is assumed that the fractured clay, the unfractured clay and the aquifer shown in Figure 12.16 have the properties defined in Table 12.7. For the purpose of calculating the Darcy velocity, the harmonic mean, $\bar{k}$, of hydraulic conductivity of the various units, and the gradient can be calculated as discussed in section 5.2.

In the natural setting (i.e. before construction of a landfill), there could be either upward or downward gradients from the aquifer to the groundwater table. In any event, the construction of a landfill can change the flow regime in the vicinity of the landfill. In some cases, the elevation of the base of the landfill and the layout of the leachate underdrain system can be designed to provide groundwater gradients into the landfill from the underlying aquifer, creating a hydraulic trap which will restrict the outward migration of contaminants to outward diffusion which can occur in opposition to the inward velocity (e.g. Chapter 10), and migration will be primarily through the matrix of the aquitard.

If there is a hydraulic trap, then, under the

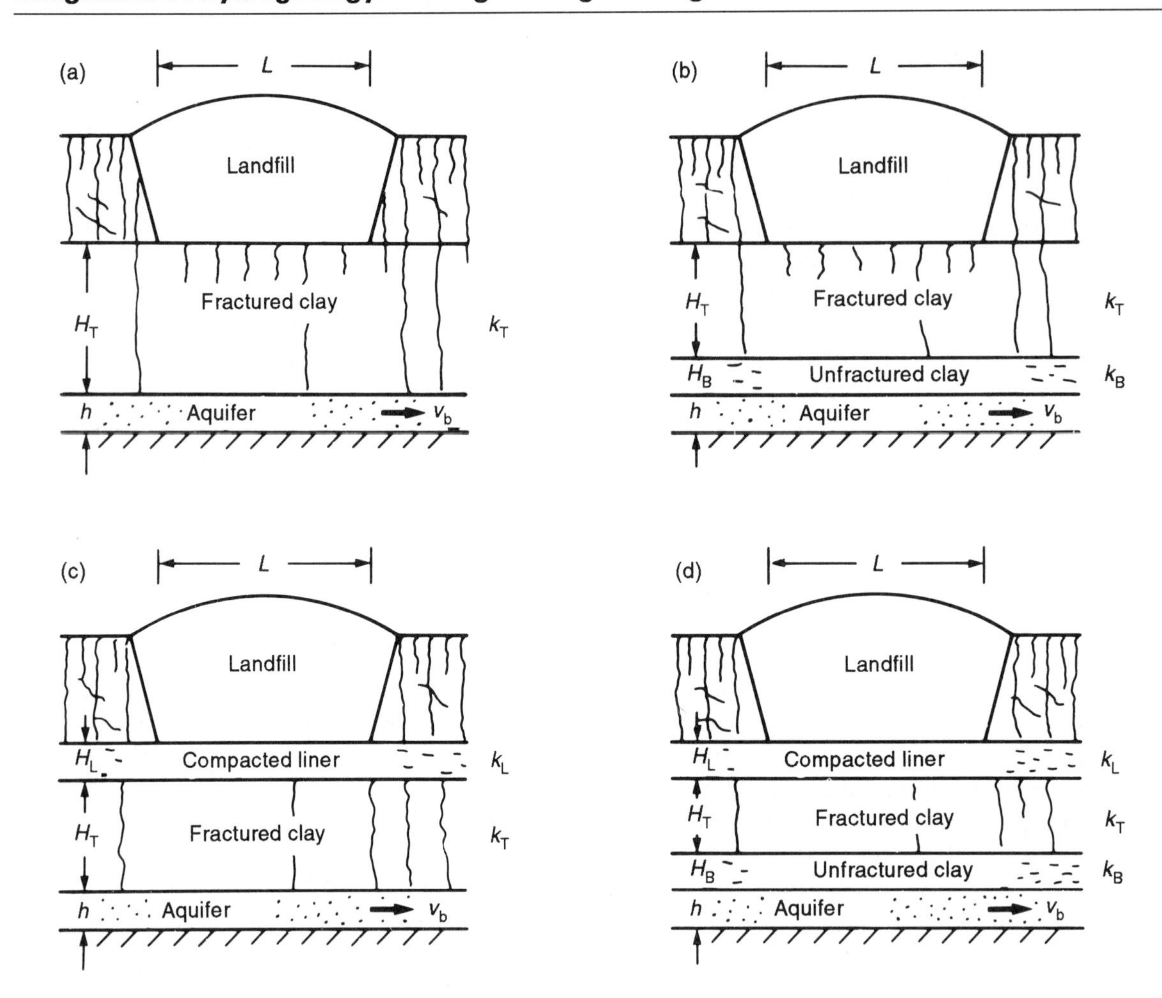

Figure 12.16 Cases considered in example D.

most adverse conditions, the advective flow into the landfill will be entirely through the fractures, and so migration through the matrix would be by pure diffusion. This can be modeled as diffusive transport through this matrix, without the need to model fractures. To illustrate this, consider contaminant migration from a landfill where the characteristics of the leachate and collection system are as discussed in section 12.6 (example D) for case (i) (i.e. $H_r = 10$ m; $c_0 = 1000$ mg/l; $q_0 = 0.15$ m/a). It is assumed that the base of the landfill is separated from a 1 m thick underlying aquifer by 4 m of fractured clay (i.e. $H_T = 4$ m, as in Figure 12.16(a); see Table 12.7, cases (0), (i) for a full set of parameters).

The concentration of chloride in the aquifer due to pure diffusion from the landfill was calculated assuming the flow in the aquifer is sufficiently small as to be negligible (i.e. $v_b = 0$) as shown by curve (0i) in Figure 12.17. This represents a worst case impact on the aquifer,

Table 12.7 Summary of parameters considered in example D

Quantity	*Value*
Length of landfill, L (m)	200
Reference height of leachate, H_r (m)	(i) 10; (ii) 10; (iii) 5; (iv) 5; (v) variable
Initial concentration, c_0 (mg/l)	(i) 1000; (ii) 1000; (iii) 1000; (iv) 2000; (v) 1000
Downward Darcy velocity, v_a (m/a)	(0) 0; (1) 0.05; (2) 0.01; (3) 0.005; (4) 0.005; (5) 0.003
Thickness of fractured clay, H_T (m)	(0) 4; (1) 4; (2) 4; (3) 3; (5) 2
Porosity of clay matrix, n_m (–)	0.4
Hydraulic conductivity of fractured clay, k_T (m/s)	1×10^{-9}
Thickness of unfractured clay, H_B (m)	(0) 0; (1) 0; (2) 0; (3) 0; (4) 1; (5) 1
Hydraulic conductivity of unfractured clay, k_B (m/s)	2×10^{-10}
Thickness of clay liner, H_L (m)	(0) 0; (1) 0; (2) 0; (3) 1; (4) 0; (5) 1
Hydraulic conductivity of liner, k_L (m/s)	2×10^{-10}
Total head drop between base of landfill and the aquifer, Δh (m)	(0) 0; (1) 6.35; (2) 1.27; (3) 1.27; (4) 1.27; (5) 1.27
Thickness of underlying aquifer, h (m)	1
Porosity of aquifer, n_b (–)	0.3
Horizontal Darcy velocity in aquifer, v_b (m/a)	$v_b = Lv_a$
Diffusion coefficient in matrix, D_m (m^2/a)	0.02
Retardation coefficient for matrix R_m (–)	1
Coefficient of hydrodynamic dispersion along fractures, D (m^2/a)	0.06
Fracture spacing,[a] $2H_1 = 2H_2$ (m)	1 (unless otherwise noted)
Fracture opening size,[a] $2h_1 = 2h_2$ (μm)	10 (unless otherwise noted)

Note: number in parentheses (e.g. (1), (i)) indicates the case; the number following the parentheses gives the value of parameter used for that case. Where only one value is given, it is the value used for all cases.
[a]Assuming orthogonal fractures at equal spacings $2H_1$ and $2H_2$ and with equal fracture opening sizes $2h_1$ and $2h_2$.

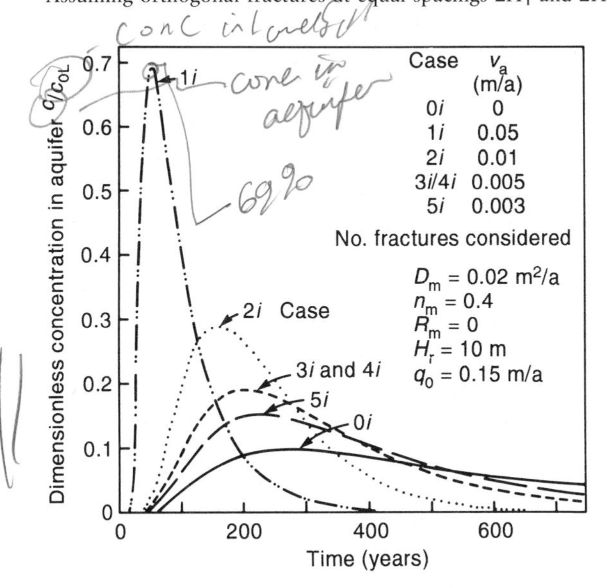

Figure 12.17 Variation in chloride concentration in the aquifer assuming migration through the matrix of the clay (neglecting fractures): example D. (Modified from Rowe, 1991a.)

since any horizontal flow (i.e. $v_b > 0$) beneath the site would imply some dilution of contaminant in the aquifer. From this result for case (0i), it is evident that with a working leachate collection system, the first arrival of chloride in the aquifer at 1% of the initial leachate value occurs after about 70 years. The concentration then increases until a peak impact of just under 100 mg/l is reached after about 275 years. Assuming that the initial background concentration of chloride in the aquifer is negligible, this increase of 100 mg/l would meet Ontario's 'Reasonable Use' Guidelines (MoEE, 1993a). It should be noted that the concentration calculated here is directly beneath the landfill, and hence the impact at the site boundary would be expected to be less (Chapter 10). Thus for the purposes of this discussion it will be assumed that with an operational leachate collection system this site is acceptable.

The foregoing assessment that this site is acceptable, based on the natural hydrogeology combined with the provision of a state-of-the-art leachate collection system assumes that the leachate collection system will operate for the contaminating lifespan of the landfill. An estimate of the contaminating lifespan can be made from equation 12.4b, assuming no attenuation in the aquitard, by calculating the time required until the concentration in the leachate, c_{0L}, is less than the maximum increase in chloride permitted in the aquifer: if $c_{0L}/c_0 = 0.125$; $H_r = 10$ m; $q_0 = 0.15$ m/a; then $t = -H_r/q_0 \ln (c_{0L}/c_0) = 139$ years. This implies a contaminating lifespan of up to 140 years.

The question which then arises is as to whether the leachate underdrain system will function for 140 years, and what the impact of a failure of the leachate collection system would be prior to this time, particularly in view of the earlier projection (section 12.5) of a service life of 40–100 years for a state-of-the-art primary leachate collection system.

The following subsections will focus on this question and the implications that fracturing of the clay might have on the determination of an answer.

(a) Example involving a barrier consisting of 4 m of fractured clay–till

Suppose that the landfill is being designed to operate with a hydraulic trap and that under these circumstances the impact is acceptable (e.g. for case (0i), as discussed above). The hydraulic trap operates for a period of time t_0, after which it fails, and a downward gradient develops due to the increased leachate level in the landfill relative to the potentiometric surface in the underlying aquifer. The level of mounding in the landfill will depend on the dimensions of the landfill, the location of perimeter drains and the location of contingency leachate wells which may be installed in the landfill to control the level of mounding (Rowe and Nadarajah, 1994b). The failure of the leachate collection system will take a period of time, and the consequent gradual change in advective velocity can be modeled (e.g. using the program POLLUTE). However, for simplicity of presentation, the change in velocity is considered to occur quickly at time t_0 which approximately corresponds to the mean time between when the failure began to occur and when the full downward gradient was developed.

Once downward gradients develop, the hydraulic conductivity and fracturing of the underlying clay become critical. Suppose, for the sake of discussion, that the hydraulic conductivity of the fractured clay had been determined from a pumping test to be 10^{-9} m/s (section 3.2.3) and that, due to the nature of the field test, this incorporates the effect of any fractures on the bulk hydraulic conductivity. The Darcy velocity is the product of the bulk hydraulic conductivity and the gradient. For determining the gradient, it is initially assumed that the failed leachate level corresponds to a difference in head, Δh, of 6.35 m between the base of the landfill and the aquifer (case (1)).

Given the bulk hydraulic conductivity of the fractured unweathered clay, there are two possible bounding situations with regard to the effect of the fractures. On the one hand, the fractures may not be significant conduits for contaminant transport, and contaminant migration may simply occur through the matrix of the clay; this will be modeled as a conventional porous medium, without explicitly considering the fractures (e.g. as in Chapter 10). On the other hand, the fractures may control migration, and it can be assumed that all migration to the aquifer occurs through the fractures and none through the matrix (although attenuation may still occur due to matrix diffusion from the fractures into the adjacent clay) as discussed in Chapter 11. Many situations will lie between these bounding cases, but by modeling these cases it is possible to obtain a reasonable engineering estimate of potential impact.

Figure 12.17 shows the results obtained

assuming contaminant migration through the matrix for the five failure cases being considered as part of example D, using the program POLLUTE and the theory presented in Chapter 7 for unfractured material. Figure 12.18 shows the corresponding results considering contaminant migration through fractured media, and were obtained using the theory presented in Chapter 7 for fractured material. For the failure cases, the time shown in the figures (e.g. Figures 12.17 and 12.18) represents the time after failure (i.e. after time t_0, as defined earlier).

Case (1i) corresponds to a major failure of the leachate collection system and a landfill underlain by 4 m of fractured material (Figure 12.16(a) and Table 12.7). Assuming migration through the matrix of the clay, Figure 12.17 shows first arrival of contaminant in the aquifer (at 1% of the source value at the time the failure occurred) within 20 years following the failure if it has not already arrived due to pure diffusion – case 0i; this comment applies to each reference to first arrival in the following discussion. The

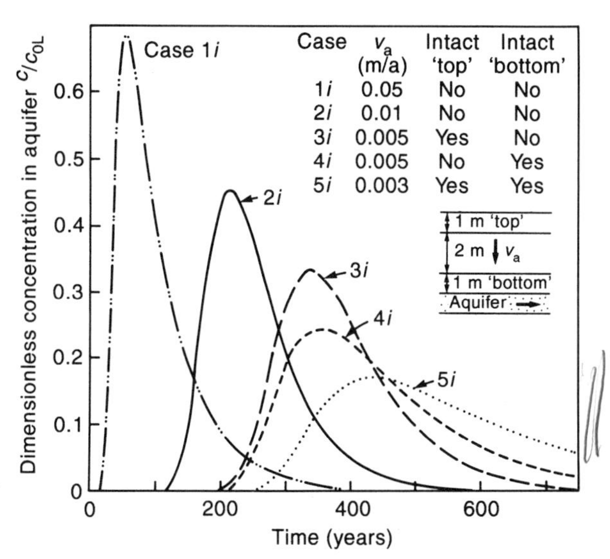

Figure 12.18 Variation in chloride concentration in the aquifer assuming migration along fractures: example D. (Modified from Rowe, 1991a.)

concentration in the aquifer increases with time, reaching a peak value of about 69% of the value in the leachate at the time of failure, approximately 52 years after the failure occurred. Assuming migration along the fractures (with diffusion into the adjacent matrix), Figure 12.18 also shows first arrival of contaminant in the aquifer within 20 years of the failure, rapidly increasing to a peak value of about 69% of the leachate concentration at failure after 54 years. In both cases, the concentration decreases after the peak value has been reached.

Comparison of the results for case (1i) in Figures 12.17 and 12.18 shows very similar results, irrespective of whether it is assumed that migration is through the matrix or through the fractures, and the attenuation factor is 0.69 (i.e. the peak concentration in the aquifer is 0.69 times the value in the leachate at the time of failure). The spacing of fractures does have some influence on the results; however, irrespective of which analysis is used, there will be a substantial impact about 50 years after failure.

If it is assumed that the maximum allowable increase in concentration in the aquifer is 125 mg/l for chloride, then the results for case (1i) can be used in conjunction with equation 12.4b (or Figure 12.14) to estimate how long the leachate collection system would have to work before a failure and mounding to the level implied by case (1i) could be allowed to occur. For example, taking the allowable increase in concentration to be $c_a = 125$ mg/l and an attenuation factor $a_T = (c/c_{0L})_{max} = 0.69$ (based on Figures 12.17 and 12.18), the allowable value in the leachate at the time of failure, c_{0L}, may be calculated by dividing the allowable concentration c_a by the attenuation factor a_T:

$$c_{0L} = \frac{c_a}{a_T}$$

$$= \frac{125}{0.69} = 181 \text{ mg/l} \qquad (12.6)$$

This result (and similar results for other levels of mounding) can be used to construct an envelope of trigger levels at which control measures would be required. Thus case (1i) corresponds to a differential head $\Delta h = 6.35$ m and a maximum allowable concentration in the leachate of 181 mg/l. This is plotted in Figure 12.19 (which will be discussed subsequently). Assuming landfill conditions (i) ($c_0 = 1000$ mg/l, $H_r = 10$ m, $q_0 = 0.15$ m/a), the time at which this failure could occur can be estimated from equation 12.4b:

$$t = \frac{-H_r}{q_0} \ln\left(\frac{c_{0L}(t)}{c_0}\right)$$

$$= \frac{-10}{0.15} \ln\left(\frac{181}{1000}\right) = 114 \text{ years} \qquad (12.7)$$

If the failure occurs prior to this time, then the eventual impact on the aquifer would be unacceptable, based on the assumed conditions in this example.

As indicated in sections 12.4 and 12.5, it may be unrealistic to expect the leachate underdrain

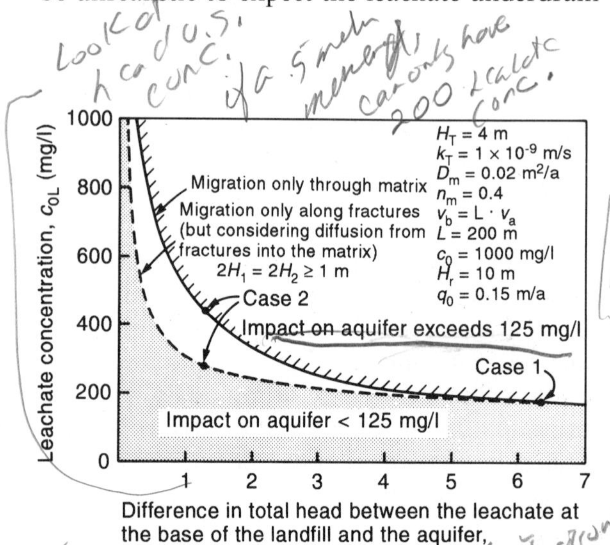

Figure 12.19 Trigger concentration levels for leachate pumping at various levels of mounding for the hypothetical case being considered: example D.

system to be fully functioning for 114 years, and so these results indicate that some measures would have to be taken to decrease impact on the underlying aquifer if a failure of the underdrain system were to occur in less than 114 years.

One option would be to install leachate wells and lower the level of leachate mounding (Rowe and Nadarajah, 1994b). To illustrate the potential effect for the fractured system shown in Figure 12.16(a), it is assumed in case (2i) that the difference in head between the leachate level at the base of the landfill and the value in the aquifer is 1.27 m (i.e. one-fifth of that assumed in case (1i)); all other parameters are identical. Figures 12.17 and 12.18 show the results assuming that the contaminant moves through the matrix and the fractures respectively. In this case, the assumption made concerning the mechanism of transport has a greater effect on the results than was the situation for case (1i). Referring to Figure 12.17, it is seen that for migration purely through the matrix, contaminant reaches this aquifer at the 1% level within 45 years, and the concentration increases to a peak value of about $0.29c_{0L}$ (i.e. the attenuation factor $a_T = 0.29$) of the leachate concentration at failure, about 160 years after failure of the leachate collection system.

Referring to Figure 12.18, it is seen that, assuming migration through the fractures, but considering attenuation due to matrix diffusion, gives a much later first arrival in the aquifer at the 1% level (after about 125 years), but in this case the peak impact is $0.45c_{0L}$ (i.e. $a_T = 0.45$) about 215 years after failure. If the fracture spacing is not less than the 1 m value assumed in this calculation, then the lower and upper bound on the peak impact (and attenuation factors) for the combination of parameters associated with case (2i) are 0.29 (Figure 12.17) and 0.45 (Figure 12.18) times the value in the leachate at failure. Thus, in order for the impact on the aquifer not to exceed 125 mg/l, the concentration in the leachate at failure would have to be less than 277 mg/l, based on a_T = 0.45 (Figure 12.18), or less than 430 mg/l, based on a_T = 0.29 (Figure 12.17). These are plotted for Δh = 1.27 m in Figure 12.19. Using equation 12.4b or Figure 12.14, it can be shown that to meet these requirements, the hydraulic trap must be maintained for between 56 and 85 years, after which the mound would need to be controlled to give a head difference of $\Delta h \approx$ 1.27 m (or less).

If it is considered to be unreasonable to expect that the leachate collection underdrain will maintain the hydraulic trap this long, then either the level of leachate mounding would have to be further reduced (thereby decreasing the head difference, Δh, and hence the downward gradient and Darcy velocity) or some other engineering would be required. By repeating the calculations discussed above for different assumed levels of leachate mounding (and considering the corresponding difference in head between the waste and the aquifer), an envelope of trigger conditions can be constructed, as shown in Figure 12.19. By monitoring the leachate levels and concentrations, and comparing with the results shown in Figure 12.19, it would be possible to determine whether supplementary leachate control (e.g. leachate wells) would be required for the case being considered here. If the combination of leachate mounding and concentration plots below the dotted curve, then the impact on the aquifer is expected to be less than 125 mg/l, and acceptable for this case. If the combination of mounding and leachate concentration plots above the solid curve, then future unacceptable impact may be anticipated, unless some leachate control measures are taken. The zone between the dashed and solid curves represents the range of variability associated with the extent to which contaminant migrates through the fractures and through the matrix of the fractured clay. It would be conservative to use the lower curve as the trigger for leachate control measures.

The practicality of leachate sump wells controlling the leachate mound after failure of the collection system also needs to be carefully considered. The lower the hydraulic conductivity of the waste, the more wells that will be required and the less practical this option becomes. Generally, the larger the landfill (and, in particular, the thicker the waste), the lower will be the hydraulic conductivity of the waste and the less practical will be leachate wells.

Thus, although under operating conditions, this design may be successful, it relies very heavily on the long-term maintenance of very low leachate levels. It may be argued that in terms of long-term potential impact and contaminating lifespan, this landfill design, involving only a fractured clay barrier, should not be accepted for the size of landfill considered here.

(b) Example involving a partly fractured and partly unfractured natural barrier

In many practical situations the unweathered fractured clay–till may be underlain by an unfractured clay layer, as indicated in Figure 12.16(b). To illustrate the effect of this layer, case (4i) is examined, in which it is assumed that this lower layer is 1 m thick with a hydraulic conductivity of 2×20^{-10} m/s. If the hydraulic trap fails at some time, t_0, and a leachate mound develops to give a head difference Δh = 1.27 m between the leachate level and the potentiometric surface in the aquifer, the harmonic mean hydraulic conductivity can be used to calculate a downward Darcy velocity of 0.005 m/a for this case. The corresponding calculated contaminant impact on the aquifer is given by curve (4i) in Figures 12.17 and 12.18 for the cases where migration in the fractured clay is through the matrix and through the fractures, respectively.

When considering migration along the fractures for case (4i), it is assumed that when contaminant reaches the bottom of the fractures it will spread out and migrate evenly through the intact layer; this is equivalent to assuming that there is a thin permeable layer between the fractured clay and the unfractured clay. Based on the results given in Figures 12.17 and 12.18, the peak impact for case (4i) lies between $0.19c_{0L}$ and $0.24c_{0L}$ and, based on equation 12.4b or Figure 12.14, the length of time that the leachate collection system must operate in order to keep the impact on the aquifer to less than 125 mg/l is between 30 and 44 years.

12.7.2 Natural barrier and compacted clay liner

In the previous subsection, consideration was given to 4 m of fractured clay with hydraulic conductivity of 1×10^{-9} m/s. The attenuation characteristics of this deposit can be improved by removing the top 1 m of the fractured clay and replacing it with a compacted clay liner having a hydraulic conductivity of 2×10^{-10} m/s. Provided that the geotechnical studies show that it is suitable, the liner might be constructed by recompacting the clay (at an appropriate water content). This corresponds to the situation shown in Figure 12.16(c). Assuming the same level of leachate mounding as in case (2i), the results for case (3i) shown in Figures 12.17 and 12.18 indicate that the clayey liner increases the time to first arrival in the aquifer, the time to peak impact and the magnitude of the peak impact. The peak impact evident from Figures 12.17 and 12.18 was $0.19c_{0L}$ and $0.33c_{0L}$ respectively.

The results presented in Figure 12.18 considered diffusion through the matrix of the 1 m thick liner, and then transport along the fractures with diffusion from the fractures into the matrix, in the lower 3 m of the deposit. This analysis assumes that once the contaminant breaks through the liner it can move to the fractures; this is equivalent to assuming that there is a thin permeable layer between the liner and the fractured clay (e.g. a thin sand layer). If this permeable zone does not exist, then the migration can be expected to be much closer to those given in Figure 12.17 than those in Figure 12.18 for this case.

Based on the results presented in Figures 12.17 and 12.18 for case (3i) and equation 12.4b (or Figure 12.14) in order to keep the impact on the underlying aquifer to less than 125 mg/l, it would be necessary for the leachate underdrain system to function between 30 and 65 years, provided that when the failure did occur, leachate pumping was initiated to ensure that the difference in head between the landfill and the aquifer did not exceed the 1.27 m value assumed in the analysis.

Assuming that there is a 1 m thick intact layer beneath the fractured layer, it is also of some interest to see what would be the effect of removing and recompacting 1 m of fractured till beneath the base of the landfill, as shown in Figure 12.16(d). Analyses were performed, assuming the parameters given for case (5i), as given in Table 12.7. The results are shown in Figures 12.17 and 12.18. Assuming that migration is only through the matrix (i.e. that the contaminant cannot spread out and move down through the fractures in the layer between the liner and the unfractured clay), it is found that the peak impact is about $0.15c_{0L}$ at about 225 years after the failure of the underdrain system. If one considers migration only through the fractures in the fractured layer (as in Figure 12.18), then the peak impact is slightly higher, at $0.17c_{0L}$ about 430 years after failure of the underdrain. Although the magnitude of impact is very similar, the earlier impact time (Figure 12.17) is likely to be more realistic for the situation shown in Figure 12.16(d) because the intact layer above and below the fractured zone is likely to force most of the contaminant migration to occur through the matrix of the fractured clay rather than through the fractures. Based on these results, the hydraulic trap would be required for between 12 and 20 years, after which the mound would need to be held at or below 1.27 m to keep the increase in chloride in the aquifer to below 125 mg/l. This may be compared to the period of 56 to 85 years calculated for case (1i). The difference reflects the different attenuation capacity of the barriers shown in Figures 12.16(a) and (d).

(a) Effect of leachate characteristics

In section 12.6, it was shown that the characteristics of the landfill, such as the height of waste, proportion of a given contaminant in the waste and the volume of leachate generated per unit time, would influence the decrease in concentration in the landfill. As discussed above, consideration of contaminating lifespan also involves consideration of the attenuation which can occur between the landfill and any critical underlying receptor (e.g. aquifer). The preceding discussion has focused on landfill conditions (i) (as defined in Table 12.7). To illustrate the effect of other assumptions concerning the leachate and waste characteristics (cases (ii)–(v)), calculations were performed for case (5) (considering the worst situation involving migration through the fractures in the fractured zone). The results are shown in Figure 12.20. Considering the peak impact for case (i) of about $0.17c_{0L}$, equation 12.4b indicates that the leachate underdrain system would have to work for only about 20 years. Landfill conditions (ii) and (iii) give an identical response and a peak impact of less than $0.1c_{0L}$. For these conditions it would not be necessary to maintain the hydraulic trap and, even if a head difference Δh of 1.27 m, developed immediately on completion of the landfill, the chloride impact on the aquifer would be less than 100 mg/l and, for these assumed conditions, would be acceptable.

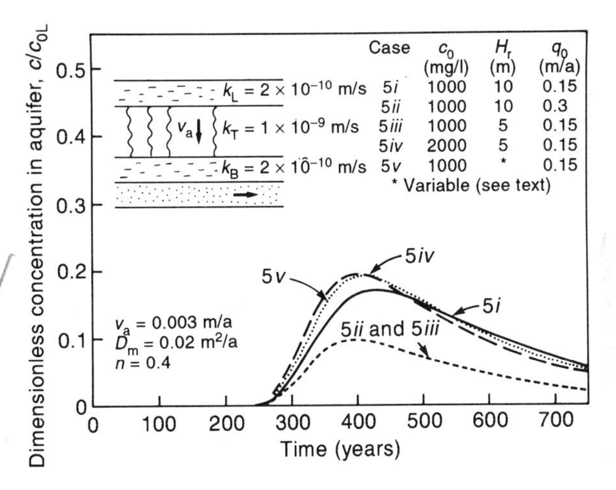

Figure 12.20 Effect of landfill source characterization on calculated impact in the underlying aquifer (case 5, considering fractures): example D.

Landfill conditions (iv), involving a peak concentration of 2000 mg/l (rather than 1000 mg/l assumed for the other cases) but the same total mass of chloride as for case (i), gives a slightly larger peak impact of about $0.19c_{0L}$, and based on the parameters for this case the hydraulic trap would need to function for less than 40 years before mounding to give Δh = 1.27 m could be permitted, i.e. for a concentration in the aquifer of less than 125 mg/l we require $c_{0L} \leqslant 125/0.19 = 658$ mg/l, and hence from equation 12.4b,

$$t = \frac{-5}{0.15} \ln\left(\frac{658}{2000}\right) = 37 \text{ years} \approx 40 \text{ years}$$

Landfill conditions (v) involve identical assumptions regarding the peak leachate concentration, mass of contaminant and infiltration as conditions (i), however it is assumed in case (v) that the source concentration remains constant for 30 years before it begins to decrease (see section 12.6 for a discussion of this). Examination of Figure 12.20 shows that case (5v) gives an earlier and somewhat higher impact on the aquifer than did case (5i). Thus in this case the assumption of a constant source for 30 years, followed by a decrease, gives a shorter contaminating lifespan but, for the same flow conditions, a higher impact than case (5i), where the source concentration begins to increase immediately. While this is true for this case, the conclusion should not be generalized. For example, if there was a secondary leachate collection system, the landfill conditions modeled in case (5v) may result in smaller impacts than case (5i) because of the reduced contaminating lifespan. Thus a sensitivity study is required on a site/case-specific basis to assess which of a group of reasonable landfill assumptions gives the greatest impact.

These results illustrate that consideration should be given to the characteristics of the landfill when assessing potential impact, and when developing trigger plots such as those shown in Figure 12.19.

(b) Contaminating lifespan

The contaminating lifespan of the facility will be controlled by the level to which leachate can mound (in the absence of control measures) and the level of passive attenuation available in the system (i.e. the attenuation which can occur without human intervention). For this landfill, which is 200 m from toe to toe, the level of mounding that could be expected to occur for a hydraulic conductivity of the waste of 10^{-6} m/s is less than 7 m above the elevation of the perimeter drains, with an average value of 5.4 m (section 2.4.3). Assuming that the base of the landfill is 3.2 m below the perimeter drains, this gives an average height of leachate mound on the landfill base of 5.4 + 3.2 = 8.6 m. Taking the potentiometric surface in the aquifer to be 1 m above the base, this gives a head difference of $\Delta h = 8.6 - 1 = 7.6$ m. Based on the harmonic mean hydraulic conductivity (3.3×10^{-10} m/s) for the system shown in Figure 12.16(d), this then gives a downward Darcy velocity of 0.02 m/a and a maximum attenuation factor (which can be calculated in the same manner as that already discussed from a plot similar to that shown in Figure 12.18) of $a_{\mathrm{T}} = 0.5$. Thus the contaminating lifespan with respect to impact on the aquifer is obtained by first calculating the minimum release concentration $c_{0\mathrm{L}}$, as indicated in equation 12.6:

$$c_{\mathrm{L}} = \frac{c_{\mathrm{a}}}{a_{\mathrm{T}}} = \frac{125}{0.5} = 250 \text{ years}$$

and hence the contamination lifespan is given by equation 12.7:

$$t = \frac{-H_{\mathrm{r}}}{q_0} \ln\left(\frac{c_{0\mathrm{L}}}{c_0}\right)$$

$$= \frac{-10}{0.15} \ln\left(\frac{250}{1000}\right) = 92 \text{ years}$$

This implies that, based on the passive attenuation provided by the natural aquitard and compacted clay liner shown in Figure 12.16(d), leachate mounding to the maximum level expected for waste having a hydraulic conductivity of 10^{-6} m/s would have an impact of less than 125 mg/l with respect to chloride provided it is not allowed to develop fully for approximately 90 years.

A lower assumed hydraulic conductivity of the waste would give a higher calculated leachate mound. The physical thickness of the waste places an upper limit on the magnitude of the leachate mound that can develop. Similarly, the infiltration places an upper limit on the exfiltration through the base of the landfill.

The foregoing assessment of contaminating lifespan considered only chloride. Similar calculations may be performed for other critical contaminants (dichloromethane, benzene etc.). The contaminant lifespan is taken as the length of time estimated for the most critical contaminant.

12.7.3 Composite liner systems and secondary leachate collection

As discussed in section 12.3.2, a composite primary liner (geomembrane and clay) combined with a secondary leachate collection system has the potential to substantially decrease contaminant impact on an underlying aquifer. However, the service life of both the primary leachate collection system (intended to minimize the leachate head acting on the geomembrane liner) and the geomembrane (intended to be a barrier to advective transport) should be considered. Failure of either system will increase advective transport through the primary barrier, thereby reducing the effectiveness of the secondary leachate collection system as a diffusion barrier.

The impact of a landfill can be assessed by modeling the operating life followed by the failure of engineered components as their service life is reached (e.g. using POLLUTE). In this case an estimate of the potential impact can be obtained from the results presented in Table 12.3. In example C, case 6a, the impact was calculated assuming that the geomembrane would last for the contaminating lifespan, and the corresponding impacts are given in Table 12.4. If this geomembrane were to fail after 50 years, the peak impact on the aquifer, c_p, will be approximately given by $c_p = c_{0L}a_T$, where c_{0L} is the concentration in the leachate at the time the geomembrane fails and a_T is an attenuation factor that reflects the attenuation which can occur through the compacted clay and natural barrier. Thus if the geomembrane were to fail, then the gradients through the primary liner would be as adopted for case 1a in Table 12.3.

Assuming perfect secondary leachate collection (case 1a, Table 12.3), the attenuation factor is given by the peak concentration divided by the initial source. For chloride and dichloromethane this will be $a_T = 240/1500 = 0.16$ and $a_T = 75/1500 = 0.057$, respectively. Allowing for 50 years of decrease in leachate strength prior to median time of geomembrane failure, the concentration in the landfill for case 1a would be 800 (based on equation 12.4a), and so the peak impact on the aquifer would be expected to be given by $c_p = a_T c_{0L}$ or about $0.16 \times 800 \approx 128$ mg/l and $0.05 \times 800 \approx 40$ μg/l for chloride and dichloromethane respectively. These values are substantially less than if the geomembrane had not been used at all (case 1a, Table 12.3). Thus for this landfill and the condition of case 1a, the combination of a composite liner and secondary leachate collection system cannot be expected to result in negligible impact on the aquifer if the geomembrane in the composite liner functions for only 50 years. However, for a thicker attenuation layer, as considered in case 1b, the composite liner does provide a reduction in impact with 50 years of operation to levels that might be judged acceptable.

If, instead of considering conditions for case 1a, we consider conditions for case 2a, after failure of the geomembrane, then there is an even greater potential significance for failure of the geomembrane. Assuming that the secondary liner is of clay, then if there is failure of the geomembrane between 40 and 60 years (where failure means a substantially increased ability to transmit fluid) the attenuation factors for the hydrogeologic conditions implied by case 2a would result in estimated peak impacts of 330 mg/l and 115 μg/l for chloride and dichloromethane respectively. These impacts would generally be judged unacceptable (especially for dichloromethane).

For situations where there are downward gradients through the secondary liner, the performance of the system can be enhanced by using both a composite primary and secondary liner. Thus if the primary geomembrane and/or primary leachate collection system fail after 40–60 years, a properly designed and constructed secondary system may be expected to collect about 99% of the leachate escaping through the primary liner and hence approaches the perfect leachate collection of case 1 for so long as the secondary collection system and geomembrane function adequately. However, careful consideration must be given to the service life of these components relative to the contaminating lifespan of the facility which can be estimated on the basis of the expected maximum level of leachate mounding and the attenuation characteristics of the natural and engineered components expected to have an indefinite life (e.g. the clay, provided the design gives due consideration to the issues raised in section 2.9 and Chapters 3 and 4).

12.7.4 Hydraulic control systems

As indicated in section 12.3.3, under certain circumstances, hydraulic control systems (i.e. both natural hydraulic traps and engineered hydraulic traps) may provide a means of protecting underlying groundwater from unacceptable impact due to the landfill. The question then arises as to how long it will be necessary to maintain the hydraulic trap.

Case 8, examined in section 12.3.3, with results shown in Table 12.5, involved a 'natural' hydraulic trap and passive use of the hydraulic control layer (HCL, Figure 12.11) wherein no water was added or subtracted from the HCL during normal operations. In the event of a failure of the primary leachate collection system, the maximum leachate mound that could develop in this case is 15 m above the base of the landfill. Based on this and the hydraulic characteristics of the liner materials and head in the underlying aquifer, it is possible to estimate a contaminating lifespan of 185 years and 170 years for cases 8a and 8b, respectively. This is the period over which either the leachate mound would have to be controlled (e.g. as discussed in section 12.5) or the hydraulic control layer would have to be operated in an active mode (e.g. collection and removal of leachate). This latter option will be discussed below.

Assuming that the service life of the primary leachate collection system is 40–100 years, followed by a gradual build-up of the leachate mound to the maximum (in this case 15 m) value, one can model (e.g. using POLLUTE) the build-up of the increased contaminant transport to the HCL. Typically, the head in this unit will be controlled at some maximum value which is related to the depth of the excavation; typically the head in the HCL would be maintained at a value no more than 1–2 m below original ground surface. For situations where it is possible to maintain a flow from the aquifer to the HCL, it may be desirable to maintain the head in the HCL at a level which will induce upward flow from the aquifer, even though this may increase the downward flow through the liner. For example, suppose that for case 8a, the landfill operates as designed for 50 years. The primary leachate collection system then degrades and a leachate mound grows at the rate of 1 m every three years to a maximum value of 15 m at 95 years. Thus the head h_1 (Figure 12.11) grows from $h_1 = 10$ m at 50 years to $h_1 = 25$ m at 95 years. The contingency for failure of the leachate collection system is to maintain the average head in the HCL at about the operational level ($h_2 = 10.65$ m; Table 12.5 and Figure 12.11) this gives a constant upward Darcy flow $v_{a2} = -0.005$ m/a (Table 12.5) and a variable Darcy flux in the primary liner ranging from upward (−0.005 m/a at 50 years) to downward (+0.113 m/a at 95 years). Since the flow through the primary liner is less than the infiltration (0.15 m/a) this implies that following full mound development, some leachate is still being collected by the primary leachate collection system, or is escaping as surface seeps (that would have to be controlled).

In order to maintain the head in the HCL at the design level ($h_2 = 10.65$ m) it would be necessary to remove fluid from the HCL at a rate ranging from 0 (at 50 years) to 2.25 l/s (at 95 years) for a landfill of 60 ha. Under these conditions the volume of leachate collected from the primary collection system and HCL is no greater than that collected from the primary system (2.85 l/s) under operating conditions. The contaminant impact calculated for this mode of operation indicates a maximum increase in concentration in the aquifer of 35 mg/l and 12 μg/l for chloride and dichloromethane, respectively. These values may be compared with those given in Table 12.5 for case 8a (15 mg/l and 5 μg/l respectively) for an operating leachate collection system. Based on the calculation of a contaminating lifespan of 185 years, it is apparent that the HCL would operate in a passive mode for the first 50 years and in an active mode for the next 135 years.

The impact on the aquifer can be further reduced by active addition and removal of fluid. Alternatively, a granular layer HCL can be operated as a secondary leachate collection system once the primary system has failed. This will be discussed further in the next section in the context of the service life of both primary and secondary systems.

12.8 Modeling the service life of primary and secondary systems

The previous section focused on assessing the contaminating lifespan, and hence inferring how long the engineered system must last. Another approach to this issue is to estimate the service life of the key engineering features and then model the operation and, at the appropriate time, breakdown of these engineered features. This can be done using the theory presented in Chapter 7 (e.g. using the program POLLUTE). For example, considering the landfill shown in Figure 12.21, one can model the following.

1. The initial operation of the system is modeled as a natural hydraulic trap with water flowing upward from the aquifer through the secondary liner, the hydraulic control layer, primary liner and into the landfill. While the primary leachate collection system functions as designed, the water level in the hydraulic control layer will be controlled by the hydraulics of the system and can be readily calculated as described in Chapter 5. The hydraulic control is completely passive during this period. The system requires no human intervention except to pump the primary leachate collection system continuously.
2. After some time t_1 (e.g. 50 years in studies reported by Rowe and Fraser, 1993a,b) the service life of the primary leachate collection system is reached, and it fails to control the leachate mound at the design value. A leachate head on the liner will then build up.

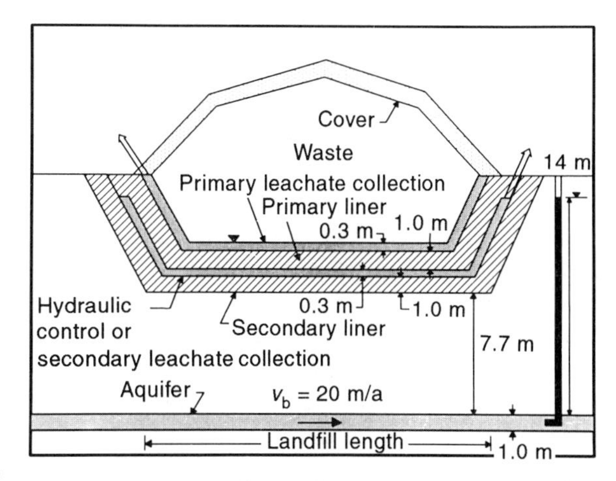

Figure 12.21 Landfill with primary and secondary leachate collection system. (Modified from Rowe and Fraser, 1993a.)

Once this is detected (either by monitoring the leachate mound in the landfill or by monitoring the hydraulic control layer), the hydraulic control layer may be pumped and operated as a secondary leachate collection system. The gradual build-up in the leachate mound can be modeled over a period $t_1 \leqslant t \leqslant t_2$. For the example shown in Figure 12.21, pumping of the secondary leachate collection system induces a natural hydraulic trap from the aquifer to the secondary collection system. There will be a strong downward gradient from the landfill through the primary liner and into the secondary system, but only diffusion from the secondary system down to the aquifer.

3. After some time t_3, the service life of the secondary leachate collection system is reached and it fails to remove all the leachate. As the hydraulic conductivity of the secondary system falls, there will be a gradual build-up in pressure in the layer over a period $t_3 \leqslant t \leqslant t_4$ and contaminant will begin to move down to the aquifer by advection as well as diffusion.

Rowe and Fraser (1993a) have examined the system shown in Figure 12.21 and have shown how the assumptions regarding the service life of the engineered collection systems can have a profound effect on the calculated impact on the underlying aquifer. This work was extended by Rowe and Fraser (1993b) who examined the effect of service life assumptions for four different landfill design options including a single (primary) leachate collection system and two different clay liner designs, a design with both primary and secondary leachate collection systems and liners, and a system involving three liners and primary, secondary and tertiary leachate collection systems.

Deterministic analyses can be readily performed as described above based on assumed service lives of the engineering systems. The greatest problem is estimating the service lives. Since there is inevitably some uncertainty regarding the service lives, one approach is to develop some distribution of service life for each system and then perform Monte Carlo simulations to establish the probability that the concentration in the aquifer will exceed some limiting allowable value. For example, Rowe and Fraser (1993a) examined a triangular distribution of service life for both the primary (PLCS) and secondary (SLCS) leachate collection systems shown in Figure 12.21. For the PLCS, they adopted minimum, mode and maximum service lives of 25, 50 and 75 years respectively. For the SLCS they adopted minimum, mode and maximum service lives of 150, 300 and 700 years respectively. Since the finite layer technique allows rapid modeling of any simulation, Monte Carlo analyses are quite feasible. Based on 5000 realizations, cumulative probability distributions for the peak concentrations in the aquifer were generated, as shown in Figure 12.22. For the particular case examined here, there is an 89% probability that the peak chloride concentration in the aquifer would meet the regulatory limits of 125 mg/l of chloride. Similar stochastic analyses have been performed for a number of different landfill designs by Rowe and Fraser (1993b).

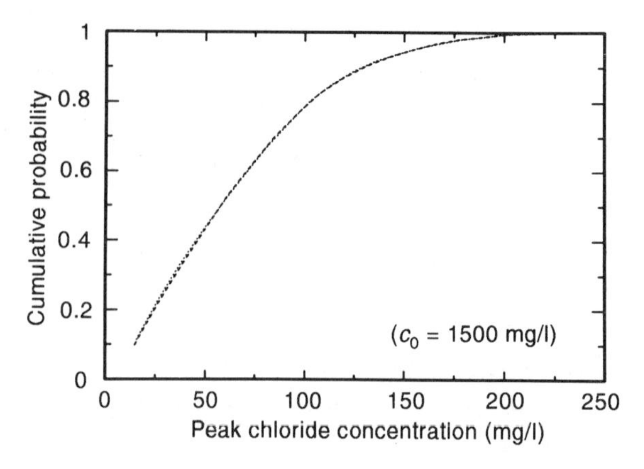

Figure 12.22 Cumulative probability curve for chloride based on Monte-Carlo analysis considering the service life of the secondary collection system (see text). (Modified from Rowe and Fraser, 1993a.)

12.9 Summary

This chapter has highlighted the fact that the safe design of waste disposal facilities requires consideration of the interaction between the natural system (hydrogeology) and the artificial system (engineering). In particular, it is important to:

1. assess the hydraulic conductivity of aquitards in the context of the proposed engineering design;
2. recognize that for highly engineered systems the hydraulic conductivity which will give the greatest potential impact is not necessarily the highest or lowest reasonable value, but may lie in between reasonable bounds;
3. give consideration to the potential effect of the proposed facility on groundwater levels and long-term performance of the landfill system (i.e. the shadow effect);
4. give consideration to the service life of each of the engineered components of the barrier system (primary and secondary leachate collection, liners, hydraulic control layers etc.) with respect to the contaminating lifespan of the landfill;

5. design a system such that there will be negligible impact on groundwater even after the service lives of the engineered components of the barrier system have been reached;
6. monitor the performance of the system and match monitored performance against that expected. Trigger conditions may be developed (e.g. level of leachate mounding and concentration of contaminants in the leachate; concentrations at volume of fluid collected from secondary collection systems) such that back-up measures can be implemented once a failure of part of the system has been detected, and before conditions have been established which will ultimately cause contamination of groundwater.

It is noted that in many systems there can be a considerable time delay (ranging from decades to centuries) between when a failure of an engineered component occurs and the subsequent contamination of an aquifer. If the failure is not detected, and appropriate back-up procedures are initiated shortly after failure and long before contamination of the aquifer, the subsequent measures required to clean up the groundwater, when the aquifer is contaminated, may be extremely time consuming and expensive.

Glossary

Adsorption The process whereby molecular attraction holds solutes to the surface of solids such as organic matter, rock or soil particles.

Advection A physical process whereby contaminants introduced into a groundwater flow system migrate in solution (as solutes) along with the movement of groundwater.

Aeration The process of exposing something to air or charging a liquid with gas.

Aerobic The biological state of living and growing in the presence of oxygen; requiring the presence of free oxygen.

After-use The use of a landfill site following completion and closure of the landfill. Typical examples of after-use include agricultural use, use for playing fields and use for golf courses.

Agglomerate To come together, form or grow into a ball or rounded mass, to cluster densely; particles of dust are brought together to form masses which are too heavy to be entrained in air.

Anaerobic The biological state of living and growing in the absence of oxygen; absence of the presence of free oxygen.

Anion A negatively charged ion (e.g. chloride, Cl^-).

Anisotropic The property of a material (e.g. hydraulic conductivity of soil) which varies with the direction of measurement at a point in the medium (e.g. vertical versus horizontal). See also **isotropic**.

Apparent density The mass of solid waste placed in a landfill divided by the volume occupied by the waste in the landfill. This is not the actual density since the mass of cover soils is not included in the calculation. This may typically be used to determine the waste capacity of a landfill.

Aquifer A geologic formation that is capable of yielding usable quantities of groundwater to wells or springs. Movement is principally in a horizontal direction through porous underground strata.

Aquitard A relatively low permeability stratum from which it is difficult to extract significant volumes of water.

Arithmetic mean Sum of a set of numbers, divided by the number of terms in the set.

Artesian pressure Water pressure in a confined aquifer with a hydraulic head above the ground surface.

Ash Inorganic residue remaining after the ignition of combustible substances; can contain bottom and/or fly ash.

Assimilative capacity A measure of the ability of a receiving body (e.g. river, lake, etc.) to render innocuous those substances deposited in it, such that the quality of the receiving body does not degrade below a predetermined level.

Attenuation The process whereby the concentrations of chemical species in groundwater or leachate are reduced as they move throughout the subsurface.

Background level Concentration of potential pollutants present in the environment prior to establishment, start-up and operation of a facility.

Base contour The contours of the bottom of the landfill.

Base flow The component of stream flow attributed to groundwater or spring contributions; the flow to which a stream will recede after a storm when surface runoff drops to zero.

Berm An artificial ridge of earth or other material used as a mitigative measure against visual and/or noise effects or, within a landfill, to contain leachate on an interim basis or provide stability to the toe of a landfill slope.

Biochemical oxygen demand (BOD) A measure of the amount of oxygen used in the biochemical oxidation of organic matter in water over a specified time under specified conditions. BOD_5 is the BOD measured in a five-day test. It is a standard test used in assessing wastewater strength.

Biodegradable The ability of a substance to be broken down physically and/or chemically by micro-organisms.

Biomass The amount of living matter in a given unit of the environment.

Blowout The upward movement of a low permeability soil due to water pressures in an underlying aquifer which exceed the weight of the overlying aquitard. Blowout can cause fracturing of the aquitard and failure of the landfill.

Borehole A hole made in a geological formation by drilling, jetting, driving or other similar techniques. It is used to determine soil and rock characteristics, and also permits the installation of a water well or an observation well for groundwater monitoring purposes.

Borrow area An area located on- or off-site from which material is extracted for use in earthworks construction (i.e. liner construction, berms, etc.) and/or operations (i.e. daily, final cover).

Buffer area The area between an emission source(s) and nearby sensitive land uses, where land-use controls are employed to minimize any significant adverse effects.

Carbon dioxide One of the principal gases which compose landfill gas.

Carcinogenic Capable of causing the cells of an organism to react in such a way as to produce cancer.

Cation exchange capacity (CEC) Reversible replacement of positively charged ions (cations) adsorbed on clays in an amount equivalent to the negative charge on the clay. Normally expressed in milliequivalents of cation/100 g of clay.

Cell With respect to a landfill site, means a deposit of waste that has been sealed by cover material so that no waste deposited in the cell is exposed to the atmosphere.

Clay Soil size particles smaller than 0.002 mm (2 μm).

Clod An artificially produced agglomeration of soil as in the process of tillage or coarse crushing prior to compaction.

Closure The completion of a landfill facility including construction of the final cover, grassing and preparation of the site for its after-use (including removal of all facilities used during the construction of the landfill and not required for the subsequent after-use).

Compaction Reduction in bulk of fill by rolling, tamping or other mechanical means.

Compaction curve The curve showing the relationship between dry unit weight (density) and the water content of a soil for a given compactive effort.

Compaction test A laboratory compaction procedure whereby a soil at a known water content is placed in a specific manner into a mold of given dimensions, subjected to a compactive effort of controlled magnitude, and the resulting unit weight determined. The procedure is repeated for various water contents sufficient to establish a relation between water content and unit weight.

Compost A relatively stable mixture of decomposed organic waste materials, generally used to condition and fertilize soil.

Composting An aerobic process involving the biological stabilization of organic matter by micro-organisms. Generally comprises spreading or windrowing the organic waste which is sometimes mixed with a bulking agent to maximize air contact.

Concentration The relative fraction of one substance in another, normally expressed in mass percent, volume percent or as mass/volume.

Concentration gradient The change in chemical concentration per unit distance in a given direction.

Confined aquifer An aquifer which is overlain by an aquitard.

Confining layer A body of geologic materials (aquitard) in the subsurface which is of sufficiently low hydraulic conductivity to limit significantly the flow of groundwater into or out of an adjacent aquifer.

Contaminant Any solid or liquid resulting directly or indirectly from human activities that may cause an adverse effect on the environment.

Contaminating lifespan The period of time during which the landfill will produce contaminants at levels that could have unacceptable impact if they were discharged into the surrounding environment.

Contingency plan An organized, planned and coordinated course of action to follow in case of any unexpected failure in the design of a waste management facility. A contingency plan is considered to be a backup measure only and proposed measures with a high level of probability of implementation are not contingency plans.

Continuous sampling The collection of cores of soil (overburden) during the drilling of a borehole. A continuous soil sample is collected from the surface to the bottom of the hole.

Correlation A statement of the kind and degree of a relationship between two or more variables.

Cover (daily and intermediate) Material that is placed on the waste during construction of the landfill to minimize impacts due to: the blowing away of waste, birds, vermin and odor.

Cover (final) Soil (and sometimes geosynthetics) placed over the waste after completion (of a portion) of the landfill. This represents the final surface of the landfill and is intended to (a) control the infiltration of water into the landfill and (b) provide a 'pleasing' appearance while containing the waste.

Decompose Separate into its elements.

Density The ratio of the mass of a substance to its volume.

Dichloromethane (**DCM**) A volatile, chlorinated hydrocarbon used in paint removers and chemical processing. Also known as methylene chloride; CH_2Cl_2. Boiling point $\simeq 40°C$.

Diffusion Migration of molecules or ions in air, water or a solid as a result of their own random movements from a region of higher to a region of lower concentration. Diffusion can occur in the absence of any bulk air or water movement.

Dilution Increasing the proportion of solvent to solute in any solution and thereby decreasing the concentration of solute per unit volume.

Dispersion See **hydrodynamic dispersion**.

Disposal Means the discharge, deposit, injection, dumping, filling or placing of solid waste into or on any land or water.

Disposal facility A collection of equipment and associated land area which serves to receive waste and dispose of it. The facility may have available one, many, or all of a large number of disposal methods.

Disposal site Includes the fill area and the buffer area within the limit of solid waste.

Dissolved oxygen (**DO**) The quantity of oxygen dissolved in a certain volume of water.

Dissolved solids The anhydrous residues of the dissolved constituents in water.

Diurnal Occurring during a single day, and recurring daily.

Domestic waste Solid nonhazardous waste generated from households. Also referred to as residential waste or municipal solid waste (MSW). It does not include liquid waste or hazardous waste.

Domestic well A water well used for private household or farm supplies.

Dry density The mass of mineral matter divided by the total volume it is within.

Dry of optimum A soil compacted 'dry of optimum' is compacted at a water content less than the optimum water content for that soil.

Dry unit weight The weight (force) of mineral matter per unit total volume.

Evaporation The physical transformation of a liquid to a gas at any temperature below its boiling point.

Evapotranspiration The combined loss of water from soil and plant surfaces by direct evaporation and by transpiration.

Exceedence An occurrence during which a regulatory limit is exceeded.

Field capacity Quantity of water held by soil or compacted solid waste where application of additional water will cause it to drain to underlying material.

Final closure Operational and engineering measures that are taken to ensure a former landfill operation will remain environmentally safe and acceptable.

Fissure A narrow opening, cleft or crevice.

Footprint The area of the site within which solid waste will be placed.

Geocomposite Manufactured, assembled material using at least one geosynthetic product among the components.

Geogrid Planar, polymetric structure consisting of a regular, open network of integrally connected tensile elements and whose openings are much larger than its constituents, used for reinforcement in geotechnical and civil engineering applications.

Geomembrane (GM) A relatively impermeable, polymetric sheet used as a liquid and vapor barrier in geotechnical and civil engineering applications.

Geomesh A geonet whose constituent elements are chemically or thermally bonded.

Geometric mean hydraulic conductivity (permeability) The average permeability of randomly distributed test results as defined by

$$k_m = {}^n\sqrt{k_1 \times k_2 \times k_3 \times \ldots k_n}$$

where n is the number of test data and k_n is the individual test result.

Geonet Planar, polymetric structure consisting of a regular, dense network of integrally connected overlapping ribs, used for liquid and vapor transmission in geotechnical and civil engineering applications.

Geospacer Three-dimensional, polymetric space layer, made of a cuspated sheet, monofilaments or any other structure, used in geotechnical and civil engineering applications.

Geosynthetic A polymetric material, synthetic or natural, used in geotechnical and civil engineering applications.

Geosynthetic clay liner (GCL) A low permeability sheet constructed from a thin layer of clay bonded between two layers of geosynthetic which is used as a liquid and vapor barrier in geotechnical and civil engineering applications.

Geotextile (GT) Planar, polymetric (synthetic or natural) textile material, which may be woven, nonwoven or knitted, used in geotechnical and civil engineering applications.

Glaciolacustrine Fine-grained sediments that have settled from suspension in water bodies resulting from the melting of glaciers.

Groundwater Water occurring in a zone of saturation (complete or partial) in a soil or rock and which flows in response to gravitational forces.

Groundwater discharge area An area where water in the saturated zone discharges or flows out of the ground surfaces.

Groundwater flow path The directions in which groundwater flows within an aquifer.

Groundwater recharge area An area where precipitation infiltrates downward through the soil to the water table or saturated zone.

Halogenated Refers to compounds containing fluorine, chlorine, bromine, iodine or astatine.

Hardness A characteristic of water, imparted mainly by salts of calcium and magnesium that causes curdling of soap, deposition of scale in boilers and sometimes objectionable taste.

Harmonic mean hydraulic conductivity (permeability) Representative hydraulic conductivity, $\bar{k}$, of a layered system for flow normal to a layered medium:

$$\bar{k} = \sum_{i=1}^{n} L_i / \sum_{i=1}^{n} (L_i/k_i)$$

where n is the number of test data, k_i is the individual test result and L_i is the length of the flowpath through medium i.

Head on liner The vertical distance between the top of the landfill liner and the top of the zone of saturation of leachate (also referred to as mound height).

Heavy metals Metallic elements such as mercury, chromium, lead, cadmium and arsenic with atomic weights greater than 23 (sodium).

Hectare A unit of area in the metric system equal to 10 000 m^2 (approximately 2.47 acres).

Heterogeneous In hydrogeology, the property (e.g. hydraulic conductivity) of the medium which varies with the location within the medium.

Homogeneous In hydrogeology, the property (e.g. hydraulic conductivity) of the medium which does not vary with the location within the medium.

Hydraulic conductivity (k) The ability of soil or rock to transmit water. The higher the hydraulic conductivity, the greater the ability to transmit water. Sometimes referred to as permeability.

Hydraulic control layer (HCL) A saturated, permeable (usually coarse stone) engineered layer which may be pressurized (either naturally or by external introduction of water) to control the hydraulic gradients across a clayey barrier (liner). May be used to induce an inward hydraulic gradient across a clayey liner and hence create a hydraulic trap.

Hydraulic gradient (i) The change in head per unit of distance in a given direction.

Hydraulic trap A term used to describe a landfill design where water flow is into the landfill and hence resists the outward movement of contaminants.

Hydrocarbon Any of the family of compounds containing hydrogen and carbon in various combinations.

Hydrodynamic dispersion Spreading of contaminant due to the combined effects of mechanical dispersion and diffusion.

Hydrogeology The study of the occurrence, movement and chemistry of groundwater in relation to the geologic environment.

Hydrology The science of dealing with the properties, distribution and movement of waters of the earth.

Hydrometer analysis A laboratory procedure for determining the distribution of particle sizes smaller than 75 mm in a soil sample by means of a sedimentation process which relates the rate of fall of soil particles in a water bath to the particle diameter.

Impact The predicted effect of influence on public health and safety or the environment caused by the introduction of a proposed environmental undertaking. An impact may be positive or negative.

Impermeable Adjective used to indicate that a soil, rock, geomembrane etc. has a very low capacity to transmit fluid, i.e. having a very low permeability.

Impervious Incapable of being passed through by moisture or chemicals.

Industrial waste Nonhazardous solid waste generated by business and industry. Collection (and sometimes disposal) is usually the responsibility of the industrial generator. It does not include liquid waste or hazardous waste.

Inert fill waste Uncontaminated earth or rock fill, also referred to as backfilling waste. These wastes typically contain no soluble or decomposable chemical substances.

Infiltration Water that penetrates the soil from the ground surface. Often used with reference to the water passing through a landfill cover (**percolation**).

Inorganic water Chemical substances of mineral origin, not containing carbon-to-carbon bonding.

***In-situ* density** The density of a soil sample in the field.

***In-situ* water content** The water content of a soil sample in the field.

Instantaneous Happening with no delay; immediate.

Institutional waste Nonhazardous solid waste generated by schools, hospitals, nursing homes etc. It does not include liquid waste or hazardous waste.

Isotropic The property of a material (e.g. hydraulic conductivity) which is independent of the direction of measurement at a point in the medium. See also **anisotropic**.

Kneading compaction Compaction of clayey soil using a technique which remolds and works the clay, and does not just attempt to compress it by pressure.

Landfill A land disposal site employing an engineered method of disposing wastes on land in a manner that minimizes environmental hazards by spreading wastes in thin layers, compacting the wastes to the smallest practical volume and applying cover materials at the end of each operating day.

Landfill gas (LFG) The mixture of gases generated by the decomposition of organic wastes.

Landfilling The disposal of waste by deposit, under controlled conditions on land, or land covered by water, and includes compaction of the waste into a cell and covering the waste with cover materials at regular intervals.

Leach To dissolve out by the action of a percolating liquid.

Leachate A liquid produced from a landfill that contains dissolved, suspended and/or microbial contaminants (see **contaminant**) from solid waste.

Leachate collection system (LCS) An engineered system designed to collect and remove leachate from the landfill.

Lift The vertical thickness of a compacted volume of solid waste and the cover material immediately above it. Usually a lift represents the quantity of waste disposed in a landfilling cell in one day.

Limit of landfill/solid waste A line delineating the limit within which municipal solid waste is contained.

Liner A relatively thin structure of compacted natural clayey soil or manufactured material (e.g. geomembranes, geosynthetic clay liners) which serves as a barrier to control the amount of leachate that reaches or mixes with groundwater in landfills, lagoons etc.

Lysimeter A sampling instrument used to monitor and measure the quantity or rate of water movement through soil, natural or artificial liners, or to collect percolated water for analyses.

Macropores Pore sizes larger than the dominant pore size of a soil. May include compaction-induced fractures and openings due to chemical shrinkage.

Mean A number or quantity contained within the range of a set of numbers or quantities and representative, by some method, of each of the set.

Mechanical dispersion Spreading of contaminant in groundwater due to small-scale variations in groundwater velocities.

Metabolic Refers to the exchange of matter and energy between an organism and its environment and the transformation of this matter and energy within the organism.

Methane An odorless, colorless, nonpoisonous and explosive gas when mixed with air or oxygen in certain proportions. It is one of the two principal gases which compose landfill gas.

Micron A unit of measurement equal to one thousandth of a millimeter (10^{-6} m).

Mitigation Any action with the intent to lessen or moderate potential negative effects; refers to methods that may be used to prevent, avoid or reduce the severity of risks, impacts or service and cost concerns.

Molecular diffusion See **diffusion**.

Monitoring program A program designed to test on-site and off-site effects of landfills. Such a program may be carried out over the operational life of a landfill and for several decades following closure.

Monitoring well A well used to obtain water samples for water quality analysis or to measure groundwater levels.

Moraine Commonly a ridge-like deposition of glacial material formed at the edge of a receding glacier.

Municipal solid waste (MSW) Consists of domestic waste and commercial and industrial wastes of similar composition in any combination or proportion but does not include liquid waste or hazardous waste.

Municipal waste landfill A waste disposal site operating as a landfill site and authorized to accept for landfill only domestic waste or only domestic waste in combination with either or both commercial waste or industrial waste but not including liquid waste or hazardous waste.

Observation well A well installed to enable the measurement of the groundwater level, the sampling of groundwater and testing of the characteristics of the surrounding overburden or bedrock.

Off-site Any site that does not meet the definition of on-site.

On-site On the same property as the proposed or existing disposal facility.

Optimum water content (ω_{opt}) The water content at the peak dry density for a given compaction energy.

Organic matter Substances containing carbon in their molecular structure.

Overburden The soil and fragmented rock materials which lie above solid bedrock.

Oxidation A loss of electrons normally involving the combination of oxygen with another element to form one or more new substances.

Peds See **soil peds**.

Perched water table Groundwater lying above a low permeability layer and separate from and above the water table.

Percolation A term applying to the downward movement of water through soil and especially through a landfill cover (see also **infiltration**).

Permeability The capacity of a porous medium to transmit a liquid or gas; see **hydraulic conductivity**.

Permeable Adjective usually used to indicate that a soil or rock has a relatively high capacity to transmit fluid, i.e. having a relatively high permeability.

pH The measure of acidity or alkalinity, measured on a logarithmic scale from 0 to 14. Neutral water, for example, has a pH value of 7, an acidic solution less than 7 and an alkaline solution greater than 7.

Phenols Organic compounds that contain a hydroxyl group (OH) bound directly to a carbon atom in a benzene ring.

Physiographic region Areas with recognizable local landform patterns.

Plume That portion of the groundwater beneath and in the vicinity of the landfill site where contaminant concentrations exceed certain specified limits. The limits may be defined on the basis of background water quality, drinking water quality standards, or other appropriate standards.

Pore Volume (PV) The volume of fluid required to occupy the pore space in a soil.

Porosity The ratio of the volume of pores of a material to the total volume.

Porous Permeable by fluids.

Potentiometric surface The contours of hydraulic head in a confined aquifer.

PPB/ppb Parts per billion (mass of substance (μg)/mass of solution (kg)).

PPM/ppm Parts per million (mass of substance (mg)/mass of solution (kg)).

Precipitation 1. A physical/chemical phenomenon in which dissolved chemical species in solution (e.g. metals) are transformed into a solid phase (precipitate) which can subsequently be separated from the solution by physical means. 2. Moisture which falls to the earth's surface as rain and snow.

Pressure head A measurement of pressure in a fluid system expressed as the height of an enclosed column of fluid which can be balanced by the pressure in the system.

Pumping test A test performed on a well to determine characteristics of the aquifer and/or adjacent aquitard.

Purge well A pumping well installed for the purpose of extracting and controlling the movement of contaminated groundwater.

Recharge The entry of infiltration into the saturated groundwater zone together with the associated flow away from the water table within the saturated zone. Generally, an area where water is added to the groundwater system by virtue of the infiltration of precipitation or surface water, and subsequently moves downward to the water table, is referred to as a recharge area.

Reference height of leachate (H_r) A measure of the mass of a contaminant of interest within a landfill. It is the ratio of the mass of a given contaminant per unit area divided by the peak leachate concentration for this contaminant; hence it has dimensions of length. It generally does not correspond to an actual level of leachate in a landfill.

Refuse See **solid waste.**

Residential waste Nonhazardous solid waste generated by households.

Runoff The portion of precipitation or irrigation water that drains over the land as surface flow.

Saturated zone The subsurface region below the water table where the soil pores are completely filled with water (saturated) and the moisture content equals the porosity.

Saturation The amount of moisture in the voids of a medium, equal to the volumetric moisture content divided by the porosity. The saturation ranges from 0 (dry) to 1 (or 100%) (completely saturated).

Scenario A possible course of action or events.

Sensitivity analysis An evaluation conducted to assess the impact of changes in the values of specific parameters.

Site boundary criteria The groundwater quality criteria that are used to assess the hydrogeological performance of the site.

Soil peds Units of soil structure such as granules or crumbs; in this text considered to be smaller than clods.

Solid waste All nonhazardous solid materials which are discarded as waste and are not recyclable or reusable; includes daily and interim cover soil that is utilized as part of the landfilling operation (also referred to as refuse or waste).

Solid waste disposal area The area of a municipal solid waste disposal site set aside for landfilling (also referred to as limit of landfill/solid waste).

Solid waste management The systematic control of the storage, collection, transportation, processing and disposal of solid domestic/commercial and nonhazardous industrial waste.

Solvent A substance capable of, or used in, dissolving or dispersing one or more other substances.

Sorption Processes whereby a contaminant is removed from solution. These processes may include cation exchange or partitioning of organic compounds onto solid organic matter. See also **adsorption.**

Standard Proctor compaction A laboratory compaction procedure used to determine the relationship between water content and dry unit weight of soils (compaction curve) compacted in a 101.6 mm or 152.4 mm (4 in to 6 in) diameter mold with a 24.4 N (5.5 lbf) rammer dropped from a height of 305 mm (12 in) producing a compactive effort of 600 kNm/m^3 (12 400 ft lbf/ft^3).

Subsidence The process of settling.

Surface drainage The overland movement of surface water.

Surface water Lakes, bays, ponds, springs, rivers, streams, creeks, estuaries, marshes, inlets, canals, and all other bodies of surface water, natural or artificial, public or private.

Surfactant A surface-active chemical agent.

Surficial At or near the surface. Generally applied to soil and/or bedrock at or near the earth's surface.

Suspension A mixture containing solid particles which are distributed throughout the fluid. The particles will ultimately settle out under the force of gravity.

Till An unsorted mixture of clay, silt, sand, pebbles, cobbles and/or boulders deposited directly by glacial ice.

Tipping face Unloading area for vehicles that are delivering waste to the current active area of landfilling.

Topsoil The uppermost layer of organic-rich soil which is capable of supporting good plant growth.

Transpiration A form of evaporation where growing plants give up water to the air.

Unit weight Weight per unit volume (with this definition, the use of the term weight means force).

Variability A quantity susceptible of fluctuating in value or magnitude under different conditions.

Vertical expansion The expansion of a landfill operation upward from the existing landscape (as opposed to horizontal expansion onto adjacent land not currently landfilled).

VOC Volatile organic compound.

Volatile Describes a substance that evaporates at a low temperature.

Volatility The property of a substance or substances to convert into vapor or gas without chemical damage.

Volatilized Alternative term for 'evaporated'.

Volumetric water content The mass of moisture divided by the total mass in a given volume.

Waste Solid and nonhazardous garbage or refuse which no longer serves any useful purposes in its present form and is discarded by its owner.

Water level The measurement of the top of groundwater. The water level is reported as geodetic elevation to provide a common and comparative reference point.

Watershed The total land area above a given point on a stream or watershed that contributes runoff to that point.

Water table The surface of underground, gravity-controlled water; the surface of an unconfined aquifer at which pore water is at atmospheric pressure. It is generally located at the top of the zone of saturation in an unconfined aquifer.

Well nest A group of two or more observation wells found in one location and installed at different depths to monitor different geologic formations.

Wetlands Lands that are seasonally or permanently covered by shallow water; also lands where the water table is close to, or at the surface. In either case the presence of abundant water has caused the formation of hydric soils and has favored the dominance of either hydrophytic or water-tolerant plants.

Wet of optimum A soil compacted 'wet of optimum' is compacted at a water content higher than the optimum water content for that soil.

Notation

A_0	area through which contaminant is being transported [L^2]
A, B	constants used in solving equations
b	parameter representing the rate of sorption [–]
$c = c(t) = c(z,t)$ $= c(x,z,t)$ $= c(x,y,z,t)$	concentration at depth z, time t [ML^{-3}]
$\bar{c} = \bar{c}(s)$	Laplace transform of concentration c
C	Fourier transform of concentration c
c_r	rate of increase in concentration with time [$ML^{-3}T^{-1}$]
c_f	concentration of contaminant at a point (x,y,z) in a fracture at time t [ML^{-3}]
c_m	concentration of contaminant at a point (x,y,z) in the matrix of a fractured medium at time t [ML^{-3}]
c_I	initial (e.g. background) concentration in soil/rock [ML^{-3}]
c_0	initial source (e.g. leachate) concentration or maximum concentration in the source [ML^{-3}]
$c_{0L}(t)$	concentration in source at time t [ML^{-3}]
c_T	concentration at top of soil/rock mass [ML^{-3}]
$c_{b(max)}$	maximum value of impact at a specified point in an aquifer [ML^{-3}]
$c_b = c_b(x,t)$	concentration at bottom of soil/rock mass (e.g. in a base aquifer) [ML^{-3}]
c_v	hydraulic diffusivity – coefficient of consolidation [L^2T^{-1}]
d	diameter of piezometer tube in Chapter 3
D	coefficient of hydrodynamic dispersion ($D = D_e + D_{md}$) [L^2T^{-1}]
D^*	retarded coefficient of hydrodynamic dispersion ($D^* = D/R$) [L^2T^{-1}]
D_e	effective diffusion coefficient through soil/rock (coefficient of molecular diffusion through soil or rock) [L^2T^{-1}]
D_{md}	coefficient of mechanical dispersion ($D_{md} = \alpha v$) [L^2T^{-1}]
D_0	free solution diffusion coefficient
D_p	porous media diffusion coefficient, $D_p = W_T D_0 = nD_e$
$D_x = D_{xx}$, $D_y = D_{yy}$, $D_z = D_{zz}$	coefficient of hydrodynamic dispersion in the x-, y- and z-directions (cartesian coordinates) (principal components of the dispersivion tensor) [L^2T^{-1}]
D_{1x}, D_{1y}	coefficient of hydrodynamic dispersion in x- and y-direction respectively for fracture set 1 [L^2T^{-1}]

D_{2x}, D_{2y}	coefficient of hydrodynamic dispersion in x- and y-direction respectively for fracture set 2 [L^2T^{-1}]
D_{3x}, D_{3y}	coefficient of hydrodynamic dispersion in x- and y-direction respectively for fracture set 3 [L^2T^{-1}]
D_H, D_v	horizontal and vertical coefficients of hydrodynamic dispersion in an aquifer ($D_H = D_x$; $D_v = D_z$) [L^2T^{-1}]
D_{H1}, D_{v1}, D_{H2}, D_{v2}	horizontal and vertical dispersivity in aquifer 1 and 2 respectively [L^2T^{-1}]
D_{ax}, D_{ay}, D_{az}	product of proportion of open area and coefficient of hydrodynamic dispersion in x-, y- and z-directions respectively (e.g. $D_{ax} = n_0D_{1x}$) [L^2T^{-1}]
e	void ratio in Chapter 4 [–]
erf(x)	error function
erfc(x)	complementary error function, erfc(x) = 1 − erf(x)
f	mass flux (mass transported per unit area per unit time) [$ML^{-2}T^{-1}$]
$\bar{f} = \bar{f}(s)$	Laplace transform of flux f
F	Fourier transform of flux f
$f_x = f_x(x,t) = f(x,y,z,t)$	mass flux transported in the cartesian x-direction at position (x,y,z) and time t [$ML^{-2}T^{-1}$]
$f_y = f_y(x,t) = f_y(x,y,z,t)$	mass flux transported in the cartesian y-direction at position (x,y,z) and time t [$ML^{-2}T^{-1}$]
$f_z = f_z(x,t) = f_z(x,y,z,t)$	mass flux transported in the cartesian z-direction at position (x,y,z) and time t [$ML^{-2}T^{-1}$]
f_{1x}, f_{2x}	component of flux in first set of fissures [$ML^{-2}T^{-1}$]
f_{1y}, f_{2y}	component of flux in second set of fissures [$ML^{-2}T^{-1}$]
f_{av}	average mass flux transported over a time period t [ML^{-2}]
$f_b(\tau) = f_b(c,\tau)$	vertical flux into a base aquifer ($z = H$) at time τ [$ML^{-2}T^{-1}$]
f_{oc}	organic carbon content (e.g. of a soil). [–]
$\dot{g}$	rate at which contaminant is adsorbed onto fracture walls per unit volume of material (matrix and fissures)
h	thickness of base aquifer [L]
h_1, h_2	thickness of aquifers 1 and 2 [L]
h_1,h_2,h_3	half the fracture opening size (i.e. width of fracture opening) in fracture sets 1, 2 and 3 respectively [L]
h_{max}	maximum height of mounding between leachate drains [L]
h_{ave}	average height of mounding between leachate drains [L]
h_p	pressure head [L]
H	thickness of soil deposit [L]
H_1, H_2	thickness of aquitards 1 and 2 [L]
H_1, H_2	piezometer head at times t_1, t_2 in Chapter 3 [L]
H_1, H_2, H_3	half the fracture spacing between fractures in fracture sets 1, 2 and 3 respectively [L]
H_f	equivalent height of leachate ($H_f = H_rq_a/q_0$) [L]
H_r	reference height of leachate [L]
H_w	average height (thickness) of waste (excluding all cover) [L]
$\bar{i}$	hydraulic gradient [–]

i, j, k	indices used in series summation or to denote one layer relative to another
k	hydraulic conductivity (permeability) [LT^{-1}]
$\bar{k}$	harmonic mean hydraulic conductivity [LT^{-1}]
k_h	hydraulic conductivity in the horizontal direction [LT^{-1}]
k_v	hydraulic conductivity in the vertical direction [LT^{-1}]
k_w	hydraulic conductivity of waste
K, K_d	partitioning or distribution coefficient [L^3M^{-1}]
K_f	fracture distribution coefficient (mass of solute adsorbed on fracture walls per unit area of surface per unit concentration of solution in the fracture [L]
K_m	partitioning or distribution coefficient in the matrix material of fractured media [L^3M^{-1}]
K_{oc}	octanol-carbon partitioning coefficient ($K_d = K_{oc} f_{oc}$) [ML^{-3}]
L	length of landfill (in direction parallel to the horizontal flow in the underlying aquifer) [L]
L	length of piezometer intake in Chapter 3 [L]
l	spacing between drains in a landfill
m_{TC}	total initial mass of a contaminant species in waste [M]
m_{1c}	increase in mass deposited in landfill up to time t [M]
m_t	mass of contaminant in source at time t [M]
m_a	mass of contaminant transported from a contaminant source (e.g. landfill) by advection [M]
m_d	mass of contaminant transported from a contaminant source (e.g. landfill) by diffusion [M]
m_0	initial mass of a given contaminant in the source (e.g. landfill) that could be released for transport or collection. It includes contaminant in solid form that may eventually be dissolved but excludes contaminant in solid form which is never expected to be dissolved or contaminant that is or will be released in the gas phase [M]
m	mass of contaminant transported into the soil/rock [M]
m	$(k_h/k_v)^{1/2}$ in Chapter 3
n	porosity [–]
n_b	porosity of base aquifer [–]
n_m	matrix porosity (primary porosity) in a fracture system [–]
n_0	proportion of open area perpendicular to flow in a fracture system $n_0 = h_1/H_1 + h_2/H_2$ [–]
n_f	fracture porosity (secondary porosity) [–]
p	proportion of a contaminant species of the total dry mass of waste [–]
q_a	normalized rate of contaminant flux into soil ($q_a = f/c$) [LT^{-1}]
q_c	volume of leachate collected per unit area of landfill [LT^{-1}]
$\dot{q}$	rate at which contaminants are transported into the matrix per unit volume of material (matrix and fissures)
q_0	steady-state infiltration into a landfill (volume of leachate generated per unit area of landfill) [LT^{-1}]
R	radius of screen or well in Chapter 3 [L]

$R = R_m$ — retardation coefficient ($R = 1 + \varrho K/n$) (R_m is for matrix material in a fracture system) [–]

r — distance from pumped well to piezometer in Chapter 3

$\dot{r}$ — rate at which contaminant is being 'injected' into the fracture system per unit volume of matrix and fissures

S — mass of solute removed from solution per unit mass of solid [–]

S — storativity in Chapter 3 [–]

S_m — solid-phase concentration corresponding to all available sorption sites being occupied [–]

S'_s — specific storage ($S_s = m_v \gamma_w$) [T^{-1}] [L^{-1}]

s — Laplace transform variable

t — time [T]

t_D, t'_D — dimensionless time factor for aquifer and aquitard in Chapter 3 [–]

t_{max} — time at which maximum impact occurs at a specified point in an aquifer [T]

T — transmissivity [L^{2}T^{-1}]

T_0 — basic time lag in Chapter 3 [T]

v — average linearized groundwater velocity (seepage velocity; groundwater velocity) [LT^{-1}]

v_a — Darcy velocity (also called discharge velocity and Darcy flux ($v_a = nv$) [LT^{-1}]

v^* — retarded groundwater velocity ($v^* = v/R$) [LT^{-1}]

v_x, v_y, v_z — components of groundwater velocity in the x, y and z cartesian directions [LT^{-1}]

v_{1x}, v_{1y} — groundwater velocity in x- and y-direction in fracture set 1 [LT^{-1}]

v_{2x}, v_{2y} — groundwater velocity in x- and y-direction in fracture set 2 (LT^{-1}]

v_{3x}, v_{3y} — groundwater velocity in x- and y-direction in fracture set 3 [LT^{-1}]

v_b — horizontal Darcy velocity (Darcy flux) in a base aquifer [LT^{-1}]

v_{b1}, v_{b2} — horizontal Darcy velocity in aquifer 1 and 2, respectively [LT^{-1}]

W — width of landfill (e.g. perpendicular to the direction of flow in the base aquifer) [L]

W_T — complex tortuosity factor (Chapter 6)

x, y, z — cartesian directions

α — dispersivity [L]

α_H, α_v — dispersivity in horizontal and vertical directions [L]

α, β — variables used in derivation of solution to the contaminant transport equations

α_j, β_k, γ_l — variable used in derivation of solution to contaminant transport equation in the matrix of fracture media

γ_w — unit weight of water [ML^{-2}T^{-1}]

Γ_R — l n2/(radioactive decay half-life) [T^{-1}]

Γ_B — ln 2/(biological decay half-life) [T^{-1}]

Γ_S — sink term for removal of fluid. Equal to the volume of fluid removed per unit volume of soil per unit time [T^{-1}]

Δ — surface area per unit volume of fracture media ($\Delta = 1/H_1 + 1/H_2 + 1/H_3$) [L^{-1}]

θ — volumetric water content (= porosity of a saturated soil)

λ	first order decay constant which has components $\lambda = \Gamma_R + \Gamma_B + \Gamma_S$ due to radioactive and biological decay and fluid withdrawal, respectively $[T^{-1}]$
ϱ	dry density of soil/rock $[ML^{-3}]$
ϱ_m	dry density of soil/rock forming the matrix material for a fractured medium $[ML^{-3}]$
ϱ_{dw}	density of waste $[ML^{-3}]$
σ	total stress $[ML^{-1}T^{-2}]$
σ'	effective stress $[ML^{-1}T^{-2}]$
σ'_p	pre-consolidation pressure $[ML^{-1}T^{-2}]$
τ	dummy time variable – used in integrations with respect to time [T]
ϕ	total head [L]
ω	water content of soil [–]
ω_{opt}	optimum water content of soil [–]
Ω	dimensionless infiltration coefficient for waste, $\Omega = q_0/k$ [–]

Specific solution for matrix diffusion: one-dimensional, two-dimensional or three-dimensional conditions

The analysis presented in section 7.6 is dependent on the quantity $\bar{\eta}$ introduced in equation 7.65. The quantity reflects the diffusion of contaminant from the fractures and into the matrix of the adjacent intact porous media. This quantity will depend on the nature of the fracturing and can be derived separately for one-dimensional, two-dimensional and three-dimensional conditions as follows.

Consider a typical unit of the matrix. If advection in the matrix can be neglected, the matrix concentration, c_m, satisfies the equation

$$n_m D_m \nabla^2 c_m = (n_m + \varrho_m K_m)\frac{\partial c_m}{\partial t} + \lambda_m c_m \qquad (C.1)$$

where the subscript m indicates a property of the matrix and n, D, K and λ are the porosity, diffusion coefficient, distribution coefficient and first order decay constant (for radioactive or biological decay) in the matrix. The concentration at the surface of the matrix will be equal to that in the fracture system, c_m, so that $c_m = c_f$ on the surface of the matrix.

It will be assumed that initially the matrix is uncontaminated, so that

$$c_m = 0 \quad \text{when } t = 0 \qquad (C.2)$$

The Laplace transform of equation C.3 is then

$$n_m D_m \nabla^2 \bar{c}_m = n_m R_m (s + \Lambda_m)\bar{c}_m \qquad (C.3)$$

where

$$R_m = 1 + \frac{\varrho K_m}{n_m}$$

$$\Lambda_m = \frac{\lambda_m}{R_m}$$

Equation C.3 is to be solved subject to the boundary conditions that $c_m = c_f$ on the surface Σ_m of the matrix.

The quantity of primary interest is

$$\dot{q} = -\frac{\int f_{mn}\, d\Sigma_m}{\int dV_m}$$

where f_{mn} denotes the component of flux normal to the surface and V_m the volume occupied by the matrix.

It thus follows from equation C.3 and Gauss's divergence theorem that

$$\bar{\dot{q}} = \frac{\int n_m D_m \nabla^2 \bar{c}_m\, dV_m}{\int dv_m} = n_m R_m (s + \Lambda_m) \frac{\int \bar{c}_m\, dV_m}{\int \bar{c}_f\, dV_m}\, \bar{c}_f \qquad \text{(C.4)}$$

Referring to equation 7.65, $\bar{\eta}$ is given by

$$\bar{\eta} = n_m R_m \frac{\int \bar{c}_m\, dV_m}{\int \bar{c}_f dV_m} \qquad \text{(C.5)}$$

where

$$\bar{c}_m = \bar{c}_f \qquad z = \pm H_1$$

It now follows that

$$\bar{c}_m = \bar{c}_f \frac{\cosh(\mu z)}{\cosh(\mu H_1)}$$

where

$$\mu^2 = \frac{R_m}{D_m}(s + \Lambda_m)$$

Hence

$$\bar{\eta} = n_m R_m \frac{\tanh(\mu H_1)}{\mu H_1} \qquad \text{(C.6a)}$$

If the solution for $\bar{c}_m$ is developed as a Fourier series, it is found that $\bar{\eta}$ can also be expressed in the form

$$\bar{\eta} = n_m R_m \left[1 - 2\sum_j \frac{s + \lambda_m}{s + \lambda_m + \alpha_j^2 (D_m/R_m)} \times \frac{1}{(\alpha_j H_1)^2}\right] \qquad \text{(C.6b)}$$

where $\alpha_j = (j - \frac{1}{2})\pi/H_1$, $j = 1, 2, \ldots, \infty$.

Case 1

First, suppose there is only a single set of fractures (set 1), then because of the assumed conditions in the landfill there is no variation of concentration with respect to y. Also, the variation of concentration is likely to be relatively slow along the fissures (x-direction) when compared with the variation between adjacent fissures (z-direction); thus to sufficient accuracy the concentration within the fissures satisfies the equation

$$D_m \frac{\partial^2 \bar{c}_m}{\partial z^2} = R_m (s + \Lambda_m) \bar{c}_m$$

Case 2

Suppose now that there are two sets of fractures (sets 1 and 2); considering diffusive transport into the matrix in the z- and y-directions, the matrix concentration $\bar{c}_m$ can be written in the form

$$\bar{c}_m = \bar{c}_f \left[1 + \sum_{j,k} X_{jk} \cos(\alpha_j z) \cos(\beta_k y)\right]$$

with

$$\alpha_j = (j - \tfrac{1}{2})\pi/H_1,\ j = 1, 2, \ldots, \infty$$

$$\beta_k = (k - ½)\pi/H_2,\ k = 1, 2, \ldots, \infty$$

so that

$$\bar{\eta} = n_m R_m \left[1 + \sum_{j,k} X_{jk} \frac{\sin(\alpha_j H_1)}{\alpha_j H_1} \frac{\sin(\beta_k H_2)}{\beta_k H_2}\right]$$

where

$$X_{jk} = -4 \frac{\sin(\alpha_j H_1)}{\alpha_j H_1} \frac{\sin(\beta_k H_2)}{\beta_k H_2} \times \frac{(s + \Lambda_m)R_m}{R_m(s + \Lambda_m) + D_m(\alpha_j^2 + \beta_k^2)} \quad \text{(C.7a)}$$

and so

$$\bar{\eta} = n_m R_m \left[1 - 4 \sum_{j,k} \times \frac{s + \Lambda_m}{s + \Lambda_m + (\alpha_j^2 + \beta_j^2)(D_m/R_m)} \times \frac{1}{(\alpha_j H_1)^2} \frac{1}{(\beta_k H_2)^2}\right] \quad \text{(C.7b)}$$

Case 3

Finally, if there are three sets of fissures, and it is observed that the transport distances of interest along the fracture system are large compared to the dimensions of the block then it is found that the solution has the form

$$\bar{c}_m = \bar{c}_f \left[1 + \sum_{j,k,l} X_{jkl} \cos(\alpha_j z)\cos(\beta_k y)\cos(\gamma_l x)\right]$$

It follows in a similar manner to the previous case that

$$\bar{\eta} = n_m R_m \left[1 - 8 \sum_{j,k,l} \times \frac{s + \Lambda_m}{s + \Lambda_m + (D_m/R_m)\,(\alpha_j^2 + \beta_j^2 + \gamma_l^2)} \times \frac{1}{(\alpha_j H_1)^2} \frac{1}{(\beta_k H_2)^2} \frac{1}{(\gamma_l H_3)^2}\right]$$

where the definitions of α_j, β_k are as described previously and

$$\gamma_l = (l - ½)\pi/H_3,\ l = 1, 2, - \infty \quad \text{(C.8)}$$

References

Abramowitz, M. and Stegun, I. (1964) *Handbook of Mathematical Functions*, National Bureau of Standards Applied Mathematics, US Government Printing Offices, Washington, DC, Ser. 55.

Acar, Y.B. and Seals, R.K. (1984) Clay barrier technology for shallow land waste disposal facilities. *Hazardous Waste*, **1**(2), 167–81.

Acar, Y.B. and Haider, L. (1990) Transport of low concentration contaminants in saturated earthen barriers. *Journal of Geotechnical Engineering*, **116** (7), pp. 1031–52.

Acar, Y.B., Hamidon, A., Field, S.D. and Scott, L. (1985) The effects of organic fluids on hydraulic conductivity of compacted kaolinite, in *Hydraulic Barriers in Soil and Rock* (eds A.I. Johnson, R.K. Frobel, N.J. Cavalli and C.B. Pettersson), ASTM STP 874, American Society for Testing and Materials, Philadelphia, pp. 171–87.

Allan, M.B. (1984) Why upwinding is reasonable, in *Proceedings 5th International Conference of Finite Elements in Water Resources*, pp. 13–23.

Al-Niami, A.N.S. and Rushton, K.R. (1977) Analysis of flow against dispersion in porous media. *Journal of Hydrology*, **33**, 87–97.

Anderson, M.P. (1979) Using models to simulate the movement of contaminants through groundwater flow systems. *CRC Critical Reviews in Environmental Control*, **9**, 97–156.

Appelt, H., Holtzclaw, K. and Pratt, P.F. (1975) Effect of anion exclusion on the movement of chloride through soils. *Soil Science Society of America Proceedings*, **39**, 264–67.

Bailey, S.W. (1980) Summary of recommendations of AIPEA nomenclature committee. *Clay Minerals*, **5**, 85.

Banerji, S.K., Piontek, K. and O'Connor, J.T. (1986) Pentachlorophenol adsorption on soils and its potential for migration into groundwater, in *Hazardous and Industrial Solid Waste Testing and Disposal* (eds D. Lorenzen, R.A. Conway, L.P. Jackson *et al.*), Vol. 6, ASTM Special Technical Publication 933, pp. 120–39.

Barker, J.A. (1982) Laplace transform solutions for solute transport in fissured aquifers. *Advanced Water Research*, **5**(2), 98–104.

Barker, J.A. and Foster, S.S.D. (1981) A diffusion exchange model for solute movement in fissured porous rock. *Quarterly Journal of Engineering in Geology, London*, **14**, 17–24.

Barker, J.F. (1992) The persistence of aromatic hydrocarbons in various groundwater environments. Research paper, Waterloo Centre for Groundwater Research.

Barker, J.F., Cherry, J.A., Carey, D.A. and Mattes, M.E. (1987) Hazardous organic chemicals in groundwater at Ontario landfills, in *Proceedings 1987 Technology Transfer Conference*, Part C, Paper C3.

Barone, F.S. (1990) Determination of diffusion and adsorption coefficients for some contaminants in clayey soil and rock; laboratory determination and field evaluation. PhD thesis, University of Western Ontario, London, Ontario, Canada, 325 pp.

Barone, F.S., Costa, J.M.A., King, K.S. *et al.* (1993) Chemical and mineralogical assessment of in situ clay liner Keele Valley Landfill, Maple, Ontario in *Proceedings Joint CSCE–ASCE National Conference on Environmental Engineering* (eds R.N. Yong, J. Hadjinicolaou and A.M.O. Mohamed), pp. 1563–72.

Barone, F.S., Rowe, R.K. and Quigley, R.M. (1990) Laboratory determination of chloride diffusion coefficient in an intact shale. *Canadian Geotechnical Journal*, **27**, 177–84.

Barone, F.S., Rowe, R.K. and Quigley, R.M. (1992a) Estimation of chloride diffusion coefficient and tortuosity factor for mudstone. *Journal of Geotechnical Engineering, ASCE*, **118**, 1031–46.

Barone, F.S., Rowe, R.K. and Quigley, R.M. (1992b) A laboratory estimation of diffusion and adsorption coefficients for several volatile organics in a natural clayey soil. *Journal of Contaminant Hydrogeology*, **10**, 225–50.

Barone, F.S., Yanful, E.K., Quigley, R.M. and Rowe, R.K. (1989) Effect of multiple contaminant migration on diffusion and adsorption of some domestic waste contaminants in a natural clayey soil. *Canadian Geotechnical Journal*, **26**(2), 189–98.

Bass, J.M., Ehrenfeld, J.R. and Valentine, J.N. (1984) Potential clogging of landfill drainage systems. *US Environmental Protection Agency Project Summary EPA-600/S2-83-109*, Feb.

References

Bear, J. (1979) *Hydraulics of Groundwater*, McGraw-Hill, New York.

Benson, C.H. and Daniel, D.E. (1990) Influence of clods on hydraulic conductivity of compacted clay. *Journal of the Geotechnical Engineering Division, ASCE*, **116**(8), 1231–48.

Biener, E. and Sasse, T. (1993) Construction and rehabilitation of landfill shafts, in *Proceedings Fourth International Landfill Symposium*, Cagliari, Italy, pp. 451–60.

Biot, M.A. (1941) General theory of three-dimensional consolidation. *Journal of Applied Physics*, **12**, pp. 155–64.

Bonaparte, R., Ah-Line, A.M., Charron, R. and Tisinger, L. (1988) Survivability and durability of a nonwoven geotextile, in *Proceedings Geosynthetics for Soil Improvement*, Nashville, pp. 68–91.

Booker, J.R. and Rowe, R.K. (1987) One dimensional advective–dispersive transport into a deep layer having a variable surface concentration. *International Journal for Numerical and Analytical Methods in Geomechanics*, **11**(2), 131–42.

Bove, J.A. (1990) Direct shear friction testing for geosynthetics in waste containment, in *Geosynthetic Testing for Waste Containment Applications* (ed. R.M. Koerner), ASTM 1081, American Society for Testing and Materials, Philadelphia, pp. 241–56.

Bowders, J.J. and Daniel, D.E. (1987) Hydraulic conductivity of compacted clay to dilute organic chemicals. *Journal of Geotechnical Engineering, ASCE*, **113**(12), 1432–48.

Bowders, J.J., Daniel, D.E., Broderick, G.P. and Liljestrand, H.M. (1986) Methods for testing the compatibility of clay liners with landfill leachate, in *Hazardous and Industrial Solid Waste Testing: Fourth Symposium* (eds J.K. Petros, W.J. Lacy and R.A. Conway), ASTM STP 886, American Society for Testing and Materials, Philadelphia, pp. 233–50.

Bracci, G., Giardi, M. and Paci, B. (1991) The problem of clay liners testing in landfills, in *Proceedings Third International Landfill Symposium*, Cagliari, Italy, pp. 679–89.

Brand, E.W. and Premchitt, J. (1982) Response characteristics of cylindrical piezometers. *Geotechnique*, **32**(3), pp. 203–16.

Brebbia, C.A. and Dominguez, J. (1989) *Boundary Elements: an Introductory Course*, Computational Mechanics Publications, McGraw-Hill.

Brebbia, C.A. and Skerget, P. (1984) Diffusion–convection problems using boundary elements, in *Proceedings 5th International Conference on Finite Elements in Water Resources*, Vermont, pp. 747–68.

Brown, K.W. and Anderson, D.C. (1983) Effects of organic solvents on the permeability of clay soils. *EPA-600/2-83-016*, US Environmental Protection Agency, Cincinnati, OH.

Brown, K.W., Green, J.W. and Thomas, J.C. (1983) The influence of selected organic liquids on the permeability of clay liners, in *Proceedings, Ninth Annual Research Symposium on Land Disposal, Incineration, and Treatment of Hazardous Waste*, Ft Mitchell, KY, May 2–4, *EPA-600/9-83-018*, US Environmental Protection Agency, Cincinnati, OH, pp. 114–25.

Brown, D.W., Thomas, J.C. and Green, J.W. (1984) Permeability of compacted soils to solvents, mixtures and petroleum products, in *Proceedings Tenth Annual Research Symposium on Land Disposal of Hazardous Waste*, Ft Mitchell, KY, April 3–5, *EPA 600/9-84-007*, US Environmental Protection Agency, Cincinnati, OH, pp. 124–37.

Brune, M., Ramke, H.G., Collins, H.J. and Hanert, H.H. (1991) Incrustation processes in drainage systems of sanitary landfills, in *Proceedings Third International Landfill Symposium*, Cagliari, Italy, pp. 999–1035.

Cancelli, A. and Cazzuffi, D. (1987) Permittivity of geotextiles in presence of water and pollutant fluids, in *Proceedings of Geosynthetics 87*, New Orleans, pp. 471–81.

Cancelli, A., Cazzuffi, D. and Rimoldi, P. (1987) Geocomposite drainage systems: mechanical properties and discharge capacity evaluation, in *Proceedings of Geosynthetics '87*, New Orleans, pp. 393–404.

Carslaw, H.S. and Jaegar, J.C. (1948) *Operational Methods in Applied Mathematics*, 2nd edn, Oxford University Press, London.

Carslaw, H.S. and Jaegar, J.C. (1959) *Conduction of Heat in Solids*, 2nd edn, Clarendon Press, Oxford.

Cazzuffi, D., Cossu, R., Ferruti, L. and Lavagnolo, C. (1991) Efficiency of geotextiles and geocomposites in landfill drainage systems, in *Proceedings Third International Landfill Symposium*, Cagliari, Italy, pp. 759–80.

Cheremisinoff, P. and Morresi, A. (1976) *Energy From Solid Waste*, Marcel Dekker, New York.

Chian, E.S.K. (1977) Stability of organic matter in landfill leachates. *Water Research*, **11**, 225–32.

Christopher, B.R. and Holtz, R.D. (1984) *Geotextile Engineering Manual*. US DoT, FHWA Contract No. DTFH 61-80-C-00094.

Collins, H.J. (1991) Influences of recycling household refuse upon sanitary landfills, in *Proceedings Third International Landfill Symposium*, Cagliari, Italy, pp. 1111–23.

Collins, H.J. (1993) Impact of temperature inside the landfill on the behaviour of the barrier system. *Proceedings Fourth International Landfill Symposium*, Cagliari, Italy, pp. 417–32.

Cooper, H.H. and Jacob, C.E. (1946) A generalized graphical method for evaluating formation constants and summarizing well field history. *Transactions of the American Geophysical Union*, **27**, 526–34.

Cope, F.W. (1987) Design of waste containment structures (synthetic liner-based systems), in *Geotechnical Practice for Waste Disposal '87* (ed. R.W. Woods), ASCE, Geotechnical Special Publication No. 13, pp. 1–20.

Crooks, V.E. and Quigley, R.M. (1984) Saline leachate migration through clay: A comparative laboratory and field investigation. *Canadian Geotechnical Journal*, **21**(2), 349–62.

Daniel, D.E. (1984) Predicting hydraulic conductivity of clay liners. *Journal of the Geotechnical Engineering Division, ASCE*, **110**(2), 285–300.

Daniel, D.E. (1987) Earthen liners for land disposal facilities, in *Geotechnical Practice for Waste Disposal '87*, ASCE, Geotechnical Special Publication No. 13, pp. 21–39.

Daniel, D.E. (1989) In situ hydraulic conductivity tests for compacted clay. *Journal of Geotechnical Engineering*, **115**(9), 1205–26.

Daniel, D. and Liljestrand, H.M. (1984) Effects of landfill leachates on natural clay liner systems. Report to Chemical Manufacturers Association, Washington, DC.

Daniel, D.E. and Trautwein, S.J. (1986) Field permeability test for earthen liners, in *Proceedings In Situ '86 ASCE Specialty Conference*, Blacksburg, Virginia, June 1986, pp. 146–60.

Daniel, D.E., Trautwein, S.J., Boynton, S.S. and Foreman, D.E. (1984) Permeability testing with flexible-wall permeameters. *Geotechnical Testing Journal (GTJODJ)*, 7(3), pp. 113–22.

D'Astous, A.Y., Ruland, W.W., Bruce, J.R.G. *et al.* (1989) Fracture effects in the shallow groundwater zone in weathered Sarnia-area clay. *Canadian Geotechnical Journal*, **26**(1), 43–56.

Day, M.J. (1977) Analysis of movement and hydrochemistry of groundwater in the fractured clay and till deposits of the Winnipeg area, Manitoba. MSc thesis, University of Waterloo.

Day, S.R. and Daniel, D.E. (1985) Hydraulic conductivity of two prototype clay liners. *Journal of Geotechnical Engineering, ASCE*, **111**(8), 957–70. (Also discussion in *Journal of Geotechnical Engineering*, 1987, **113**(7), 796–819.)

Dematrocopoulis, A.C., Korfiatis, G.P., Bourodimos, E.L. and Nawy, E.G. (1984) Modelling for design of landfill bottom liners. *Journal of the Environmental Engineering Division, ASCE*, **110**(6), 1084.

Desaulniers, D.D., Cherry, J.A. and Fritz, P. (1981) Origin, age and movement of pore water in argillaceous quaternary deposits at four sites in southwestern Ontario. *Journal of Hydrology*, **50**, 231–57.

de Smedt (1981) Solute transfer through unsaturated porous media. *Quality of Groundwater – Studies in Environmental Science* (eds W. Duigvenbooden and P. Glasbergen), Elsevier Scientific Publishing Co., Amsterdam, Vol. 17, pp. 1011–16.

Domenico, P.A. and Schwartz, F.W. (1990) *Physical and Chemical Hydrogeology*, John Wiley, New York.

Dunn, R.J. and Mitchell, J.K. (1984) Fluid conductivity testing of fine-grained soils. *Journal of the Geotechnical Engineering Division, ASCE,* **110**(10), 1648–65.

Dutt, G.R. and Low, P.F. (1962) Diffusion of alkali chlorides in clay–water systems. *Soil Science*, **93**, 233–40.

Ehrig, H-J. (1989) Leachate quality, in *Sanitary Landfilling* (eds T.H. Christensen, R. Cossu and R. Stegmann), Academic Press, London, Chapter 4.2.

Ehrig, H-J. and Scheelhaase, T. (1993) Pollution potential and long term behaviour of sanitary landfills, in *Proceedings Fourth International Landfill Symposium*, Cagliari, Italy, pp. 1204–25.

Elrick, D.E., Smiles, D.E., Baumgartner, N. and Groenevelt, P.H. (1976) Coupling phenomena in saturated homo-ionic and montmorillonite: I. *Soil Science Society America Proceedings*, **40**, 490–91.

Elsbury, B.R., Daniel, D.E., Sraders, G.A. and Anderson, D.C. (1990) Lessons learned from compacted clay liners. *Journal of the Geotechnical Engineering Division, ASCE*, **116**(11), 1641–60.

Environmental Protection Agency (1983) Lining of waste impoundment and disposal facilities. *SW-870*, Office of Solid Waste and Emergency Response, Washington, DC 20460.

Erdelyi, A., Magnus, W., Oberhetting, F. and Tucomi, F.G. (1954) *Tables of Integral Transforms*, Vol. 1, McGraw-Hill, New York.

Farquhar, G.J. (1987) Leachate: production and characterization, in *Proceedings Canadian Society of Civil Engineers Centennial Conference*, Montreal, pp. 1–25.

Faure, Y., Gourc, J.P., Brocher, P. and Rollin, A.L. (1986) Soil–geotextile interaction in filter systems, in *3rd International Conference on Geotextiles*, Vienna, pp. 1207–12.

Fein, J. and Yu, P. (1988) Bioremediation technology for treatment of industrial pollutants: microbiological considerations, in *Proceedings Technology Transfer Conference 1988*, MOE Ontario, pp. 81–105.

Fernandez, F. and Quigley, R.M. (1985) Hydraulic conductivity of natural clays permeated with simple liquid hydrocarbons. *Canadian Geotechnical Journal*, **22**, 205–14.

Fernandez, F. and Quigley, R.M. (1987) Effect of viscosity on the hydraulic conductivity of clayey soils permeated with water-soluble organics, preprint, in *40th Canadian Geotechnical Conference*, Regina, Sask., pp. 313–20.

Fernandez, F. and Quigley, R.M. (1988a) Viscosity and dielectric constant controls on the hydraulic conductivity of clayey soils permeated with water-soluble organics. *Canadian Geotechnical Journal*, **25**, 582–89.

Fernandez, F. and Quigley, R.M. (1988b) Effects of increasing amounts of non-polar organic liquids in domestic waste leachate on the hydraulic conductivity of clay liners in southwestern Ontario, in *Proceedings Technological Transfer Conference No. 8*, Session C, Ontario MOE, pp. 55–79.

Fernandez, F. and Quigley, R.M. (1989) Organic liquids and the hydraulic conductivity of barrier clays. *Proceedings 12th International Conference on Soil Mechanics and Foundation Engineering*, Rio de Janeiro, Brazil, pp. 1867–70.

Fernandez, F. and Quigley, R.M. (1991) Controlling the destructive effects of clay–organic liquid interactions by application of effective stresses. *Canadian Geotechnical Journal*, **28**, 388–98.

Foreman, D. and Daniel, D.E. (1986) Permeation of compacted clay with organic chemicals. *ASCE Journal of Geotechnical Engineering*, **112**, 669–81.

Foster, S.S.D. (1975) The chalk groundwater tritium anomaly – a possible explanation. *Journal of Hydrology*, **25**, 159–65.

Freeze, R.A. and Cherry, J.A. (1979) *Groundwater*, Prentice-Hall, Englewood Cliffs, NJ.

Fried, J.J. (1976) *Groundwater Pollution*, Elsevier, New York.

Frind, E.O. (1987) Modelling of contaminant transport in groundwater. An overview, in *The Canadian Society for Civil Engineering Centennial Symposium on Management of Waste Contamination of Groundwater*, Montreal, May. 35 pp.

Frind, E.O. and Hokkanen, G.E. (1987) Simulation of the Borden plume using the alternating direction Galerkin technique. *Water Resources Research*, **23**(5), pp. 918–30.

Frind, E.O., Sudicky, E.A. and Schellenberg, S.L. (1987) Micro-scale modelling in the study of plume evolution in heterogeneous media. *Stochastic Hydrology and Hydraulics*, **1**, 263–79.

Fritz, P., Matthess, G. and Brown, R.M. (1976) Deuterium and oxygen-18 as indicators of leachwater movement from a sanitary landfill, in *Interpretation of Environmental Isotope and Hydrochemical Data in Groundwater Hydrology*, International Atomic Energy Agency, Vienna, pp. 131–42.

Fullerton, D.S. (1980) Preliminary correlation of post-Erie interstadial events (16,000–10,000 radiocarbon years before present), Central and Eastern Great Lakes Region, Hudson, Champlain and St. Lawrence Lowlands United States and Canada. *US GS Professional Paper 1089*, accompanied by two charts.

Gardner, W.R. (1958) Some steady-state solutions of the unsaturated moisture flow equation with applications to evaporation from a water table. *Soil Science*, **85**(4), pp. 228–32.

Gartner Lee Ltd. (1986) Burlington regional landfill plume delineation report for the regional municipality of Halton. (Also Halton regional landfill – Burlington 1986 monitoring report.)

Gartner Lee Ltd. (1988) Halton regional landfill – Burlington 1987 monitoring report. Report *GLL87-224*, submitted to the region of Halton, March.

Gartung, E., Pruhs, H. and Nowack, F. (1993) Measurements on vertical shafts in landfills. *Proceedings Fourth International Landfill Symposium*, Cagliari, Italy, pp. 461–68.

Gaudet, J.P., Jegat, H., Vachand, G. and Wierenga, P.J. (1977) Solute transfer, with exchange between mobile and stagnant water, through unsaturated sand. *Soil Science Society of America Journal*, **41**, 665–71.

Gelhar, L.W. and Axness, C.L. (1983) Three-dimensional stochastic analysis of macrodispersion in aquifers. *Water Resources Research*, **15**(6), 1387–97.

Gelhar, L.W., Mantoglou, A., Welty, C. and Rehfeldt, K.R. (1985) A review of field-scale physical solute transport processes in saturated and unsaturated porous media. *Electric Power Research Institute EPRI EA-4190 Project 2485–5*.

Geoservices (1987a) Background document on bottom liner performance in double-lined landfills and surface impoundments. *EPA/530-SW-87.013*, US Environmental Protection Agency, Cincinnati, OH.

Geoservices (1987b) Background document on proposed liner and leak detection rule. *EPA/530-SW-87.015*, US Environmental Protection Agency, Cincinnati, OH.

Gerhardt, R.A. (1984) Landfill leachate migration and attenuation in the unsaturated zone in layered and nonlayered coarse-grained soils. *Groundwater Monitoring Review*, **4**(2), pp. 56–65.

Geusebroek, H.J.L. and Luning, L. (1993) Innovative landfill design in a former sand quarry. *Proceedings Fourth International Landfill Symposium*, Cagliari, Italy, pp. 443–50.

Gillham, R.W. and Cherry, J.A. (1982) Contaminant migration in saturated unconsolidated geologic deposits. *Special Paper 189*, Geophysical Society of America, pp. 31–62.

Gillham, R.W., Robin, M.J.L., Dytynyshyn, D.J. and Johnston, H.M. (1984) Diffusion of nonreactive and reactive solutes through fine-grained barrier materials. *Canadian Geotechnical Journal*, **21**, 541–50.

Giroud, J.P. (1982) Filter criteria for geotextiles. *Proceedings 2nd International Conference on Geotextiles*, Las Vegas, USA, Vol. 1, pp. 103–09.

Giroud, J.P. (1987) Tomorrow's designs for geotextile applications in *Geotextile Testing and the Design Engineer* (ed. J.E. Fluet, Jr), ASTM STP 952, American Society for Testing and Materials, Philadelphia, pp. 145–58.

Giroud, J.P. (1994) Quantification of geosynthetic

behaviour, in: *Proceedings of the 5th International Conference on Geotextiles, Geomembranes and Related Products*, Singapore, Vol. 4, pp. 3–27.

Giroud, J.P. and Bonaparte, R. (1989) Leakage through liners constructed with geomembranes – Part I, Geomembrane liners; Part II, Composite liners. *Geotextiles and Geomembranes*, **8**, 27–68; 71–112.

Giroud, J.P., Badu-Tweneboah, K. and Bonaparte, R. (1992) Rate of leakage through a composite liner due to geomembrane defects. *Geotextiles and Geomembranes*, **11**(1), 1–29.

Giroud, J.P., Bonaparte, R., Beech, J.F. and Gross, B.A. (1990) Design of soil layer – geosynthetic systems overlying voids. *Geotextiles and Geomembranes*, **9**, 11–50.

Goldman, L.J., Greenfield, L.I., Damle, A.S. *et al.* (1990) *Clay Liners for Waste Management Facilities, Design, Construction and Evaluation.* Noyes Data Corporation for US EPA, New Jersey, USA.

Goodall, D.E. and Quigley, R.M. (1977) Pollutant migration from two sanitary landfill sites near Sarnia, Ontario. *Canadian Geotechnical Journal*, **14**, 223–36.

Gordon, M.E., Huebner, P.M. and Miazga, T.J. (1989) Hydraulic conductivity of three landfill clay liners. *Journal of Geotechnical Engineering, ASCE*, **115**(8), 1148–62.

Green, W.J., Lee, G.F. and Jones, R.A. (1981) Clay-soils permeability and hazardous waste storage. *J. Water Pollution Control Federation, Washington, DC*, **53**(8), 1347–54.

Griffin, R.A., Cartwright, K., Shimp, N.F. *et al.* (1976) Attenuation of pollutants in municipal landfill leachate by clay minerals: Part 1 – Column leaching and field verification. *Illinois State Geological Survey, Environmental Geology Notes, No. 78.*

Griffin, R.A. and Shimp, N.F. (1978) Attenuation of pollutants in municipal landfill leachate by clay minerals. Municipal Environmental Research Lab., US EPA, Report no. 600/14, Cincinnati, OH.

Griffin, R.A., Frost, R.R., Au, A.K. *et al.* (1977) Attenuation of pollutants in municipal landfill leachate by clay minerals: Part 2 – Heavy-metal adsorption. *Illinois State Geological Survey, Environmental Geology Notes, No. 79.*

Griffin, R.A., Sack, W.A., Roy, W.R. (1986) Batch type 24 hour distribution ratio for contaminant adsorption by soil materials, in *Hazardous and Industrial Solid Waste Testing and Disposal*, Vol. 6, ASTM, Special Technical Publication 933, pp. 390–408.

Grim, R.E. (1953) *Clay Mineralogy*, McGraw-Hill.

Grim, R.E. (1962) *Applied Clay Mineralogy*, McGraw-Hill.

Grisak, G.E. and Pickens, J.F. (1980) Solute transport through fractured media, 1, The effect of matrix diffusion. *Water Resources Research*, **16**(4), pp. 719–30.

Grisak, G.E., Pickens, J.F. and Cherry, J.A. (1980) Solute transport through fractured media, 2, Column study of fractured till. *Water Resources Research*, **16**, 731–39.

Gross, B.A., Bonaparte, R. and Giroud, J.P. (1990) Evaluation of flow from landfill leachate detection layers, in *Fourth International Conference on Geotextiles and Geomembranes*, The Hague, pp. 481–86.

Grouch, S.L. and Starfield, A.M. (1983) *Boundary Element Methods in Solid Mechanics*, Allen & Unwin, London.

Gschwend, P.M. and Wu, S. (1985) On the constancy of sediment–water partition coefficients of hydrophobic pollutants. *Journal of Environmental Science and Technology*, **19**, 90–96.

Ham, R.K. (1976) Solid waste degradation due to shredders and sludge addition, in *Gas and Leachate From Landfills: Formation, Collection and Treatment, EPA 600/9-76-004 PB 25116*, US Environmental Protection Agency, Cincinnati, OH.

Ham, R.K. in Lockland & Assoc. (1978) Final Report to EPA. Contract #68-03-2536. Recovery, processing and utilization of gas from sanitary landfills, US EPA.

Ham, R.K., Hekimian, K., Ketter, S. *et al.* (1979) *Recovery, Processing and Utilization of Gas from Sanitary Landfills, EPA-600/2-79-001*, US Environmental Protection Agency, Cincinnati, OH.

Hamilton, N.F. and Dylingowski, P.J. (1989) Bioslime control on geosynthetics, in *Proceedings Durability and Aging of Geosynthetics* (ed. R.M. Koerner), Elsevier Applied Science, London, pp. 278–92.

Hammoud, A. (1989) A theoretical examination of contaminant in a regularly fractured media. MESc thesis, University of Western Ontario.

Hantush, M.S. (1956) Analysis of data from pumping tests in leaky aquifers, *Transactions of the American Geophysical Union*, **37**, 702–14.

Hantush, M.S. (1960) Modification of the theory of leaky aquifers. *Journal of Geophysical Research*, **65**, 3713–25.

Harr, M.E. (1962) *Groundwater and Seepage*, McGraw-Hill, New York, pp. 43–44.

Herzog, B.L. and Morse, W.J. (1986) Hydraulic conductivity at a hazardous waste disposal site: Comparison of laboratory and field determined values. *Waste Management and Research*, **4**, 177–87.

Herzog, B.L., Griffin, R.A., Stohr, C.J. *et al.* (1989) Investigation of failure mechanisms and migration of organic contaminant at Wilsonville, Illinois. Spring 1989 *Groundwater Monitoring Review*, pp. 82–89.

Hewetson, J.P. (1985) An investigation of the groundwater zone in fractured shale at a landfill. MESc thesis, University of Waterloo.

Hillaire-Marcel, C.M. (1979) *Les mers post-glacieres au Quebec: Quelques aspects*, published Doctorat d'Etat thesis, Université Pierre et Marie Curie, Paris, 2 volumes.

Hoek, E. and Bray, J.W. (1981) *Rock Slope Engineering*, 3rd edn, The Institute of Mining and Metallurgy, London, p. 133.

Howard, P.H. (1989) *Handbook of Environmental Fate and Exposure Data for Organic Chemicals*, Lewis Publishers, Michigan.

Hughes, J.W. and Monteleone, M.J. (1987) Geomembrane/synthesized leachate compatibility testing, in *Geotechnical and Geohydrological Aspects of Waste Management* (eds D.J.A. van Zyl, J.D. Nelson, S.R. Abt and T.A. Shepherd), Fort Collins, CO, Lewis Publishers Inc., Michigan, pp. 35–50.

Hughes, G.M., Landon, R.A. and Farvolden, R.N. (1971) Hydrogeology of solid waste disposal sites in northeastern Illinois. *Report SW-12d*, US Environmental Protection Agency.

Huyakorn, P.S., Lester, B.H. and Mercer, J.W. (1983) An efficient finite element technique for modelling transport in fractured porous media, 1, Single species transport. *Water Resources Research*, **19**(3), 841–54.

Hvorslev, M.J. (1951) Time lag and soil permeability in groundwater observations. *Bulletin 36*, United States Army, Corps of Engineers, Waterways Experiment Station, Vicksburg, MS.

Hwu, B., Koerner, R.M. and Sprague, C.J. (1990) Geotextile intrusion into geonets, in *Fourth International Conference on Geotextiles*, The Hague, pp. 351–56.

International Environment Reporter (1981) EEC Drinking Water Directive, January 14, 151:0706–151: 0712.

Javendel, I., Christine, D. and Tsang, C. (1984) TDAST – analytic solution of 2D flow, in *Groundwater Transport: Handbook of Mathematical Models*, American Geophysical Union, Washington, DC, 228 pp.

Jewell, R.A. (1991) Application of revised design charts for steep reinforced slopes. *Geotextiles and Geomembranes*, **10**, 203–33.

Karickhoff, S.W., Brown, D.S. and Scott, T.A. (1979) Sorption of hydrophobic pollutants on natural sediments. *Water Research*, **13**, 241–8.

Kemper, J.M. and Smith, R.B. (1980) Leachate and gas production by landfilled processed municipal wastes, in *University of Wisconsin Municipal and Industrial Waste Research and Practice 3rd Conference*, September 10.

Kemper, W.D. and van Schaik, J.C. (1966) Diffusion of salts in clay–water systems. *Soil Science Society of America Proceedings*, **30**, 534–40.

King, K.S., Quigley, R.M., Fernandez, F. *et al.* (1993) Hydraulic conductivity and diffusion monitoring of the Keele Valley landfill liner, Maple, Ontario. *Canadian Geotechnical Journal*, **30**.

Kirk, D. and Law, M. (1985) Codisposal of industrial and municipal waste, in *Proceedings MOE Technology Transfer Conference*, Toronto.

Klute, A. and Letey, J. (1958) The dependence of ionic diffusion on the moisture content of nonadsorbing porous media. *Soil Science Society of America Proceedings*, **22**, 213–15.

Koerner, G.R. and Koerner, R.M. (1989) Biological clogging in leachate collection systems, in *Proceedings Durability and Aging of Geosynthetics* (ed. R.M. Koerner), Elsevier Applied Science, London, pp. 260–77.

Koerner, G.R. and Koerner, R.M. (1990a) Biological activity and remediation involving geotextile landfill leachate filters, in *Symposium on Geosynthetic Testing for Waste Containment Applications*, ASTM STP 1081, American Society for Testing and Materials, Philadelphia.

Koerner, G.R. and Koerner, R.M. (1990b) The installation survivability of geotextiles and geogrids, in *Fourth International Conference on Geotextiles*, The Hague, pp. 597–602.

Koerner, R.M. (1990a) Preservation of the environment via geosynthetic containment systems, in *Proceedings of the 4th International Conference on Geotextiles and Geomembranes*, Vol. III, The Hague, pp. 975–88.

Koerner, R.M. (ed.) (1990b) *Durability and Aging of Geosynthetics*. Elsevier Applied Science, London.

Koerner, R.M. (1990c) *Designing with Geosynthetics*, Prentice-Hall, Englewood Cliffs, NJ.

Koerner, R.M. and Hwu, B. (1989) *Behaviour of Double Geonet Drainage Systems*. Geotechnical Fabrics Report, IFAI Publication, pp. 39–44.

Koerner, R.M., Bove, J.A. and Martin, J.P. (1984) Water and air transmissivity of geotextiles. *Geotextiles and Geomembranes*, 1, 57–73.

Koerner, R.M., Halse, Y.H. and Lord, A.E. Jr (1990) Long-term durability and aging of geomembranes, in *Waste Containment Systems: Construction, Regulation, and Performance* (ed. R. Bonaparte), ASCE Special Technical Publication No. 26, New York, pp. 106–34.

Koerner, R.M., Hsuan, Y. and Lord, A.E. (1993) Remaining technical barriers to obtaining general acceptance of geosynthetics. *Geotextiles and Geomembranes*, **12**(1), 1–53.

Koerner, R.M., Lucianai, V.A., Freese, J.S. and Carroll, R.G., Jr (1986) Prefabricated drainage composites: evaluation and design guidelines, in *Third International Conference on Geotextiles*, Vienna, pp. 551–56.

Krug, T.A. and McDougall, S. (1988) Preliminary

assessment of a microfiltration/reverse osmosis process for the treatment of landfill leachate, in *Proceedings Technology Transfer Conference, 1988*, MOE Ontario, pp. 117–32.

Lahti, L.R., King, K.S., Reades, D.W. and Bacopoulos, A. (1987) Quality assurance monitoring of a large clay liner, in *Proceedings of ASCE Specialty Conference on Geotechnical Aspects of Waste Disposal, '87*, Ann Arbor, pp. 640–54.

Lai, T.M. and Mortland, M.M. (1962) Self-diffusion of exchangeable cations in bentonite, in *Clays and Clay Minerals, 9th Conference*, Pergamon Press, New York, pp. 229–47.

Lambe, T.W. (1956) The storage of oil in an earth reservoir. *Journal of Boston Society of Civil Engineers*, **43**, 179–241.

Lambe, T.W. (1958) The engineering behaviour of compacted clay. *Journal Soil Mechanics and Foundation Engineering Division, ASCE*, **84**(SM2), Paper 1655, 35 pp.

Lambe, T.W. (1960) Compacted clay – a symposium. *Transactions of the ASCE*, **125**, Part I, 681–756. (Two papers – Compacted clay: Structure; and Compacted clay; Engineering behaviour plus discussions.)

Lambe, T.W. and Martin, R.T. (1955) Composition and engineering properties of soil III, in *Proceedings 34th Annual Meeting of the Highway Research Board*, USA, pp. 566–82.

Lambe, T.W. and Whitman, R.E. (1979) *Soil Mechanics*, SI version, John Wiley.

Landva, A.O. and Clark, J.I. (1990) Geotechnics of waste fills, in *Geotechnics of Waste Fills* (eds A.O. Landva and G. Knowles), ASTM STP1070, pp. 86–103.

Landva, A.O. and Knowles, G. (eds) (1990) *Geotechnics of Waste Fills – Theory and Practice*, ASTM STP1070, Baltimore.

Lapidus, L. and Amundson, N.R. (1952) Mathematics of adsorption in beds, VI. The effect of longitudinal diffusion in ion exchange and chromatographic columns. *Journal of Physical Chemistry*, **56**(8), 984–88.

Lerman, A. (1979) *Geochemical Processes – Water and Sediment Environments*. Wiley Interscience, New York.

Lewis, C.F.M. (1969). Late quaternary history of lake levels in the Huron and Erie Basin, in *Proceedings 12th Conference on Great Lakes Research*, International Association for Great Lakes Research, pp. 250–70.

Lindstrom, F.T., Haque, R., Freed, V.H. and Boersma, L. (1967) Theory of the movement of some herbicides in soils – linear diffusion and convection of chemicals in soils. *Environmental Science and Technology*, **1**(7), 561–65.

Lord, A.E., Koerner, R.M. and Swan, J.R. (1988) Chemical mass transport measurement to determine flexible membrane liner lifetime. *Geotechnical Testing Journal, ASTM*, 83–91.

Lu, J.C.S., Eichenberger, B. and Stearns, R.J. (1985) Leachate from municipal landfills, production and management. *Pollution Technology Review No. 119*, Noyes Publications, Park Ridge, NJ.

Lu, J.C.S., Morrison, R.D. and Stearns, R.J. (1981) Leachate production and management from municipal landfills: Summary and assessment, in *Proceedings of the Seventh Annual Research Symposium, Land Disposal: Municipal Solid Waste*, EPA-600/9-81-002a, US Environmental Protection Agency, Cincinnati, OH, pp. 1–17.

Luber, M. (1992) Diffusion of chlorinated organic compounds through synthetic landfill liners. Report, Department of Earth Sciences, University of Waterloo.

Lundell, C.M. and Menoff, S.D. (1989) The use of geosynthetics as drainage media at solid waste landfills, in *Proceedings Geosynthetics '89*, San Diego, pp. 10–17.

Matich, M.A.J. and Tao, W.F. (1984) A new concept of waste disposal, in *Proceedings of the CEO–SOS Seminar on Design and Construction of Municipal and Industrial Waste Disposal Facilities*, Toronto, pp. 43–60.

McBean, E.A., Mosher, F.R. and Rovers, F.A. (1993) Reliability based design for leachate collection systems, in *Proceedings Fourth International Landfill Symposium*, Cagliari, Italy, pp. 431–41.

McBean, E.A., Poland, R., Rovers, F.A. and Crutcher, A.J. (1982) Leachate collection design for contaminant landfills. *Journal of Environmental Engineering Division, ASCE*, **108**, 204.

McGinley, P.M. and Kmet, P. (1984) Formation, characteristics, treatment and disposal of leachate from municipal solid waste landfills. Special report, Bureau of Solid Waste Management, Wisconsin Department of Natural Resources, Madison, Wisconsin.

McKay, L.D. and Trudell, M.R. (1989) The sorption of trichloroethylene in clayey till. Symposium of Groundwater Contamination, Saskatoon, Saskatchewan, June 14–15. Sponsored by Ground-Water Division, National Hydrology Research Institute, Paper E-40.

Mesri, G. and Olson, R.E. (1971) Mechanisms controlling the permeability of clays. *Clays and Clay Minerals*, **19**, 151–58.

Michaels, A.S. and Lin, C.S. (1954) The permeability of kaolinite. *Industrial and Engineering Chemistry*, **46**, 1239–46.

Mineralogical Society (1980) *Crystal Structures of Clay Minerals and Their X-ray Identification* (eds G.W. Brindley and G. Brown), Mineralogical Society Monograph No. 5.

Ministry of National Health and Welfare (1978) Guidelines for Canadian Drinking Water Quality. ISBN: 0 660 10429 6 (updated 1988).

Mitchell, J.K. (1993) *Fundamentals of Soil Behaviour*, 2nd edn, John Wiley.

Mitchell, J.K. and Madsen, F.T. (1987) Chemical effects on clay hydraulic conductivity, in *Geotechnical Practice for Waste Disposal '87* (ed. R.D. Woods), American Society of Civil Engineers, Geotechnical Special Publication No. 13, pp. 87–116.

Mitchell, J.K., Hooper, D.R. and Campanella, R.G. (1965) Permeability of compacted clay. *Journal of the Geotechnical Engineering Division, ASCE*, **91**(SM4), 41–65.

Mitchell, J.K., Seed, R.B. and Seed, H.B. (1990a) Kettleman Hills waste landfill slope failure I: Liner system properties. *Journal of Geotechnical Engineering, ASCE*, **116**(4), 647–68.

Mitchell, J.K., Seed, R.B. and Seed, H.B. (1990b) Stability considerations in the design and construction of lined waste repositories, in *Geotechnics of Waste Fills – Theory and Practice* (eds A.O. Landva and G. Knowles), ASTM STP1070, pp. 207–24.

MoEE (1993a) Ontario Policy, Incorporation of the reasonable use concept into MOEE groundwater management activities. Policy 15-08, Ministry of the Environment and Energy, March 1993.

MoEE (1993b) Engineered facilities at landfills that receive municipal and non-hazardous wastes. Policy 14-15, Ministry of the Environment and Energy, March 1993.

Moore, C.A. (1983) *Landfill and Surface Impoundment Performance Evaluation Manual, EPA SW 869*, US Environmental Protection Agency, Cincinnati, OH.

Moore, I.D. (1993) HDPE pipe for burial under landfills, in *Proceedings Fourth International Landfill Symposium* Cagliari, Italy, pp. 1473–82.

Moreno, L. and Rasmuson, A. (1986) Contaminant transport through a fractured porous rock: Impact of the inlet boundary condition on the concentration profile in the rock matrix. *Water Resources Research*, **22**(12), 1728–30.

Morrison, R.T. and Boyd, R.N. (1983) *Organic Chemistry*. Allyn & Bacon, Toronto, 1370 pp.

Mucklow, J.P. (1990) Phenol migration from landfills by diffusion in natural clayey soils. MSc thesis, University of Western Ontario, London, Ontario, Canada, 234 pp.

Myrand, D., Gillham, R.W., Cherry, J.A. and Johnson, R.L. (1987) Diffusion of volatile organic compounds in natural clay deposits. Department of Earth Sciences Report, University of Waterloo, Waterloo, Ontario, Canada, 33 pp.

Neretnieks, I. (1980) Diffusion in the rock matrix: An important factor in radio nuclide retardation. *Journal of Geophysical Research*, **85**(B8), 4379–97.

Neuman, S.P. and Witherspoon, P.A. (1972) Field determination of the hydraulic properties of leaky multiple aquifer systems. *Water Resources Research*, **8**(5), 1284–98.

Nkedi-Kissa, P., Rao, P.S.C. and Hornsby, A.G. (1985) Influence of organic cosolvents on sorption of hydrophobic organic chemicals by soils. *Journal of Environmental Science and Technology*, **19**, 975–79.

Novakowski, K.S. (1993) Interpretation of the transient flow rate obtained from constant-head tests conducted in situ in clays. *Canadian Geotechnical Journal*, **30**(4), 600–06.

Ogata, A. (1970) Theory of dispersivity in a granular medium. US Geological Survey, Professions Paper.

Ogata, A. and Banks, R.B. (1961) A solution of the differential equation of longitudinal dispersion in porous media. US Geological Survey, Professions Paper 411-A.

Ogunbadejo, T.A. (1973) Physico-chemistry of weathered clay crust formation. PhD thesis, University of Western Ontario, London, Ontario.

Olsen, H.W. (1966) Darcy's law in saturated kaolinite. *Water Resources Research*, **2**(2), 287–95.

Page, L.M., Raila, S.J. and Woliner, N.R. (1982) Investigation of leachate contamination of ground and surface waters at the Fresh Kills landfill, New York City, in *Proceedings 5th Madison Conference of Applied Research and Practice on Municipal and Industrial Waste*, University of Wisconsin, pp. 175–209.

Park, G.S. (1986) Transport principles – solution. Diffusion and permeation in polymer membranes, in *Synthetic Membranes: Science, Engineering, and Applications* (eds P.M. Bangay *et al.*), Reidel Publ. Co.

Perkins, T.K. and Johnston, D.C. (1963) A review of diffusion and dispersion in porous media. *Society of Petroleum Engineering Journal*, 3(1), 70–84.

Porter, L.K., Kempter, W.D., Jackson, R.J. and Stewart, B.A. (1960) Chloride diffusion in soils as influenced by moisture content. *Soil Science Society of America Proceedings*, **24**, 460–63.

Poulos, H.G. and Davis, E.H. (1980) *Pile Foundation Analysis and Design*, John Wiley, New York.

Press, W.H., Flannery, B.P., Teukolsky, S.A. and Vetterling, W.T. (1986) *Numerical Recipes – The Art of Scientific Computing*, Cambridge University Press, Cambridge.

Puig, J., Govy, J-L. and Labrove, L. (1986) Ferric clogging of drains, in *Proceedings 3rd International Conference on Geotextiles*, Vienna, pp. 1179–84.

Quigley, R.M. (1991) Chemical assessment of an insitu clayey liner – Keele Valley landfill, Maple, Ontario (December 8, 1988 exhumation). Final Report to Golder Associates Ltd., Mississauga, Ontario.

Quigley, R.M. and Crooks, V.E. (1983) Chemical

profiles in soft clays and the role of long-term diffusion, in *Geological Environment and Soil Properties* (ed. R.N. Yong), ASCE Special Publication (unnumbered), pp. 5–18.

Quigley, R.M. and Fernandez, F. (1987) Effects of increasing amounts of nonpolar organic liquids in domestic waste leachate on the hydraulic conductivity of clay liners in southwestern Ontario. Geotechnical research report *GEOT-13-87*, UWO.

Quigley, R.M. and Fernandez, F. (1989) Clay/organic interactions and their effect on the hydraulic conductivity of barrier clays, in *Contaminant Transport in Groundwater* (eds H.E. Kobus and W. Kinzelbach), pp. 117–24. Proceedings of an international symposium, Stuttgart, 4–6 April 1989. 500 pp., Hfl. 205/US$120.00. A.A. Balkema, Old Post Road, Brookfield, Vermont 05036 (telephone: 802-276-3162; telefax: 802-276-3837).

Quigley, R.M. and Fernandez, F. (1992) Organic liquid interactions with water–wet barrier clays, in *Subsurface Contamination by Immiscible Fluids* (ed. K.U. Weyer), Proceedings International Conference on Subsurface Contamination by Immiscible Fluids, Calgary, Alberta, April 1990, Balkema, Rotterdam, pp. 49–56.

Quigley, R.M. and Rowe, R.K. (1986) Leachate migration through clay below a domestic waste landfill, Sarnia, Ontario, Canada: Chemical interpretation and modelling philosophies, in *Hazardous and Industrial Solid Waste Testing and Disposal*, Vol. 6 (eds D. Lorenzen, R.A. Conway, L.P. Jackson *et al.*), ASTM STP 933 American Society for Testing and Materials, Philadelphia, pp. 93–103.

Quigley, R.M., Crooks, V.E. and Yanful, E. (1984) Contaminant migration through clay below a domestic waste landfill site, Sarnia, Ontario, Canada, in *Proceedings International Groundwater Symposium on Groundwater Resources Utilization and Contaminant Hydrogeology*, Montreal, May, Vol. II, pp. 499–506.

Quigley, R.M., Fernandez, F., Yanful, E. *et al.* (1987) Hydraulic conductivity of contaminated natural clay directly beneath a domestic landfill. *Canadian Geotechnical Journal*, **24**(3), pp. 377–83.

Quigley, R.M., Fernandez, F. and Rowe, R.K. (1988) Clayey barrier assessment for impoundment of domestic waste leachate (southern Ontario) including clay–leachate compatibility by hydraulic conductivity testing. *Canadian Geotechnical Journal*, **25**(B), 574–81.

Quigley, R.M., Gwyn, Q.H.J., White, O.L. *et al.* (1983) Leda clay from deep boreholes at Hawkesbury, Ontario. Part I: Geology and geotechnique. *Canadian Geotechnical Journal*, **20**(2), 288–98.

Quigley, R.M., Mucklow, J.P. and Yanful, E.K. (1990) Contaminant migration by diffusion at the Confederation Road landfill, Sarnia, Ontario, in *Engineering in our Environment, Proceedings of 1990 Conference of Canadian Society for Civil Engineering*, Hamilton, May, pp. 876–92.

Quigley, R.M., Yanful, E.K. and Fernandez, F. (1987) Ion transfer by diffusion through clayey barriers, in *Geotechnical Practice for Waste Disposal '87* (ed. R.D. Woods), ASCE, Geotechnical Special Publication No. 13, pp. 137–58.

Quigley, R.M., Yanful, E.K. and Fernandez, F. (1990) Biological factors influencing laboratory and field diffusion, in *Microbiology in Civil Engineering* (ed. P. Howsam), Proceedings Federation of European Microbiological Societies Symposium, Cranfield Institute of Technology, UK, September 3–5, pp. 261–73.

Ramke, H-G. and Brune, M. (1990) Untersuchungen zur Funktionsfähigkeit von Entwässerungsschichten in Deponiebasisabdichtungssystemen, Abschlussbericht Bundesminister für Forschung und Technologie. FKZ BMFT 145 0457 3.

Randolph, M.F. and Booker, J.R. (1982) Analysis of seepage into a cylindrical permeameter, in *Proceedings Fourth International Conference on Numerical Methods in Geomechanics*, Edmonton, Ontario, pp. 349–58.

Rao, P.S.C., Green, R.E., Balasubramanian, V. and Kanehio, Y. (1974) Field study of solute movement in a highly aggregated oxisol with intermittent flooding: II Picloram, *Journal Environmental Quality*, **3**, 197–202.

Reades, D.W. and Thompson, C.D. (1984) Quality control testing and monitoring of performance of clay till liner, State 1, Keele Valley landfill, Maple, Ontario, in *Proceedings of the CEO–CGS Seminar on Design and Construction of Municipal and Industrial Waste Disposal Facilities*, Toronto, pp. 135–46.

Reades, D.W., King, K.S., Benda, E. *et al.* (1989) The results of on-going monitoring of the performance of a low permeability clay liner, Keele Valley landfill, Maple, Ontario, in *Proceedings Focus Conference on Eastern Regional Ground Water Issues*, National Water Well Association, Kitchener, Ontario, pp. 79–91.

Reades, D.W., Lahti, L.R., Quigley, R.M. and Bacopoulos, A. (1990) Detailed case history of clay liner performance, in *Waste Containment Systems: Construction, Regulation, and Performance* (ed. R. Bonaparte), ASCE Special Technical Publication No. 26, New York, pp. 156–74.

Reades, D.W., Poland, R.J., Kelly, G. and King, S. (1987) Discussion of, Hydraulic conductivity of two prototype clay liners, by Day and Daniel. *Journal of Geotechnical Engineering*, **113**(7), pp. 809–13.

Reitzel, S.F. (1990) The temporal characterization of municipal solid waste. MASc thesis, University of Waterloo, Waterloo, Ontario, Canada.

Richardson, G.N. and Koerner, R.M. (1988) Geosynthetic design guidance for hazardous waste landfill cells and surface impoundments. *EPA/600/52-87/097*, US Environmental Protection Agency, Cincinnati, OH.

Rollin, A.L. and Denis, R. (1987) Geosynthetic filtration in landfill design, in *Proceedings Geosynthetics '87*, pp. 456–70.

Rollin, A.L. and Lombard, G. (1988) Mechanisms affecting long-term filtration behaviour of geotextiles. *Geotextiles and Geomembranes*, 7, 119–45.

Rollin, A.L. and Rigo, J-M. (eds) (1991) *Geomembranes: Identification and Performance Testing*, Chapman & Hall, London.

Rowe, R.K. (1987) Pollutant transport through barriers, in *Geotechnical Practice for Waste Disposal '87* (ed. R.D. Woods), ASCE Special Geotechnical Publication No. 13, pp. 159–81.

Rowe, R.K. (1988) Contaminant migration through groundwater: The role of modelling in the design of barriers. *Canadian Geotechnical Journal*, **25**(4), 778–98.

Rowe, R.K. (1991a) Contaminant impact assessment and the contaminating lifespan of landfills. *Canadian Journal of Civil Engineering*, **18**(2), 244–53.

Rowe, R.K. (1991b) Some considerations in the design of barrier systems, in *Proceedings, First Canadian Conference on Environmental Geotechnics*, Montreal, May, pp. 157–64.

Rowe, R.K. (1991c) Leachate detection of hydraulic control: possible roles for a granular layer beneath landfill liners, in *Proceedings 3rd International Symposium on Sanitary Landfills*, Sardinia, Italy, pp. 979–987.

Rowe, R.K. (1991d) Environmental geotechnology: Some pertinent considerations, keynote paper, in: *Proceedings of the 7th International Conference of the International Association for Computer Methods and Advances in Geomechanics*, Cairns, 6–10 May 1991 Vol. 1, pp. 35–48. Beer, G., Booker, J.R. and Carter, J.P. (eds.), 3 volumes, Hfl. 480/US$275.00. A.A. Balkema, Old Post Road, Brookfield, Vermont 05036 (telephone: 802-276-3162; telefax: 802-276-3837).

Rowe, R.K. (1992) Integration of hydrogeology and engineering in the design of waste management sites, in *Proceedings of the International Association of Hydrogeologists Conference on 'Modern Trends in Hydrogeology'*, Hamilton, Ontario, pp. 7–21.

Rowe, R.K. (1993) Some challenging applications of geotextiles in filtration and drainage, in *Geotextiles in Filtration and Drainage*, Thomas Telford, London, pp. 1–12.

Rowe, R.K. and Badv, K. (1994a) Chloride migration through clay underlain by fine sand. Geotechnical Research Centre Report, *GEOT-11-94*, Faculty of Engineering Science, University of Western Ontario, London, Canada.

Rowe, R.K. and Badv, K. (1994b) Diffusion controlled and advective–diffusive contaminant migration through unsaturated coarse sand and fine gravel. Geotechnical Research Centre Report, *GEOT-12-94*, Faculty of Engineering Science, University of Western Ontario, London, Canada.

Rowe, R.K. and Badv, K. (1994c) Contaminant transport through a clayey liner underlain by an unsaturated stone collection layer. Geotechnical Research Centre Report, *GEOT-13-94*, Faculty of Engineering Science, University of Western Ontario, London, Canada.

Rowe, R.K. and Barone, F.S. (1991) Diffusion tests for chloride and dichloromethane in Halton Till. Report of the Geotechnical Research Centre, University of Western Ontario, London, Ontario, Canada.

Rowe, R.K. and Booker, J.R. (1983) SFIN – a finite element analysis program for single contaminant migration under 1D conditions. Geotechnical Research Centre, University of Western Ontario, London, Ontario.

Rowe, R.K. and Booker, J.R. (1985a) 1-D pollutant migration in soils of finite depth. *Journal of Geotechnical Engineering, ASCE*, **111**(GT4), 479–99.

Rowe, R.K. and Booker, J.R. (1985b) 2D pollutant migration in soils of finite depth. *Canadian Geotechnical Journal*, **22**(4), pp. 429–36.

Rowe, R.K. and Booker, J.R. (1986) A finite layer technique for calculating three-dimensional pollutant migration in soil. *Geotechnique*, **36**(2), 205–14.

Rowe, R.K. and Booker, J.R. (1987) An efficient analysis of pollutant migration through soil, in *Numerical Methods in Transient and Coupled Systems* (eds R.W. Lewis, E. Hinton, P. Bettess and B.A. Schrefler), John Wiley, pp. 13–42.

Rowe, R.K. and Booker, J.R. (1988a). Modelling of contaminant movement through fractured or jointed media with parallel fractures, in *Proceedings of 6th International Conference on Numerical Methods in Geomechanics*, Innsbruck, April, 855–62.

Rowe, R.K. and Booker, J.R. (1988b) MIGRATE – analysis of 2D pollutant migration in a non-homogeneous soil system: Users manual. Report number *GEOP-1-88*, Geotechnical Research Centre, University of Western Ontario, London, Ontario.

Rowe, R.K. and Booker, J.R. (1989a) A semi-analytic model for contaminant migration in a regular two or three dimensional fractured network: Conservative contaminants. *International Journal for Numerical and Analytical Methods in Geomechanics*, **13**, pp. 531–50.

Rowe, R.K. and Booker, J.R. (1989b) Analysis of contaminant transport through fractured rock at an Ontario landfill, in *Proceedings of the 3rd International Symposium on Numerical Methods in Geomechanics*, Niagara Falls, pp. 383–90.

Rowe, R.K. and Booker, J.R. (1990a) Contaminant

migration through fractured till into an underlying aquifer. *Canadian Geotechnical Journal*, **27**, 484–95.

Rowe, R.K. and Booker, J.R. (1990b). A semi-analytic model for contaminant migration in a regular two or three dimensional fractured network: Reactive contaminants. *International Journal for Numerical and Analytical Methods in Geomechanics*, **14**, 401–25.

Rowe, R.K. and Booker, J.R. (1991a). Modelling of 2D contaminant migration in a layered and fractured zone beneath landfills. *Canadian Geotechnical Journal*, **28**(3), 338–52.

Rowe, R.K. and Booker, J.R. (1991b). Pollutant migration through a liner underlain by fractured soil. *Journal of Geotechnical Engineering, ASCE*, **118**(7), 1031–46.

Rowe, R.K. and Booker, J.R. (1994). Program POLLUTE, Geotechnical Research Centre, University of Western Ontario Report. © 1983, 1990, 1994. Distributed by GAEA Environmental Engineering Ltd., 30 Frontenac Road, London, Ontario, Canada.

Rowe, R.K. and Fraser, M.J. (1993a) Long-term behaviour of engineered barrier systems, in *Proceedings, Fourth International Landfill Symposium*, Cagliari, Italy, pp. 397–406.

Rowe, R.K. and Fraser, M.J. (1993b) Service life of barrier systems in the assessment of contaminant impact, in *Joint CSCE–ASCE National Conference on Environmental Engineering*, Montreal, July, Vol. 2, pp. 1217–24.

Rowe, R.K. and Giroud, J-P. (1994) Quality assurance of barrier systems for landfills. *IGS News*, **10**(1), pp. 6–8.

Rowe, R.K. and Nadarajah, P. (1993) Evaluation of the hydraulic conductivity of aquitards. *Canadian Geotechnical Journal*, **30**(5), 781–800.

Rowe, R.K. and Nadarajah, P. (1994a) An analytical method for predicting the velocity field beneath landfills. Geotechnical Research Centre Report, *GEOT-14-94*, University of Western Ontario, London, Canada.

Rowe, R.K. and Nadarajah, P. (1994b) Design of leachate pumping well fields in landfills. Geotechnical Research Centre Report, *GEOT-15-94*, University of Western Ontario, London, Canada.

Rowe, R.K. and San, K.W. (1992) Effect of source characteristics of landfills on environmental impact, in *Proceedings of International Association of Hydrogeologists Conference on Modern Trends in Hydrogeology*, Hamilton, Canada, pp. 249–61.

Rowe, R.K. and San, K.W. (1994) Effect of leachate mounding and sand lenses on contaminant migration from a landfill cell. *Computers and Geotechnics* **16**(3), pp. 173–204.

Rowe, R.K. and Sawicki, D. (1992) Modelling of a natural diffusion profile and the implications for landfill design, in *Proceedings of 4th International Symposium on Numerical Methods in Geomechanics* (eds G.N. Pande and S. Pietruszak), Swansea, pp. 481–89.

Rowe, R.K., Booker, J.R. and Fraser, M.J. (1994) *POLLUTE v6 and POLLUTE-GUI User's Guide*, GAEA Environmental Engineering Ltd., London, Ontario.

Rowe, R.K., Caers, C.J., Booker, J.R. and Crooks, V.E. (1985) Pollutant migration through clay soils, in *Proceedings 11th International Conference on Soil Mechanics and Foundation Engineering*, San Francisco, pp. 1293–98.

Rowe, R.K., Caers, C.J. and Barone, F. (1988) Laboratory determination of diffusion and distribution coefficients of contaminants using undisturbed soil. *Canadian Geotechnical Journal*, **25**, pp. 108–18.

Rowe, R.K., Caers, C.J. and Chan, C. (1993) Evaluation of a compacted till liner test pad constructed over a granular subliner contingency layer. *Canadian Geotechnical Journal*, **30**(4), 667–89.

Ruland, W.W. (1988) Fracture depth and active groundwater flow in clayey till in Lambton County, Ontario. MSc thesis, The University of Waterloo.

Schubert, W.R., Harrington, T.J. and Finno, R.J. (1984) Glacial clay liners in waste disposal practice, in *Environmental Engineering Specialty Conference, American Society of Civil Engineers*, pp. 36–41.

Schwarzenbach, R.P. and Westall, J. (1981) Transport of non-polar organic compounds from surface water to groundwater. Laboratory sorption studies. *Environmental Science and Technology*, Vol. **15**(11), pp. 1360–67.

Seed, R.B., Mitchell, J.K. and Seed, H.B. (1990) Kettleman Hills waste landfill slope failure II: Stability analyses. *Journal of Geotechnical Engineering, ASCE*, **116**(4), 669–89.

Selim, H.M. and Mansell, R.S. (1976) Analytical solution of the equation for transport of reactive solute. *Water Resources Research*, **12**(3), 528–32.

Shackelford, C.D. and Daniel, D.E. (1991) Diffusion in saturated soil II: Results for compacted clay. *Journal of Geotechnical Engineering, ASCE*, **117**(3), pp. 485–506.

Sherard, J.L., Decker, R.S. and Ryker, N.L. (1972) Piping – earth dams of dispersive clay, in *Proceedings ASCE Specialty Conference on the Performance of Earth and Earth Supported Structures*, Purdue University, West Lafayette, Indiana, pp. 589–626.

Smith, D.W., Rowe, R.K. and Booker, J.R. (1992) Contaminant transport and non-equilibrium sorption, in *Proceedings of the 4th International Symposium on Numerical Methods in Geomechanics* (eds G.N. Pande and S. Pietruszak), Swansea, pp. 509–18.

Smith, D.W., Rowe, R.K. and Booker, J.R. (1993) The analysis of pollutant migration through soil with linear hereditary time dependent sorption. *International Journal for Analytical and Numerical Methods in Geomechanics*, **17**(4), 255–74.

Sudicky, E.A. (1986) A natural gradient experiment on solute transport in a sand aquifer: Spatial variability of hydraulic conductivity and its role in the dispersion process. *Water Resources Research*, **22**(13), pp. 2069–82.

Sudicky, E.A. (1990) The Laplace Transform Galerkin Technique for efficient time-continuous solution of solute transport in double-porosity media. *Geoderma*, **46**, 209–32.

Sudicky, E.A. and Frind, E.O. (1982) Contaminant transport in fractured porous media: analytical solutions for a system of parallel fractures. *Water Resources Research*, **18**(6), pp. 1634–42.

Talbot, A. (1979) The accurate numerical integration of Laplace transforms. *Journal of Institute Mathematics Applications*, **23**, 97–120.

Tang, D.H., Frind, E.O. and Sudicky, E.A. (1981) Contaminant transport in fractured porous media: analytical solution for a single fracture. *Water Resources Research*, **17**(3), 555–64.

Tavenas, F., Tremblay, M. and Leroueil, S. (1983) Mesure in situ de la perméabilité des argiles. Symposium international sur la reconnaissance des sols et des roches par essais en place, Paris. *Bulletin of the International Association of Engineering Geology*, Nos. 26–27, 509–15.

Tavenas, F., Diene, M. and Leroueil, S. (1990) Analysis of the in situ constant-head permeability test in clays. *Canadian Geotechnical Journal*, **27**, 305–14.

Tavenas, F., Tremblay, M., Larouche, G. and Leroueil, S. (1986) In situ measurement of permeability in soft clays, in *ASCE Specialty Conference, In Situ '86*, Blacksburg, pp. 1034–48.

Thomas, G.W. and Swoboda, A.R. (1970) Anion exclusion effects on chloride movement in soils. *Soil Science*, **110**, 163–66.

Travis, C.C. and Lanz, M.L. (1990). Estimating the mean of data sets with non-detectable values. *Environmental Science and Technology*, **24**(7), 961–62.

Uchrin, C.G. and Katz, J. (1986) Sorption kinetics of competing organic substances on New Jersey coastal plain aquifier solids, in *Hazardous and Industrial Solid Waste Testing and Disposal* (eds D. Lorenzen, R. A. Conway, L.P. Jackson *et al.*) Vol. 6, ASTM Special Technical Publication 933, pp. 140–50.

Usher, S.J. and Cherry, J.A. (1988). Evaluation of a two layer aquitard through the use of a one-dimensional analytical solution of the transient head distribution, in *41st Canadian Geotechnical Conference*, Waterloo, Ontario.

US EPA (1985) Compatibility test for wastes and geomembrane liners, Method 9090, US Environmental Protection Agency, Cincinnati, OH.

US EPA (1988) Guide to technical resources for the design of land disposal facilities. *EPA/625/6-88/018*, US Environmental Protection Agency, Cincannati, OH.

US EPA (1987) Office of Drinking Water, March 31.

US EPA (1989) Requirements for hazardous waste landfill design, construction and closure. *EPA/625/4-89/022*, US Environmental Protection Agency, Cincinnati, OH.

Van Genuchten, M.Th. (1978). Calculating the unsaturated hydraulic conductivity with a new closed form analytical model. Research Report 78-WR-08, Department of Civil Engineering, Princeton University, Princeton, NJ.

Van Olphen, H. (1977) *An Introduction to Clay Colloid Chemistry*, 2nd edn, John Wiley.

Vogel, T.M., Criddle, C.S. and McCarty, P.L. (1987) Transformations of halogenated diphatic compounds. *Environmental Science and Technology*, **2**(8), 722–36.

Voice, T.C. and Weber, W.J. (1983) Sorption of hydrophobic compounds by sediments, soils and suspended solids – I Theory and background. *Water Research*, **17**(10), 1433–41.

Voice, T.C., Rice, C.P. and Weber, W.J. (1983) Effect of solids concentration on the sorptive partitioning of hydrophobic pollutants in aquatic systems. *Journal of Environmental Science Technology*, **17**, 513–18.

Wigh, R.J. (1979) Boone County Field Site Interim Report, EPA-600/2-79-058.

WHO (1984) Guidelines for Drinking Water Quality, Vol. 1: Recommendations. World Health Organization, ISBN: 92 4 154168 7.

Williams, N., Giroud, J.P. and Bonaparte, R. (1984) Properties of plastic nets for liquid and gas drainage associated with geomembranes. *International Conference on Geomembranes*, Denver, pp. 399–404.

Wong, J. (1977) The design of a system for collecting leachate from a lined landfill site. *Water Resources Research*, **13**(2).

Yanful, E.K. (1993) Oxygen diffusion through soil covers on sulphidic mill tailings. *ASCE Journal of Geotechnical Engineering*, **119**(8), 1207–28.

Yanful, E.K. and Quigley, R.M. (1986) Heavy metal deposition at the clay/waste interface of a landfill site, Sarnia, Ontario , in *Proceedings 3rd Canadian Hydrogeological Conference*, Saskatoon, April, pp. 35–42.

Yanful, E.K. and Quigley, R.M. (1990) Tritium, oxygen-18 and deuterium diffusion at the Confederation Road landfill site, Sarnia, Ontario, Canada. *Canadian Geotechnical Journal*, **27**(3), 271–75.

Yanful, E.K., Fernandez, F. and Quigley, R.M. (1987) The chemical instability of domestic waste leachate and its geotechnical implications. The University of Western Ontario, Faculty of Engineering Science Research Report *GEOT-15-87*.

Yanful, E.K., Nesbitt, H.W. and Quigley, R.M. (1988a) Heavy metal migration at a landfill site, Sarnia, Ontario, Canada – I: Thermodynamic assessment and chemical interpretations. *Applied Geochemistry*, **3**, 523–33.

Yanful, E.K., Quigley, R.M. and Nesbitt, W. (1988b) Heavy metal migration at a landfill site. Part II: Metal partitioning and geotechnical implications. *Applied Geochemistry*, **3**, 623–29.

Yeh, G.T. (1984) Solution of contaminant transport equations using an orthogonal upstream weighting finite element scheme. *Proceedings 5th International Conference on Finite Elements in Water Resources*, Vermont, pp. 285–93.

Yong, R.N. and Sheremata, T.W. (1991) Effect of chloride ions on adsorption of cadmium from a landfill leachate. *Canadian Geotechnical Journal*, **28**(3), 378–87.

Yong, R.N. and Warkentin, B.P. (1975) *Soil Properties and Behaviour*, 2nd edn, Elsevier Scientific.

Zagorski, G.A. and Wayne, M.H. (1990) Geonet seams. *Geotextiles and Geomembranes*, **9**, 487–99.

Zehnder, A.J.B. (1978) Ecology of methane formation, in *Water Pollution Microbiology*, Vol. 2 (ed. R. Mitchell), John Wiley, New York, pp. 349–76.

Zienkiewicz, O.C. (1977) *The Finite Element Method*, McGraw-Hill, UK.

Author index

Subject index